Multivariate Statistics

Old School

Mathematical and methodological introduction to multivariate statistical analytics, including linear models, principal components, covariance structures, classification, and clustering, providing background for machine learning and big data study, with R

John I. Marden
Department of Statistics
University of Illinois at Urbana-Champaign

Email: multivariate@stat.istics.net

URL: http://stat.istics.net/Multivariate

Typeset using the memoir package [Madsen and Wilson, 2015] with LaTeX [LaTeX Project Team, 2015]. The faces in the cover image were created using the faces routine in the R package aplpack [Wolf, 2014].

Preface

This text was developed over many years while teaching the graduate course in multivariate analysis in the Department of Statistics, University of Illinois at Urbana-Champaign. Its goal is to teach the basic mathematical grounding that Ph. D. students need for future research, as well as cover the important multivariate techniques useful to statisticians in general.

There is heavy emphasis on multivariate normal modeling and inference, both theory and implementation. Several chapters are devoted to developing linear models, including multivariate regression and analysis of variance, and especially the "both-sides models" (i.e., generalized multivariate analysis of variance models), which allow modeling relationships among variables as well as individuals. Growth curve and repeated measure models are special cases.

Inference on covariance matrices covers testing equality of several covariance matrices, testing independence and conditional independence of (blocks of) variables, factor analysis, and some symmetry models. Principal components is a useful graphical/exploratory technique, but also lends itself to some modeling.

Classification and clustering are related areas. Both attempt to categorize individuals. Classification tries to classify individuals based upon a previous sample of observed individuals and their categories. In clustering, there is no observed categorization, nor often even knowledge of how many categories there are. These must be estimated from the data.

Other useful multivariate techniques include biplots, multidimensional scaling, and canonical correlations.

The bulk of the results here are mathematically justified, but I have tried to arrange the material so that the reader can learn the basic concepts and techniques while plunging as much or as little as desired into the details of the proofs. Topic- and level-wise, this book is somewhere in the convex hull of the classic book by Anderson [2003] and the texts by Mardia, Kent, and Bibby [1979] and Johnson and Wichern [2007], probably closest in spirit to Mardia, Kent and Bibby.

The material assumes the reader has had mathematics up through calculus and linear algebra, and statistics up through mathematical statistics, e.g., Hogg, McKean, and Craig [2012], and linear regression and analysis of variance, e.g., Weisberg [2013].

In a typical semester, I would cover Chapter 1 (introduction, some graphics, and principal components); go through Chapter 2 fairly quickly, as it is a review of mathematical statistics the students should know, but being sure to emphasize Section 2.3.1 on means and covariance matrices for vectors and matrices, and Section 2.5 on conditional probabilities; go carefully through Chapter 3 on the multivariate normal, and Chapter 4 on setting up linear models, including the both-sides model; cover most of Chapter 5 on projections and least squares, though usually skipping 5.7.1 on the proofs of the QR and Cholesky decompositions; cover Chapters 6 and 7 on estimation and testing in the both-sides model; skip most of Chapter 8, which has many technical proofs, whose results are often referred to later; cover most of Chapter 9, but usually skip the exact likelihood ratio test in a special case (Section 9.4.1), and Sections 9.5.2 and 9.5.3 with details about the Akaike information criterion; cover Chapters 10 (covariance models), 11 (classifications), and 12 (clustering) fairly thoroughly; and make selections from Chapter 13, which presents more on principal components, and introduces singular value decompositions, multidimensional scaling, and canonical correlations.

A path through the book that emphasizes methodology over mathematical theory would concentrate on Chapters 1 (skip Section 1.8), 4, 6, 7 (skip Sections 7.2.5 and 7.5.2), 9 (skip Sections 9.3.4, 9.5.1. 9.5.2, and 9.5.3), 10 (skip Section 10.4), 11, 12 (skip Section 12.4), and 13 (skip Sections 13.1.5 and 13.1.6). The more data-oriented exercises come at the end of each chapter's set of exercises.

One feature of the text is a fairly rigorous presentation of the basics of linear algebra that are useful in statistics. Sections 1.4, 1.5, 1.6, and 1.8 and Exercises 1.9.1 through 1.9.13 cover idempotent matrices, orthogonal matrices, and the spectral decomposition theorem for symmetric matrices, including eigenvectors and eigenvalues. Sections 3.1 and 3.3 and Exercises 3.7.6, 3.7.12, 3.7.16 through 3.7.20, and 3.7.24 cover positive and nonnegative definiteness, Kronecker products, and the Moore-Penrose inverse for symmetric matrices. Chapter 5 covers linear subspaces, linear independence, spans, bases, projections, least squares, Gram-Schmidt orthogonalization, orthogonal polynomials, and the QR and Cholesky decompositions. Section 13.1.3 and Exercise 13.4.3 look further at eigenvalues and eigenspaces, and Section 13.3 and Exercise 13.4.12 develop the singular value decomposition.

Practically all the calculations and graphics in the examples are implemented using the statistical computing environment R [R Development Core Team, 2015]. Throughout the text we have scattered some of the actual R code we used. Many of the data sets and original R functions can be found in the R package `msos` [Marden and Balamuta, 2014], thanks to the much appreciated efforts of James Balamuta. For other material we refer to available R packages.

I thank Michael Perlman for introducing me to multivariate analysis, and his friendship and mentorship throughout my career. Most of the ideas and approaches in this book got their start in the multivariate course I took from him forty years ago. I think they have aged well. Also, thanks to Steen Andersson, from whom I learned a lot, including the idea that one should define a model before trying to analyze it.

This book is dedicated to Ann.

Contents

Preface iii

Contents v

1 A First Look at Multivariate Data 1
- 1.1 The data matrix 1
  - 1.1.1 Example: Planets data 2
- 1.2 Glyphs 2
- 1.3 Scatter plots 3
  - 1.3.1 Example: Fisher-Anderson iris data 5
- 1.4 Sample means, variances, and covariances 6
- 1.5 Marginals and linear combinations 8
  - 1.5.1 Rotations 10
- 1.6 Principal components 10
  - 1.6.1 Biplots 13
  - 1.6.2 Example: Sports data 13
- 1.7 Other projections to pursue 15
  - 1.7.1 Example: Iris data 17
- 1.8 Proofs 18
- 1.9 Exercises 20

2 Multivariate Distributions 27
- 2.1 Probability distributions 27
  - 2.1.1 Distribution functions 27
  - 2.1.2 Densities 28
  - 2.1.3 Representations 29
  - 2.1.4 Conditional distributions 30
- 2.2 Expected values 32
- 2.3 Means, variances, and covariances 33
  - 2.3.1 Vectors and matrices 34
  - 2.3.2 Moment generating functions 35
- 2.4 Independence 35
- 2.5 Additional properties of conditional distributions 37
- 2.6 Affine transformations 40

2.7 Exercises 41

3 The Multivariate Normal Distribution 49
3.1 Definition 49
3.2 Some properties of the multivariate normal 51
3.3 Multivariate normal data matrix 52
3.4 Conditioning in the multivariate normal 55
3.5 The sample covariance matrix: Wishart distribution 57
3.6 Some properties of the Wishart 59
3.7 Exercises 60

4 Linear Models on Both Sides 69
4.1 Linear regression 69
4.2 Multivariate regression and analysis of variance 72
4.2.1 Examples of multivariate regression 72
4.3 Linear models on both sides 77
4.3.1 One individual 77
4.3.2 IID observations 78
4.3.3 The both-sides model 81
4.4 Exercises 82

5 Linear Models: Least Squares and Projections 87
5.1 Linear subspaces 87
5.2 Projections 89
5.3 Least squares 90
5.4 Best linear unbiased estimators 91
5.5 Least squares in the both-sides model 93
5.6 What is a linear model? 94
5.7 Gram-Schmidt orthogonalization 95
5.7.1 The QR and Cholesky decompositions 97
5.7.2 Orthogonal polynomials 99
5.8 Exercises 101

6 Both-Sides Models: Estimation 109
6.1 Distribution of $\widehat{\beta}$ 109
6.2 Estimating the covariance 109
6.2.1 Multivariate regression 109
6.2.2 Both-sides model 111
6.3 Standard errors and t-statistics 111
6.4 Examples 112
6.4.1 Mouth sizes 112
6.4.2 Using linear regression routines 114
6.4.3 Leprosy data 115
6.4.4 Covariates: Leprosy data 115
6.4.5 Histamine in dogs 117
6.5 Submodels of the both-sides model 118
6.6 Exercises 120

7 Both-Sides Models: Hypothesis Tests on β 125
7.1 Approximate χ^2 test 125

7.1.1 Example: Mouth sizes 126
7.2 Testing blocks of β are zero 126
7.2.1 Just one column: F test 128
7.2.2 Just one row: Hotelling's T^2 128
7.2.3 General blocks 129
7.2.4 Additional test statistics 129
7.2.5 The between and within matrices 130
7.3 Examples 132
7.3.1 Mouth sizes 132
7.3.2 Histamine in dogs 133
7.4 Testing linear restrictions 134
7.5 Model selection: Mallows' C_p 135
7.5.1 Example: Mouth sizes 137
7.5.2 Mallows' C_p verification 139
7.6 Exercises 141

8 Some Technical Results **145**
8.1 The Cauchy-Schwarz inequality 145
8.2 Conditioning in a Wishart 146
8.3 Expected value of the inverse Wishart 147
8.4 Distribution of Hotelling's T^2 148
8.4.1 A motivation for Hotelling's T^2 149
8.5 Density of the multivariate normal 150
8.6 The QR decomposition for the multivariate normal 151
8.7 Density of the Wishart 153
8.8 Exercises 154

9 Likelihood Methods **161**
9.1 Likelihood 161
9.2 Maximum likelihood estimation 161
9.3 The MLE in the both-sides model 162
9.3.1 Maximizing the likelihood 162
9.3.2 Examples 164
9.3.3 Calculating the estimates 166
9.3.4 Proof of the MLE for the Wishart 168
9.4 Likelihood ratio tests 168
9.4.1 The LRT in the both-sides model 169
9.5 Model selection: AIC and BIC 170
9.5.1 BIC: Motivation 171
9.5.2 AIC: Motivation 173
9.5.3 AIC: Multivariate regression 174
9.5.4 Example: Skulls 175
9.5.5 Example: Histamine 179
9.6 Exercises 181

10 Models on Covariance Matrices **187**
10.1 Testing equality of covariance matrices 188
10.1.1 Example: Grades data 189
10.1.2 Testing the equality of several covariance matrices 190
10.2 Testing independence of two blocks of variables 190

10.2.1 Example: Grades data 191
10.2.2 Example: Testing conditional independence 192
10.3 Factor analysis 194
10.3.1 Estimation 195
10.3.2 Describing the factors 197
10.3.3 Example: Grades data 198
10.4 Some symmetry models 202
10.4.1 Some types of symmetry 203
10.4.2 Characterizing the structure 205
10.4.3 Maximum likelihood estimates 205
10.4.4 Hypothesis testing and model selection 207
10.4.5 Example: Mouth sizes 207
10.5 Exercises 210

11 Classification **215**
11.1 Mixture models 215
11.2 Classifiers 217
11.3 Fisher's linear discrimination 219
11.4 Cross-validation estimate of error 221
11.4.1 Example: Iris data 222
11.5 Fisher's quadratic discrimination 225
11.5.1 Example: Iris data, continued 226
11.6 Modifications to Fisher's discrimination 227
11.7 Conditioning on X: Logistic regression 227
11.7.1 Example: Iris data 229
11.7.2 Example: Spam 230
11.8 Trees 233
11.8.1 CART 235
11.9 Exercises 240

12 Clustering **245**
12.1 K-means 246
12.1.1 Example: Sports data 246
12.1.2 Silhouettes 247
12.1.3 Plotting clusters in one and two dimensions 248
12.1.4 Example: Sports data, using R 250
12.2 K-medoids 252
12.3 Model-based clustering 254
12.3.1 Example: Automobile data 254
12.3.2 Some of the models in mclust 257
12.4 An example of the EM algorithm 259
12.5 Soft K-means 260
12.5.1 Example: Sports data 261
12.6 Hierarchical clustering 261
12.6.1 Example: Grades data 262
12.6.2 Example: Sports data 263
12.7 Exercises 265

13 Principal Components and Related Techniques **267**
13.1 Principal components, redux 267

13.1.1 Example: Iris data 268
13.1.2 Choosing the number of principal components 270
13.1.3 Estimating the structure of the component spaces 271
13.1.4 Example: Automobile data 273
13.1.5 Principal components and factor analysis 276
13.1.6 Justification of the principal component MLE 278
13.2 Multidimensional scaling 280
13.2.1 $\boldsymbol{\Delta}$ is Euclidean: The classical solution 280
13.2.2 $\boldsymbol{\Delta}$ may not be Euclidean: The classical solution 282
13.2.3 Non-metric approach 283
13.2.4 Examples: Grades and sports 283
13.3 Canonical correlations 284
13.3.1 Example: Grades 287
13.3.2 How many canonical correlations are positive? 288
13.3.3 Partial least squares 290
13.4 Exercises 290

A Extra R Routines **295**
A.1 Entropy 295
A.1.1 negent: Estimate negative entropy 296
A.1.2 negent2D: Maximize negentropy for two dimensions 296
A.1.3 negent3D: Maximize negentropy for three dimensions 297
A.2 Both-sides model 297
A.2.1 bothsidesmodel: Calculate the least squares estimates 298
A.2.2 reverse.kronecker: Reverse the matrices in a Kronecker product . 299
A.2.3 bothsidesmodel.mle: Calculate the maximum likelihood estimates 299
A.2.4 bothsidesmodel.chisquare: Test subsets of $\boldsymbol{\beta}$ are zero 300
A.2.5 bothsidesmodel.hotelling: Test blocks of $\boldsymbol{\beta}$ are zero 300
A.2.6 bothsidesmodel.lrt: Test subsets of $\boldsymbol{\beta}$ are zero 301
A.2.7 Helper functions 302
A.3 Classification 302
A.3.1 lda: Linear discrimination 302
A.3.2 qda: Quadratic discrimination 302
A.3.3 predict_qda: Quadratic discrimination prediction 303
A.4 Silhouettes for K-means clustering 303
A.4.1 silhouette.km: Calculate the silhouettes 303
A.4.2 sort_silhouette: Sort the silhouettes by group 303
A.5 Patterns of eigenvalues 304
A.5.1 pcbic: BIC for a particular pattern 304
A.5.2 pcbic.stepwise: Choose a good pattern 304
A.5.3 Helper functions 305
A.6 Function listings 305

Bibliography **317**

Author Index **325**

Subject Index **329**

Chapter 1

A First Look at Multivariate Data

In this chapter, we try to give a sense of what multivariate data sets look like, and introduce some of the basic matrix manipulations needed throughout these notes. Chapters 2 and 3 lay down the distributional theory. Linear models are probably the most popular statistical models ever. With multivariate data, we can model relationships between individuals or between variables, leading to what we call "both-sides models," which do both simultaneously. Chapters 4 through 8 present these models in detail. The linear models are concerned with means. Before turning to models on covariances, Chapter 9 briefly reviews likelihood methods, including maximum likelihood estimation, likelihood ratio tests, and model selection criteria (Bayes and Akaike). Chapter 10 looks at a number of models based on covariance matrices, including equality of covariances, independence and conditional independence, factor analysis, and other structural models. Chapter 11 deals with classification, in which the goal is to find ways to classify individuals into categories, e.g., healthy or unhealthy, based on a number of observed variable. Chapter 12 has a similar goal, except that the categories are unknown and we seek to group individuals based on just the observed variables. Finally, Chapter 13 explores principal components, which we first see in Section 1.6. It is an approach for reducing the number of variables, or at least find a few interesting ones, by searching through linear combinations of the observed variables. Multidimensional scaling has a similar objective, but tries to exhibit the individual data points in a low-dimensional space while preserving the original inter-point distances. Canonical correlations has two sets of variables, and finds linear combinations of the two sets to explain the correlations between them.

On to the data.

1.1 The data matrix

Data generally will consist of a number of variables recorded on a number of individuals, e.g., heights, weights, ages, and sexes of a sample of students. Also, generally, there will be n individuals and q variables, and the data will be arranged in an $n \times q$

data matrix, with rows denoting individuals and the columns denoting variables:

$$\mathbf{Y} = \begin{array}{c} \\ \text{Individual 1} \\ \text{Individual 2} \\ \vdots \\ \text{Individual } n \end{array} \begin{array}{c} \begin{array}{cccc} \text{Var 1} & \text{Var 2} & \cdots & \text{Var } q \end{array} \\ \begin{pmatrix} y_{11} & y_{12} & \cdots & y_{1q} \\ y_{21} & y_{22} & \cdots & y_{2q} \\ \vdots & \vdots & \ddots & \vdots \\ y_{n1} & y_{n2} & \cdots & y_{nq} \end{pmatrix} \end{array}. \tag{1.1}$$

Then y_{ij} is the value of the variable j for individual i. Much more complex data structures exist, but this course concentrates on these straightforward data matrices.

1.1.1 Example: Planets data

Six astronomical variables are given on each of the historical nine planets (or eight planets, plus Pluto). The variables are (average) distance in millions of miles from the Sun, length of day in Earth days, length of year in Earth days, diameter in miles, temperature in degrees Fahrenheit, and number of moons. The data matrix:

	Dist	Day	Year	Diam	Temp	Moons
Mercury	35.96	59.00	88.00	3030	332	0
Venus	67.20	243.00	224.70	7517	854	0
Earth	92.90	1.00	365.26	7921	59	1
Mars	141.50	1.00	687.00	4215	−67	2
Jupiter	483.30	0.41	4332.60	88803	−162	16
Saturn	886.70	0.44	10759.20	74520	−208	18
Uranus	1782.00	0.70	30685.40	31600	−344	15
Neptune	2793.00	0.67	60189.00	30200	−261	8
Pluto	3664.00	6.39	90465.00	1423	−355	1

(1.2)

The data can be found in Wright [1997], for example.

1.2 Glyphs

Graphical displays of univariate data, that is, data on one variable, are well-known: histograms, stem-and-leaf plots, pie charts, box plots, etc. For two variables, scatter plots are valuable. It is more of a challenge when dealing with three or more variables.

Glyphs provide an option. A little picture is created for each individual, with characteristics based on the values of the variables. **Chernoff's faces** [Chernoff, 1973] may be the most famous glyphs. The idea is that people intuitively respond to characteristics of faces, so that many variables can be summarized in a face.

Figure 1.1 exhibits faces for the nine planets. We use the faces routine by H. P. Wolf in the R package aplpack, Wolf [2014]. The distance the planet is from the sun is represented by the height of the face (Pluto has a long face), the length of the planet's day by the width of the face (Venus has a wide face), etc. One can then cluster the planets. Mercury, Earth and Mars look similar, as do Saturn and Jupiter. These face plots are more likely to be amusing than useful, especially if the number of individuals is large. A **star plot** is similar. Each individual is represented by a q-pointed star, where each point corresponds to a variable, and the distance of the

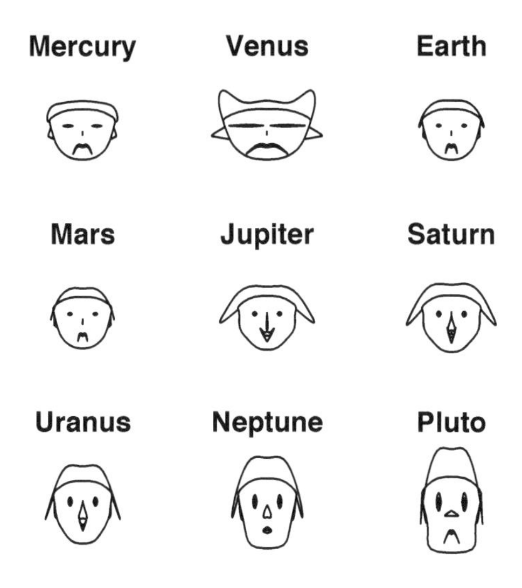

Figure 1.1: Chernoff's faces for the planets. Each feature represents a variable. For these data, *distance = height of face, day = width of face, year = shape of face, diameter = height of mouth, temperature = width of mouth, moons = curve of smile.*

point from the center is based on the variable's value for that individual. See Figure 1.2.

1.3 Scatter plots

Two-dimensional scatter plots can be enhanced by using different symbols for the observations instead of plain dots. For example, different colors could be used for different groups of points, or glyphs representing other variables could be plotted. Figure 1.2 plots the planets with the logarithms of day length and year length as the axes, where the stars created from the other four variables are the plotted symbols. Note that the planets pair up in a reasonable way. Mercury and Venus are close, both in terms of the scatter plot and in the look of their stars. Similarly, Earth and Mars pair up, as do Jupiter and Saturn, and Uranus and Neptune. See Listing 1.1 for the R code.

A **scatter plot matrix** arranges all possible two-way scatter plots in a $q \times q$ matrix. These displays can be enhanced with **brushing**, in which individual points or groups of points can be selected in one plot, and be simultaneously highlighted in the other plots.

Listing 1.1: R code for the star plot of the planets, Figure 1.2. The data are in the matrix planets. The first statement normalizes the variables to range from 0 to 1. The ep matrix is used to place the names of the planets. Tweaking is necessary, depending on the size of the plot.

```
p <- apply(planets,2,function(z) (z-min(z))/(max(z)-min(z)))
x <- log(planets[,2])
y <- log(planets[,3])
ep <- rbind(c(-.3,.4),c(-.5,.4),c(.5,0),c(.5,0),c(.6,-1),c(-.5,1.4),
        c(1,-.6),c(1.3,.4),c(1,-.5))
symbols(x,y,stars=p[,-(2:3)],xlab='log(day)',ylab='log(year)',inches=.4)
text(x+ep[,1],y+ep[,2],labels=rownames(planets),cex=.7)
```

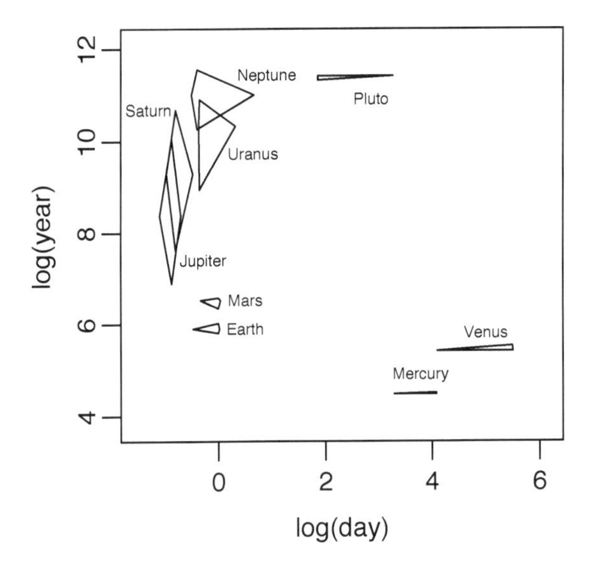

Figure 1.2: Scatter plot of log(day) versus log(year) for the planets, with plotting symbols being stars created from the other four variables: distance, diameter, temperature, moons.

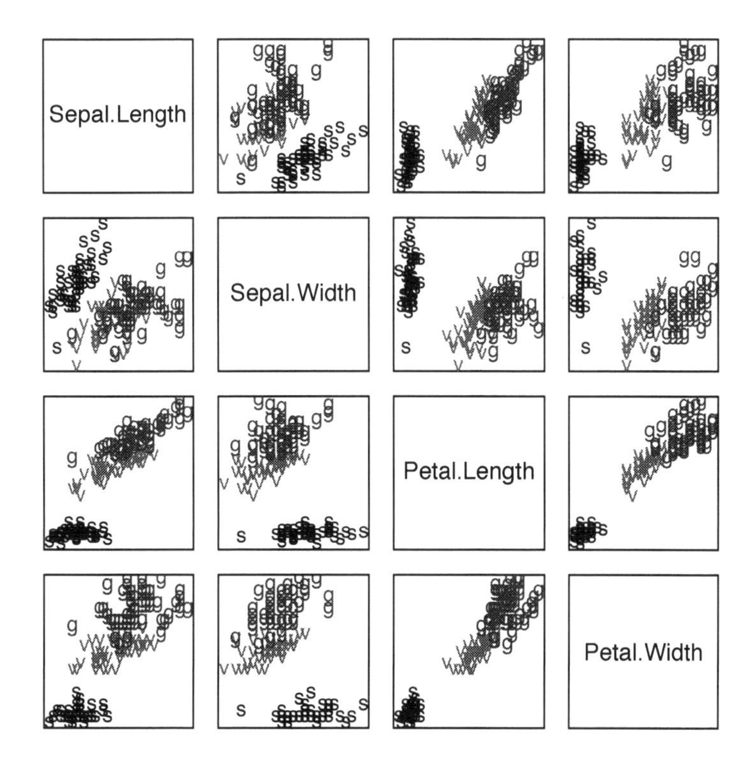

Figure 1.3: A scatter plot matrix for the Fisher-Anderson iris data. In each plot, "s" indicates setosa plants, "v" indicates versicolor, and "r" indicates virginica.

1.3.1 Example: Fisher-Anderson iris data

The most famous data set in multivariate analysis is the iris data analyzed by Fisher [1936] based on data collected by Anderson [1935]. See also Anderson [1936]. There are fifty specimens each of three species of iris: setosa, versicolor, and virginica. There are four variables measured on each plant: sepal length, sepal width, petal length, and pedal width. Thus $n = 150$ and $q = 4$. Figure 1.3 contains the corresponding scatter plot matrix, with species indicated by letter. We used the R function pairs. Note that the three species separate fairly well, with setosa especially different from the other two.

As a preview of classification (Chapter 11), Figure 1.4 uses faces to exhibit five observations from each species, and five random observations without their species label. Can you guess which species each is? See page 20. The setosas are not too difficult, since they have small faces, but distinguishing the other two can be a challenge.

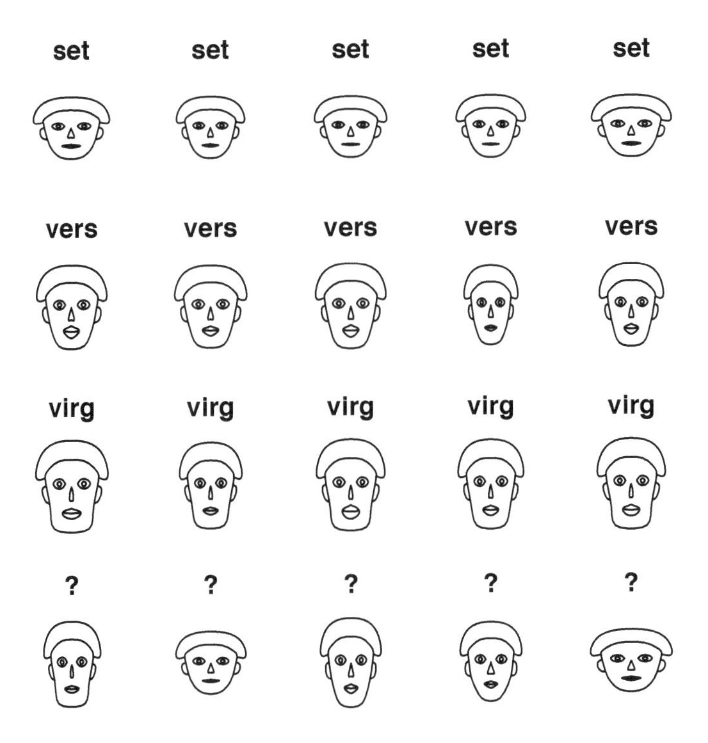

Figure 1.4: Chernoff's faces for five specimens from each iris species, plus five from unspecified species. Here, "set" indicates setosa plants, "vers" indicates versicolor, and "virg" indicates virginica.

1.4 Sample means, variances, and covariances

For univariate values $x_1, \ldots, x_n$, the **sample mean** is

$$\overline{x} = \frac{1}{n} \sum_{i=1}^{n} x_i, \tag{1.3}$$

and the **sample variance** is

$$s_x^2 = s_{xx} = \frac{1}{n} \sum_{i=1}^{n} (x_i - \overline{x})^2. \tag{1.4}$$

Note the two notations: The s_x^2 is most common when dealing with individual variables, but the s_{xx} transfers better to multivariate data. Often one is tempted to divide by $n-1$ instead of n. That's fine, too. With a second set of values $z_1, \ldots, z_n$, we have the **sample covariance** between the x_i's and z_i's to be

$$s_{xz} = \frac{1}{n}\sum_{i=1}^{n}(x_i - \overline{x})(z_i - \overline{z}). \tag{1.5}$$

So the covariance between the x_i's and themselves is the variance, which is to say that $s_x^2 = s_{xx}$. The **sample correlation coefficient** is a normalization of the covariance that ranges between -1 and $+1$, defined by

$$r_{xz} = \frac{s_{xz}}{s_x s_z} \tag{1.6}$$

provided both variances are positive. (See Corollary 8.1.) In a scatter plot of x versus z, the correlation coefficient is $+1$ if all the points lie on a line with positive slope, and -1 if they all lie on a line with negative slope.

For a data matrix $\mathbf{Y}$ (1.1) with q variables, there are q means:

$$\overline{y}_j = \frac{1}{n}\sum_{i=1}^{n} y_{ij}. \tag{1.7}$$

Placing them in a row vector, we have

$$\overline{\mathbf{y}} = (\overline{y}_1, \ldots, \overline{y}_q). \tag{1.8}$$

The $n \times 1$ **one** vector is $\mathbf{1}_n = (1, 1, \ldots, 1)'$, the vector of all 1's. Then the mean vector (1.8) can be written

$$\overline{\mathbf{y}} = \frac{1}{n}\mathbf{1}_n'\mathbf{Y}. \tag{1.9}$$

To find the variances and covariances, we first have to subtract the means from the individual observations in $\mathbf{Y}$: change y_{ij} to $y_{ij} - \overline{y}_j$ for each i, j. That transformation can be achieved by subtracting the $n \times q$ matrix $\mathbf{1}_n\overline{\mathbf{y}}$ from $\mathbf{Y}$ to get the matrix of deviations. Using (1.9), we can write

$$\mathbf{Y} - \mathbf{1}_n\overline{\mathbf{y}} = \mathbf{Y} - \mathbf{1}_n\frac{1}{n}\mathbf{1}_n'\mathbf{Y} = (\mathbf{I}_n - \frac{1}{n}\mathbf{1}_n\mathbf{1}_n')\mathbf{Y} \equiv \mathbf{H}_n\mathbf{Y}. \tag{1.10}$$

There are two important matrices in that formula: The $n \times n$ **identity matrix** $\mathbf{I}_n$,

$$\mathbf{I}_n = \begin{pmatrix} 1 & 0 & \ldots & 0 \\ 0 & 1 & \ldots & 0 \\ \vdots & \vdots & \ddots & \vdots \\ 0 & 0 & \ldots & 1 \end{pmatrix}, \tag{1.11}$$

and the $n \times n$ **centering matrix** $\mathbf{H}_n$,

$$\mathbf{H}_n = \mathbf{I}_n - \frac{1}{n}\mathbf{1}_n\mathbf{1}_n' = \begin{pmatrix} 1 - \frac{1}{n} & -\frac{1}{n} & \ldots & -\frac{1}{n} \\ -\frac{1}{n} & 1 - \frac{1}{n} & \ldots & -\frac{1}{n} \\ \vdots & \vdots & \ddots & \vdots \\ -\frac{1}{n} & -\frac{1}{n} & \ldots & 1 - \frac{1}{n} \end{pmatrix}. \tag{1.12}$$

The identity matrix leaves any vector or matrix alone, so if $\mathbf{A}$ is $n \times m$, then $\mathbf{A} = \mathbf{I}_n\mathbf{A} = \mathbf{A}\mathbf{I}_m$, and the centering matrix subtracts the column mean from each element in $\mathbf{H}_n\mathbf{A}$. Similarly, $\mathbf{A}\mathbf{H}_m$ results in the row mean being subtracted from each element.

For an $n \times 1$ vector $\mathbf{x}$ with mean $\bar{x}$, and $n \times 1$ vector $\mathbf{z}$ with mean $\bar{z}$, we can write

$$\sum_{i=1}^{n}(x_i - \bar{x})^2 = (\mathbf{x} - \bar{x}\mathbf{1}_n)'(\mathbf{x} - \bar{x}\mathbf{1}_n) \tag{1.13}$$

and

$$\sum_{i-1}^{n}(x_i - \bar{x})(z_i - \bar{z}) = (\mathbf{x} - \bar{x}\mathbf{1}_n)'(\mathbf{z} - \bar{z}\mathbf{1}_n). \tag{1.14}$$

Thus taking the deviations matrix in (1.10), $(\mathbf{H}_n\mathbf{Y})'(\mathbf{H}_n\mathbf{Y})$ contains all the $\sum(y_{ij} - \bar{y}_j)(y_{ik} - \bar{y}_k)$'s. We will call that matrix the **sum of squares and cross-products matrix**. Notice that

$$(\mathbf{H}_n\mathbf{Y})'(\mathbf{H}_n\mathbf{Y}) = \mathbf{Y}'\mathbf{H}_n'\mathbf{H}_n\mathbf{Y} = \mathbf{Y}'\mathbf{H}_n\mathbf{Y}. \tag{1.15}$$

What happened to the $\mathbf{H}_n$'s? First, $\mathbf{H}_n$ is clearly symmetric, so that $\mathbf{H}_n' = \mathbf{H}_n$. Then notice that $\mathbf{H}_n\mathbf{H}_n = \mathbf{H}_n$. Such a matrix is called **idempotent**, that is, a square matrix

$$\mathbf{A} \text{ is idempotent if } \mathbf{A}\mathbf{A} = \mathbf{A}. \tag{1.16}$$

Dividing the sum of squares and cross-products matrix by n gives the **sample variance-covariance matrix**, or more simply **sample covariance matrix**:

$$\mathbf{S} = \frac{1}{n}\,\mathbf{Y}'\mathbf{H}_n\mathbf{Y} = \begin{pmatrix} s_{11} & s_{12} & \cdots & s_{1q} \\ s_{21} & s_{22} & \cdots & s_{2q} \\ \vdots & \vdots & \ddots & \vdots \\ s_{q1} & s_{q2} & \cdots & s_{qq} \end{pmatrix}, \tag{1.17}$$

where s_{jj} is the sample variance of the jth variable (column), and s_{jk} is the sample covariance between the jth and k^{th} variables. (When doing inference later, we may divide by $n - df$ instead of n for some "degrees-of-freedom" integer df.)

1.5 Marginals and linear combinations

A natural first stab at looking at data with several variables is to look at the variables one at a time, so with q variables, one would first make q histograms, or box plots, or whatever suits one's fancy. Such techniques are based on **marginals**, that is, based on subsets of the variables rather than all variables at once as in glyphs. One-dimensional marginals are the individual variables, two-dimensional marginals are the pairs of variables, three-dimensional marginals are the sets of three variables, etc.

Consider one-dimensional marginals. It is easy to construct the histograms, say. But why be limited to the q variables? Functions of the variables can also be histogrammed, e.g., weight $\div$ height. The number of possible functions one could imagine is vast. One convenient class is the set of *linear* transformations, that is, for some constants $b_1, \ldots, b_q$, a new variable is $W = b_1Y_1 + \cdots + b_qY_q$, so the transformed data consist of $w_1, \ldots, w_n$, where

$$w_i = b_1y_{i1} + \cdots + b_qy_{iq}. \tag{1.18}$$

Placing the coefficients into a column vector $\mathbf{b} = (b_1, \ldots, b_q)'$, we can write

$$\mathbf{W} \equiv \begin{pmatrix} w_1 \\ w_2 \\ \vdots \\ w_n \end{pmatrix} = \mathbf{Yb}, \tag{1.19}$$

transforming the original data matrix to another one, albeit with only one variable.

Now there is a histogram for each vector $\mathbf{b}$. A one-dimensional **grand tour** runs through the vectors $\mathbf{b}$, displaying the histogram for $\mathbf{Yb}$ as it goes. (See Asimov [1985] and Buja and Asimov [1986] for general grand tour methodology.) Actually, one does not need all $\mathbf{b}$, e.g., the vectors $\mathbf{b} = (1,2,5)'$ and $\mathbf{b} = (2,4,10)'$ would give the same histogram. Just the scale of the horizontal axis on the histograms would be different. One simplification is to look at only the $\mathbf{b}$'s with norm 1. That is, the **norm** of a vector $\mathbf{x} = (x_1, \ldots, x_q)'$ is

$$\|\mathbf{x}\| = \sqrt{x_1^2 + \cdots + x_q^2} = \sqrt{\mathbf{x}'\mathbf{x}}, \tag{1.20}$$

so one would run through the $\mathbf{b}$'s with $\|\mathbf{b}\| = 1$. Note that the one-dimensional marginals are special cases: take

$$\mathbf{b}' = (1,0,\ldots,0), (0,1,0,\ldots,0), \ldots, \text{ or } (0,0,\ldots,1). \tag{1.21}$$

Scatter plots of two linear combinations are more common. That is, there are two sets of coefficients (b_{1j}'s and b_{2j}'s), and two resulting variables:

$$\begin{aligned} w_{i1} &= b_{11}y_{i1} + b_{21}y_{i2} + \cdots + b_{q1}y_{iq}, \text{ and} \\ w_{i2} &= b_{12}y_{i1} + b_{22}y_{i2} + \cdots + b_{q2}y_{iq}. \end{aligned} \tag{1.22}$$

In general, the data matrix generated from p linear combinations can be written

$$\mathbf{W} = \mathbf{YB}, \tag{1.23}$$

where $\mathbf{W}$ is $n \times p$, and $\mathbf{B}$ is $q \times p$ with column k containing the coefficients for the kth linear combination. As for one linear combination, the coefficient vectors are taken to have norm 1, i.e., $\|(b_{1k}, \ldots, b_{qk})\| = 1$, which is equivalent to having all the diagonals of $\mathbf{B}'\mathbf{B}$ being 1.

Another common restriction is to have the linear combination vectors be **orthogonal**, where two column vectors $\mathbf{b}$ and $\mathbf{c}$ are orthogonal if $\mathbf{b}'\mathbf{c} = 0$. Geometrically, orthogonality means the vectors are perpendicular to each other. One benefit of restricting to orthogonal linear combinations is that one avoids scatter plots that are highly correlated but not meaningfully so, e.g., one might have w_1 be *Height* + *Weight*, and w_2 be $.99 \times$ *Height* + $1.01 \times$ *Weight*. Having those two highly correlated does not tell us anything about the data set. If the columns of $\mathbf{B}$ are orthogonal to each other, as well as having norm 1, then

$$\mathbf{B}'\mathbf{B} = \mathbf{I}_p. \tag{1.24}$$

A set of norm 1 vectors that are mutually orthogonal are said to be **orthonormal**.

Return to $q = 2$ *orthonormal* linear combinations. A two-dimensional grand tour plots the two variables as the $q \times 2$ matrix $\mathbf{B}$ runs through all the matrices with a pair of orthonormal columns.

1.5.1 Rotations

If the **B** in (1.24) is $q \times q$, i.e., there are as many orthonormal linear combinations as variables, then **B** is an **orthogonal matrix**.

Definition 1.1. *A $q \times q$ matrix* **G** *is* ***orthogonal*** *if*

$$\mathbf{G}'\mathbf{G} = \mathbf{G}\mathbf{G}' = \mathbf{I}_q. \tag{1.25}$$

Note that the definition says that the columns are orthonormal, and the rows are orthonormal. In fact, the rows are orthonormal if and only of the columns are (if the matrix is square), since then $\mathbf{G}'$ is the inverse of **G**. Hence the middle equality in (1.25) is not strictly needed in the definition.

Think of the data matrix **Y** being the set of n points in q-dimensional space. For orthogonal matrix **G**, what does the set of points $\mathbf{W} = \mathbf{YG}$ look like? It looks exactly like **Y**, but rotated or flipped. Think of a pinwheel turning, or a chicken on a rotisserie, or the earth spinning around its axis or rotating about the sun. Figure 1.5 illustrates a simple rotation of two variables. In particular, the norms of the points in **Y** are the same as in **W**, so each point remains the same distance from 0.

Rotating point clouds for three variables work by first multiplying the $n \times 3$ data matrix by a 3×3 orthogonal matrix, then making a scatter plot of the first two resulting variables. By running through the orthogonal matrices smoothly, one gets the illusion of three dimensions. See the discussion immediately above Exercise 1.9.23 for some suggestions on software for real-time rotations.

1.6 Principal components

The grand tours and rotating point clouds described in the last two subsections do not have mathematical objectives, that is, one just looks at them to see if anything interesting pops up. In **projection pursuit** [Huber, 1985], one looks for a few (often just one or two) (orthonormal) linear combinations that maximize a given objective function. For example, if looking at just one linear combination, one may wish to find the one that maximizes the variance of the data, or the skewness or kurtosis, or one whose histogram is most bimodal. With two linear combinations, one may be after clusters of points, high correlation, curvature, etc.

Principal components are the orthonormal combinations that maximize the variance. They predate the term *projection pursuit* by decades [Pearson, 1901], and are the most commonly used. The idea behind them is that variation is information, so if one has several variables, one wishes the linear combinations that capture as much of the variation in the data as possible. You have to decide in particular situations whether variation is the important criterion. To find a column vector **b** to maximize the sample variance of $\mathbf{W} = \mathbf{Yb}$, we could take **b** infinitely large, which yields infinite variance. To keep the variance meaningful, we restrict to vectors **b** of unit norm.

For q variables, there are q principal components: The first has the maximal variance any one linear combination (with norm 1) can have, the second has the maximal variance among linear combinations orthogonal to the first, etc. The technical definition for a data matrix is below. First, we note that for a given $q \times p$ matrix **B**, the mean and variance of the elements in the linear transformation $\mathbf{W} = \mathbf{YB}$ are easily obtained from the mean and covariance matrix of $\mathbf{Y}$ using (1.8) and (1.15):

$$\overline{\mathbf{w}} = \frac{1}{n}\,\mathbf{1}_n'\mathbf{W} = \frac{1}{n}\,\mathbf{1}_n'\mathbf{Y}\mathbf{B} = \overline{\mathbf{y}}\mathbf{B}, \tag{1.26}$$

by (1.9), and

$$\mathbf{S_W} = \frac{1}{n}\,\mathbf{W}'\mathbf{H}_n\mathbf{W} = \frac{1}{n}\,\mathbf{B}'\mathbf{Y}'\mathbf{H}_n\mathbf{Y}\mathbf{B} = \mathbf{B}'\mathbf{S}\mathbf{B}, \tag{1.27}$$

where $\mathbf{S}$ is the covariance matrix of $\mathbf{Y}$ in (1.17). In particular, for a column vector $\mathbf{b}$, the sample variance of $\mathbf{Yb}$ is $\mathbf{b}'\mathbf{Sb}$. Thus the principal components aim to maximize $\mathbf{g}'\mathbf{Sg}$ for $\mathbf{g}$'s of unit length.

Definition 1.2. *Suppose* $\mathbf{S}$ *is the sample covariance matrix for the* $n \times q$ *data matrix* $\mathbf{Y}$*. Let* $\mathbf{g}_1, \ldots, \mathbf{g}_q$ *be an orthonormal set of* $q \times 1$ *vectors such that*

$$\begin{aligned}
&\mathbf{g}_1 \textit{ is any } \mathbf{g} \textit{ that maximizes } \mathbf{g}'\mathbf{Sg} \textit{ over } \|\mathbf{g}\| = 1;\\
&\mathbf{g}_2 \textit{ is any } \mathbf{g} \textit{ that maximizes } \mathbf{g}'\mathbf{Sg} \textit{ over } \|\mathbf{g}\| = 1, \mathbf{g}'\mathbf{g}_1 = 0;\\
&\mathbf{g}_3 \textit{ is any } \mathbf{g} \textit{ that maximizes } \mathbf{g}'\mathbf{Sg} \textit{ over } \|\mathbf{g}\| = 1, \mathbf{g}'\mathbf{g}_1 = \mathbf{g}'\mathbf{g}_2 = 0;\\
&\qquad\vdots\\
&\mathbf{g}_q \textit{ is any } \mathbf{g} \textit{ that maximizes } \mathbf{g}'\mathbf{Sg} \textit{ over } \|\mathbf{g}\| = 1, \mathbf{g}'\mathbf{g}_1 = \cdots = \mathbf{g}'\mathbf{g}_{q-1} = 0.
\end{aligned} \tag{1.28}$$

Then $\mathbf{Yg}_i$ *is the* i^{th} ***sample principal component****,* $\mathbf{g}_i$ *is its* ***loading vector****, and* $l_i \equiv \mathbf{g}_i'\mathbf{Sg}_i$ *is its sample variance.*

Because the function $\mathbf{g}'\mathbf{Sg}$ is continuous in $\mathbf{g}$, and the maximizations are over compact sets, these principal components always exist. They may not be unique, although for sample covariance matrices, if $n \geq q$, they almost always are unique, up to sign. See Section 13.1 and Exercise 13.4.3 for further discussion.

By the construction in (1.28), we have that the sample variances of the principal components are ordered as

$$l_1 \geq l_2 \geq \cdots \geq l_q. \tag{1.29}$$

What is not as obvious, but quite important, is that the principal components are uncorrelated, as in the next lemma, proved in Section 1.8.

Lemma 1.1. *The* $\mathbf{S}$ *and* $\mathbf{g}_1, \ldots, \mathbf{g}_q$ *in Definition 1.2 satisfy*

$$\mathbf{g}_i'\mathbf{Sg}_j = 0 \textit{ for } i \neq j. \tag{1.30}$$

Now $\mathbf{G} \equiv (\mathbf{g}_1, \ldots, \mathbf{g}_q)$ is an orthogonal matrix, and the matrix of principal components is

$$\mathbf{W} = \mathbf{YG}. \tag{1.31}$$

Equations (1.29) and (1.30) imply that the sample covariance matrix, say $\mathbf{L}$, of $\mathbf{W}$ is diagonal, with the l_i's on the diagonal. Hence by (1.27),

$$\mathbf{S_W} = \mathbf{G}'\mathbf{SG} = \mathbf{L} = \begin{pmatrix} l_1 & 0 & \cdots & 0 \\ 0 & l_2 & \cdots & 0 \\ \vdots & \vdots & \ddots & \vdots \\ 0 & 0 & \cdots & l_q \end{pmatrix} \Rightarrow \mathbf{S} = \mathbf{GLG}'. \tag{1.32}$$

Thus we have the following.

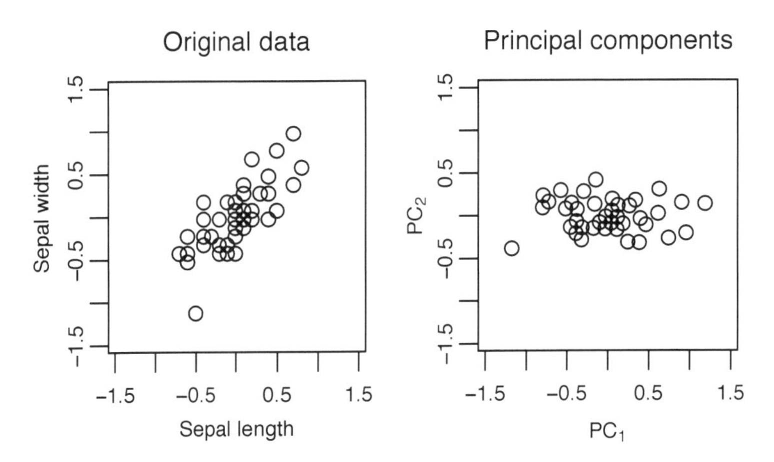

Figure 1.5: The sepal length and sepal width for the setosa iris data. The first plot is the raw data, centered. The second shows the two principal components.

Theorem 1.1 (The spectral decomposition theorem for symmetric matrices). *If* $\mathbf{S}$ *is a symmetric* $q \times q$ *matrix, then there exists a* $q \times q$ *orthogonal (1.25) matrix* $\mathbf{G}$ *and a* $q \times q$ *diagonal matrix* $\mathbf{L}$ *with diagonals* $l_1 \geq l_2 \geq \cdots \geq l_q$ *such that*

$$\mathbf{S} = \mathbf{G}\mathbf{L}\mathbf{G}'. \tag{1.33}$$

Although we went through the derivation with $\mathbf{S}$ being a covariance matrix, all we really need for this theorem is that $\mathbf{S}$ be symmetric. The $\mathbf{g}_i$'s and l_i's have mathematical names, too: Eigenvectors and eigenvalues.

Definition 1.3 (Eigenvalues and eigenvectors). *Suppose* $\mathbf{A}$ *is a* $q \times q$ *matrix. Then* λ *is an* ***eigenvalue*** *of* $\mathbf{A}$ *if there exists a nonzero* $q \times 1$ *vector* $\mathbf{u}$ *such that* $\mathbf{A}\mathbf{u} = \lambda\mathbf{u}$. *The vector* $\mathbf{u}$ *is the corresponding* ***eigenvector***. *Similarly,* $\mathbf{u} \neq \mathbf{0}$ *is an eigenvector if there exists an eigenvalue to which it corresponds.*

A little linear algebra shows that, indeed, each $\mathbf{g}_i$ is an eigenvector of $\mathbf{S}$ corresponding to l_i. Hence the following:

Symbol	Principal components	Spectral decomposition
l_i	Variance	Eigenvalue
$\mathbf{g}_i$	Loadings	Eigenvector

(1.34)

Figure 1.5 plots the principal components for the $q = 2$ variables sepal length and sepal width for the fifty iris observations of the species setosa. The data has been centered, so that the means are zero. The variances of the two original variables are 0.124 and 0.144, respectively. The first graph shows the two variables are highly correlated, with most of the points lining up near the 45° line. The principal component loading matrix $\mathbf{G}$ rotates the points approximately 45° clockwise as in the second graph, so that the data are now most spread out along the horizontal axis (variance is 0.234), and least along the vertical (variance is 0.034). The two principal components are also, as it appears, uncorrelated.

Best K components

In the process above, we found the principal components one by one. It may be that we would like to find the rotation for which the first K variables, say, have the maximal sum of variances. That is, we wish to find the orthonormal set of $q \times 1$ vectors $\mathbf{b}_1, \ldots, \mathbf{b}_K$ to maximize

$$\mathbf{b}_1' \mathbf{S} \mathbf{b}_1 + \cdots + \mathbf{b}_K' \mathbf{S} \mathbf{b}_K. \tag{1.35}$$

Fortunately, the answer is the same, i.e., take $\mathbf{b}_i = \mathbf{g}_i$ for each i, the principal component loadings. See Proposition 1.1 in Section 1.8. Section 13.1 explores principal components further.

1.6.1 Biplots

When plotting observations using the first few principal component variables, the relationship between the original variables and principal components is often lost. An easy remedy is to rotate and plot the original axes as well. Imagine in the original data space, in addition to the observed points, one plots arrows of length λ along the axes. That is, the arrows are the line segments

$$\mathbf{a}_i = \{(0, \ldots, 0, c, 0, \ldots, 0)' \mid 0 < c < \lambda\} \quad \text{(the } c \text{ is in the } i^{th} \text{ slot)}, \tag{1.36}$$

where an arrowhead is added at the non-origin end of the segment. If $\mathbf{Y}$ is the matrix of observations, and $\mathbf{G}_1$ the matrix containing the first p loading vectors, then

$$\widehat{\mathbf{X}} = \mathbf{Y}\mathbf{G}_1. \tag{1.37}$$

We also apply the transformation to the arrows:

$$\widehat{\mathbf{A}} = (\mathbf{a}_1, \ldots, \mathbf{a}_q)\mathbf{G}_1. \tag{1.38}$$

The plot consisting of the points $\widehat{\mathbf{X}}$ and the arrows $\widehat{\mathbf{A}}$ is then called the **biplot**. See Gabriel [1981]. The points of the arrows in $\widehat{\mathbf{A}}$ are just

$$\lambda \mathbf{I}_q \mathbf{G}_1 = \lambda \mathbf{G}_1, \tag{1.39}$$

so that in practice all we need to do is for each axis, draw an arrow pointing from the origin to $\lambda\times$ (the ith row of $\mathbf{G}_1$). The value of λ is chosen by trial-and-error, so that the arrows are amidst the observations. Notice that the components of these arrows are proportional to the loadings, so that the length of the arrows represents the weight of the corresponding variables on the principal components.

1.6.2 Example: Sports data

Louis Roussos asked $n = 130$ people to rank seven sports, assigning #1 to the sport they most wish to participate in, and #7 to the one they least wish to participate in. The sports are baseball, football, basketball, tennis, cycling, swimming and jogging. Here are a few of the observations:

Obs i	BaseB	FootB	BsktB	Ten	Cyc	Swim	Jog
1	1	3	7	2	4	5	6
2	1	3	2	5	4	7	6
3	1	3	2	5	4	7	6
⋮	⋮	⋮	⋮	⋮	⋮	⋮	⋮
129	5	7	6	4	1	3	2
130	2	1	6	7	3	5	4

(1.40)

E.g., the first person likes baseball and tennis, but not basketball or jogging (too much running?).

We find the principal components. The data is in the matrix sportsranks. It is easier to interpret the plot if we reverse the ranks, so that 7 is best and 1 is worst, then center the variables. The function eigen calculates the eigenvectors and eigenvalues of its argument, returning the results in the components vectors and values, respectively:

```
y <− 8−sportsranks
y <− scale(y,scale=F) # Centers the columns
eg <− eigen(var(y))
```

The function prcomp can also be used. The eigenvalues (variances) are

j	1	2	3	4	5	6	7
l_j	10.32	4.28	3.98	3.3	2.74	2.25	0

(1.41)

The first eigenvalue is 10.32, quite a bit larger than the second. The second through sixth are fairly equal, so it may be reasonable to look at just the first component. (The seventh eigenvalue is 0, but that follows because the rank vectors all sum to $1 + \cdots + 7 = 28$, hence exist in a six-dimensional space.)

We create the biplot using the first two dimensions. We first plot the people:

```
ev <− eg$vectors
w <− y%*%ev # The principal components
r <− range(w)
plot(w[,1:2],xlim=r,ylim=r,xlab=expression('PC'[1]),ylab=expression('PC'[2]))
```

The biplot adds in the original axes. Thus we want to plot the seven ($q = 7$) points as in (1.39), where $\mathbf{G}_1$ contains the first two eigenvectors. Plotting the arrows and labels:

```
arrows(0,0,5*ev[,1],5*ev[,2])
text(7*ev[,1:2],labels=colnames(y))
```

The constants "5" (which is the λ) and "7" were found by trial and error so that the graph, Figure 1.6, looks good. We see two main clusters. The left-hand cluster of people is associated with the team sports' arrows (baseball, football and basketball), and the right-hand cluster is associated with the individual sports' arrows (cycling, swimming, jogging). Tennis is a bit on its own, pointing south.

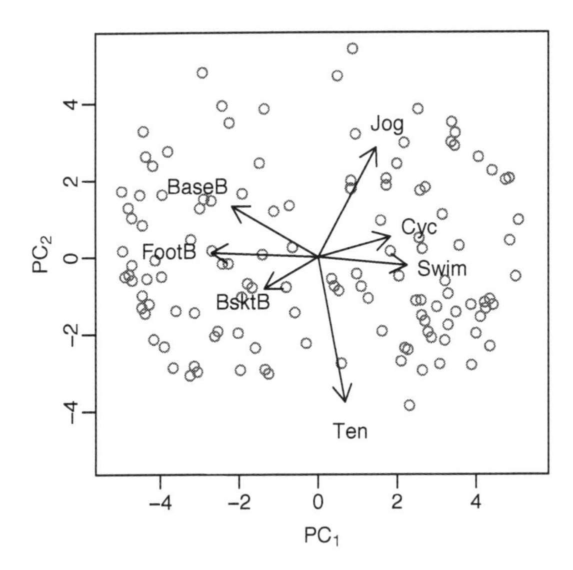

Figure 1.6: Biplot of the sports data, using the first two principal components.

1.7 Other projections to pursue

Principal components can be very useful, but you do have to be careful. For one, they depend crucially on the scaling of your variables. For example, suppose the data set has two variables, height and weight, measured on a number of adults. The variance of height, in inches, is about 9, and the variance of weight, in pounds, is 900 ($= 30^2$). One would expect the first principal component to be close to the weight variable, because that is where the variation is. On the other hand, if height were measured in millimeters, and weight in tons, the variances would be more like 6000 (for height) and 0.0002 (for weight), so the first principal component would be essentially the height variable. In general, if the variables are not measured in the same units, it can be problematic to decide what units to use for the variables. See Section 13.1.1. One common approach is to divide each variable by its standard deviation, so that the resulting variables all have variance 1.

Another caution is that the linear combination with largest variance is not necessarily the most interesting, e.g., you may want one that is maximally correlated with another variable, or that distinguishes two populations best, or that shows the most clustering.

Popular objective functions to maximize, other than variance, are skewness, kurtosis, and negative entropy. The idea is to find projections that are not normal (in the sense of the normal distribution). The hope is that these will show clustering or some other interesting feature.

Skewness measures a certain lack of symmetry, where one tail is longer than the other. For a sample $x_1, \ldots, x_n$, it is measured by the normalized sample third central (meaning subtract the mean) moment:

$$\text{Skewness} = \frac{\sum_{i=1}^{n}(x_i - \overline{x})^3/n}{(\sum_{i=1}^{n}(x_i - \overline{x})^2/n)^{3/2}}. \tag{1.42}$$

Positive values indicate a longer tail to the right, and negative to the left. Kurtosis is the normalized sample fourth central moment:

$$\text{Kurtosis} = \frac{\sum_{i=1}^{n}(x_i - \overline{x})^4/n}{(\sum_{i=1}^{n}(x_i - \overline{x})^2/n)^{2}} - 3. \tag{1.43}$$

The "−3" is there so that exactly normal data will have kurtosis 0. A variable with low kurtosis is more "boxy" than the normal. One with high kurtosis tends to have thick tails and a pointy middle. (A variable with low kurtosis is *platykurtic,* and one with high kurtosis is *leptokurtic,* from the Greek: *kyrtos* = curved, *platys* = flat, like a platypus, and *lepto* = thin.) Bimodal distributions often have low kurtosis.

Entropy

(You may wish to look through Section 2.1 before reading this section.) The entropy of a random variable Y with pdf $f(y)$ is

$$\text{Entropy}(f) = -E_f[\log(f(Y))]. \tag{1.44}$$

Entropy is supposed to measure lack of structure, so that the larger the entropy, the more diffuse the distribution is. For the normal, we have that

$$\text{Entropy}(N(\mu,\sigma^2)) = E_{N(\mu,\sigma^2)}\left[\log(\sqrt{2\pi\sigma^2}) + \frac{(Y-\mu)^2}{2\sigma^2}\right] = \frac{1}{2}\,(1+\log(2\pi\sigma^2)). \tag{1.45}$$

Note that it does not depend on the mean μ, and that it increases without bound as σ^2 increases. Thus maximizing entropy unrestricted is not an interesting task. However, one can imagine maximizing entropy for a given mean and variance, which leads to the next lemma, to be proved in Section 1.8.

Lemma 1.2. *The* $N(\mu,\sigma^2)$ *uniquely maximizes the entropy among all pdf's with mean* μ *and variance* σ^2.

Thus a measure of non-normality of g is its entropy subtracted from that of the normal with the same variance. Since there is a negative sign in front of the entropy of g, this difference is called **negentropy** defined for any g as

$$\text{Negent}(g) = \frac{1}{2}\,(1+\log(2\pi\sigma^2)) - \text{Entropy}(g), \quad \text{where } \sigma^2 = Var_g[Y]. \tag{1.46}$$

This value is known as the Kullback-Leibler distance, or discrimination information, from g to the normal density. See Kullback and Leibler [1951]. With data, the pdf g is unknown, so that the negentropy must be estimated.

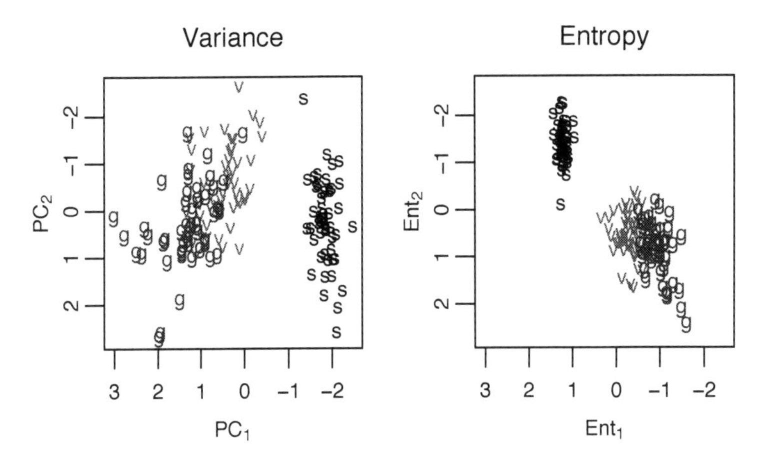

Figure 1.7: Projection pursuit for the iris data. The first plot is based on maximizing the variances of the projections, i.e., principal components. The second plot maximizes estimated entropies.

1.7.1 Example: Iris data

Consider the first three variables of the iris data (sepal length, sepal width, and petal length), normalized so that each variable has mean zero and variance one. We find the first two principal components, which maximize the variances, and the first two components that maximize the estimated entropies, defined as in Definition 1.28, but with estimated entropy of $\mathbf{Yg}$ substituted for the variance $\mathbf{g}'\mathbf{S}\mathbf{g}$. The table (1.47) contains the loadings for the variables. Note that the two objective functions do produce different projections. The first principal component weights equally on the two length variables, while the first entropy variable is essentially petal length.

	Variance		Entropy	
	$\mathbf{g}_1$	$\mathbf{g}_2$	$\mathbf{g}_1^*$	$\mathbf{g}_2^*$
Sepal length	0.63	0.43	0.08	0.74
Sepal width	−0.36	0.90	0.00	−0.68
Petal length	0.69	0.08	−1.00	0.06

(1.47)

Figure 1.7 graphs the results. The plots both show separation between setosa and the other two species, but the principal components plot has the observations more spread out, while the entropy plot shows the two groups much tighter.

The matrix iris has the iris data, with the first four columns containing the measurements, and the fifth specifying the species. The observations are listed with the fifty setosas first, then the fifty versicolors, then the fifty virginicas. To find the principal components for the first three variables, we use the following:

```
y <− scale(as.matrix(iris[,1:3]))
g <− eigen(var(y))$vectors
pc <− y%*%g
```

The first statement centers and scales the variables. The plot of the first two columns of pc is the first plot in Figure 1.7. The procedure we used for entropy is negent3D in Listing A.3, explained in Appendix A.1. The code is

```
gstar <- negent3D(y,nstart=10)$vectors
ent <-y%*%gstar
```

To create plots like the ones in Figure 1.7, use

```
par(mfrow=c(1,2))
sp <- rep(c('s','v','g'),c(50,50,50))
plot(pc[,1:2],pch=sp) # pch specifies the characters to plot.
plot(ent[,1:2],pch=sp)
```

1.8 Proofs

Proof of the principal components result, Lemma 1.1

The idea here is taken from Davis and Uhl [1999]. Consider the $\mathbf{g}_1, \ldots, \mathbf{g}_q$ as defined in (1.28). Take $i < j$, and for angle θ, let

$$h(\theta) = \mathbf{g}(\theta)'\mathbf{S}\mathbf{g}(\theta) \text{ where } \mathbf{g}(\theta) = \cos(\theta)\mathbf{g}_i + \sin(\theta)\mathbf{g}_j. \tag{1.48}$$

Because the $\mathbf{g}_i$'s are orthogonal,

$$\|\mathbf{g}(\theta)\| = 1 \text{ and } \mathbf{g}(\theta)'\mathbf{g}_1 = \cdots = \mathbf{g}(\theta)'\mathbf{g}_{i-1} = 0. \tag{1.49}$$

According to the ith stage in (1.28), $h(\theta)$ is maximized when $\mathbf{g}(\theta) = \mathbf{g}_i$, i.e., when $\theta = 0$. The function is differentiable, hence its derivative must be zero at $\theta = 0$. To verify (1.30), differentiate:

$$\begin{aligned} 0 &= \frac{d}{d\theta} h(\theta)|_{\theta=0} \\ &= \frac{d}{d\theta}(\cos^2(\theta)\mathbf{g}_i'\mathbf{S}\mathbf{g}_i + 2\sin(\theta)\cos(\theta)\mathbf{g}_i'\mathbf{S}\mathbf{g}_j + \sin^2(\theta)\mathbf{g}_j'\mathbf{S}\mathbf{g}_j)|_{\theta=0} \\ &= 2\mathbf{g}_i'\mathbf{S}\mathbf{g}_j. \end{aligned} \tag{1.50}$$

□

Best K components

We next consider finding the set $\mathbf{b}_1, \ldots, \mathbf{b}_K$ of orthonormal vectors to maximize the sum of variances, $\sum_{i=1}^K \mathbf{b}_i'\mathbf{S}\mathbf{b}_i$, as in (1.35). It is convenient here to have the next definition.

Definition 1.4 (Trace). *The* ***trace*** *of an $m \times m$ matrix* $\mathbf{A}$ *is the sum of its diagonals,* $\text{trace}(\mathbf{A}) = \sum_{i=1}^m a_{ii}$.

Thus if we let $\mathbf{B} = (\mathbf{b}_1, \ldots, \mathbf{b}_K)$, we have that

$$\sum_{i=1}^K \mathbf{b}_i'\mathbf{S}\mathbf{b}_i = \text{trace}(\mathbf{B}'\mathbf{S}\mathbf{B}). \tag{1.51}$$

Proposition 1.1. ***Best K components.*** *Suppose* $\mathbf{S}$ *is a* $q \times q$ *covariance matrix, and define* $\mathcal{B}_K$ *to be the set of* $q \times K$ *matrices with orthonormal columns,* $1 \leq K \leq q$*. Then*

$$\max_{\mathbf{B} \in \mathcal{B}_K} \text{trace}(\mathbf{B}'\mathbf{S}\mathbf{B}) = l_1 + \cdots + l_K, \tag{1.52}$$

which is achieved by taking $\mathbf{B} = (\mathbf{g}_1, \ldots, \mathbf{g}_K)$*, where* $\mathbf{g}_i$ *is the* i^{th} *principal component loading vector for* $\mathbf{S}$*, and* l_i *is the corresponding variance.*

Proof. Let $\mathbf{G}\mathbf{L}\mathbf{G}'$ be the spectral decomposition of $\mathbf{S}$ as in (1.33). Set $\mathbf{A} = \mathbf{G}'\mathbf{B}$, so that $\mathbf{A}$ is also in $\mathcal{B}_K$. Then $\mathbf{B} = \mathbf{G}\mathbf{A}$, and

$$\begin{aligned} \text{trace}(\mathbf{B}'\mathbf{S}\mathbf{B}) &= \text{trace}(\mathbf{A}'\mathbf{G}'\mathbf{S}\mathbf{G}\mathbf{A}) \\ &= \text{trace}(\mathbf{A}'\mathbf{L}\mathbf{A}) \\ &= \sum_{i=1}^{q} [(\sum_{j=1}^{K} a_{ij}^2) l_i] \\ &= \sum_{i=1}^{q} c_i l_i, \end{aligned} \tag{1.53}$$

where the a_{ij}'s are the elements of $\mathbf{A}$, and $c_i = \sum_{j=1}^{K} a_{ij}^2$. Because the columns of $\mathbf{A}$ have norm one, and the rows of $\mathbf{A}$ have norms less than or equal to one,

$$\sum_{i=1}^{q} c_i = \sum_{j=1}^{K} [\sum_{i=1}^{q} a_{ij}^2] = K \text{ and } c_i \leq 1. \tag{1.54}$$

Since $l_1 \geq l_2 \geq \cdots \geq l_q$, to maximize (1.53) under those constraints on the c_i's, we try to make the earlier c_i's as large as possible. Thus we set $c_1 = \cdots = c_K = 1$ and $c_{K+1} = \cdots = c_q = 0$. The resulting value of (1.53) is then $l_1 + \cdots + l_K$. Note that taking $\mathbf{A}$ with $a_{ii} = 1$, $i = 1, \ldots, K$, and 0 elsewhere (so that $\mathbf{A}$ consists of the first K columns of $\mathbf{I}_q$), achieves that maximum. With that $\mathbf{A}$, we have that $\mathbf{B} = (\mathbf{g}_1, \ldots, \mathbf{g}_K)$. □

Proof of the entropy result, Lemma 1.2

Let f be the $N(\mu, \sigma^2)$ density, and g be any other pdf with mean μ and variance σ^2. Then

$$\begin{aligned} \text{Entropy}(f) - \text{Entropy}(g) &= -\int f(y) \log(f(y)) dy + \int g(y) \log(g(y)) dy \\ &= \int g(y) \log(g(y)) dy - \int g(y) \log(f(y)) dy \\ &\quad + \int g(y) \log(f(y)) dy - \int f(y) \log(f(y)) dy \\ &= -\int g(y) \log(f(y)/g(y)) dy \\ &\quad + E_g \left[\log(\sqrt{2\pi\sigma^2}) + \frac{(Y-\mu)^2}{2\sigma^2} \right] \\ &\quad - E_f \left[\log(\sqrt{2\pi\sigma^2}) + \frac{(Y-\mu)^2}{2\sigma^2} \right] \qquad (1.55) \\ &= E_g[-\log(f(Y)/g(Y))]. \qquad (1.56) \end{aligned}$$

The last two terms in (1.55) are equal, since Y has the same mean and variance under f and g.

At this point we need an important inequality about convexity, to whit, what follows is a definition and lemma.

Definition 1.5 (Convexity). *The real-valued function h, defined on an interval $\mathbf{I} \subset \mathbb{R}$, is* ***convex*** *if for each $x_0 \in \mathbf{I}$, there exists an a_0 and b_0 such that*

$$h(x_0) = a_0 + b_0 x_0 \text{ and } h(x) \geq a_0 + b_0 x \text{ for } x \neq x_0. \tag{1.57}$$

The function is ***strictly convex*** *if the inequality is strict in (1.57).*

The line $a_0 + b_0 x$ is the tangent line to h at x_0. Convex functions have tangent lines that are below the curve, so that convex functions are "bowl-shaped." The next lemma is proven in Exercise 1.9.15.

Lemma 1.3 (Jensen's inequality). *Suppose W is a random variable with finite expected value. If $h(w)$ is a convex function, then*

$$E[h(W)] \geq h(E[W]), \tag{1.58}$$

where the left-hand expectation may be infinite. Furthermore, the inequality is strict if $h(w)$ is strictly convex and W is not constant, that is, $P[W = c] < 1$ for any c.

One way to remember the direction of the inequality is to imagine $h(w) = w^2$, in which case (1.58) states that $E[W^2] \geq E[W]^2$, which we already know because $Var[X] = E[W^2] - E[W]^2 \geq 0$.

Now back to (1.56). The function $h(w) = -\log(w)$ is strictly convex, and if g is not equivalent to f, $W = f(Y)/g(Y)$ is not constant. Jensen's inequality thus shows that

$$\begin{aligned} E_g[-\log(f(Y)/g(Y))] &> -\log\left(E[f(Y)/g(Y)]\right) \\ &= -\log\left(\int (f(y)/g(y))g(y)dy\right) \\ &= -\log\left(\int f(y)dy\right) = -\log(1) = 0. \end{aligned} \tag{1.59}$$

Putting (1.56) and (1.59) together yields

$$\text{Entropy}(N(0,\sigma^2)) - \text{Entropy}(g) > 0, \tag{1.60}$$

which completes the proof of Lemma 1.2. □

Answers: The question marks in Figure 1.4 are, respectively, virginica, setosa, virginica, versicolor, and setosa.

1.9 Exercises

Exercise 1.9.1. Let $\mathbf{H}_n$ be the centering matrix in (1.12). (a) What is $\mathbf{H}_n \mathbf{1}_n$? (b) Suppose $\mathbf{x}$ is an $n \times 1$ vector whose elements sum to zero. What is $\mathbf{H}_n \mathbf{x}$? (c) Show that $\mathbf{H}_n$ is idempotent (1.16).

Exercise 1.9.2. Define the matrix $\mathbf{J}_n = (1/n)\mathbf{1}_n\mathbf{1}_n'$, so that $\mathbf{H}_n = \mathbf{I}_n - \mathbf{J}_n$. (a) What does $\mathbf{J}_n$ do to a vector? (That is, what is $\mathbf{J}_n\mathbf{a}$?) (b) Show that $\mathbf{J}_n$ is idempotent. (c) Find the spectral decomposition (1.33) for $\mathbf{J}_n$ explicitly when $n = 3$. [Hint: In $\mathbf{G}$, the first column (eigenvector) is proportional to $\mathbf{1}_3$. The remaining two eigenvectors can be any other vectors such that the three eigenvectors are orthonormal. Once you have a $\mathbf{G}$, you can find the $\mathbf{L}$.] (d) Find the spectral decomposition for $\mathbf{H}_3$. [Hint: Use the same eigenvectors as for $\mathbf{J}_3$, but in a different order.] (e) What do you notice about the eigenvalues for these two matrices?

Exercise 1.9.3. A covariance matrix has *intraclass correlation structure* if all the variances are equal, and all the covariances are equal. So for $n = 3$, it would look like

$$\mathbf{A} = \begin{pmatrix} a & b & b \\ b & a & b \\ b & b & a \end{pmatrix}. \tag{1.61}$$

Find the spectral decomposition for this type of matrix. [Hint: Use the $\mathbf{G}$ in Exercise 1.9.2, and look at $\mathbf{G}'\mathbf{A}\mathbf{G}$. You may have to reorder the eigenvectors depending on the sign of b.]

Exercise 1.9.4. Show (1.24): if the $q \times p$ matrix $\mathbf{B}$ has orthonormal columns, then $\mathbf{B}'\mathbf{B} = \mathbf{I}_p$.

Exercise 1.9.5. Suppose $\mathbf{Y}$ is an $n \times q$ data matrix, and $\mathbf{W} = \mathbf{Y}\mathbf{G}$, where $\mathbf{G}$ is a $q \times q$ orthogonal matrix. Let $\mathbf{y}_1, \ldots, \mathbf{y}_n$ be the rows of $\mathbf{Y}$, and similarly $\mathbf{w}_i$'s be the rows of $\mathbf{W}$. (a) Show that the corresponding points have the same length: $\|\mathbf{y}_i\| = \|\mathbf{w}_i\|$. (b) Show that the distances between the points have not changed: $\|\mathbf{y}_i - \mathbf{y}_j\| = \|\mathbf{w}_i - \mathbf{w}_j\|$, for any i, j.

Exercise 1.9.6. Show that the construction in Definition 1.1 implies that the variances satisfy (1.30): $l_1 \geq l_2 \geq \cdots \geq l_q$.

Exercise 1.9.7. Suppose that the columns of $\mathbf{G}$ constitute the principal component loading vectors for the sample covariance matrix $\mathbf{S}$. Show that $\mathbf{g}_i'\mathbf{S}\mathbf{g}_i = l_i$ and $\mathbf{g}_i'\mathbf{S}\mathbf{g}_j = 0$ for $i \neq j$, as in (1.30), implies (1.32): $\mathbf{G}'\mathbf{S}\mathbf{G} = \mathbf{L}$.

Exercise 1.9.8. Verify (1.49) and (1.50).

Exercise 1.9.9. In (1.53), show that $\text{trace}(\mathbf{A}'\mathbf{L}\mathbf{A}) = \sum_{i=1}^{q}[(\sum_{j=1}^{K} a_{ij}^2)l_i]$.

Exercise 1.9.10. This exercise is to show that the eigenvalue matrix of a covariance matrix $\mathbf{S}$ is unique. Suppose $\mathbf{S}$ has two spectral decompositions, $\mathbf{S} = \mathbf{G}\mathbf{L}\mathbf{G}' = \mathbf{H}\mathbf{M}\mathbf{H}'$, where $\mathbf{G}$ and $\mathbf{H}$ are orthogonal matrices, and $\mathbf{L}$ and $\mathbf{M}$ are diagonal matrices with nonincreasing diagonal elements. Use Proposition 1.1 on both decompositions of $\mathbf{S}$ to show that for each $K = 1, \ldots, q$, $l_1 + \cdots + l_K = m_1 + \cdots + m_K$. Thus $\mathbf{L} = \mathbf{K}$.

Exercise 1.9.11. Suppose $\mathbf{Y}$ is a data matrix, and $\mathbf{Z} = \mathbf{Y}\mathbf{F}$ for some orthogonal matrix $\mathbf{F}$, so that $\mathbf{Z}$ is a rotated version of $\mathbf{Y}$. Show that the variances of the principal components are the same for $\mathbf{Y}$ and $\mathbf{Z}$. (This result should make intuitive sense.) [Hint: Find the spectral decomposition of the covariance of $\mathbf{Z}$ from that of $\mathbf{Y}$, then note that these covariance matrices have the same eigenvalues.]

Exercise 1.9.12. Show that in the spectral decomposition (1.33), each l_i is an eigenvalue, with corresponding eigenvector $\mathbf{g}_i$, i.e., $\mathbf{S}\mathbf{g}_i = l_i\mathbf{g}_i$.

Exercise 1.9.13. Suppose λ is an eigenvalue of the covariance matrix $\mathbf{S}$. Show that λ must equal one of the l_i's in the spectral decomposition of $\mathbf{S}$. [Hint: Let $\mathbf{u}$ be an eigenvector corresponding to λ. Show that λ is also an eigenvalue of $\mathbf{L}$, with corresponding eigenvector $\mathbf{v} = \mathbf{G}'\mathbf{u}$, so that $l_i v_i = \lambda v_i$ for each i. Why does that fact lead to the desired result?]

Exercise 1.9.14. Verify the expression for $\int g(y) \log(f(y))dy$ in (1.55).

Exercise 1.9.15. Consider the setup in Jensen's inequality, Lemma 1.3. (a) Show that if h is convex, $E[h(W)] \geq h(E[W])$. [Hint: Set $x_0 = E[W]$ in Definition 1.5.] (b) Suppose h is strictly convex. Give an example of a random variable W for which $E[h(W)] = h(E[W])$. (c) Show that if h is strictly convex and W is not constant, that $E[h(W)] > E[W]$.

Exercise 1.9.16 (Spam). In the Hewlett-Packard spam data, $n = 4601$ emails were classified according to whether they were spam, where "0" means not spam, "1" means spam. Fifty-seven explanatory variables based on the content of the emails were recorded, including various word and symbol frequencies. The emails were sent to George Forman (not the boxer) at Hewlett-Packard labs, hence emails with the words "George" or "hp" would likely indicate non-spam, while "credit" or "!" would suggest spam. The data were collected by Hopkins et al. [1999], and are in the data matrix Spam. (They are also in the R data frame spam from the ElemStatLearn package [Halvorsen, 2012], as well as at the UCI Machine Learning Repository [Lichman, 2013].)

Based on an email's content, is it possible to accurately guess whether it is spam or not? Here we use Chernoff's faces. Look at the faces of some emails known to be spam and some known to be non-spam (the "training data"). Then look at some randomly chosen faces (the "test data"). E.g., to have twenty observations known to be spam, twenty known to be non-spam, and twenty test observations, use the following R code:

```
x0 <- Spam[Spam[,'spam']==0,] # The non-spam
x1 <- Spam[Spam[,'spam']==1,] # The spam
train0 <- x0[1:20,]
train1 <- x1[1:20,]
test <- rbind(x0[-(1:20),],x1[-(1:20),])[sample(1:4561,20),]
```

Based on inspecting the training data, try to classify the test data. How accurate are your guesses? The faces program uses only the first fifteen variables of the input matrix, so you should try different sets of variables. For example, for each variable find the value of the t-statistic for testing equality of the spam and email groups, then choose the variables with the largest absolute t's.

Exercise 1.9.17 (Spam). Continue with the spam data from Exercise 1.9.16. (a) Plot the variances of the explanatory variables (the first 57 variables) versus the index (i.e., the x-axis has $(1, 2, \ldots, 57)$, and the y-axis has the corresponding variances.) You might not see much, so repeat the plot, but taking logs of the variances. What do you see? Which three variables have the largest variances? (b) Find the principal components using just the explanatory variables. Plot the eigenvalues versus the index. Plot the log of the eigenvalues versus the index. What do you see? (c) Look at the loadings for the first three principal components. (E.g., if spamload contains the loadings (eigenvectors), then you can try plotting them using matplot(1:57,spamload[,1:3]).) What is the main feature of the loadings? How do they relate to your answer in part (a)?

(d) Now scale the explanatory variables so each has mean zero and variance one: `spamscale <- scale(Spam[,1:57])`. Find the principal components using this matrix. Plot the eigenvalues versus the index. What do you notice, especially compared to the results of part (b)? (e) Plot the loadings of the first three principal components obtained in part (d). How do they compare to those from part (c)? Why is there such a difference?

Exercise 1.9.18 (Sports data). Consider the Louis Roussos sports data described in Section 1.6.2. Use faces to cluster the observations. Use the raw variables, or the principal components, and try different orders of the variables (which maps the variables to different sets of facial features). After clustering some observations, look at how they ranked the sports. Do you see any pattern? Were you able to distinguish between people who like team sports versus individual sports? Those who like (dislike) tennis? Jogging?

Exercise 1.9.19 (Election). The data set election has the results of the first three US presidential races of the 2000's (2000, 2004, 2008). The observations are the 50 states plus the District of Columbia, and the values are the $(D-R)/(D+R)$ for each state and each year, where D is the number of votes the Democrat received, and R is the number the Republican received. (a) Without scaling the variables, find the principal components. What are the first two principal component loadings measuring? What is the ratio of the standard deviation of the first component to the second's? (c) Plot the first versus second principal components, using the states' two-letter abbreviations as the plotting characters. (They are in the vector stateabb.) Make the plot so that the two axes cover the same range. (d) There is one prominent outlier. What is it, and for which variable is it mostly outlying? (e) Comparing how states are grouped according to the plot and how close they are geographically, can you make any general statements about the states and their voting profiles (at least for these three elections)?

Exercise 1.9.20 (Painters). The data set painters has ratings of 54 famous painters. It is in the MASS package [Venables and Ripley, 2002]. See Davenport and Studdert-Kennedy [1972] for a more in-depth discussion. The R help file says

> The subjective assessment, on a 0 to 20 integer scale, of 54 classical painters. The painters were assessed on four characteristics: composition, drawing, colour and expression. The data is due to the Eighteenth century art critic, de Piles.

The fifth variable gives the school of the painter, using the following coding:

> A: Renaissance; B: Mannerist; C: Seicento; D: Venetian; E: Lombard; F: Sixteenth Century; G: Seventeenth Century; H: French

Create the two-dimensional biplot for the data. Start by turning the data into a matrix, then centering both dimensions, then scaling:

```
x <- scale(as.matrix(painters[,1:4]),scale=F)
x <- t(scale(t(x),scale=F))
x <- scale(x)
```

Use the fifth variable, the painters' schools, as the plotting character, and the four rating variables as the arrows. Interpret the two principal component variables. Can you make any generalizations about which schools tend to rate high on which scores?

Exercise 1.9.21 (Cereal). Chakrapani and Ehrenberg [1981] analyzed people's attitudes towards a variety of breakfast cereals. The data matrix cereal is 8×11, with rows corresponding to eight cereals, and columns corresponding to potential attributes about cereals. The attributes: Return (a cereal one would come back to), tasty, popular (with the entire family), digestible, nourishing, natural flavor, affordable, good value, crispy (stays crispy in milk), fit (keeps one fit), and fun (for children). The original data consisted of the percentage of subjects who thought the given cereal possessed the given attribute. The present matrix has been doubly centered, so that the row means and columns means are all zero. (The original data can be found in the S-Plus® [TIBCO Software Inc., 2009] data set cereal.attitude.) Create the two-dimensional biplot for the data with the cereals as the points (observations), and the attitudes as the arrows (variables). What do you see? Are there certain cereals/attributes that tend to cluster together? (You might want to look at the Wikipedia entry [Wikipedia, 2015] on breakfast cereals.)

Exercise 1.9.22 (Decathlon). The decathlon data set has scores on the top 24 men in the decathlon (a set of ten events) at the 2008 Olympics. The scores are the numbers of points each participant received in each event, plus each person's total points. See [Wikipedia, 2014a]. Create the biplot for these data based on the first ten variables (i.e., do not use their total scores). Doubly center, then scale, the data as in Exercise 1.9.20. The events should be the arrows. Do you see any clustering of the events? The athletes?

The remaining questions require software that will display rotating point clouds of three dimensions, and calculate some projection pursuit objective functions. The Spin program at `http://istics.net/Spin` is sufficient for our purposes. GGobi [Cook and Swayne, 2007] has an excellent array of graphical tools for interactively exploring multivariate data. See also Wickham et al. [2011] and the spin3R routine in the R package aplpack [Wolf, 2014].

Exercise 1.9.23 (Iris). Consider the three variables X = sepal length, Y = petal length, and Z = petal width in the Fisher-Anderson iris data. (a) Look at the data while rotating. What is the main feature of these three variables? (b) Scale the data so that the variables all have the same sample variance. (The Spin program automatically performs the scaling.) For various objective functions (variance, skewness, kurtosis, negative kurtosis, negentropy), find the rotation that maximizes the function. (That is, the first component of the rotation maximizes the criterion over all rotations. The second then maximizes the criterion for components orthogonal to the first. The third component is then whatever is orthogonal to the first two.) Which criteria are most effective in yielding rotations that exhibit the main feature of the data? Which are least effective? (c) Which of the original variables are most prominently represented in the first two components of the most effective rotations?

Exercise 1.9.24 (Automobiles). The data set cars [Consumers' Union, 1990] contains $q = 11$ size measurements on $n = 111$ models of automobile. The original data can be found in the S-Plus® [TIBCO Software Inc., 2009] data frame cu.dimensions. In cars, the variables have been normalized to have medians of 0 and median absolute deviations (MAD) of 1.4826 (the MAD for a $N(0, 1)$). Inspect the three-dimensional data set consisting of the variables length, width, and height. (In the Spin program, the data set is called "Cars.") (a) Find the linear combination with the largest variance.

What is the best linear combination? (Can you interpret it?) What is its variance? Does the histogram look interesting? (b) Now find the linear combination to maximize negentropy. What is the best linear combination, and its entropy? What is the main feature of the histogram? (c) Find the best two linear combinations for entropy. What are they? What feature do you see in the scatter plot?

Exercise 1.9.25 (RANDU). *RANDU* [IBM, 1970] is a venerable, fast, efficient, and very flawed random number generator. See Dudewicz and Ralley [1981] for a thorough review of old-time random number generators. For given "seed" x_0, RANDU produces x_{i+1} from x_i via

$$x_{i+1} = (65539\, x_i) \mod 2^{31}. \tag{1.62}$$

The "random" Uniform(0,1) values are then $u_i = x_i/2^{31}$. The R data set `randu` is based on a sequence generated using RANDU, where each of $n = 400$ rows is a set of $p = 3$ consecutive u_i's. Rotate the data, using objective criteria if you wish, to look for significant nonrandomness in the data matrix. If the data are really random, the points should uniformly fill up the three-dimensional cube. What feature do you see that reveals the nonrandomness?

The data sets Example 1, Example 2, ..., Example 5 are artificial three-dimensional point clouds. The goal is to rotate the point clouds to reveal their structures.

Exercise 1.9.26. Consider the Example 1 data set. (a) Find the first two principal components for these data. What are their variances? (b) Rotate the data. Are the principal components unique? (c) Find the two-dimensional plots based on maximizing the skewness, kurtosis, negative kurtosis, and negentropy criteria. What do you see? What does the histogram for the linear combination with the largest kurtosis look like? Is it "pointy"? What does the histogram for the linear combination with the most negative kurtosis look like? Is it "boxy"? (d) Describe the three-dimensional structure of the data points. Do the two-dimensional plots in part (c) give a good idea of the three-dimensional structure?

Exercise 1.9.27. This question uses the Example 2 data set. (a) What does the histogram for the linear combination with the largest variance look like? (b) What does the histogram for the linear combination with the largest negentropy look like? (c) Describe the three-dimensional object.

Exercise 1.9.28. For each of Example 3, 4, and 5, try to guess the shape of the cloud of data points based on just the 2-way scatter plots. Then rotate the points enough to convince yourself of the actual shape.

Chapter 2

Multivariate Distributions

This chapter reviews the elements of distribution theory that we need, especially for vectors and matrices. (Classical multivariate analysis is basically linear algebra, so everything we do eventually gets translated into matrix equations.) See any good mathematical statistics book such as Hogg, McKean, and Craig [2012], Bickel and Doksum [2000], or Lehmann and Casella [2003] for a more comprehensive treatment.

2.1 Probability distributions

We will deal with random variables and finite collections of random variables. A random variable X has range or **space** $\mathcal{X} \subset \mathbb{R}$, the real line. A collection of random variables is a finite set of random variables. They could be arranged in any convenient way, such as a row or column vector, matrix, triangular array, or three-dimensional array, and will often be indexed to match the arrangement. The default arrangement will be to index the random variables by $1, \ldots, N$, so that the collection is $\mathbf{X} = (X_1, \ldots, X_N)$, considered as a row vector. The space of $\mathbf{X}$ is $\mathcal{X} \subset \mathbb{R}^N$, N-dimensional Euclidean space. A probability distribution P for a random variable or collection of random variables specifies the chance that the random object will fall in a particular subset of its space. That is, for $A \subset \mathcal{X}$, $P[A]$ is the probability that the random $\mathbf{X}$ is in A, also written $P[\mathbf{X} \in A]$. In principle, to describe a probability distribution, one must specify $P[A]$ for all subsets A. (Technically, all "measurable" subsets, but we will not worry about measurability.) Fortunately, there are easier ways. We will use densities, but the main method will be to use **representations**, by which we mean describing a collection of random variables $\mathbf{Y}$ in terms of another collection $\mathbf{X}$ for which we already know the distribution, usually through a function, i.e., $\mathbf{Y} = g(\mathbf{X})$.

2.1.1 Distribution functions

The **distribution function** for the probability distribution P for the collection $\mathbf{X} = (X_1, \ldots, X_N)$ of random variables is the function

$$\begin{aligned} F &: \mathbb{R}^N \to [0,1] \\ F(x_1, x_2, \ldots, x_N) &= P[X_1 \le x_1, X_2 \le x_2, \ldots, X_N \le x_N]. \end{aligned} \tag{2.1}$$

Note that it is defined on all of $\mathbb{R}^N$, not just the space of $\mathbf{X}$. It is nondecreasing, and continuous from the right, in each x_i. The limit as all $x_i \to -\infty$ is zero, and as all $x_i \to \infty$, the limit is one. The distribution function uniquely defines the distribution, though we will not find much use for it.

2.1.2 Densities

A collection of random variables $\mathbf{X}$ is said to have a **density with respect to Lebesgue measure on** $\mathbb{R}^N$, if there is a nonnegative function $f(\mathbf{x})$,

$$f : \mathcal{X} \longrightarrow [0, \infty), \tag{2.2}$$

such that for any $A \subset \mathcal{X}$,

$$\begin{aligned} P[A] &= \int_A f(\mathbf{x}) d\mathbf{x} \\ &= \int \int \cdots \int_A f(x_1, \ldots, x_N) dx_1 \cdots dx_N. \end{aligned} \tag{2.3}$$

The second line is there to emphasize that we have a multiple integral. (The Lebesgue measure of a subset $\mathbf{A}$ of $\mathbb{R}^N$ is the integral $\int_{\mathbf{A}} d\mathbf{x}$, i.e., as if $f(\mathbf{x}) = 1$ in (2.3). Thus if $N = 1$, the Lebesgue measure of a line segment is its length. In two dimensions, the Lebesgue measure of a set is its area. For $N = 3$, it is the volume.)

We will call a density f as in (2.3) the "pdf," for "probability density function." (Densities can also be defined with respect to other measures than Lebesgue, but for our purposes, pdf means the dominating measure is Lebesgue.) Because $P[\mathbf{X} \in \mathcal{X}] = 1$, the integral of the pdf over the entire space $\mathcal{X}$ must be 1. Random variables or collections that have pdf's are **continuous** in the sense that their distribution functions are continuous in $\mathbf{x}$. (There are continuous distributions that do not have pdf's, such as the uniform distribution on the unit circle.)

If $\mathbf{X}$ does have a pdf, then it can be obtained from the distribution function in (2.1) by differentiation:

$$f(x_1, \ldots, x_N) = \frac{\partial^N}{\partial x_1 \cdots \partial x_N} F(x_1, \ldots, x_N). \tag{2.4}$$

If the space $\mathcal{X}$ is a countable (which includes finite) set, then its probability can be given by specifying the probability of each individual point. The **probability mass function** f, or "pmf," with

$$f : \mathcal{X} \longrightarrow [0, 1], \tag{2.5}$$

is given by

$$f(\mathbf{x}) = P[\mathbf{X} = \mathbf{x}] = P[\{\mathbf{x}\}]. \tag{2.6}$$

The probability of any subset A is the sum of the probabilities of the individual points in A,

$$P[A] = \sum_{\mathbf{x} \subset A} f(\mathbf{x}). \tag{2.7}$$

Such an $\mathbf{X}$ is called **discrete**. (A pmf is also a density, but with respect to counting measure on $\mathcal{X}$, not Lebesgue measure.)

Not all random variables are either discrete or continuous, and especially a collection of random variables could have some discrete and some continuous members. In such cases, the probability of a set is found by integrating over the continuous parts and summing over the discrete parts. For example, suppose our collection is a $1 \times N$ vector combining two other collections, i.e.,

$$\mathbf{W} = (\mathbf{X}, \mathbf{Y}) \text{ has space } \mathcal{W}, \ \mathbf{X} \text{ is } 1 \times N_x \text{ and } \mathbf{Y} \text{ is } 1 \times N_y, \ N = N_x + N_y. \tag{2.8}$$

For a subset $A \subset \mathcal{W}$, define the **marginal space** by

$$\mathcal{X}^A = \{\mathbf{x} \in \mathbb{R}^{N_x} \mid (\mathbf{x}, \mathbf{y}) \in A \ \text{ for some } \mathbf{y}\}, \tag{2.9}$$

and the **conditional space given X** $= \mathbf{x}$ by

$$\mathcal{Y}_{\mathbf{x}}^A = \{\mathbf{y} \in \mathbb{R}^{N_y} \mid (\mathbf{x}, \mathbf{y}) \in A\}. \tag{2.10}$$

Suppose $\mathbf{X}$ is discrete and $\mathbf{Y}$ is continuous. Then $f(\mathbf{x}, \mathbf{y})$ is a mixed-type density for the distribution of $\mathbf{W}$ if for any $A \subset \mathcal{W}$,

$$P[A] = \sum_{\mathbf{x} \in \mathcal{X}^A} \int_{\mathcal{Y}_{\mathbf{x}}^A} f(\mathbf{x}, \mathbf{y}) d\mathbf{y}. \tag{2.11}$$

We will use the generic term "density" to mean pdf, pmf, or the mixed type of density in (2.11). There are other types of densities, but we will not need to deal with them.

2.1.3 Representations

Representations are very useful, especially when no pdf exists. For example, suppose $\mathbf{Y} = (Y_1, Y_2)$ is uniform on the unit circle, by which we mean $\mathbf{Y}$ has space $\mathcal{Y} = \{\mathbf{y} \in \mathbb{R}^2 \mid \|\mathbf{y}\| = 1\}$, and it is equally likely to be any point on that circle. There is no pdf, because the area of the circle in $\mathbb{R}^2$ is zero, so the integral over any subset of $\mathcal{Y}$ of any function is zero. The distribution can be thought of in terms of the angle $\mathbf{y}$ makes with the x-axis, that is, $\mathbf{y}$ is equally likely to be at any angle. Thus we can let $X \sim \text{Uniform}(0, 2\pi]$: X has space $(0, 2\pi]$ and pdf $f_X(x) = 1/(2\pi)$. Then we can define

$$\mathbf{Y} = (\cos(X), \sin(X)). \tag{2.12}$$

In general, suppose we are given the distribution for $\mathbf{X}$ with space $\mathcal{X}$ and function g,

$$g : \mathcal{X} \longrightarrow \mathcal{Y}. \tag{2.13}$$

Then for any $B \subset \mathcal{Y}$, we can define the probability of $\mathbf{Y}$ by

$$P[\mathbf{Y} \in B] = P[g(\mathbf{X}) \in B] = P[\mathbf{X} \in g^{-1}(B)]. \tag{2.14}$$

We know the final probability because $g^{-1}(B) \subset \mathcal{X}$.

One special type of function yields **marginal distributions**, analogous to the marginals in Section 1.5, that picks off some of the components. Consider the setup in (2.8). The marginal function for $\mathbf{X}$ simply chooses the $\mathbf{X}$ components:

$$g(\mathbf{x}, \mathbf{y}) = \mathbf{x}. \tag{2.15}$$

The space of $\mathbf{X}$ is then given by (2.9) with $A = \mathcal{W}$, i.e.,

$$\mathcal{X} \equiv \mathcal{X}^{\mathcal{W}} = \{\mathbf{x} \in \mathbb{R}^{N_x} \mid (\mathbf{x}, \mathbf{y}) \in \mathcal{W} \text{ for some } \mathbf{y}\}. \tag{2.16}$$

If $f(\mathbf{x}, \mathbf{y})$ is the density for $(\mathbf{X}, \mathbf{Y})$, then the density of $\mathbf{X}$ can be found by "integrating (or summing) out" the $\mathbf{y}$. That is, if f is a pdf, then $f_{\mathbf{X}}(\mathbf{x})$ is the pdf for $\mathbf{X}$, where

$$f_{\mathbf{X}}(\mathbf{x}) = \int_{\mathcal{Y}_{\mathbf{x}}} f(\mathbf{x}, \mathbf{y}) d\mathbf{y}, \tag{2.17}$$

and

$$\mathcal{Y}_{\mathbf{x}} = \mathcal{Y}_{\mathbf{x}}^{\mathcal{W}} = \{\mathbf{y} \in \mathbb{R}^{N_y} \mid (\mathbf{x}, \mathbf{y}) \in \mathcal{W}\} \tag{2.18}$$

is the conditional space (2.10) with $A = \mathcal{W}$. If $\mathbf{y}$ has some discrete components, then they are summed in (2.17).

Note that we can find the marginals of any subset, not just sets of consecutive elements. E.g., if $\mathbf{X} = (X_1, X_2, X_3, X_4, X_5)$, we can find the marginal of (X_2, X_4, X_5) by integrating out the X_1 and X_3.

Probability distributions can also be represented through conditioning, discussed in the next section.

2.1.4 Conditional distributions

The conditional distribution of one or more variables given another set of variables is central to multivariate analysis. E.g., what is the distribution of health measures given diet, smoking, and ethnicity? We start with the two collections of variables $\mathbf{Y}$ and $\mathbf{X}$, each of which may be a random variable, vector, matrix, etc. We want to make sense of the notion

$$\text{Conditional distribution of } \mathbf{Y} \text{ given } \mathbf{X} = \mathbf{x}, \text{ written } \mathbf{Y} \mid \mathbf{X} = \mathbf{x}. \tag{2.19}$$

What this means is that for each fixed value $\mathbf{x}$, there is a possibly different distribution for $\mathbf{Y}$.

Very generally, such conditional distributions will exist, though they may be hard to figure out, even what they mean. In the discrete case, the concept is straightforward, and by analogy the case with densities follows. For more general situations, we will use properties of conditional distributions rather than necessarily specifying them.

We start with the $(\mathbf{X}, \mathbf{Y})$ as in (2.8), and assume we have their **joint distribution** P. The word "joint" is technically unnecessary, but helps to emphasize that we are considering the two collections together. The joint space is $\mathcal{W}$, and let $\mathcal{X}$ denote the marginal space of $\mathbf{X}$ as in (2.16), and for each $\mathbf{x} \in \mathcal{X}$, the conditional space of $\mathbf{Y}$ given $\mathbf{X} = \mathbf{x}$, $\mathcal{Y}_{\mathbf{x}}$, is given in (2.18). For example, if the space $\mathcal{W} = \{(x, y) \mid 0 < x < y < 1\}$, then $\mathcal{X} = (0, 1)$, and for $x \in \mathcal{X}$, $\mathcal{Y}_x = (x, 1)$.

Next, given the joint distribution of $(\mathbf{X}, \mathbf{Y})$, we define the conditional distribution (2.19) in the discrete, then pdf, cases.

Discrete case

For sets A and B, the conditional probability of A given B is defined as

$$P[A \mid B] = \frac{P[A \cap B]}{P[B]} \quad \text{if} \quad B \neq \emptyset. \tag{2.20}$$

If B is empty, then the conditional probability is not defined since we would have $\frac{0}{0}$. For a discrete pair $(\mathbf{X}, \mathbf{Y})$, let $f(\mathbf{x}, \mathbf{y})$ be the pmf. Then the conditional distribution of

$\mathbf{Y}$ given $\mathbf{X} = \mathbf{x}$ can be specified by

$$P[\mathbf{Y} = \mathbf{y} \mid \mathbf{X} = \mathbf{x}], \text{ for } \mathbf{x} \in \mathcal{X},\ \mathbf{y} \in \mathcal{Y}_{\mathbf{x}}, \tag{2.21}$$

at least if $P[\mathbf{X} = \mathbf{x}] > 0$. The expression in (2.21) is, for fixed $\mathbf{x}$, the conditional pmf for $\mathbf{Y}$:

$$\begin{aligned} f_{\mathbf{Y}|\mathbf{X}}(\mathbf{y} \mid \mathbf{x}) &= P[\mathbf{Y} = \mathbf{y} \mid \mathbf{X} = \mathbf{x}] \\ &= \frac{P[\mathbf{Y} = \mathbf{y} \text{ and } \mathbf{X} = \mathbf{x}]}{P[\mathbf{X} = \mathbf{x}]} \\ &= \frac{f(\mathbf{x}, \mathbf{y})}{f_{\mathbf{X}}(\mathbf{x})},\ \ \mathbf{y} \in \mathcal{Y}_{\mathbf{x}}, \end{aligned} \tag{2.22}$$

if $f_{\mathbf{X}}(\mathbf{x}) > 0$, where $f_{\mathbf{X}}(\mathbf{x})$ is the marginal pmf of $\mathbf{X}$ from (2.17) with sums.

Pdf case

In the discrete case, the restriction that $P[\mathbf{X} = \mathbf{x}] > 0$ is not worrisome, since the chance is 0 we will have an $\mathbf{x}$ with $P[\mathbf{X} = \mathbf{x}] = 0$. In the continuous case, we cannot follow the same procedure, since $P[\mathbf{X} = \mathbf{x}] = 0$ for *all* $\mathbf{x} \in \mathcal{X}$. However, if we have pdf's, or general densities, we can analogize (2.22) and declare that the conditional density of $\mathbf{Y}$ given $\mathbf{X} = \mathbf{x}$ is

$$f_{\mathbf{Y}|\mathbf{X}}(\mathbf{y} \mid \mathbf{x}) = \frac{f(\mathbf{x}, \mathbf{y})}{f_{\mathbf{X}}(\mathbf{x})},\ \ \mathbf{y} \in \mathcal{Y}_{\mathbf{x}}, \tag{2.23}$$

if $f_{\mathbf{X}}(\mathbf{x}) > 0$. In this case, as in the discrete one, the restriction that $f_{\mathbf{X}}(\mathbf{x}) > 0$ is not worrisome, since the set on which $\mathbf{X}$ has density zero has probability zero. It turns out that the definition (2.23) is mathematically legitimate.

The $\mathbf{Y}$ and $\mathbf{X}$ can be very general. Often, both will be functions of a collection of random variables, so that we may be interested in conditional distributions of the type

$$g(\mathbf{Y}) \mid h(\mathbf{X}) = \mathbf{z} \tag{2.24}$$

for some functions g and h.

Reconstructing the joint distribution

Note that if we are given the marginal space and density for $\mathbf{X}$, and the conditional spaces and densities for $\mathbf{Y}$ given $\mathbf{X} = \mathbf{x}$, then we can reconstruct the joint space and joint density:

$$\mathcal{W} = \{(\mathbf{x}, \mathbf{y}) \mid \mathbf{y} \in \mathcal{Y}_{\mathbf{x}}, \mathbf{x} \in \mathcal{X}\} \text{ and } f(\mathbf{x}, \mathbf{y}) = f_{\mathbf{Y}|\mathbf{X}}(\mathbf{y} \mid \mathbf{x}) f_{\mathbf{X}}(\mathbf{x}). \tag{2.25}$$

Thus another way to represent a distribution for $\mathbf{Y}$ is to specify the conditional distribution given each $\mathbf{X} = \mathbf{x}$, and the marginal of $\mathbf{X}$. The marginal distribution of $\mathbf{Y}$ is then found by first finding the joint as in (2.25), then integrating out the $\mathbf{x}$:

$$f_{\mathbf{Y}}(\mathbf{y}) = \int_{\mathcal{X}_{\mathbf{y}}} f_{\mathbf{Y}|\mathbf{X}}(\mathbf{y} \mid \mathbf{x}) f_{\mathbf{X}}(\mathbf{x}) d\mathbf{x}. \tag{2.26}$$

2.2 Expected values

Means, variances, and covariances (Section 1.4) are key sample quantities in describing data. Similarly, they are important for describing random variables. These are all expected values of some function, defined next.

Definition 2.1 (Expected value). *Suppose* $\mathbf{X}$ *has space* $\mathcal{X}$, *and consider the real-valued function* g,

$$g : \mathcal{X} \longrightarrow \mathbb{R}. \tag{2.27}$$

If $\mathbf{X}$ *has pdf* f, *then the expected value of* $g(\mathbf{X})$, $E[g(\mathbf{X})]$, *is*

$$E[g(\mathbf{X})] = \int_{\mathcal{X}} g(\mathbf{x})f(\mathbf{x})d\mathbf{x} \tag{2.28}$$

if the integral converges. If $\mathbf{X}$ *has pmf* f, *then*

$$E[g(\mathbf{X})] = \sum_{\mathbf{x}\in\mathcal{X}} g(\mathbf{x})f(\mathbf{x}) \tag{2.29}$$

if the sum converges.

As in (2.11), if the collection is $(\mathbf{X}, \mathbf{Y})$, where $\mathbf{X}$ is discrete and $\mathbf{Y}$ is continuous, and $f(\mathbf{x}, \mathbf{y})$ is its mixed-type density, then for function $g(\mathbf{x}, \mathbf{y})$,

$$E[g(\mathbf{X}, \mathbf{Y})] = \sum_{\mathbf{x}\in\mathcal{X}} \int_{\mathcal{Y}_\mathbf{x}} g(\mathbf{x}, \mathbf{y})f(\mathbf{x}, \mathbf{y})d\mathbf{y}. \tag{2.30}$$

if everything converges. (The spaces are defined in (2.16) and (2.18).)

Expected values for representations cohere is the proper way, that is, if $\mathbf{Y}$ is a collection of random variables such that $\mathbf{Y} = h(\mathbf{X})$, then for a function g,

$$E[g(\mathbf{Y})] = E[g(h(\mathbf{X}))], \tag{2.31}$$

if the latter exists. Thus we often can find the expected values of functions of $\mathbf{Y}$ based on the distribution of $\mathbf{X}$. See Exercise 2.7.5 for the discrete case.

Conditioning

If $(\mathbf{X}, \mathbf{Y})$ has a joint distribution, then we can define the conditional expectation of $g(\mathbf{Y})$ given $\mathbf{X} = \mathbf{x}$ to be the regular expected value of $g(\mathbf{Y})$, but we use the conditional distribution $\mathbf{Y} \mid \mathbf{X} = \mathbf{x}$. In the pdf case, we write

$$E[g(\mathbf{Y}) \mid \mathbf{X} = \mathbf{x}] = \int_{\mathcal{Y}_\mathbf{x}} g(\mathbf{y})f_{\mathbf{Y}|\mathbf{X}}(\mathbf{y}|\mathbf{x})d\mathbf{y} \equiv e_g(\mathbf{x}). \tag{2.32}$$

Note that the conditional expectation is a function of $\mathbf{x}$. We can then take the expected value of that, using the marginal distribution of $\mathbf{X}$. We end up with the same result (if we end up with anything) as taking the usual expected value of $g(\mathbf{Y})$. That is,

$$E[g(\mathbf{Y})] = E[E[g(\mathbf{Y}) \mid \mathbf{X} = \mathbf{x}]]. \tag{2.33}$$

There is a bit of a notational glitch in the formula, since the inner expected value is a function of $\mathbf{x}$, a constant, and we really want to take the expected value over $\mathbf{X}$. We cannot just replace $\mathbf{x}$ with $\mathbf{X}$, however, because then we would have the undesired

$E[g(\mathbf{Y}) \mid \mathbf{X} = \mathbf{X}]$. So a more precise way to express the result is to use the $e_g(\mathbf{x})$ in (2.32), so that

$$E[g(\mathbf{Y})] = E[e_g(\mathbf{X})]. \tag{2.34}$$

This result holds in general. It is not hard to see in the pdf case:

$$\begin{aligned}
E[e_g(\mathbf{X})] &= \int_{\mathcal{X}} e_g(\mathbf{x}) f_{\mathbf{X}}(\mathbf{x}) d\mathbf{x} \\
&= \int_{\mathcal{X}} \int_{\mathcal{Y}_{\mathbf{x}}} g(\mathbf{y}) f_{\mathbf{Y}|\mathbf{X}}(\mathbf{y}|\mathbf{x}) d\mathbf{y} f_{\mathbf{X}}(\mathbf{x}) d\mathbf{x} && \text{by (2.32)} \\
&= \int_{\mathcal{X}} \int_{\mathcal{Y}_{\mathbf{x}}} g(\mathbf{y}) f(\mathbf{x}, \mathbf{y}) d\mathbf{y} d\mathbf{x} && \text{by (2.25)} \\
&= \int_{\mathcal{W}} g(\mathbf{y}) f(\mathbf{x}, \mathbf{y}) d\mathbf{x} d\mathbf{y} && \text{by (2.25)} \\
&= E[g(\mathbf{Y})].
\end{aligned} \tag{2.35}$$

A useful corollary is the **total probability formula**: For $B \subset \mathcal{Y}$, if $\mathbf{X}$ has a pdf,

$$P[\mathbf{Y} \in B] = \int_{\mathcal{X}} P[\mathbf{Y} \in B \mid \mathbf{X} = \mathbf{x}] f_{\mathbf{X}}(\mathbf{x}) d\mathbf{x}. \tag{2.36}$$

If $\mathbf{X}$ has a pmf, then we sum. The formula follows by taking g to be the indicator function I_B, given as

$$I_B(\mathbf{y}) = \begin{cases} 1 & \text{if} \quad \mathbf{y} \in B, \\ 0 & \text{if} \quad \mathbf{y} \notin B. \end{cases} \tag{2.37}$$

2.3 Means, variances, and covariances

Means, variances, and covariances are particular expected values. For a collection of random variables $\mathbf{X} = (X_1, \ldots, X_N)$, the **mean** of X_j is its expected value, $E[X_j]$. (Throughout this section, we will be acting as if the expected values exist. So if $E[X_j]$ doesn't exist, then the mean of X_j doesn't exist, but we might not explicitly mention that.) Often the mean is denoted by μ, so that $E[X_j] = \mu_j$.

The **variance** of X_j, often denoted σ_j^2 or σ_{jj}, is

$$\sigma_{jj} = Var[X_j] = E[(X_j - \mu_j)^2]. \tag{2.38}$$

The **covariance** between X_j and X_k is defined to be

$$\sigma_{jk} = Cov[X_j, X_k] = E[(X_j - \mu_j)(X_k - \mu_k)]. \tag{2.39}$$

Their **correlation coefficient** is

$$Corr[X_j, X_k] = \rho_{jk} = \frac{\sigma_{jk}}{\sqrt{\sigma_{jj}\sigma_{kk}}}, \tag{2.40}$$

if both variances are positive. Compare these definitions to those of the sample analogs, (1.3), (1.4), (1.5), and (1.6). So, e.g., $Var[X_j] = Cov[X_j, X_j]$.

The **mean of the collection X** is the corresponding collection of means. That is,

$$\boldsymbol{\mu} = E[\mathbf{X}] = (E[X_1], \ldots, E[X_N]). \tag{2.41}$$

2.3.1 Vectors and matrices

If a collection has a particular structure, then its mean has the same structure. That is, if $\mathbf{X}$ is a row vector as in (2.41), then $E[\mathbf{X}] = (E[X_1], \ldots, E[X_N])$. If $\mathbf{X}$ is a column vector, so is its mean. Similarly, if $\mathbf{W}$ is an $n \times p$ matrix, then so is its mean. That is,

$$E[\mathbf{W}] = E\left[\begin{pmatrix} W_{11} & W_{12} & \cdots & W_{1p} \\ W_{21} & W_{22} & \cdots & W_{2p} \\ \vdots & \vdots & \ddots & \vdots \\ W_{n1} & W_{n2} & \cdots & W_{np} \end{pmatrix}\right]$$
$$= \begin{pmatrix} E[W_{11}] & E[W_{12}] & \cdots & E[W_{1p}] \\ E[W_{21}] & E[W_{22}] & \cdots & E[W_{2p}] \\ \vdots & \vdots & \ddots & \vdots \\ E[W_{n1}] & E[W_{n2}] & \cdots & E[W_{np}] \end{pmatrix}. \tag{2.42}$$

Turning to variances and covariances, first suppose that $\mathbf{X}$ is a vector (row or column). There are N variances and $\binom{N}{2}$ covariances among the X_j's to consider, recognizing that $Cov[X_j, X_k] = Cov[X_k, X_j]$. By convention, we will arrange them into a matrix, the **variance-covariance matrix**, or simply **covariance matrix** of $\mathbf{X}$:

$$\boldsymbol{\Sigma} = Cov[\mathbf{X}]$$
$$= \begin{pmatrix} Var[X_1] & Cov[X_1, X_2] & \cdots & Cov[X_1, X_N] \\ Cov[X_2, X_1] & Var[X_2] & \cdots & Cov[X_2, X_N] \\ \vdots & \vdots & \ddots & \vdots \\ Cov[X_N, X_1] & Cov[X_N, X_2] & \cdots & Var[X_N] \end{pmatrix}, \tag{2.43}$$

so that the elements of $\boldsymbol{\Sigma}$ are the σ_{jk}'s. Compare this arrangement to that of the sample covariance matrix (1.17). If $\mathbf{X}$ is a row vector, and $\boldsymbol{\mu} = E[\mathbf{X}]$, a convenient expression for its covariance is

$$Cov[\mathbf{X}] = E\left[(\mathbf{X} - \boldsymbol{\mu})'(\mathbf{X} - \boldsymbol{\mu})\right]. \tag{2.44}$$

Similarly, if $\mathbf{X}$ is a column vector, $Cov[\mathbf{X}] = E[(\mathbf{X} - \boldsymbol{\mu})(\mathbf{X} - \boldsymbol{\mu})']$.

Now suppose $\mathbf{X}$ is a matrix as in (2.42). Notice that individual components have double subscripts: X_{ij}. We need to decide how to order the elements in order to describe its covariance matrix. We will use the convention that the elements are strung out by row, so that $\text{row}(\mathbf{X})$ is the $1 \times N$ vector, $N = np$, given by

$$\begin{aligned} \text{row}(\mathbf{X}) = (&X_{11}, X_{12}, \cdots, X_{1p}, \\ &X_{21}, X_{22}, \cdots, X_{2p}, \\ &\cdots \\ &X_{n1}, X_{n2}, \cdots, X_{np}). \end{aligned} \tag{2.45}$$

Then $Cov[\mathbf{X}]$ is defined to be $Cov[\text{row}(\mathbf{X})]$, which is an $(np) \times (np)$ matrix.

One more covariance: The covariance between two vectors is defined to be the matrix containing all the individual covariances of one variable from each vector.

That is, if $\mathbf{X}$ is $1 \times p$ and $\mathbf{Y}$ is $1 \times q$, then the $p \times q$ matrix of covariances is

$$Cov[\mathbf{X}, \mathbf{Y}] = E[(\mathbf{X} - E[\mathbf{X}])'(\mathbf{Y} - E[\mathbf{Y}])]$$

$$= \begin{pmatrix} Cov[X_1, Y_1] & Cov[X_1, Y_2] & \cdots & Cov[X_1, Y_q] \\ Cov[X_2, Y_1] & Cov[X_2, Y_2] & \cdots & Cov[X_2, Y_q] \\ \vdots & \vdots & \ddots & \vdots \\ Cov[X_p, Y_1] & Cov[X_p, Y_2] & \cdots & Cov[X_p, Y_q] \end{pmatrix}. \tag{2.46}$$

2.3.2 Moment generating functions

The **moment generating function** (**mgf** for short) of $\mathbf{X}$ is a function from $\mathbb{R}^N \rightarrow [0, \infty]$ given by

$$M_{\mathbf{X}}(\mathbf{t}) = M_{\mathbf{X}}(t_1, \ldots, t_N) = E\left[e^{t_1 X_1 + \cdots + t_N X_N}\right] = E\left[e^{\mathbf{X}\mathbf{t}'}\right] \tag{2.47}$$

for $\mathbf{t} = (t_1, \ldots, t_N)$. It is very useful in distribution theory, especially convolutions (sums of independent random variables), asymptotics, and for generating moments. The main use we have is that the mgf determines the distribution:

Theorem 2.1 (Uniqueness of mgf). *If for some $\epsilon > 0$,*

$$M_{\mathbf{X}}(\mathbf{t}) < \infty \textit{ and } M_{\mathbf{X}}(\mathbf{t}) = M_{\mathbf{Y}}(\mathbf{t}) \textit{ for all } \mathbf{t} \textit{ such that } \|\mathbf{t}\| < \epsilon, \tag{2.48}$$

then $\mathbf{X}$ and $\mathbf{Y}$ have the same distribution.

See Ash [1970] for an approach to proving this result. The mgf does not always exist, that is, often the integral or sum defining the expected value diverges. That is ok, as long as it is finite for $\mathbf{t}$ in a neighborhood of $\mathbf{0}$. If one knows complex variables, the **characteristic function** is handy because it always exists. It is defined as $\phi_{\mathbf{X}}(\mathbf{t}) = E[\exp(i\mathbf{X}\mathbf{t}')]$.

If a distribution's mgf is finite when $\|\mathbf{t}\| < \epsilon$ for some $\epsilon > 0$, then all of its moments are finite, and can be calculated via differentiation:

$$E[X_1^{k_1} \cdots X_N^{k_N}] = \frac{\partial^K}{\partial t_1^{k_1} \cdots \partial t_N^{k_N}} M_{\mathbf{X}}(\mathbf{t}) \Big|_{\mathbf{t}=\mathbf{0}}, \tag{2.49}$$

where the k_i are nonnegative integers, and $K = k_1 + \cdots + k_N$. See Exercise 2.7.22.

2.4 Independence

Two sets of random variables are **independent** if the values of one set do not affect the values of the other. More precisely, suppose the collection is $(\mathbf{X}, \mathbf{Y})$ as in (2.8), with space $\mathcal{W}$. Let $\mathcal{X}$ and $\mathcal{Y}$ be the marginal spaces (2.16) of $\mathbf{X}$ and $\mathbf{Y}$, respectively. First, we need the following:

Definition 2.2. *If $A \subset \mathbb{R}^K$ and $B \subset \mathbb{R}^L$, then $A \times B$ is a **rectangle**, the subset of $\mathbb{R}^{K+L}$ given by*

$$A \times B = \{(\mathbf{x}, \mathbf{y}) \in \mathbb{R}^{K+L} \mid \mathbf{x} \in A \textit{ and } \mathbf{y} \in B\}. \tag{2.50}$$

Now for the main definition.

Definition 2.3. *Given the setup above, the collections* $\mathbf{X}$ *and* $\mathbf{Y}$ *are* ***independent*** *if* $\mathcal{W} = \mathcal{X} \times \mathcal{Y}$, *and for every* $A \subset \mathcal{X}$ *and* $B \subset \mathcal{Y}$,

$$P[(\mathbf{X}, \mathbf{Y}) \in A \times B] = P[\mathbf{X} \in A]P[\mathbf{Y} \in B]. \tag{2.51}$$

In the definition, the left-hand side uses the joint probability distribution for $(\mathbf{X}, \mathbf{Y})$, and the right-hand side uses the marginal probabilities for $\mathbf{X}$ and $\mathbf{Y}$, respectively.

If the joint collection $(\mathbf{X}, \mathbf{Y})$ has density f, then $\mathbf{X}$ and $\mathbf{Y}$ are independent if and only if $\mathcal{W} = \mathcal{X} \times \mathcal{Y}$, and

$$f(\mathbf{x}, \mathbf{y}) = f_{\mathbf{X}}(\mathbf{x}) f_{\mathbf{Y}}(\mathbf{y}) \text{ for all } \mathbf{x} \in \mathcal{X} \textit{ and } \mathbf{y} \in \mathcal{Y}, \tag{2.52}$$

where $f_{\mathbf{X}}$ and $f_{\mathbf{Y}}$ are the marginal densities (2.17) of $\mathbf{X}$ and $\mathbf{Y}$, respectively. (Technically, (2.52) only has to hold with probability one. Also, except for sets of probability zero, the requirements (2.51) or (2.52) imply that $\mathcal{W} = \mathcal{X} \times \mathcal{Y}$, so that the requirement we place on the spaces is redundant. But we keep it for emphasis.)

A useful result is that $\mathbf{X}$ and $\mathbf{Y}$ are independent if and only if

$$E[g(\mathbf{X})h(\mathbf{Y})] = E[g(\mathbf{X})]E[h(\mathbf{Y})] \tag{2.53}$$

for all functions $g : \mathbf{X} \to \mathbb{R}$ and $h : \mathbf{Y} \to \mathbb{R}$ with finite expectation.

The last expression can be used to show that independent variables have covariance equal to 0. If X and Y are independent random variables with finite expectations, then

$$\begin{aligned} Cov[X, Y] &= E[(X - E[X])(Y - E[Y])] \\ &= E[(X - E[X])]\, E[(Y - E[Y])] \\ &= 0. \end{aligned} \tag{2.54}$$

The second equality uses (2.53), and the final equality uses that $E[X - E[X]] = E[X] - E[X] = 0$. Be aware that the reverse is *not* true, that is, variables can have 0 covariance but still not be independent.

If the collections $\mathbf{X}$ and $\mathbf{Y}$ are independent, then $Cov[X_k, Y_l] = 0$ for all k, l, so that

$$Cov[(\mathbf{X}, \mathbf{Y})] = \begin{pmatrix} Cov[\mathbf{X}] & \mathbf{0} \\ \mathbf{0} & Cov[\mathbf{Y}] \end{pmatrix}, \tag{2.55}$$

at least if the covariances exist. (Throughout this book, "$\mathbf{0}$" represents a matrix of zeroes, its dimension implied by the context.)

Collections $\mathbf{Y}$ and $\mathbf{X}$ are independent if and only if the conditional distribution of $\mathbf{Y}$ given $\mathbf{X} = \mathbf{x}$ does not depend on $\mathbf{x}$. If $(\mathbf{X}, \mathbf{Y})$ has a pdf or pmf, this property is easy to see. If $\mathbf{X}$ and $\mathbf{Y}$ are independent, then $\mathcal{Y}_{\mathbf{x}} = \mathcal{Y}$ since $\mathcal{W} = \mathcal{X} \times \mathcal{Y}$, and by (2.23) and (2.52),

$$f_{\mathbf{Y}|\mathbf{X}}(\mathbf{y} \mid \mathbf{x}) = \frac{f(\mathbf{x}, \mathbf{y})}{f_{\mathbf{X}}(\mathbf{x})} = \frac{f_{\mathbf{Y}}(\mathbf{y}) f_{\mathbf{X}}(\mathbf{x})}{f_{\mathbf{X}}(\mathbf{x})} = f_{\mathbf{Y}}(\mathbf{y}), \tag{2.56}$$

so that the conditional distribution does not depend on $\mathbf{x}$. On the other hand, if the conditional distribution does not depend on $\mathbf{x}$, then the conditional space and pdf cannot depend on $\mathbf{x}$, in which case they are the marginal space and pdf, so that $\mathcal{W} = \mathcal{X} \times \mathcal{Y}$ and

$$\frac{f(\mathbf{x},\mathbf{y})}{f_{\mathbf{X}}(\mathbf{x})} = f_{\mathbf{Y}}(\mathbf{y}) \Longrightarrow f(\mathbf{x},\mathbf{y}) = f_{\mathbf{X}}(\mathbf{x}) f_{\mathbf{Y}}(\mathbf{y}). \tag{2.57}$$

So far, we have treated independence of just two sets of variables. Everything can be easily extended to any finite number of sets. That is, suppose $\mathbf{X}_1, \ldots, \mathbf{X}_S$ are collections of random variables, with N_s and $\mathcal{X}_s$ being the dimension and space for $\mathbf{X}_s$, and $\mathbf{X} = (\mathbf{X}_1, \ldots, \mathbf{X}_S)$, with dimension $N = N_1 + \cdots + N_S$ and space $\mathcal{X}$.

Definition 2.4. *Given the setup above, the collections $\mathbf{X}_1, \ldots, \mathbf{X}_S$ are **mutually independent** if $\mathcal{X} = \mathcal{X}_1 \times \cdots \times \mathcal{X}_S$, and for every set of subsets $A_s \subset \mathcal{X}_s$,*

$$P[(\mathbf{X}_1, \ldots, \mathbf{X}_S) \in A_1 \times \cdots \times A_S] = P[\mathbf{X}_1 \in A_1] \cdots P[\mathbf{X}_S \in A_S]. \tag{2.58}$$

In particular, $\mathbf{X}_1, \ldots, \mathbf{X}_S$ being mutually independent implies that every pair $\mathbf{X}_i, \mathbf{X}_j$ $(i \neq j)$ is independent. The reverse need not be true, however, that is, each pair could be independent without having all mutually independent. Analogs of the equivalences in (2.52) to (2.53) hold for this case, too. E.g., $\mathbf{X}_1, \ldots, \mathbf{X}_S$ are mutually independent if and only if

$$E[g_1(\mathbf{X}_1) \cdots g_S(\mathbf{X}_S)] = E[g_1(\mathbf{X}_1)] \cdots E[g_S(\mathbf{X}_S)] \tag{2.59}$$

for all functions $g_s : \mathcal{X}_s \rightarrow \mathbb{R}$, $s = 1, \ldots, S$, with finite expectation.

A common situation is that the individual random variables X_i's in $\mathbf{X}$ are mutually independent. Then, e.g., if there are densities,

$$f(x_1, \ldots, x_N) = f_1(x_1) \cdots f_N(x_N), \tag{2.60}$$

where f_j is the density of X_j. Also, if the variances exist, the covariance matrix is diagonal:

$$Cov[\mathbf{X}] = \begin{pmatrix} Var[X_1] & 0 & \cdots & 0 \\ 0 & Var[X_2] & \cdots & 0 \\ \vdots & \vdots & \ddots & \vdots \\ 0 & \cdots & 0 & Var[X_N] \end{pmatrix}. \tag{2.61}$$

2.5 Additional properties of conditional distributions

The properties that follow are straightforward to prove in the discrete case. They still hold for the continuous and more general cases, but are not always easy to prove.

Plug-in formula

Suppose the collection of random variables is given by $(\mathbf{X}, \mathbf{Y})$, and we are interested in the conditional distribution of the function $g(\mathbf{X}, \mathbf{Y})$ given $\mathbf{X} = \mathbf{x}$. Then

$$g(\mathbf{X}, \mathbf{Y}) \mid \mathbf{X} = \mathbf{x} \ =^{\mathcal{D}} \ g(\mathbf{x}, \mathbf{Y}) \mid \mathbf{X} = \mathbf{x}. \tag{2.62}$$

That is, the conditional distribution of $g(\mathbf{X},\mathbf{Y})$ given $\mathbf{X} = \mathbf{x}$ is the same as that of $g(\mathbf{x},\mathbf{Y})$ given $\mathbf{X} = \mathbf{x}$. (The "$=^{\mathcal{D}}$" means "equal in distribution.") Furthermore, if $\mathbf{Y}$ and $\mathbf{X}$ are independent, we can take off the conditional part at the end of (2.62):

$$\mathbf{X} \text{ and } \mathbf{Y} \text{ independent} \Longrightarrow g(\mathbf{X},\mathbf{Y}) \mid \mathbf{X} = \mathbf{x} \ =^{\mathcal{D}} \ g(\mathbf{x},\mathbf{Y}). \tag{2.63}$$

Exercise 2.7.6 considers the discrete case.

This property may at first seem so obvious to be meaningless, but it can be very useful. For example, suppose X and Y are independent $N(0,1)$'s, and $g(X,Y) = X+Y$, so we wish to find $X+Y \mid X = x$. The official way is to let $W = X+Y$, and $Z = X$, and use the transformation of variables to find the space and pdf of (W,Z). One can then figure out $\mathcal{W}_z$, and use the formula (2.23). Instead, using the plug-in formula with independence (2.63), we have that

$$X+Y \mid X = x \ =^{\mathcal{D}} \ x+Y, \tag{2.64}$$

which we immediately realize is $N(x,1)$.

Conditional independence

Given a set of three collections, $(\mathbf{X},\mathbf{Y},\mathbf{Z})$, $\mathbf{X}$ are $\mathbf{Y}$ are said to be *conditionally independent given* $\mathbf{Z} = \mathbf{z}$ if

$$P[(\mathbf{X},\mathbf{Y}) \in A \times B \mid \mathbf{Z} = \mathbf{z}] = P[\mathbf{X} \in A \mid \mathbf{Z} = \mathbf{z}]P[\mathbf{Y} \in B \mid \mathbf{Z} = \mathbf{z}], \tag{2.65}$$

for sets $A \subset \mathcal{X}_{\mathbf{z}}$ and $B \subset \mathcal{Y}_{\mathbf{z}}$ as in (2.51). If further $\mathbf{X}$ is independent of $\mathbf{Z}$, then $\mathbf{X}$ is independent of the combined $(\mathbf{Y},\mathbf{Z})$. See Exercise 2.7.7.

Dependence on x only through a function

If the conditional distribution of $\mathbf{Y}$ given $\mathbf{X} = \mathbf{x}$ depends on $\mathbf{x}$ only through the function $h(\mathbf{x})$, then that conditional distribution is the same as the conditional distribution given $h(\mathbf{X}) = h(\mathbf{x})$. Symbolically, if $\mathbf{v} = h(\mathbf{x})$,

$$\mathbf{Y} \mid \mathbf{X} = \mathbf{x} \ =^{\mathcal{D}} \ \mathbf{Y} \mid h(\mathbf{X}) = \mathbf{v}. \tag{2.66}$$

As an illustration, suppose (X,Y) is uniformly distributed over the unit disk, so that the pdf is $f(x,y) = 1/\pi$ for $x^2+y^2 < 1$. Then it can be shown (Exercise 2.7.8) that

$$Y \mid X = x \sim \text{Uniform}(-\sqrt{1-x^2}, \sqrt{1-x^2}). \tag{2.67}$$

Note that the distribution depends on x only through $h(x) = x^2$, so that, e.g., conditioning on $X = 1/2$ is the same as conditioning on $X = -1/2$. The statement (2.66) then yields

$$Y \mid X^2 = v \sim \text{Uniform}(-\sqrt{1-v}, \sqrt{1-v}). \tag{2.68}$$

That is, we have managed to turn a statement about conditioning on X to one about conditioning on X^2.

Variance decomposition

The formula (2.34) shows that the expected value of $g(\mathbf{Y})$ is the expected value of the conditional expected value, $e_g(\mathbf{X})$. A similar formula holds for the variance, but it is not simply that the variance is the expected value of the conditional variance. Using the well-known identity $Var[Z] = E[Z^2] - E[Z]^2$ on $Z = g(\mathbf{Y})$, as well as (2.34) on $g(\mathbf{Y})$ and $g(\mathbf{Y})^2$, we have

$$\begin{aligned} Var[g(\mathbf{Y})] &= E[g(\mathbf{Y})^2] - E[g(\mathbf{Y})]^2 \\ &= E[e_{g^2}(\mathbf{X})] - E[e_g(\mathbf{X})]^2. \end{aligned} \tag{2.69}$$

The identity holds conditionally as well, i.e.,

$$\begin{aligned} v_g(\mathbf{x}) = Var[g(\mathbf{Y}) \mid \mathbf{X} = \mathbf{x}] &= E[g(\mathbf{Y})^2 \mid \mathbf{X} = \mathbf{x}] - E[g(\mathbf{Y}) \mid \mathbf{X} = \mathbf{x}]^2 \\ &= e_{g^2}(\mathbf{x}) - e_g(\mathbf{x})^2. \end{aligned} \tag{2.70}$$

Taking expected value over $\mathbf{X}$ in (2.70), we have

$$E[v_g(\mathbf{X})] = E[e_{g^2}(\mathbf{X})] - E[e_g(\mathbf{X})^2]. \tag{2.71}$$

Comparing (2.69) and (2.71), we see the difference lies in where the square is in the second terms. Thus

$$\begin{aligned} Var[g(\mathbf{Y})] &= E[v_g(\mathbf{X})] + E[e_g(\mathbf{X})^2] - E[e_g(\mathbf{X})]^2 \\ &= E[v_g(\mathbf{X})] + Var[e_g(\mathbf{X})], \end{aligned} \tag{2.72}$$

now using the identity on $e_g(\mathbf{X})$. Thus the variance of $g(\mathbf{Y})$ equals the variance of the conditional expected value plus the expected value of the conditional variance.

For a collection $\mathbf{Y}$ of random variables, where

$$e_{\mathbf{Y}}(\mathbf{x}) = E[\mathbf{Y} \mid \mathbf{X} = \mathbf{x}] \;\text{ and }\; v_{\mathbf{Y}}(\mathbf{x}) = Cov[\mathbf{Y} \mid \mathbf{X} = \mathbf{x}], \tag{2.73}$$

(2.72) extends to

$$Cov[\mathbf{Y}] = E[v_{\mathbf{Y}}(\mathbf{X})] + Cov[e_{\mathbf{Y}}(\mathbf{X})]. \tag{2.74}$$

See Exercise 2.7.12.

Bayes theorem

Bayes formula reverses conditional distributions, that is, it takes the conditional distribution of $\mathbf{Y}$ given $\mathbf{X}$, and the marginal of $\mathbf{X}$, and returns the conditional distribution of $\mathbf{X}$ given $\mathbf{Y}$. Bayesian inference is based on this formula, starting with the distribution of the data given the parameters, and a marginal ("prior") distribution of the parameters, and producing the conditional distribution ("posterior") of the parameters given the data. Inferences are then based on this posterior, which is the distribution one desires because the data are observed while the parameters are not.

Theorem 2.2 (Bayes). *In the setup of (2.8), suppose that the conditional density of $\mathbf{Y}$ given $\mathbf{X} = \mathbf{x}$ is $f_{\mathbf{Y}|\mathbf{X}}(\mathbf{y} \mid \mathbf{x})$, and the marginal density of $\mathbf{X}$ is $f_{\mathbf{X}}(\mathbf{x})$. Then for $(\mathbf{x}, \mathbf{y}) \in \mathcal{W}$, the conditional density of $\mathbf{X}$ given $\mathbf{Y} = \mathbf{y}$ is*

$$f_{\mathbf{X}|\mathbf{Y}}(\mathbf{x} \mid \mathbf{y}) = \frac{f_{\mathbf{Y}|\mathbf{X}}(\mathbf{y} \mid \mathbf{x}) f_{\mathbf{X}}(\mathbf{x})}{\int_{\mathcal{X}_{\mathbf{y}}} f_{\mathbf{Y}|\mathbf{X}}(\mathbf{y} \mid \mathbf{z}) f_{\mathbf{X}}(\mathbf{z}) d\mathbf{z}}. \tag{2.75}$$

Proof. From (2.23) and (2.25),

$$f_{\mathbf{X}|\mathbf{Y}}(\mathbf{x} \mid \mathbf{y}) = \frac{f(\mathbf{x}, \mathbf{y})}{f_{\mathbf{Y}}(\mathbf{y})} = \frac{f_{\mathbf{Y}|\mathbf{X}}(\mathbf{y} \mid \mathbf{x}) f_{\mathbf{X}}(\mathbf{x})}{f_{\mathbf{Y}}(\mathbf{y})}. \tag{2.76}$$

By (2.26), using **z** for **x**, to avoid confusion with the **x** in (2.76),

$$f_{\mathbf{Y}}(\mathbf{y}) = \int_{\mathcal{X}_{\mathbf{y}}} f_{\mathbf{Y}|\mathbf{X}}(\mathbf{y} \mid \mathbf{z}) f_{\mathbf{X}}(\mathbf{z}) d\mathbf{z}, \tag{2.77}$$

which, substituted in the denominator of (2.76), shows (2.75). □

2.6 Affine transformations

In Section 1.5, linear combinations of the data were used heavily. Here we consider the distributional analogs of linear functions, or their extensions, **affine transformations**. For a single random variable X, an affine transformation is $a + bX$ for constants a and b. Equation (2.82) is an example of an affine transformation with two random variables.

More generally, an affine transformation of a collection of N random variables **X** is a collection of M random variables **Y** where

$$Y_j = a_j + b_{j1}X_1 + \cdots + b_{jN}X_N, \quad j = 1, \ldots, M, \tag{2.78}$$

the a_j's and b_{jk}'s being constants. Note that marginals are examples of affine transformations: the a_j's are 0, and most of the b_{jk}'s are 0, and a few are 1. Depending on how the elements of **X** and **Y** are arranged, affine transformations can be written as a matrix equation. For example, if **X** and **Y** are row vectors, and **B** is $M \times N$, then

$$\mathbf{Y} = \mathbf{a} + \mathbf{X}\mathbf{B}', \tag{2.79}$$

where **B** is the matrix of b_{jk}'s, and $\mathbf{a} = (a_1, \ldots, a_M)$. If **X** and **Y** are column vectors, then the equation is $\mathbf{Y} = \mathbf{a}' + \mathbf{B}\mathbf{X}$. For an example using matrices, suppose **X** is $n \times p$, **C** is $m \times n$, **D** is $q \times p$, and **A** is $m \times q$, and

$$\mathbf{Y} = \mathbf{A} + \mathbf{C}\mathbf{X}\mathbf{D}'. \tag{2.80}$$

Then **Y** is an $m \times q$ matrix, each of whose elements is some affine transformation of the elements of **X**. The relationship between the b_{jk}'s and the elements of **C** and **D** is somewhat complicated but could be made explicit, if desired. Look ahead to (3.32d), if interested.

Expectations are linear, that is, for any random variables (X, Y), and constant c,

$$E[cX] = cE[X] \quad \text{and} \quad E[X + Y] = E[X] + E[Y], \tag{2.81}$$

which can be seen from (2.28) and (2.29) by the linearity of integrals and sums. Considering any constant a as a (nonrandom) random variable, with $E[a] = a$, (2.81) can be used to show, e.g.,

$$E[a + bX + cY] = a + bE[X] + cE[Y]. \tag{2.82}$$

The mean of an affine transformation is the affine transformation of the mean. This property follows from (2.81) as in (2.82), i.e., for (2.78),

$$E[Y_j] = a_j + b_{j1}E[X_1] + \cdots + b_{jN}E[X_N], \quad j = 1, \ldots, M. \tag{2.83}$$

If the collections are arranged as vectors or matrices, then so are the means, so that for the row vector (2.79) and matrix (2.80) examples, one has, respectively,

$$E[\mathbf{Y}] = \mathbf{a} + E[\mathbf{X}]\mathbf{B}' \quad \text{and} \quad E[\mathbf{Y}] = \mathbf{A} + \mathbf{C}E[\mathbf{X}]\mathbf{D}'. \tag{2.84}$$

The covariance matrix of $\mathbf{Y}$ can be obtained from that of $\mathbf{X}$. It is a little more involved than for the means, but not too bad, at least in the vector case. Suppose $\mathbf{X}$ and $\mathbf{Y}$ are row vectors, and (2.79) holds. Then from (2.44),

$$\begin{aligned}
Cov[\mathbf{Y}] &= E\left[(\mathbf{Y} - E[\mathbf{Y}])'(\mathbf{Y} - E[\mathbf{Y}])\right] \\
&= E\left[(\mathbf{a} + \mathbf{XB}' - (\mathbf{a} + E[\mathbf{X}]\mathbf{B}'))'(\mathbf{a} + \mathbf{XB}' - (\mathbf{a} + E[\mathbf{X}]\mathbf{B}'))\right] \\
&= E\left[(\mathbf{XB}' - E[\mathbf{X}]\mathbf{B}')'(\mathbf{XB}' - E[\mathbf{X}]\mathbf{B}')\right] \\
&= E\left[\mathbf{B}(\mathbf{X} - E[\mathbf{X}])'(\mathbf{X} - E[\mathbf{X}])\mathbf{B}'\right] \\
&= \mathbf{B}E\left[(\mathbf{X} - E[\mathbf{X}])'(\mathbf{X} - E[\mathbf{X}])\right]\mathbf{B}' \quad \text{by second part of (2.84)} \\
&= \mathbf{B}Cov[\mathbf{X}]\mathbf{B}'.
\end{aligned} \tag{2.85}$$

Compare this formula to the sample version in (1.27). Though modest looking, the formula $Cov[\mathbf{XB}'] = \mathbf{B}Cov[\mathbf{X}]\mathbf{B}'$ is extremely useful. It is often called a "sandwich" formula, with the $\mathbf{B}$ as the slices of bread. The formula for column vectors is the same. Compare this result to the familiar one from univariate analysis: $Var[a + bX] = b^2 Var[X]$.

For matrices, we again will wait. (We are waiting for Kronecker products as in Definition 3.5, in case you are wondering.)

2.7 Exercises

Exercise 2.7.1. Consider the pair of random variables (X, Y), where X is discrete and Y is continuous. Their space is

$$\mathcal{W} = \{(x, y) \mid x \in \{1, 2, 3\} \;\&\; 0 < y < x\}, \tag{2.86}$$

and their mixed-type density is

$$f(x, y) = \frac{x + y}{21}. \tag{2.87}$$

Let $A = \{(x, y) \in \mathcal{W} \mid y \leq x/2\}$. (It is a good idea to sketch $\mathcal{W}$ and A.) (a) Find $\mathcal{X}^A$. (b) Find $\mathcal{Y}_x^A$ for each $x \in \mathcal{X}^A$. (c) Find $P[A]$. (d) Find the marginal density and space of X. (e) Find the marginal space of Y. (f) Find the conditional space of X given Y, $\mathcal{X}_y$, for each y. (Do it separately for $y \in (0, 1)$, $y \in [1, 2)$ and $y \in [2, 3)$.) (g) Find the marginal density of Y.

Exercise 2.7.2. Given the setup in (2.8) through (2.10), show that for $A \in \mathcal{W}$,

$$A = \{(\mathbf{x}, \mathbf{y}) \mid \mathbf{x} \in \mathcal{X}^A \text{ and } \mathbf{y} \in \mathcal{Y}_{\mathbf{x}}^A\} = \{(\mathbf{x}, \mathbf{y}) \mid \mathbf{y} \in \mathcal{Y}^A \text{ and } \mathbf{x} \in \mathcal{X}_{\mathbf{y}}^A\}. \tag{2.88}$$

Exercise 2.7.3. Verify (2.17), that is, given $B \subset \mathcal{X}$, show that

$$P[X \in B] = \int_B \left[\int_{\mathcal{Y}_\mathbf{x}} f(\mathbf{x}, \mathbf{y}) d\mathbf{y} \right] d\mathbf{x}. \tag{2.89}$$

[Hint: Show that for $A = \{(\mathbf{x}, \mathbf{y}) \mid \mathbf{x} \in B \text{ and } \mathbf{y} \in \mathcal{Y}_\mathbf{x}\}$, $\mathbf{x} \in B$ if and only if $(\mathbf{x}, \mathbf{y}) \in A$, so that $P[\mathbf{X} \in B] = P[(\mathbf{X}, \mathbf{Y}) \in A]$. Then note that the latter probability is $\int_A f(\mathbf{x}, \mathbf{y}) d\mathbf{x} d\mathbf{y}$, which with some interchanging equals the right-hand side of (2.89).]

Exercise 2.7.4. Show that $\mathbf{X}$ and $\mathbf{Y}$ are independent if and only if $E[g(\mathbf{X})h(\mathbf{Y})] = E[g(\mathbf{X})]E[h(\mathbf{Y})]$ as in (2.53) for all g and h with finite expectations. You can assume densities exist, i.e., (2.52). [Hint: To show independence implies (2.53), write out the sums/integrals. For the other direction, consider indicator functions for g and h as in (2.37).]

Exercise 2.7.5. Prove (2.31), $E[g(\mathbf{Y})] = E[g(h(\mathbf{X}))]$ for $\mathbf{Y} = h(\mathbf{X})$, in the discrete case. [Hint: Start by writing

$$f_\mathbf{Y}(\mathbf{y}) = P[\mathbf{Y} = \mathbf{y}] = P[h(\mathbf{X}) = \mathbf{y}] = \sum_{\mathbf{x} \in \mathcal{X}_\mathbf{y}} f_\mathbf{X}(\mathbf{x}), \tag{2.90}$$

where $\mathcal{X}_\mathbf{y} = \{\mathbf{x} \in \mathcal{X} \mid h(\mathbf{x}) = \mathbf{y}\}$. Then

$$E[g(\mathbf{Y})] = \sum_{\mathbf{y} \in \mathcal{Y}} g(\mathbf{y}) \sum_{\mathbf{x} \in \mathcal{X}_\mathbf{y}} f_\mathbf{X}(\mathbf{x}) = \sum_{\mathbf{y} \in \mathcal{Y}} \sum_{\mathbf{x} \in \mathcal{X}_\mathbf{y}} g(\mathbf{y}) f_\mathbf{X}(\mathbf{x}). \tag{2.91}$$

In the inner summation in the final expression, $h(\mathbf{x})$ is always equal to $\mathbf{y}$. (Why?) Substitute $h(\mathbf{x})$ for $\mathbf{y}$ in the g, then. Now the summand is free of $\mathbf{y}$. Argue that the double summation is the same as summing over $\mathbf{x} \in \mathcal{X}$, yielding $\sum_{\mathbf{x} \in \mathcal{X}} g(h(\mathbf{x})) f_\mathbf{X}(\mathbf{x}) = E[g(h(\mathbf{X}))]$.]

Exercise 2.7.6. (a) Prove the plugin formula (2.62) in the discrete case. [Hint: For $\mathbf{z}$ in the range of g, write $P[g(\mathbf{X}, \mathbf{Y}) = \mathbf{z} \mid \mathbf{X} = \mathbf{x}] = P[g(\mathbf{X}, \mathbf{Y}) = \mathbf{z} \text{ and } \mathbf{X} = \mathbf{x}] / P[\mathbf{X} = \mathbf{x}]$, then note that in the numerator, the $\mathbf{X}$ can be replaced by $\mathbf{x}$.] (b) Prove (2.63). [Hint: Follow the proof in part (a), then note the two events $g(\mathbf{x}, \mathbf{Y}) = \mathbf{z}$ and $\mathbf{X} = \mathbf{x}$ are independent.]

Exercise 2.7.7. Suppose $(\mathbf{X}, \mathbf{Y}, \mathbf{Z})$ has a discrete distribution, $\mathbf{X}$ and $\mathbf{Y}$ are conditionally independent given $\mathbf{Z}$ (as in (2.65)), and $\mathbf{X}$ and $\mathbf{Z}$ are independent. Show that $\mathbf{X}$ is independent of $(\mathbf{Y}, \mathbf{Z})$. [Hint: Use the total probability formula (2.36) on $P[\mathbf{X} \in A \text{ and } (\mathbf{Y}, \mathbf{Z}) \in B]$, conditioning on $\mathbf{Z}$. Then argue that the summand can be written

$$\begin{aligned} P[\mathbf{X} \in A \text{ and } (\mathbf{Y}, \mathbf{Z}) \in B \mid \mathbf{Z} = \mathbf{z}] &= P[\mathbf{X} \in A \text{ and } (\mathbf{Y}, \mathbf{z}) \in B \mid \mathbf{Z} = \mathbf{z}] \\ &= P[\mathbf{X} \in A \mid \mathbf{Z} = \mathbf{z}] P[(\mathbf{Y}, \mathbf{z}) \in B \mid \mathbf{Z} = \mathbf{z}]. \end{aligned} \tag{2.92}$$

Use the independence of $\mathbf{X}$ and $\mathbf{Z}$ on the first probability in the final expression, and bring it out of the summation.]

Exercise 2.7.8. Prove (2.67). [Hint: Find $\mathcal{Y}_x$ and the marginal $f_X(x)$.]

Exercise 2.7.9. Suppose $\mathbf{Y} = (Y_1, Y_2, Y_3, Y_4)$ is multinomial with parameters n and $\mathbf{p} = (p_1, p_2, p_3, p_4)$. Thus n is a positive integer, the p_i's are positive and sum to 1, and the Y_i's are nonnegative integers that sum to n. The pmf is

$$f(\mathbf{y}) = \binom{n}{y_1, y_2, y_3, y_4} p_1^{y_1} \cdots p_4^{y_4}, \tag{2.93}$$

where $\binom{n}{y_1,y_2,y_3,y_4} = n!/(y_1! \cdots y_4!)$. Consider the conditional distribution of (Y_1, Y_2) given $(Y_3, Y_4) = (c, d)$. (a) What is the conditional space of (Y_1, Y_2) given $(Y_3, Y_4) = (c, d)$? Give Y_2 as a function of Y_1, c, and d. What is the conditional range of Y_1? (b) Write the conditional pmf of (Y_1, Y_2) given $(Y_3, Y_4) = (c, d)$, and simplify noting that

$$\binom{n}{y_1, y_2, c, d} = \binom{n}{n-c-d, c, d} \binom{n-c-d}{c, d} \tag{2.94}$$

What is the conditional distribution of $Y_1 \mid (Y_3, Y_4) = (c, d)$? (c) What is the conditional distribution of Y_1 given $Y_3 + Y_4 = a$?

Exercise 2.7.10. Prove (2.44). [Hint: Write out the elements of the matrix $(\mathbf{X} - \boldsymbol{\mu})'(\mathbf{X} - \boldsymbol{\mu})$, then use (2.42).]

Exercise 2.7.11. Suppose $\mathbf{X}$, $1 \times N$, has finite covariance matrix. Show that $Cov[\mathbf{X}] = E[\mathbf{X}'\mathbf{X}] - E[\mathbf{X}]'E[\mathbf{X}]$.

Exercise 2.7.12. (a) Prove the variance decomposition holds for the $1 \times q$ vector $\mathbf{Y}$, as in (2.74). (b) Write $Cov[Y_i, Y_j]$ as a function of the conditional quantities

$$Cov[Y_i, Y_j \mid \mathbf{X} = \mathbf{x}],\ E[Y_i \mid \mathbf{X} = \mathbf{x}],\ \text{and}\ E[Y_j \mid \mathbf{X} = \mathbf{x}]. \tag{2.95}$$

Exercise 2.7.13. The **Beta-binomial**(n, α, β) distribution is a mixture of binomial distributions. That is, suppose Y given $P = p$ is Binomial(n, p) ($f_Y(y) = \binom{n}{y} p^y (1-p)^{n-y}$ for $y = 0, 1, \ldots, n$), and P is (marginally) Beta(α, β):

$$f_P(p) = \frac{\Gamma(\alpha+\beta)}{\Gamma(\alpha)\Gamma(\beta)} p^{\alpha-1}(1-p)^{\beta-1},\ p \in (0,1), \tag{2.96}$$

where Γ is the gamma function,

$$\Gamma(\alpha) = \int_0^\infty u^{\alpha-1} e^{-u} du,\ \alpha > 0. \tag{2.97}$$

(a) The conditional mean and variance of Y are np and $np(1-p)$. (Right?) The unconditional mean and variance of P are $\alpha/(\alpha+\beta)$ and $\alpha\beta/(\alpha+\beta)^2(\alpha+\beta+1)$. What are the unconditional mean and variance of Y? (b) Compare the variance of a Binomial(n, q) to that of a Beta-binomial(n, α, β), where $q = \alpha/(\alpha+\beta)$. (c) Find the joint density of (P, Y). (d) Find the pmf of the beta-binomial. [Hint: Notice that the part of the joint density depending on p looks like a beta pdf, but without the constant. Thus integrating out p yields the reciprocal of the constant.]

Exercise 2.7.14 (Bayesian inference). This question develops Bayesian inference for a binomial. Suppose

$$Y \mid P = p \sim \text{Binomial}(n, p) \ \text{and}\ P \sim \text{Beta}(\alpha_0, \beta_0), \tag{2.98}$$

that is, the probability of success P has a beta prior. (a) Show that the posterior distribution is

$$P \mid Y = y \sim \text{Beta}(\alpha_0 + y, \beta_0 + n - y). \tag{2.99}$$

The beta prior is called the **conjugate prior** for the binomial p, meaning the posterior has the same form, but with updated parameters. [Hint: Exercise 2.7.13 (d) has the joint density of (P, Y).] (b) Find the posterior mean, $E[P \mid Y = y]$. Show that it can be written as a weighted mean of the sample proportion $\widehat{p} = y/n$ and the prior mean $p_0 = \alpha_0/(\alpha_0 + \beta_0)$.

Exercise 2.7.15. Do the mean and variance formulas (2.33) and (2.72) work if g is a function of X and Y? [Hint: Consider the collection $(\mathbf{X}, \mathbf{W})$, where $\mathbf{W} = (\mathbf{X}, \mathbf{Y})$.]

Exercise 2.7.16. Suppose $h(y)$ is a histogram with K equal-sized bins. That is, we have bins $(b_{i-1}, b_i], i = 1, \ldots, K$, where $b_i = b_0 + d \times i$, d being the width of each bin. Then

$$h(y) = \begin{cases} p_i/d & \text{if } b_{i-1} < y \leq b_i,\ i = 1, \ldots, K \\ 0 & \text{if } y \notin (b_0, b_K], \end{cases} \tag{2.100}$$

where the p_i's are probabilities that sum to 1. Suppose Y is a random variable with pdf h. For $y \in (b_0, b_K]$, let $\mathcal{I}(y)$ be y's bin, i.e., $\mathcal{I}(y) = i$ if $b_{i-1} < y \leq b_i$. (a) What is the distribution of the random variable $\mathcal{I}(Y)$? Find its mean and variance. (b) Find the mean and variance of $b_{\mathcal{I}(Y)} = b_0 + d\mathcal{I}(Y)$. (c) What is the conditional distribution of Y given $\mathcal{I}(Y) = i$, for each $i = 1, \ldots, K$? [It is uniform. Over what range?] Find the conditional mean and variance. (d) Show that unconditionally,

$$E[Y] = b_0 + d(E[\mathcal{I}] - 1/2) \text{ and } Var[Y] = d^2(Var[\mathcal{I}] + 1/12). \tag{2.101}$$

(e) Recall the entropy in (1.44). Note that for our pdf, $h(Y) = p_{\mathcal{I}(Y)}/d$. Show that

$$\text{Entropy}(h) = -\sum_{i=1}^{K} p_i \log(p_i) + \log(d), \tag{2.102}$$

and for the negentropy in (1.46),

$$\text{Negent}(h) = \frac{1}{2}\left(1 + \log\left(2\pi\left(Var[\mathcal{I}] + 1/12\right)\right)\right) + \sum_{i=1}^{K} p_i \log(p_i). \tag{2.103}$$

Exercise 2.7.17. Suppose Y is a random variable with finite variance, and one wishes to guess the value of Y by the constant c, using the **least squares** criterion. That is, we want to choose c to minimize $E[(Y - c)^2]$. (a) What is the minimizing constant c? (b) Using that c, what is the value of $E[(Y - c)^2]$?

Exercise 2.7.18. Suppose for random vector (X, Y), one observes $X = x$, and wishes to guess the value of Y by $h(x)$, say, using the least squares criterion: Choose h to minimize $E[q(X, Y)]$, where $q(X, Y) = (Y - h(X))^2$. This h is called the **regression function** of Y on X. Assume all the relevant means and variances are finite. (a) Write $E[q(X, Y)]$ as the expected value of the conditional expected value conditioning on $X = x$, $e_q(x)$. For fixed x, note that $h(x)$ is a scalar, hence one can minimize $e_q(x)$ over $h(x)$ using differentiation. What $h(x)$ achieves the minimum conditional expected value of q? (b) Show that the h found in part (a) minimizes the unconditional expected value $E[q(X, Y)]$. (c) Find the value of $E[q(X, Y)]$ for the minimizing h.

Exercise 2.7.19. Continue with Exercise 2.7.18, but this time restrict h to be a linear function, $h(x) = \alpha + \beta x$. Thus we wish to find α and β to minimize $E[(Y - \alpha - \beta X)^2]$. The minimizing function is the **linear regression function** of Y on X. (a) Find the α and β to minimize $E[(Y - \alpha - \beta X)^2]$. [You can differentiate that expected value directly, without worrying about conditioning.] (b) Find the value of $E[(Y - \alpha - \beta X)^2]$ for the minimizing α and β.

Exercise 2.7.20. Suppose $\mathbf{Y}$ is $1 \times q$ and $\mathbf{X}$ is $1 \times p$, $E[\mathbf{X}] = \mathbf{0}$, $Cov[\mathbf{X}] = \mathbf{I}_p$, $E[\mathbf{Y} \mid \mathbf{X} = \mathbf{x}] = \boldsymbol{\mu} + \mathbf{x}\boldsymbol{\beta}$ for some $p \times q$ matrix $\boldsymbol{\beta}$, and $Cov[\mathbf{Y} \mid \mathbf{X} = \mathbf{x}] = \boldsymbol{\Psi}$ for some $q \times q$ diagonal matrix $\boldsymbol{\Psi}$. Thus the Y_i's are conditionally uncorrelated given $\mathbf{X} = \mathbf{x}$. Find the unconditional $E[\mathbf{Y}]$ and $Cov[\mathbf{Y}]$. (The covariance matrix of $\mathbf{Y}$ has a factor-analytic structure, which we will see in Section 10.3. The X_i's are factors that explain the correlations among the Y_i's. Typically, the factors are not observed.)

Exercise 2.7.21. Suppose $\mathbf{Y}_1, \ldots, \mathbf{Y}_q$ are independent $1 \times p$ vectors, where $\mathbf{Y}_i$ has moment generating function $M_i(\mathbf{t}), i = 1, \ldots, q$, all of which are finite for $\|\mathbf{t}\| < \epsilon$ for some $\epsilon > 0$. Show that the moment generating function of $\mathbf{Y}_1 + \cdots + \mathbf{Y}_q$ is $M_1(\mathbf{t}) \cdots M_q(\mathbf{t})$. For which $\mathbf{t}$ is this moment generating function finite?

Exercise 2.7.22. Prove (2.49). It is legitimate to interchange the derivatives and expectation, and to set $\mathbf{t} = \mathbf{0}$ within the expectation, when $\|\mathbf{t}\| < \epsilon$. [Extra credit: Prove that those operations are legitimate.]

Exercise 2.7.23. The **cumulant generating function** of $\mathbf{X}$ is defined to be $c_{\mathbf{X}}(\mathbf{t}) = \log(M_{\mathbf{X}}(\mathbf{t}))$, and, if the function is finite for $\mathbf{t}$ in a neighborhood of zero, then the $(k_1, \ldots, k_N)^{th}$ mixed cumulant is the corresponding mixed derivative of $c_{\mathbf{X}}(\mathbf{t})$ evaluated at zero. (a) For $N = 1$, find the first four cumulants, $\kappa_1, \ldots, \kappa_4$, where

$$\kappa_i = \frac{\partial}{\partial t} c_X(t) \Big|_{t=0}. \tag{2.104}$$

Show that $\kappa_3 / \kappa_2^{3/2}$ is the population analog of skewness (1.42), and κ_4 / κ_2^2 is the population analog of kurtosis (1.43), i.e.,

$$\frac{\kappa_3}{\kappa_2^{3/2}} = \frac{E[(X - \mu)^3]}{\sigma^3} \quad \text{and} \quad \frac{\kappa_4}{\kappa_2^2} = \frac{E[(X - \mu)^4]}{\sigma^4} - 3, \tag{2.105}$$

where $\mu = E[X]$ and $\sigma^2 = Var[X]$. [Write everything in terms of $E[X^k]$'s by expanding the $E[(X - \mu)^k]$'s.] (b) For general N, find the second mixed cumulants, i.e.,

$$\frac{\partial^2}{\partial t_i \partial t_j} c_{\mathbf{X}}(\mathbf{t}) \Big|_{\mathbf{t}=\mathbf{0}}, \; i \neq j. \tag{2.106}$$

Exercise 2.7.24. Suppose $\mathbf{X}$ is $1 \times p$, $\mathbf{Y}$ is $1 \times q$, $\mathbf{A}$ is $k \times p$, and $\mathbf{B}$ is $l \times q$. Use calculations as in (2.85) to show that

$$Cov[\mathbf{X}\mathbf{A}', \mathbf{Y}\mathbf{B}'] = \mathbf{A}Cov[\mathbf{X}, \mathbf{Y}]\mathbf{B}', \tag{2.107}$$

where $Cov[\mathbf{X}, \mathbf{Y}]$ is given in (2.46).

Exercise 2.7.25. A study was conducted on people near Newcastle on Tyne in 1972-74 [Appleton et al., 1996], and followed up twenty years later. We will focus on 1314 women in the study. The three variables we will consider are Z: age group (three values); X: whether they smoked or not (in 1974); and Y: whether they were still alive in 1994. Here are the frequencies:

Age group	Young (18 − 34)		Middle (35 − 64)		Old (65+)	
Smoker?	Yes	No	Yes	No	Yes	No
Died	5	6	92	59	42	165
Lived	174	213	262	261	7	28

(2.108)

(a) Treating proportions in the table as probabilities, find

$$P[Y = \text{Lived} \mid X = \text{Smoker}] \text{ and } P[Y = \text{Lived} \mid X = \text{Nonsmoker}]. \tag{2.109}$$

Who were more likely to live, smokers or nonsmokers? (b) Find $P[X = \text{Smoker} \mid Z = z]$ for z= Young, Middle, and Old. What do you notice? (c) Find

$$P[Y = \text{Lived} \mid X = \text{Smoker} \;\&\; Z = z] \tag{2.110}$$

and

$$P[Y = \text{Lived} \mid X = \text{Nonsmoker} \;\&\; Z = z] \tag{2.111}$$

for z= Young, Middle, and Old. Adjusting for age group, who were more likely to live, smokers or nonsmokers? (d) Conditionally on age, the relationship between smoking and living is negative for each age group. Is it true that marginally (not conditioning on age), the relationship between smoking and living is negative? What is the explanation? (Simpson's Paradox.)

Exercise 2.7.26. Suppose in a large population, the proportion of people who are infected with the HIV virus is $\epsilon = 1/100,000$. People can take a blood test to see whether they have the virus. The test is 99% accurate: The chance the test is positive given the person has the virus is 99%, and the chance the test is negative given the person does not have the virus is also 99%. Suppose a randomly chosen person takes the test. (a) What is the chance that this person does have the virus given that the test is positive? Is this close to 99%? (b) What is the chance that this person does have the virus given that the test is negative? Is this close to 1%? (c) Do the probabilities in (a) and (b) sum to 1?

Exercise 2.7.27. Suppose Z_1, Z_2, Z_3 are iid with $P[Z_i = -1] = P[Z_i = +1] = \frac{1}{2}$. Let

$$X_1 = Z_1 Z_2, \;\; X_2 = Z_1 Z_3, \;\; X_3 = Z_2 Z_3. \tag{2.112}$$

(a) Find the conditional distribution of $(X_1, X_2) \mid Z_1 = +1$. Are X_1 and X_2 conditionally independent given $Z_1 = +1$? (b) Find the conditional distribution of $(X_1, X_2) \mid Z_1 = -1$. Are X_1 and X_2 conditionally independent given $Z_1 = -1$? (c) Is (X_1, X_2) independent of Z_1? Are X_1 and X_2 independent (unconditionally)? (d) Are X_1 and X_3 independent? Are X_2 and X_3 independent? Are X_1, X_2 and X_3 mutually independent? (e) What is the space of (X_1, X_2, X_3)? (f) What is the distribution of $X_1 X_2 X_3$?

Exercise 2.7.28. Yes/no questions: (a) Suppose X_1 and X_2 are independent, X_1 and X_3 are independent, and X_2 and X_3 are independent. Are X_1, X_2 and X_3 mutually independent? (b) Suppose X_1, X_2 and X_3 are mutually independent. Are X_1 and X_2 conditionally independent given $X_3 = x_3$?

Exercise 2.7.29. (a) Let $U \sim$Uniform$(0,1)$, so that it has space $(0,1)$ and pdf $f_U(u) = 1$. Find its distribution function (2.1), $F_U(u)$. (b) Suppose X is a random variable with space (a,b) and pdf $f_X(x)$, where $f_X(x) > 0$ for $x \in (a,b)$. [Either or both of a and b may be infinite.] Thus the inverse function $F_X^{-1}(u)$ exists for $u \in (0,1)$. (Why?) Show that the distribution of $Y = F_X(X)$ is Uniform$(0,1)$. [Hint: For $y \in (0,1)$, write $P[Y \leq y] = P[F_X(X) \leq y] = P[X \leq F_X^{-1}(y)]$, then use the definition of F_X.] (c) Suppose $U \sim$ Uniform$(0,1)$. For the X in part (b), show that $F_X^{-1}(U)$ has the same distribution as X. [Note: This fact provides a way of generating random variables X from random uniforms.]

Exercise 2.7.30. Suppose $\mathbf{Y}$ is 1×2 with covariance matrix

$$\mathbf{\Sigma} = \begin{pmatrix} 2 & 1 \\ 1 & 2 \end{pmatrix}. \tag{2.113}$$

Let $\mathbf{W} = \mathbf{YB}'$, for

$$\mathbf{B} = \begin{pmatrix} 1 & 1 \\ 1 & c \end{pmatrix} \tag{2.114}$$

for some c. Find c so that the covariance between the two variables in $\mathbf{W}$ is zero. What are the variances of the resulting two variables?

Exercise 2.7.31. Let $\mathbf{Y}$ be a 1×4 vector with

$$Y_j = \mu_j + B + E_j,$$

where the μ_j are constants, B has mean zero and variance σ_B^2, the E_j's are independent, each with mean zero and variance σ_E^2, and B is independent of the E_j's. (a) Find the mean and covariance matrix of

$$\mathbf{X} \equiv \begin{pmatrix} B & E_1 & E_2 & E_3 & E_4 \end{pmatrix}. \tag{2.115}$$

(b) Write $\mathbf{Y}$ as an affine transformation of $\mathbf{X}$. (c) Find the mean and covariance matrix of $\mathbf{Y}$. (d) $Cov[\mathbf{Y}]$ can be written as

$$Cov[\mathbf{Y}] = a\mathbf{I}_4 + b\mathbf{1}_4\mathbf{1}_4'. \tag{2.116}$$

Give a and b in terms of σ_B^2 and σ_E^2. (e) What are the mean and covariance matrix of $\overline{Y} = (Y_1 + \cdots + Y_4)/4$?

Exercise 2.7.32. Suppose $\mathbf{Y}$ is a 5×4 data matrix, and

$$Y_{ij} = \mu + B_i + \gamma + E_{ij} \ \text{ for } \ j = 1,2, \tag{2.117}$$

$$Y_{ij} = \mu + B_i - \gamma + E_{ij} \ \text{ for } \ j = 3,4, \tag{2.118}$$

where the B_i's are independent, each with mean zero and variance σ_B^2, the E_{ij} are independent, each with mean zero and variance σ_E^2's, and the B_i's are independent of the E_{ij}'s. (Thus each row of $\mathbf{Y}$ is distributed as the vector in Extra 2.7.31, for some particular values of μ_j's.) [Note: This model is an example of a randomized block model, where the rows of $\mathbf{Y}$ represent the blocks. For example, a farm might be broken into 5 blocks, and each block split into four plots, where two of the plots

(Y_{i1}, Y_{i2}) get one fertilizer, and two of the plots (Y_{i3}, Y_{i4}) get another fertilizer.] (a) $E[\mathbf{Y}] = \mathbf{x}\boldsymbol{\beta}\mathbf{z}'$, where $\boldsymbol{\beta} = (\mu, \gamma)$. Give $\mathbf{x}$ and $\mathbf{z}'$. [The $\mathbf{x}$ and $\mathbf{z}$ contain known constants.] (b) Are the rows of $\mathbf{Y}$ uncorrelated? (c) Find $Cov[\mathbf{Y}]$. (d) Setting which parameter equal to zero guarantees that all elements of $\mathbf{Y}$ have the same mean? (e) Setting which parameter equal to zero guarantees that all elements of $\mathbf{Y}$ are uncorrelated?

Chapter 3

The Multivariate Normal Distribution

3.1 Definition

There are not very many commonly used multivariate distributions to model a data matrix $\mathbf{Y}$. The **multivariate normal** is by far the most common, at least for continuous data. Which is not to say that all data are distributed normally, nor that all techniques assume such. Rather, typically one either assumes normality, or makes few assumptions at all and relies on asymptotic results.

The multivariate normal arises from the standard normal:

Definition 3.1. *The random variable Z is **standard normal**, written $Z \sim N(0,1)$, if it has space $\mathbb{R}$ and pdf*

$$\phi(z) = \frac{1}{\sqrt{2\pi}} e^{-\frac{1}{2}z^2}. \tag{3.1}$$

It is not hard to show that if $Z \sim N(0,1)$,

$$E[Z] = 0, \;\; Var[Z] = 1, \;\; \text{and} \;\; M_Z(t) = e^{\frac{1}{2}t^2}. \tag{3.2}$$

Definition 3.2. *The collection of random variables $\mathbf{Z} = (Z_1, \ldots, Z_M)$ is a **standard normal collection** if the Z_i's are mutually independent standard normal random variables.*

Because the variables in a standard normal collection are independent, by (3.2), (2.61) and (2.59),

$$E[\mathbf{Z}] = \mathbf{0}, \;\; Cov[\mathbf{Z}] = \mathbf{I}_M, \;\; \text{and} \;\; M_{\mathbf{Z}}(\mathbf{t}) = e^{\frac{1}{2}(t_1^2+\cdots+t_M^2)} = e^{\frac{1}{2}\|\mathbf{t}\|^2}. \tag{3.3}$$

The mgf is finite for all $\mathbf{t}$.

A general multivariate normal distribution can have any (legitimate) mean and covariance, achieved through the use of affine transformations. Here is the definition.

Definition 3.3. *The collection $\mathbf{Y}$ is **multivariate normal** if it is an affine transformation of a standard normal collection.*

The mean and covariance of a multivariate normal can be calculated from the coefficients in the affine transformation. In particular, suppose $\mathbf{Z}$ is a standard normal collection represented as an $1 \times M$ row vector, and $\mathbf{Y}$ is a $1 \times N$ row vector

$$\mathbf{Y} = \boldsymbol{\mu} + \mathbf{Z}\mathbf{B}', \tag{3.4}$$

where $\mathbf{B}$ is $N \times M$ and $\boldsymbol{\mu}$ is $1 \times N$. From (3.3), (2.84) and (2.85),

$$\boldsymbol{\mu} = E[\mathbf{Y}] \text{ and } \boldsymbol{\Sigma} = Cov[\mathbf{Y}] = \mathbf{B}\mathbf{B}'. \tag{3.5}$$

The mgf is calculated, for $1 \times N$ vector $\mathbf{s}$, as

$$\begin{aligned} M_{\mathbf{Y}}(\mathbf{s}) &= E[\exp(\mathbf{Y}\mathbf{s}')] \\ &= E[\exp((\boldsymbol{\mu} + \mathbf{Z}\mathbf{B}')\mathbf{s}')] \\ &= \exp(\boldsymbol{\mu}\mathbf{s}')E[\exp(\mathbf{Z}(\mathbf{s}\mathbf{B})')] \\ &= \exp(\boldsymbol{\mu}\mathbf{s}')M_{\mathbf{Z}}(\mathbf{s}\mathbf{B}) \\ &= \exp(\boldsymbol{\mu}\mathbf{s}')\exp(\frac{1}{2}\|\mathbf{s}\mathbf{B}\|^2) \quad \text{by (3.3)} \\ &= \exp(\boldsymbol{\mu}\mathbf{s}' + \frac{1}{2}\mathbf{s}\mathbf{B}\mathbf{B}'\mathbf{s}') \\ &= \exp(\boldsymbol{\mu}\mathbf{s}' + \frac{1}{2}\mathbf{s}\boldsymbol{\Sigma}\mathbf{s}') \end{aligned} \tag{3.6}$$

The mgf depends on $\mathbf{B}$ through only $\boldsymbol{\Sigma} = \mathbf{B}\mathbf{B}'$. Because the mgf determines the distribution (Theorem 2.1), two different $\mathbf{B}$'s can produce the same distribution. That is, as long as $\mathbf{B}\mathbf{B}' = \mathbf{C}\mathbf{C}'$, the distributions of $\boldsymbol{\mu} + \mathbf{Z}\mathbf{B}'$ and $\boldsymbol{\mu} + \mathbf{Z}\mathbf{C}'$ are the same. Which is to say that the distribution of the multivariate normal depends on only the mean and covariance. Thus it is legitimate to write

$$\mathbf{Y} \sim N_N(\boldsymbol{\mu}, \boldsymbol{\Sigma}), \tag{3.7}$$

which is read "$\mathbf{Y}$ has N-dimensional multivariate normal distribution with mean $\boldsymbol{\mu}$ and covariance $\boldsymbol{\Sigma}$."

For example, consider the two matrices

$$\mathbf{B} = \begin{pmatrix} 1 & 2 & 1 \\ 0 & 3 & 4 \end{pmatrix} \text{ and } \mathbf{C} = \begin{pmatrix} \sqrt{2} & 2 \\ 0 & 5 \end{pmatrix}. \tag{3.8}$$

It is not hard to show that

$$\mathbf{B}\mathbf{B}' = \mathbf{C}\mathbf{C}' = \begin{pmatrix} 6 & 10 \\ 10 & 25 \end{pmatrix} \equiv \boldsymbol{\Sigma}. \tag{3.9}$$

Thus if the Z_i's are independent $N(0,1)$,

$$\begin{aligned} (Z_1, Z_2, Z_3)\mathbf{B}' &= (Z_1 + 2Z_2 + Z_3, 3Z_2 + 4Z_3) \\ &=^{\mathcal{D}} (\sqrt{2}Z_1 + 2Z_2, 5Z_2) \\ &= (Z_1, Z_2)\mathbf{C}', \end{aligned} \tag{3.10}$$

i.e., both vectors are $N(\mathbf{0}, \boldsymbol{\Sigma})$. Note that the two expressions are based on differing numbers of standard normals, not just different linear combinations.

Which $\boldsymbol{\mu}$ and $\boldsymbol{\Sigma}$ are legitimate parameters in (3.7)? Any $\boldsymbol{\mu} \in \mathbb{R}^N$ is. The covariance matrix $\boldsymbol{\Sigma}$ can be $\mathbf{B}\mathbf{B}'$ for any $N \times M$ matrix $\mathbf{B}$. Any such matrix $\mathbf{B}$ is considered a **square root** of $\boldsymbol{\Sigma}$. Clearly, $\boldsymbol{\Sigma}$ must be symmetric, but we already knew that. It must also be **nonnegative definite**, which we define now.

Definition 3.4. *A symmetric $q \times q$ matrix $\mathbf{A}$ is **nonnegative definite** if*

$$\mathbf{bAb}' \geq 0 \text{ for all } 1 \times q \text{ vectors } \mathbf{b}. \tag{3.11}$$

*Also, $\mathbf{A}$ is **positive definite** if*

$$\mathbf{bAb}' > 0 \text{ for all } 1 \times q \text{ vectors } \mathbf{b} \neq \mathbf{0}. \tag{3.12}$$

Note that $\mathbf{bBB}'\mathbf{b}' = \|\mathbf{bB}\|^2 \geq 0$, which means that $\mathbf{\Sigma}$ must be nonnegative definite. But from (2.85),

$$\mathbf{b\Sigma b}' = Cov[\mathbf{Yb}'] = Var[\mathbf{Yb}'] \geq 0, \tag{3.13}$$

because all variances are nonnegative. That is, any covariance matrix has to be nonnegative definite, not just multivariate normal ones.

So we know that $\mathbf{\Sigma}$ must be symmetric and nonnegative definite. Are there any other restrictions, or for any symmetric nonnegative definite matrix is there a corresponding $\mathbf{B}$? In fact, there are potentially many square roots of $\mathbf{\Sigma}$. These follow from the spectral decomposition theorem, Theorem 1.1. Because $\mathbf{\Sigma}$ is symmetric, we can write

$$\mathbf{\Sigma} = \mathbf{\Gamma\Lambda\Gamma}', \tag{3.14}$$

where $\mathbf{\Gamma}$ is orthogonal, and $\mathbf{\Lambda}$ is diagonal with diagonal elements $\lambda_1 \geq \lambda_2 \geq \cdots \geq \lambda_N$. Because $\mathbf{\Sigma}$ is nonnegative definite, the eigenvalues are nonnegative (Exercise 3.7.12), hence they have square roots. Consider

$$\mathbf{B} = \mathbf{\Gamma\Lambda}^{1/2}, \tag{3.15}$$

where $\mathbf{\Lambda}^{1/2}$ is the diagonal matrix with diagonal elements the $\lambda_j^{1/2}$'s. Then, indeed,

$$\mathbf{BB}' = \mathbf{\Gamma\Lambda}^{1/2}\mathbf{\Lambda}^{1/2}\mathbf{\Gamma}' = \mathbf{\Gamma\Lambda\Gamma}' = \mathbf{\Sigma}. \tag{3.16}$$

That is, in (3.7), $\boldsymbol{\mu}$ is unrestricted, and $\mathbf{\Sigma}$ can be any symmetric nonnegative definite matrix. Note that $\mathbf{C} = \mathbf{\Gamma\Lambda}^{1/2}\mathbf{\Psi}$ for any $N \times N$ orthogonal matrix $\mathbf{\Psi}$ is also a square root of $\mathbf{\Sigma}$. If we take $\mathbf{\Psi} = \mathbf{\Gamma}'$, then we have the **symmetric square root, $\mathbf{\Gamma\Lambda}^{1/2}\mathbf{\Gamma}'$**.

If $N = 1$, then we have a normal random variable, say Y, and $Y \sim N(\mu, \sigma^2)$ signifies that it has mean μ and variance σ^2. If $\mathbf{Y}$ is a multivariate normal collection represented as an $n \times q$ matrix, we write

$$\mathbf{Y} \sim N_{n\times q}(\boldsymbol{\mu}, \mathbf{\Sigma}) \iff \text{row}(\mathbf{Y}) \sim N_{nq}(\text{row}(\boldsymbol{\mu}), \mathbf{\Sigma}). \tag{3.17}$$

3.2 Some properties of the multivariate normal

Affine transformations of multivariate normals are also multivariate normal, because any affine transformation of a multivariate normal collection is an affine transformation of an affine transformation of a standard normal collection, and an affine transformation of an affine transformation is also an affine transformation. That is, suppose $\mathbf{Y} \sim N_q(\boldsymbol{\mu}, \mathbf{\Sigma})$, and $\mathbf{W} = \mathbf{c} + \mathbf{YD}'$ for $p \times q$ matrix $\mathbf{D}$ and $1 \times p$ vector $\mathbf{c}$. Then we know that for some $\mathbf{B}$ with $\mathbf{BB}' = \mathbf{\Sigma}$, $\mathbf{Y} = \boldsymbol{\mu} + \mathbf{ZB}'$, where $\mathbf{Z}$ is a standard normal vector. Hence

$$\mathbf{W} = \mathbf{c} + \mathbf{YD}' = \mathbf{c} + (\boldsymbol{\mu} + \mathbf{ZB}')\mathbf{D}' = \mathbf{c} + \boldsymbol{\mu}\mathbf{D}' + \mathbf{Z}(\mathbf{B}'\mathbf{D}'), \tag{3.18}$$

and as in (3.4),

$$\mathbf{W} \sim N_p(\mathbf{c} + \boldsymbol{\mu}\mathbf{D}', \mathbf{D}\mathbf{B}\mathbf{B}'\mathbf{D}') = N_p(\mathbf{c} + \boldsymbol{\mu}\mathbf{D}', \mathbf{D}\boldsymbol{\Sigma}\mathbf{D}'). \tag{3.19}$$

Of course, the mean and covariance result we already knew from (2.84) and (2.85).

Because marginals are special cases of affine transformations, marginals of multivariate normals are also multivariate normal. One needs just to pick off the appropriate means and covariances. So if $\mathbf{Y} = (Y_1, \ldots, Y_5)$ is $N_5(\boldsymbol{\mu}, \boldsymbol{\Sigma})$, and $\mathbf{W} = (Y_2, Y_5)$, then

$$\mathbf{W} \sim N_2\left((\mu_2, \mu_5), \begin{pmatrix} \sigma_{22} & \sigma_{25} \\ \sigma_{52} & \sigma_{55} \end{pmatrix}\right). \tag{3.20}$$

In Section 2.4, we showed that independence of two random variables means that their covariance is 0, but that a covariance of 0 does not imply independence. But, with multivariate normals, it does. That is, if $\mathbf{X}$ is a multivariate normal collection, and $Cov[X_j, X_k] = 0$, then X_j and X_k are independent. The next theorem generalizes this independence to sets of variables.

Theorem 3.1. *If $\mathbf{W} = (\mathbf{X}, \mathbf{Y})$ is a multivariate normal collection, then $Cov[\mathbf{X}, \mathbf{Y}] = \mathbf{0}$ (see Equation 2.46) implies that $\mathbf{X}$ and $\mathbf{Y}$ are independent.*

Proof. For simplicity, we will assume the mean of $\mathbf{W}$ is $\mathbf{0}$. Let $\mathbf{B}$ $(p \times M_1)$ and $\mathbf{C}$ $(q \times M_2)$ be matrices such that $\mathbf{B}\mathbf{B}' = Cov[\mathbf{X}]$ and $\mathbf{C}\mathbf{C}' = Cov[\mathbf{Y}]$, and $\mathbf{Z} = (\mathbf{Z}_1, \mathbf{Z}_2)$ be a standard normal collection of $M_1 + M_2$ variables, where $\mathbf{Z}_1$ is $1 \times M_1$ and $\mathbf{Z}_2$ is $1 \times M_2$. By assumption on the covariances between the X_k's and Y_l's, and properties of $\mathbf{B}$ and $\mathbf{C}$,

$$Cov[\mathbf{W}] = \begin{pmatrix} Cov[\mathbf{X}] & \mathbf{0} \\ \mathbf{0} & Cov[\mathbf{Y}] \end{pmatrix} = \begin{pmatrix} \mathbf{B}\mathbf{B}' & 0 \\ 0 & \mathbf{C}\mathbf{C}' \end{pmatrix} = \mathbf{A}\mathbf{A}', \tag{3.21}$$

where

$$\mathbf{A} = \begin{pmatrix} \mathbf{B} & 0 \\ 0 & \mathbf{C} \end{pmatrix}. \tag{3.22}$$

Which shows that $\mathbf{W}$ has distribution given by $\mathbf{Z}\mathbf{A}'$. With that representation, we have that $\mathbf{X} = \mathbf{Z}_1\mathbf{B}'$ and $\mathbf{Y} = \mathbf{Z}_2\mathbf{C}'$. Because the Z_i's are mutually independent, and the subsets $\mathbf{Z}_1$ and $\mathbf{Z}_2$ do not overlap, $\mathbf{Z}_1$ and $\mathbf{Z}_2$ are independent, which means that $\mathbf{X}$ and $\mathbf{Y}$ are independent. □

The theorem can also be proved using mgf's or pdf's. See Exercises 3.7.15 and 8.8.12.

3.3 Multivariate normal data matrix

Here we connect the $n \times q$ data matrix $\mathbf{Y}$ (1.1) to the multivariate normal. Each row of $\mathbf{Y}$ represents the values of q variables for an individual. Often, the data are modeled considering the rows of $\mathbf{Y}$ as independent observations from a population. Letting $\mathbf{Y}_i$ be the ith row of $\mathbf{Y}$, we would say that

$$\mathbf{Y}_1, \ldots, \mathbf{Y}_n \text{ are independent and identically distributed } (iid). \tag{3.23}$$

In the iid case, the vectors all have the same mean $\boldsymbol{\mu}$ and covariance matrix $\boldsymbol{\Sigma}$. Thus the mean of the entire matrix $\mathbf{M} = E[\mathbf{Y}]$ is

$$\mathbf{M} = \begin{pmatrix} \boldsymbol{\mu} \\ \boldsymbol{\mu} \\ \vdots \\ \boldsymbol{\mu} \end{pmatrix}. \tag{3.24}$$

For the covariance of the $\mathbf{Y}$, we need to string all the elements out, as in (2.45), as $(\mathbf{Y}_1, \ldots, \mathbf{Y}_n)$. By independence, the covariance between variables from different individuals is 0, that is, $Cov[Y_{ij}, Y_{kl}] = 0$ if $i \neq k$. Each group of q variables from a single individual has covariance $\boldsymbol{\Sigma}$, so that $Cov[\mathbf{Y}]$ is *block diagonal*:

$$\boldsymbol{\Omega} = Cov[\mathbf{Y}] = \begin{pmatrix} \boldsymbol{\Sigma} & \mathbf{0} & \cdots & \mathbf{0} \\ \mathbf{0} & \boldsymbol{\Sigma} & \cdots & \mathbf{0} \\ \vdots & \vdots & \ddots & \vdots \\ \mathbf{0} & \mathbf{0} & \cdots & \boldsymbol{\Sigma} \end{pmatrix}. \tag{3.25}$$

Patterned matrices such as (3.24) and (3.25) can be more efficiently represented as **Kronecker products.**

Definition 3.5. *If $\mathbf{A}$ is a $p \times q$ matrix and $\mathbf{B}$ is an $n \times m$ matrix, then the **Kronecker product** is the $(np) \times (mq)$ matrix $\mathbf{A} \otimes \mathbf{B}$ given by*

$$\mathbf{A} \otimes \mathbf{B} = \begin{pmatrix} a_{11}\mathbf{B} & a_{12}\mathbf{B} & \cdots & a_{1q}\mathbf{B} \\ a_{21}\mathbf{B} & a_{22}\mathbf{B} & \cdots & a_{2q}\mathbf{B} \\ \vdots & \vdots & \ddots & \vdots \\ a_{p1}\mathbf{B} & a_{p2}\mathbf{B} & \cdots & a_{pq}\mathbf{B} \end{pmatrix}. \tag{3.26}$$

Thus the mean in (3.24) and covariance matrix in (3.25) can be written as follows:

$$\mathbf{M} = \mathbf{1}_n \otimes \boldsymbol{\mu} \text{ and } \boldsymbol{\Omega} = \mathbf{I}_n \otimes \boldsymbol{\Sigma}. \tag{3.27}$$

Recall that $\mathbf{1}_n$ is the $n \times 1$ vector of all 1's, and $\mathbf{I}_n$ is the $n \times n$ identity matrix. Now if the rows of $\mathbf{Y}$ are iid multivariate normal, we write

$$\mathbf{Y} \sim N_{n \times q}(\mathbf{1}_n \otimes \boldsymbol{\mu}, \mathbf{I}_n \otimes \boldsymbol{\Sigma}). \tag{3.28}$$

Often the rows are independent with common covariance $\boldsymbol{\Sigma}$, but not necessarily having the same means. Then we have

$$\mathbf{Y} \sim N_{n \times q}(\mathbf{M}, \mathbf{I}_n \otimes \boldsymbol{\Sigma}). \tag{3.29}$$

We have already seen examples of linear combinations of elements in the data matrix. In (1.9) and (1.10), we had combinations of the form $\mathbf{CY}$, where the matrix multiplied $\mathbf{Y}$ on the left. The linear combinations are of the individuals within the variable, so that each variable is affected in the same way. In (1.23), and for principal components, the matrix is on the right: $\mathbf{YD}'$. In this case, the linear combinations are of the variables, with the variables for each individual affected the same way. More generally, we have affine transformations of the form (2.80),

$$\mathbf{W} = \mathbf{A} + \mathbf{CYD}'. \tag{3.30}$$

Because $\mathbf{W}$ is an affine transformation of $\mathbf{Y}$, it is also multivariate normal. When $Cov[\mathbf{Y}]$ has the form as in (3.26), then so does $Cov[\mathbf{W}]$.

Proposition 3.1. *If* $\mathbf{Y} \sim N_{n\times q}(\mathbf{M}, \mathbf{H}\otimes\mathbf{\Sigma})$ *and* $\mathbf{W} = \mathbf{A} + \mathbf{C}\mathbf{Y}\mathbf{D}'$, *where* $\mathbf{C}$ *is* $m\times n$, $\mathbf{D}$ *is* $p\times q$, $\mathbf{A}$ *is* $m\times p$, $\mathbf{H}$ *is* $n\times n$, *and* $\mathbf{\Sigma}$ *is* $q\times q$, *then*

$$\mathbf{W} \sim N_{m\times p}(\mathbf{A} + \mathbf{C}\mathbf{M}\mathbf{D}', \mathbf{C}\mathbf{H}\mathbf{C}'\otimes\mathbf{D}\mathbf{\Sigma}\mathbf{D}'). \tag{3.31}$$

The mean part follows directly from the second part of (2.84). For the covariance, we need some facts about Kronecker products, proofs of which are tedious but straightforward. See Exercises 3.7.17 to 3.7.19.

Proposition 3.2. *Presuming the matrix operations make sense and the inverses exist,*

$$\begin{aligned}
(\mathbf{A}\otimes\mathbf{B})' &= \mathbf{A}'\otimes\mathbf{B}' && (3.32a)\\
(\mathbf{A}\otimes\mathbf{B})(\mathbf{C}\otimes\mathbf{D}) &= (\mathbf{A}\mathbf{C})\otimes(\mathbf{B}\mathbf{D}) && (3.32b)\\
(\mathbf{A}\otimes\mathbf{B})^{-1} &= \mathbf{A}^{-1}\otimes\mathbf{B}^{-1} && (3.32c)\\
\text{row}(\mathbf{C}\mathbf{Y}\mathbf{D}') &= \text{row}(\mathbf{Y})(\mathbf{C}\otimes\mathbf{D})' && (3.32d)\\
\text{trace}(\mathbf{A}\otimes\mathbf{B}) &= \text{trace}(\mathbf{A})\,\text{trace}(\mathbf{B}) && (3.32e)\\
|\mathbf{A}\otimes\mathbf{B}| &= |\mathbf{A}|^b|\mathbf{B}|^a, && (3.32f)
\end{aligned}$$

where in the final equation, $\mathbf{A}$ *is* $a\times a$ *and* $\mathbf{B}$ *is* $b\times b$. *If* $Cov[\mathbf{U}] = \mathbf{A}\otimes\mathbf{B}$, *then*

$$\begin{aligned}
Var[U_{ij}] &= a_{ii}b_{jj}, \text{ more generally,} && (3.33a)\\
Cov[U_{ij}, U_{kl}] &= a_{ik}b_{jl} && (3.33b)\\
Cov[i^{th} \text{ row of } \mathbf{U}] &= a_{ii}\,\mathbf{B} && (3.33c)\\
Cov[j^{th} \text{ column of } \mathbf{U}] &= b_{jj}\,\mathbf{A}. && (3.33d)
\end{aligned}$$

To prove the covariance result in Proposition 3.1, write

$$\begin{aligned}
Cov[\mathbf{C}\mathbf{Y}\mathbf{D}'] &= Cov[\text{row}(\mathbf{Y})(\mathbf{C}\otimes\mathbf{D})'] && \text{by (3.32d)}\\
&= (\mathbf{C}\otimes\mathbf{D})Cov[\text{row}(\mathbf{Y})](\mathbf{C}\otimes\mathbf{D})' && \text{by (2.85)}\\
&= (\mathbf{C}\otimes\mathbf{D})(\mathbf{H}\otimes\mathbf{\Sigma})(\mathbf{C}'\otimes\mathbf{D}') && \text{by (3.32a)}\\
&= \mathbf{C}\mathbf{H}\mathbf{C}'\otimes\mathbf{D}\mathbf{\Sigma}\mathbf{D}' && \text{by (3.32b), twice.} \qquad (3.34)
\end{aligned}$$

One direct application of the proposition is the sample mean in the iid case (3.28), so that $\mathbf{Y} \sim N_{n\times q}(\mathbf{1}_n\otimes\boldsymbol{\mu}, \mathbf{I}_n\otimes\mathbf{\Sigma})$. Then from (1.9),

$$\overline{\mathbf{Y}} = \frac{1}{n}\mathbf{1}_n'\mathbf{Y}, \tag{3.35}$$

so we can use Proposition 3.1 with $\mathbf{C} = \frac{1}{n}\mathbf{1}_n'$, $\mathbf{D}' = \mathbf{I}_q$, and $\mathbf{A} = \mathbf{0}$. Thus

$$\overline{\mathbf{Y}} \sim N_q((\frac{1}{n}\mathbf{1}_n'\mathbf{1}_n)\otimes\boldsymbol{\mu}, (\frac{1}{n}\mathbf{1}_n')\mathbf{I}_n(\frac{1}{n}\mathbf{1}_n')'\otimes\mathbf{\Sigma}) = N_q(\boldsymbol{\mu}, \frac{1}{n}\mathbf{\Sigma}), \tag{3.36}$$

since $(1/n)\mathbf{1}_n'\mathbf{1}_n = 1$, and $c\otimes\mathbf{A} = c\mathbf{A}$ if c is a scalar. This result should not be surprising because it is the analog of the univariate result that $\overline{Y} \sim N(\mu, \sigma^2/n)$.

3.4 Conditioning in the multivariate normal

We start here with $\mathbf{X}$ being a $1 \times p$ vector and $\mathbf{Y}$ being a $1 \times q$ vector, then specialize to the data matrix case at the end of this section. If $(\mathbf{X}, \mathbf{Y})$ is multivariate normal, then the conditional distributions of $\mathbf{Y}$ given $\mathbf{X} = \mathbf{x}$ are multivariate normal as well. Let

$$(\mathbf{X}, \mathbf{Y}) \sim N_{p+q}\left((\boldsymbol{\mu}_X, \boldsymbol{\mu}_Y), \left(\begin{array}{cc} \boldsymbol{\Sigma}_{XX} & \boldsymbol{\Sigma}_{XY} \\ \boldsymbol{\Sigma}_{YX} & \boldsymbol{\Sigma}_{YY} \end{array}\right)\right). \tag{3.37}$$

Rather than diving in to joint densities, as in (2.23), we start by predicting the vector $\mathbf{Y}$ from $\mathbf{X}$ with an affine transformation. That is, we wish to find $\boldsymbol{\alpha}$, $1 \times q$, and $\boldsymbol{\beta}$, $p \times q$, so that

$$\mathbf{Y} \approx \boldsymbol{\alpha} + \mathbf{X}\boldsymbol{\beta}. \tag{3.38}$$

We use the least squares criterion, which is to find $(\boldsymbol{\alpha}, \boldsymbol{\beta})$ to minimize

$$q(\boldsymbol{\alpha}, \boldsymbol{\beta}) = E[\|\mathbf{Y} - \boldsymbol{\alpha} - \mathbf{X}\boldsymbol{\beta}\|^2]. \tag{3.39}$$

We start by noting that if the $1 \times q$ vector $\mathbf{W}$ has finite covariance matrix, then $E[\|\mathbf{W} - \mathbf{c}\|^2]$ is uniquely minimized over $\mathbf{c} \in \mathbb{R}^q$ by $\mathbf{c} = E[\mathbf{W}]$. See Exercise (3.7.21). Letting $\mathbf{W} = \mathbf{Y} - \mathbf{X}\boldsymbol{\beta}$, we have that fixing $\boldsymbol{\beta}$, $q(\boldsymbol{\alpha}, \boldsymbol{\beta})$ is minimized over $\boldsymbol{\alpha}$ by taking

$$\boldsymbol{\alpha} = E[\mathbf{Y} - \mathbf{X}\boldsymbol{\beta}] = \boldsymbol{\mu}_Y - \boldsymbol{\mu}_X\boldsymbol{\beta}. \tag{3.40}$$

Using that $\boldsymbol{\alpha}$ in (3.39), we now want to minimize

$$q(\boldsymbol{\mu}_Y - \boldsymbol{\mu}_X\boldsymbol{\beta}, \boldsymbol{\beta}) = E[\|\mathbf{Y} - (\boldsymbol{\mu}_Y - \boldsymbol{\mu}_X\boldsymbol{\beta}) - \mathbf{X}\boldsymbol{\beta}\|^2] = E[\|(\mathbf{Y} - \boldsymbol{\mu}_Y) - (\mathbf{X} - \boldsymbol{\mu}_X)\boldsymbol{\beta}\|^2] \tag{3.41}$$

over $\boldsymbol{\beta}$. Using the trick that for a row vector $\mathbf{z}$, $\|\mathbf{z}\|^2 = \text{trace}(\mathbf{z}'\mathbf{z})$, and letting $\mathbf{X}^* = \mathbf{X} - \boldsymbol{\mu}_X$ and $\mathbf{Y}^* = \mathbf{Y} - \boldsymbol{\mu}_Y$, we can write (3.41) as

$$\begin{aligned} E[\text{trace}((\mathbf{Y}^* - \mathbf{X}^*\boldsymbol{\beta})'(\mathbf{Y}^* - \mathbf{X}^*\boldsymbol{\beta}))] &= \text{trace}(E[(\mathbf{Y}^* - \mathbf{X}^*\boldsymbol{\beta})'(\mathbf{Y}^* - \mathbf{X}^*\boldsymbol{\beta}))]) \\ &= \text{trace}(E[\mathbf{Y}^{*\prime}\mathbf{Y}^*] - E[\mathbf{Y}^{*\prime}\mathbf{X}^*]\boldsymbol{\beta} \\ &\quad - \boldsymbol{\beta}'E[\mathbf{X}^{*\prime}\mathbf{Y}^*] + \boldsymbol{\beta}'E[\mathbf{X}^{*\prime}\mathbf{X}^*]\boldsymbol{\beta}) \\ &= \text{trace}(\boldsymbol{\Sigma}_{YY} - \boldsymbol{\Sigma}_{YX}\boldsymbol{\beta} - \boldsymbol{\beta}'\boldsymbol{\Sigma}_{XY} + \boldsymbol{\beta}'\boldsymbol{\Sigma}_{XX}\boldsymbol{\beta}). \end{aligned} \tag{3.42}$$

Now we complete the square. That is, we want to find $\boldsymbol{\beta}^*$ so that

$$\boldsymbol{\Sigma}_{YY} - \boldsymbol{\Sigma}_{YX}\boldsymbol{\beta} - \boldsymbol{\beta}'\boldsymbol{\Sigma}_{XY} + \boldsymbol{\beta}'\boldsymbol{\Sigma}_{XX}\boldsymbol{\beta} = (\boldsymbol{\beta} - \boldsymbol{\beta}^*)'\boldsymbol{\Sigma}_{XX}(\boldsymbol{\beta} - \boldsymbol{\beta}^*) + \boldsymbol{\Sigma}_{YY} - \boldsymbol{\beta}^{*\prime}\boldsymbol{\Sigma}_{XX}\boldsymbol{\beta}^*. \tag{3.43}$$

Matching, we must have that $\boldsymbol{\beta}'\boldsymbol{\Sigma}_{XX}\boldsymbol{\beta}^* = \boldsymbol{\beta}'\boldsymbol{\Sigma}_{XY}$, so that if $\boldsymbol{\Sigma}_{XX}$ is invertible, we need that $\boldsymbol{\beta}^* = \boldsymbol{\Sigma}_{XX}^{-1}\boldsymbol{\Sigma}_{XY}$. Then the trace of the expression in (3.43) is minimized by taking $\boldsymbol{\beta} = \boldsymbol{\beta}^*$, since that sets to 0 the part depending on $\boldsymbol{\beta}$, and you can't do better than that. Which means that (3.39) is minimized with

$$\boldsymbol{\beta} = \boldsymbol{\Sigma}_{XX}^{-1}\boldsymbol{\Sigma}_{XY}, \tag{3.44}$$

and $\boldsymbol{\alpha}$ in (3.40). The minimum of (3.39) is the trace of

$$\boldsymbol{\Sigma}_{YY} - \boldsymbol{\beta}'\boldsymbol{\Sigma}_{XX}\boldsymbol{\beta} = \boldsymbol{\Sigma}_{YY} - \boldsymbol{\Sigma}_{YX}\boldsymbol{\Sigma}_{XX}^{-1}\boldsymbol{\Sigma}_{XY}. \tag{3.45}$$

The prediction of $\mathbf{Y}$ is then $\boldsymbol{\alpha} + \mathbf{X}\boldsymbol{\beta}$. Define the **residual** to be the error in the prediction:

$$\mathbf{R} = \mathbf{Y} - \boldsymbol{\alpha} - \mathbf{X}\boldsymbol{\beta}. \tag{3.46}$$

Next step is to find the joint distribution of $(\mathbf{X}, \mathbf{R})$. Because it is an affine transformation of $(\mathbf{X}, \mathbf{Y})$, the joint distribution is multivariate normal, hence we just need to find the mean and covariance matrix. The mean of $\mathbf{X}$ we know is $\boldsymbol{\mu}_X$, and the mean of $\mathbf{R}$ is $\mathbf{0}$, from (3.40). The transform is

$$\begin{pmatrix} \mathbf{X} & \mathbf{R} \end{pmatrix} = \begin{pmatrix} \mathbf{X} & \mathbf{Y} \end{pmatrix} \begin{pmatrix} \mathbf{I}_p & -\boldsymbol{\beta} \\ 0 & \mathbf{I}_q \end{pmatrix} + \begin{pmatrix} 0 & -\boldsymbol{\alpha} \end{pmatrix}, \tag{3.47}$$

hence

$$\begin{aligned} Cov[\begin{pmatrix} \mathbf{X} & \mathbf{R} \end{pmatrix}] &= \begin{pmatrix} \mathbf{I}_p & 0 \\ -\boldsymbol{\beta}' & \mathbf{I}_q \end{pmatrix} \begin{pmatrix} \boldsymbol{\Sigma}_{XX} & \boldsymbol{\Sigma}_{XY} \\ \boldsymbol{\Sigma}_{YX} & \boldsymbol{\Sigma}_{YY} \end{pmatrix} \begin{pmatrix} \mathbf{I}_p & -\boldsymbol{\beta} \\ 0 & \mathbf{I}_q \end{pmatrix} \\ &= \begin{pmatrix} \boldsymbol{\Sigma}_{XX} & 0 \\ 0 & \boldsymbol{\Sigma}_{YY\cdot X} \end{pmatrix} \end{aligned} \tag{3.48}$$

where

$$\boldsymbol{\Sigma}_{YY\cdot X} = \boldsymbol{\Sigma}_{YY} - \boldsymbol{\Sigma}_{YX}\boldsymbol{\Sigma}_{XX}^{-1}\boldsymbol{\Sigma}_{XY}, \tag{3.49}$$

the minimizer in (3.45).

Note the zero in the covariance matrix. Because we have multivariate normality, $\mathbf{X}$ and $\mathbf{R}$ are independent, and

$$\mathbf{R} \sim N_q(\mathbf{0}, \boldsymbol{\Sigma}_{YY\cdot X}). \tag{3.50}$$

Use the plug-in formula with independence, (2.63), on $\mathbf{Y} = g(\mathbf{X}, \mathbf{R}) = \alpha + \mathbf{x}\boldsymbol{\beta} + \mathbf{R}$ to obtain

$$\mathbf{Y} \mid \mathbf{X} = \mathbf{x} \;=^{\mathcal{D}}\; \boldsymbol{\alpha} + \mathbf{x}\boldsymbol{\beta} + \mathbf{R}, \tag{3.51}$$

which leads to the next result.

Proposition 3.3. *If $(\mathbf{X}, \mathbf{Y})$ is multivariate normal as in (3.37), and $\boldsymbol{\Sigma}_{XX}$ invertible, then*

$$\mathbf{Y} \mid \mathbf{X} = \mathbf{x} \sim N(\boldsymbol{\alpha} + \mathbf{x}\boldsymbol{\beta}, \boldsymbol{\Sigma}_{YY\cdot X}), \tag{3.52}$$

where $\boldsymbol{\alpha}$ is given in (3.40), $\boldsymbol{\beta}$ is given in (3.44), and the conditional covariance matrix is given in (3.49).

The conditional distribution is particularly nice:

- It is multivariate normal;
- The conditional mean is an affine transformation of $\mathbf{x}$;
- It is *homoscedastic*, that is, the conditional covariance matrix does not depend on $\mathbf{x}$.

These properties are the typical assumptions in linear regression. See Exercise 3.7.24 for the case that $\boldsymbol{\Sigma}_{XX}$ is not invertible.

Conditioning in a multivariate normal data matrix

So far we have looked at just one $\mathbf{X}/\mathbf{Y}$ vector, whereas data will have a number of such vectors. Stacking these vectors into a data matrix, we have the distribution as in (3.29), but with an $\mathbf{X}$ matrix as well. That is, let $\mathbf{X}$ be $n \times p$ and $\mathbf{Y}$ be $n \times q$, where

$$\begin{pmatrix} \mathbf{X} & \mathbf{Y} \end{pmatrix} \sim N_{n\times(p+q)}\left(\begin{pmatrix} \mathbf{M}_X & \mathbf{M}_Y \end{pmatrix}, \mathbf{I}_n \otimes \begin{pmatrix} \mathbf{\Sigma}_{XX} & \mathbf{\Sigma}_{XY} \\ \mathbf{\Sigma}_{YX} & \mathbf{\Sigma}_{YY} \end{pmatrix}\right). \tag{3.53}$$

The conditional distribution of $\mathbf{Y}$ given $\mathbf{X} = \mathbf{x}$ can be obtained by applying Proposition 3.3 to $(\text{row}(\mathbf{X}), \text{row}(\mathbf{Y}))$, whose distribution can be written

$$\begin{aligned} \begin{pmatrix} \text{row}(\mathbf{X}) & \text{row}(\mathbf{Y}) \end{pmatrix} \sim N_{np+nq}\Big(\begin{pmatrix} \text{row}(\mathbf{M}_X) & \text{row}(\mathbf{M}_Y) \end{pmatrix}, \\ \begin{pmatrix} \mathbf{I}_n \otimes \mathbf{\Sigma}_{XX} & \mathbf{I}_n \otimes \mathbf{\Sigma}_{XY} \\ \mathbf{I}_n \otimes \mathbf{\Sigma}_{YX} & \mathbf{I}_n \otimes \mathbf{\Sigma}_{YY} \end{pmatrix}\Big). \end{aligned} \tag{3.54}$$

See Exercise 3.7.23. We again have the same $\boldsymbol{\beta}$, but $\boldsymbol{\alpha}$ is a bit expanded (it is $n \times q$):

$$\boldsymbol{\alpha} = \mathbf{M}_Y - \mathbf{M}_X\boldsymbol{\beta}, \;\; \boldsymbol{\beta} = \mathbf{\Sigma}_{XX}^{-1}\mathbf{\Sigma}_{XY}. \tag{3.55}$$

With $\mathbf{R} = \mathbf{Y} - \boldsymbol{\alpha} - \mathbf{X}\boldsymbol{\beta}$, we obtain that $\mathbf{X}$ and $\mathbf{R}$ are independent, and

$$\mathbf{Y} \mid \mathbf{X} = \mathbf{x} \sim N_{n\times q}(\boldsymbol{\alpha} + \mathbf{x}\boldsymbol{\beta}, \mathbf{I}_n \otimes \mathbf{\Sigma}_{YY\cdot X}). \tag{3.56}$$

3.5 The sample covariance matrix: Wishart distribution

Consider the iid case (3.28), $\mathbf{Y} \sim N_{n\times q}(\mathbf{1}_n \otimes \boldsymbol{\mu}, \mathbf{I}_n \otimes \mathbf{\Sigma})$. The sample covariance matrix is given in (1.17) and (1.15),

$$\mathbf{S} = \frac{1}{n}\mathbf{W}, \;\; \mathbf{W} = \mathbf{Y}'\mathbf{H}_n\mathbf{Y}, \;\; \mathbf{H}_n = \mathbf{I}_n - \frac{1}{n}\mathbf{1}_n\mathbf{1}_n'. \tag{3.57}$$

See (1.12) for the centering matrix, $\mathbf{H}_n$. Here we find the joint distribution of the sample mean $\overline{\mathbf{Y}}$ and $\mathbf{W}$. The marginal distribution of the sample mean is given in (3.36). Start by looking at the mean and the deviations together:

$$\begin{pmatrix} \overline{\mathbf{Y}} \\ \mathbf{H}_n\mathbf{Y} \end{pmatrix} = \begin{pmatrix} \frac{1}{n}\mathbf{1}_n' \\ \mathbf{H}_n \end{pmatrix}\mathbf{Y}. \tag{3.58}$$

Thus they are jointly normal. The mean of the sample mean is $\boldsymbol{\mu}$, and the mean of the deviations $\mathbf{H}_n\mathbf{Y}$ is $\mathbf{H}_n\mathbf{1}_n \otimes \boldsymbol{\mu} = \mathbf{0}$. (Recall Exercise 1.9.1.) The covariance is given by

$$\begin{aligned} Cov\left[\begin{pmatrix} \overline{\mathbf{Y}} \\ \mathbf{H}_n\mathbf{Y} \end{pmatrix}\right] &= \begin{pmatrix} \frac{1}{n}\mathbf{1}_n' \\ \mathbf{H}_n \end{pmatrix}\begin{pmatrix} \frac{1}{n}\mathbf{1}_n & \mathbf{H}_n \end{pmatrix} \otimes \mathbf{\Sigma} \\ &= \begin{pmatrix} \frac{1}{n} & \mathbf{0} \\ \mathbf{0} & \mathbf{H}_n \end{pmatrix} \otimes \mathbf{\Sigma}. \end{aligned} \tag{3.59}$$

The zeroes in the covariance show that $\overline{\mathbf{Y}}$ and $\mathbf{H}_n\mathbf{Y}$ are independent (as they are in the familiar univariate case), implying that $\overline{\mathbf{Y}}$ and $\mathbf{W}$ are independent. Also,

$$\mathbf{U} \equiv \mathbf{H}_n\mathbf{Y} \sim N(\mathbf{0}, \mathbf{H}_n \otimes \mathbf{\Sigma}). \tag{3.60}$$

Because $\mathbf{H}_n$ is idempotent, $\mathbf{W} = \mathbf{Y}'\mathbf{H}_n\mathbf{Y} = \mathbf{U}'\mathbf{U}$. At this point, instead of trying to figure out the distribution of $\mathbf{W}$, we define it to be what it is. Actually, Wishart [1928] did this a while ago. Next is the formal definition.

Definition 3.6 (Wishart distribution). *If $\mathbf{Z} \sim N_{\nu\times p}(\mathbf{0}, \mathbf{I}_\nu \otimes \mathbf{\Sigma})$, then $\mathbf{Z}'\mathbf{Z}$ is Wishart on ν degrees of freedom, with parameter $\mathbf{\Sigma}$, written*

$$\mathbf{Z}'\mathbf{Z} \sim \text{Wishart}_p(\nu, \mathbf{\Sigma}). \tag{3.61}$$

The difference between the distribution of $\mathbf{U}$ and the $\mathbf{Z}$ in the definition is the former has $\mathbf{H}_n$ where we would prefer an identity matrix. We can deal with this issue by rotating the $\mathbf{H}_n$. We need its spectral decomposition. More generally, suppose $\mathbf{J}$ is an $n \times n$ symmetric and idempotent matrix, with spectral decomposition (Theorem 1.1) $\mathbf{J} = \mathbf{\Gamma\Lambda\Gamma}'$, where $\mathbf{\Gamma}$ is orthogonal and $\mathbf{\Lambda}$ is diagonal with nondecreasing diagonal elements. Because it is idempotent, $\mathbf{JJ} = \mathbf{J}$, hence

$$\mathbf{\Gamma\Lambda\Gamma}' = \mathbf{\Gamma\Lambda\Gamma}'\mathbf{\Gamma\Lambda\Gamma}' = \mathbf{\Gamma\Lambda\Lambda\Gamma}', \tag{3.62}$$

so that $\mathbf{\Lambda} = \mathbf{\Lambda}^2$, or $\lambda_i = \lambda_i^2$ for the eigenvalues $i = 1, \ldots, n$. That means that each of the eigenvalues is either 0 or 1. If matrices $\mathbf{A}$ and $\mathbf{B}$ have the same dimensions, then

$$\text{trace}(\mathbf{AB}') = \text{trace}(\mathbf{B}'\mathbf{A}). \tag{3.63}$$

See Exercise 3.7.6. Thus

$$\text{trace}(\mathbf{J}) = \text{trace}(\mathbf{\Lambda\Gamma}'\mathbf{\Gamma}) = \text{trace}(\mathbf{\Lambda}) = \lambda_1 + \cdots + \lambda_n, \tag{3.64}$$

which is the number of eigenvalues that equal 1. Because the eigenvalues are ordered from largest to smallest, $\lambda_1 = \cdots = \lambda_{\text{trace}(\mathbf{J})} = 1$, and the rest are 0. Hence the following result.

Lemma 3.1. *Suppose $\mathbf{J}$, $n \times n$, is symmetric and idempotent. Then its spectral decomposition is*

$$\mathbf{J} = \begin{pmatrix} \mathbf{\Gamma}_1 & \mathbf{\Gamma}_2 \end{pmatrix} \begin{pmatrix} \mathbf{I}_k & \mathbf{0} \\ \mathbf{0} & \mathbf{0} \end{pmatrix} \begin{pmatrix} \mathbf{\Gamma}_1' \\ \mathbf{\Gamma}_2' \end{pmatrix} = \mathbf{\Gamma}_1\mathbf{\Gamma}_1', \tag{3.65}$$

where $k = \text{trace}(\mathbf{J})$, $\mathbf{\Gamma}_1$ is $n \times k$, and $\mathbf{\Gamma}_2$ is $n \times (n-k)$.

Now suppose

$$\mathbf{U} \sim N(\mathbf{0}, \mathbf{J} \otimes \mathbf{\Sigma}), \tag{3.66}$$

for $\mathbf{J}$ as in the lemma. Letting $\mathbf{\Gamma} = (\mathbf{\Gamma}_1, \mathbf{\Gamma}_2)$ in (3.65), we have $E[\mathbf{\Gamma}'\mathbf{U}] = \mathbf{0}$ and

$$Cov[\mathbf{\Gamma}'\mathbf{U}] = Cov\left[\begin{pmatrix} \mathbf{\Gamma}_1'\mathbf{U} \\ \mathbf{\Gamma}_2'\mathbf{U} \end{pmatrix}\right] = \begin{pmatrix} \mathbf{I}_k & \mathbf{0} \\ \mathbf{0} & \mathbf{0} \end{pmatrix} \otimes \mathbf{\Sigma}. \tag{3.67}$$

Thus $\mathbf{\Gamma}_2'\mathbf{U}$ has mean and covariance zero, hence must be zero itself (with probability one), which yields

$$\mathbf{U}'\mathbf{U} = \mathbf{U}'\mathbf{\Gamma\Gamma}'\mathbf{U} = \mathbf{U}'\mathbf{\Gamma}_1\mathbf{\Gamma}_1'\mathbf{U} + \mathbf{U}'\mathbf{\Gamma}_2\mathbf{\Gamma}_2'\mathbf{U} = \mathbf{U}'\mathbf{\Gamma}_1\mathbf{\Gamma}_1'\mathbf{U}. \tag{3.68}$$

By (3.66), and since $\mathbf{J} = \mathbf{\Gamma}_1\mathbf{\Gamma}_1'$ in (3.65),

$$\mathbf{\Gamma}_1'\mathbf{U} \sim N(\mathbf{0}, \mathbf{\Gamma}_1'\mathbf{\Gamma}_1\mathbf{\Gamma}_1'\mathbf{\Gamma}_1 \otimes \mathbf{\Sigma}) = N(\mathbf{0}, \mathbf{I}_k \otimes \mathbf{\Sigma}). \tag{3.69}$$

Now we can apply the Wishart definition (3.61) to $\mathbf{\Gamma}_1'\mathbf{U}$, to obtain the next result.

Corollary 3.1. *If* $\mathbf{U} \sim N_{n\times p}(0, \mathbf{J} \otimes \mathbf{\Sigma})$ *for idempotent* $\mathbf{J}$, *then*

$$\mathbf{U}'\mathbf{U} \sim \text{Wishart}_p(\text{trace}(\mathbf{J}), \mathbf{\Sigma}). \tag{3.70}$$

To apply the corollary to $\mathbf{W} = \mathbf{Y}'\mathbf{H}_n\mathbf{Y}$ in (3.57), by (3.60), we need only the trace of $\mathbf{H}_n$:

$$\text{trace}(\mathbf{H}_n) = \text{trace}(\mathbf{I}_n) - \frac{1}{n}\,\text{trace}(\mathbf{1}_n\mathbf{1}_n') = n - \frac{1}{n}(n) = n - 1. \tag{3.71}$$

Thus

$$\mathbf{W} \sim \text{Wishart}_q(n-1, \mathbf{\Sigma}). \tag{3.72}$$

3.6 Some properties of the Wishart

In this section we present some useful properties of the Wishart. The density is derived later in Section 8.7, and a conditional distribution is presented in Section 8.2.

Mean

Letting $\mathbf{Z}_1, \ldots, \mathbf{Z}_\nu$ be the rows of $\mathbf{Z}$ in Definition 3.6, we have that

$$\mathbf{Z}'\mathbf{Z} = \mathbf{Z}_1'\mathbf{Z}_1 + \cdots + \mathbf{Z}_\nu'\mathbf{Z}_\nu \sim \text{Wishart}_q(\nu, \mathbf{\Sigma}). \tag{3.73}$$

Each $\mathbf{Z}_i \sim N_{1\times q}(\mathbf{0}, \mathbf{\Sigma})$, so $E[\mathbf{Z}_i'\mathbf{Z}_i] = Cov[\mathbf{Z}_i] = \mathbf{\Sigma}$. Thus

$$E[\mathbf{W}] = \nu\mathbf{\Sigma}. \tag{3.74}$$

In particular, for the $\mathbf{S}$ in (3.57), because $\nu = n-1$, $E[\mathbf{S}] = ((n-1)/n)\mathbf{\Sigma}$, so that an unbiased estimator of $\mathbf{\Sigma}$ is

$$\widehat{\mathbf{\Sigma}} = \frac{1}{n-1}\,\mathbf{Y}'\mathbf{H}_n\mathbf{Y}. \tag{3.75}$$

Sum of independent Wisharts

If

$$\mathbf{W}_1 \sim \text{Wishart}_q(\nu_1, \mathbf{\Sigma}) \;\text{ and }\; \mathbf{W}_2 \sim \text{Wishart}_q(\nu_2, \mathbf{\Sigma}), \tag{3.76}$$

and $\mathbf{W}_1$ and $\mathbf{W}_2$ are independent, then $\mathbf{W}_1 + \mathbf{W}_2 \sim \text{Wishart}_q(\nu_1 + \nu_2, \mathbf{\Sigma})$. This fact can be easily shown by writing each as in (3.73), then summing.

Chi-squares

If $Z_1, \ldots, Z_\nu$ are independent $N(0, \sigma^2)$'s, then

$$W = Z_1^2 + \cdots + Z_\nu^2 \tag{3.77}$$

is said to be "chi-squared on ν degrees of freedom with scale σ^2," written $W \sim \sigma^2\chi_\nu^2$. (If $\sigma^2 = 1$, we call it just "chi-squared on ν degrees of freedom.") If $q = 1$ in the Wishart (3.73), the $\mathbf{Z}_i$'s in (3.73) are one-dimensional, i.e., $N(0, \sigma^2)$'s, hence

$$\text{Wishart}_1(\nu, \sigma^2) = \sigma^2\chi_\nu^2. \tag{3.78}$$

Linear transformations

If $\mathbf{Z} \sim N_{\nu\times q}(\mathbf{0}, \mathbf{I}_\nu \otimes \mathbf{\Sigma})$, then for $p \times q$ matrix $\mathbf{A}$, $\mathbf{Z}\mathbf{A}' \sim N_{\nu\times p}(\mathbf{0}, \mathbf{I}_\nu \otimes \mathbf{A}\mathbf{\Sigma}\mathbf{A}')$. Using the definition of Wishart (3.61),

$$\mathbf{A}\mathbf{Z}'\mathbf{Z}\mathbf{A}' \sim \text{Wishart}_p(\nu, \mathbf{A}\mathbf{\Sigma}\mathbf{A}'), \tag{3.79}$$

i.e.,

$$\mathbf{A}\mathbf{W}\mathbf{A}' \sim \text{Wishart}_p(\nu, \mathbf{A}\mathbf{\Sigma}\mathbf{A}'). \tag{3.80}$$

Marginals

Because marginals are special cases of linear transformations, central blocks of a Wishart are Wishart. E.g., if $\mathbf{W}_{11}$ is the upper-left $p \times p$ block of $\mathbf{W}$, then $\mathbf{W}_{11} \sim \text{Wishart}_p(\nu, \mathbf{\Sigma}_{11})$, where $\mathbf{\Sigma}_{11}$ is the upper-left block of $\mathbf{\Sigma}$. See Exercise 3.7.36. A special case of such marginal is a diagonal element, W_{ii}, which is $\text{Wishart}_1(\nu, \sigma_{ii})$, i.e., $\sigma_{ii}\chi^2_\nu$. Furthermore, if $\mathbf{\Sigma}$ is diagonal, then the diagonals of $\mathbf{W}$ are independent because the corresponding normals are.

3.7 Exercises

Exercise 3.7.1. Verify the calculations in (3.9).

Exercise 3.7.2. Find the matrix $\mathbf{B}$ for which $\mathbf{W} \equiv (Y_2, Y_5) = (Y_1, \ldots, Y_5)\mathbf{B}$, and verify (3.20).

Exercise 3.7.3. Verify (3.42).

Exercise 3.7.4. Suppose $\mathbf{Y} \sim N(\boldsymbol{\mu}, \mathbf{\Sigma})$, where $\mathbf{Y}$ is $1 \times q$. Let r be the number of nonzero eigenvalues of $\mathbf{\Sigma}$. Show that we can write $\mathbf{Y} = \boldsymbol{\mu} + \mathbf{Z}\mathbf{B}'$ as in (3.4), where $\mathbf{Z}$ is $1 \times r$. [Hint: Using the spectral decomposition theorem, write

$$\mathbf{\Sigma} = (\mathbf{\Gamma}_1, \mathbf{\Gamma}_2)\begin{pmatrix} \mathbf{\Lambda}_1 & \mathbf{0} \\ \mathbf{0} & \mathbf{0} \end{pmatrix}\begin{pmatrix} \mathbf{\Gamma}_1' \\ \mathbf{\Gamma}_2' \end{pmatrix} = \mathbf{\Gamma}_1\mathbf{\Lambda}_1\mathbf{\Gamma}_1', \tag{3.81}$$

where $\mathbf{\Lambda}_1$ is the $r \times r$ diagonal matrix with the positive eigenvalues, $\mathbf{\Gamma}_1$ contains the eigenvectors for the positive eigenvalues (so is $q \times r$), and $\mathbf{\Gamma}_2$ is $q \times (q - r)$. Show that $\mathbf{B}' = \mathbf{\Lambda}_1^{1/2}\mathbf{\Gamma}_1'$ works.]

Exercise 3.7.5. Verify the covariance calculation in (3.59).

Exercise 3.7.6. Suppose that $\mathbf{A}$ and $\mathbf{B}$ are both $n \times p$ matrices. Denote the elements of $\mathbf{A}$ by a_{ij}, and of $\mathbf{B}$ by b_{ij}. (a) Give the following in terms of those elements: $(\mathbf{A}\mathbf{B}')_{ii}$ (the ith diagonal element of the matrix $\mathbf{A}\mathbf{B}'$); and $(\mathbf{B}'\mathbf{A})_{jj}$ (the jth diagonal element of the matrix $\mathbf{B}'\mathbf{A}$). (b) Using the above, show that $\text{trace}(\mathbf{A}\mathbf{B}') = \text{trace}(\mathbf{B}'\mathbf{A})$.

Exercise 3.7.7. Show that in (3.69), $\mathbf{\Gamma}_1'\mathbf{\Gamma}_1 = \mathbf{I}_k$.

Exercise 3.7.8. Explicitly write the sum of $\mathbf{W}_1$ and $\mathbf{W}_2$ as in (3.76) as a sum of $\mathbf{Z}_i'\mathbf{Z}_i$'s as in (3.73).

Exercise 3.7.9. Suppose $W \sim \sigma^2\chi^2_\nu$ from (3.77), that is, $W = Z_1^2 + \cdots + Z_\nu^2$, where the Z_i's are independent $N(0, \sigma^2)$'s. This exercise shows that W has pdf

$$f_W(w \mid \nu, \sigma^2) = \frac{1}{\Gamma(\frac{\nu}{2})(2\sigma^2)^{\nu/2}} w^{\frac{\nu}{2}-1} e^{-w/(2\sigma^2)}, \; w > 0. \tag{3.82}$$

It will help to know that U has the Gamma(α, λ) density if $\alpha > 0$, $\lambda > 0$, and

$$f_U(u \mid \alpha, \lambda) = \frac{1}{\Gamma(\alpha)\lambda^\alpha} x^{\alpha-1} e^{-x/\lambda} \text{ for } x > 0. \tag{3.83}$$

The Γ function is defined in (2.97). (It is the constant needed to have the pdf integrate to one.) We'll use moment generating functions. Working directly with convolutions is another possibility. (a) Show that the moment generating function of U in (3.83) is $(1 - \lambda t)^{-\alpha}$ when it is finite. For which t is the mgf finite? (b) Let $Z \sim N(0, \sigma^2)$, so that $Z^2 \sim \sigma^2\chi^2_1$. Find the moment generating function for Z^2. [Hint: Write $E[\exp(tZ^2)]$ as an integral using the pdf of Z, then note the exponential term in the integrand looks like a normal with mean zero and some variance, but without the constant. Thus the integral over that exponential is the reciprocal of the constant.] (c) Find the moment generating function for W. (See Exercise 2.7.21.) (d) W has a gamma distribution. What are the parameters? Does this gamma pdf coincide with (3.82)? (e) [Aside] The density of Z^2 can be derived by writing

$$P[Z^2 \le w] = \int_{-\sqrt{w}}^{\sqrt{w}} f_Z(z)dz, \tag{3.84}$$

then taking the derivative. Match the result with the $\sigma^2\chi^2_1$ density found above. What is $\Gamma(\frac{1}{2})$?

Exercise 3.7.10. The balanced one-way random effects model in analysis of variance has

$$Y_{ij} = \mu + A_i + e_{ij}, \; i = 1, \ldots, g; j = 1, \ldots, r, \tag{3.85}$$

where the A_i's are iid $N(0, \sigma_A^2)$ and the e_{ij}'s are iid $N(0, \sigma_e^2)$, and the e_{ij}'s are independent of the A_i's. Let $\mathbf{Y}$ be the $g \times r$ matrix of the Y_{ij}'s. Show that

$$\mathbf{Y} \sim N_{g\times r}(\mathbf{M}, \mathbf{I}_g \otimes \mathbf{\Sigma}), \tag{3.86}$$

and give $\mathbf{M}$ and $\mathbf{\Sigma}$ in terms of the μ, σ_A^2 and σ_e^2.

Exercise 3.7.11. The double exponential random variable U has density

$$f(u) = \frac{1}{2} e^{-|u|}, \;\; u \in \mathbb{R}. \tag{3.87}$$

It has mean 0, variance 2, and moment generating function $M(t) = 1/(1 - t^2)$ for $|t| < 1$. Suppose $\mathbf{U} = (U_1, U_2)$, where U_1 and U_2 are independent double exponentials, and let

$$\mathbf{X} = \mathbf{U} \begin{pmatrix} 5 & 4 \\ 0 & 2 \end{pmatrix}. \tag{3.88}$$

(a) Find the covariance matrix of $\mathbf{X}$. (b) Find a lower triangular 2×2 matrix $\mathbf{A}$ (meaning $a_{12} = 0$) for which $Cov[\mathbf{X}] = \mathbf{A}'\mathbf{A}$. Let $\mathbf{Y} = \mathbf{U}\mathbf{A}/\sqrt{2}$. (c) Do $\mathbf{X}$ and $\mathbf{Y}$ have the same mean? (d) Do $\mathbf{X}$ and $\mathbf{Y}$ have the same covariance matrix? (e) Are $\mathbf{X}$ and $\mathbf{Y}$ both linear combinations of independent double exponentials? (f) Do $\mathbf{X}$ and $\mathbf{Y}$ have the same distribution? [Look at their moment generating functions.] (g) [Extra credit] Derive the mgf of the double exponential.

Exercise 3.7.12. Suppose $\boldsymbol{\Omega}$ is a $q \times q$ symmetric matrix with spectral decomposition (Theorem 1.1) $\boldsymbol{\Gamma}\boldsymbol{\Lambda}\boldsymbol{\Gamma}'$. (a) Show that $\boldsymbol{\Omega}$ is nonnegative definite if and only if $\lambda_i \geq 0$ for all $i = 1, \ldots, q$. [Hint: Suppose it is nonnegative definite. Let γ_i be the ith column of $\boldsymbol{\Gamma}$, and look at $\gamma_i'\boldsymbol{\Omega}\gamma_i$. What can you say about λ_i? The other way, suppose all $\lambda_i \geq 0$. Consider $\mathbf{b}\boldsymbol{\Omega}\mathbf{b}'$, and let $\mathbf{w} = \mathbf{b}\boldsymbol{\Gamma}$. Write $\mathbf{b}\boldsymbol{\Omega}\mathbf{b}'$ in terms of $\mathbf{w}$ and the λ_i.] (b) Show that $\boldsymbol{\Omega}$ is positive definite if and only if $\lambda_i > 0$ for all $i = 1, \ldots, q$. (c) Show that $\boldsymbol{\Omega}$ is invertible if and only if $\lambda_i \neq 0$ for all $i = 1, \ldots, q$. What is the spectral decomposition of $\boldsymbol{\Omega}^{-1}$ if the inverse exists?

Exercise 3.7.13. Extend Theorem 3.1: Show that if $\mathbf{W} = (\mathbf{Y}_1, \ldots, \mathbf{Y}_g)$ is a multivariate normal collection, then $Cov[\mathbf{Y}_i, \mathbf{Y}_j] = \mathbf{0}$ for each $i \neq j$ implies that $\mathbf{Y}_1, \ldots, \mathbf{Y}_g$ are mutually independent.

Exercise 3.7.14. Given the random vector (X, Y, Z), answer true or false to the following questions: (a) Pairwise independence implies mutual independence. [This question is the same as Exercise 2.7.28(a).] (b) Pairwise independence and multivariate normality implies mutual independence. (c) Conditional independence of X and Y given Z implies that X and Y are unconditionally independent. (d) (X, Y, Z) multivariate normal implies $(1, X, Y, Z)$ is multivariate normal.

Exercise 3.7.15. Let $\mathbf{X}$ be $1 \times p$ and $\mathbf{Y}$ be $1 \times q$, where

$$(\mathbf{X}, \mathbf{Y}) \sim N_{1\times(p+q)}\left((\boldsymbol{\mu}_X, \boldsymbol{\mu}_Y), \begin{pmatrix} \boldsymbol{\Sigma}_{XX} & \mathbf{0} \\ \mathbf{0} & \boldsymbol{\Sigma}_{YY} \end{pmatrix}\right), \tag{3.89}$$

so that $Cov[\mathbf{X}] = \boldsymbol{\Sigma}_{XX}$, $Cov[\mathbf{Y}] = \boldsymbol{\Sigma}_{YY}$, and $Cov[\mathbf{X}, \mathbf{Y}] = \mathbf{0}$. Using moment generating functions, show that $\mathbf{X}$ and $\mathbf{Y}$ are independent.

Exercise 3.7.16. True/false questions: (a) If $\mathbf{A}$ and $\mathbf{B}$ are identity matrices, then $\mathbf{A} \otimes \mathbf{B}$ is an identity matrix. (b) If $\mathbf{A}$ and $\mathbf{B}$ are orthogonal, then $\mathbf{A} \otimes \mathbf{B}$ is orthogonal. (c) If $\mathbf{A}$ is orthogonal and $\mathbf{B}$ is not orthogonal, then $\mathbf{A} \otimes \mathbf{B}$ is orthogonal. (d) If $\mathbf{A}$ and $\mathbf{B}$ are diagonal, then $\mathbf{A} \otimes \mathbf{B}$ is diagonal. (e) If $\mathbf{A}$ and $\mathbf{B}$ are idempotent, then $\mathbf{A} \otimes \mathbf{B}$ is idempotent. (f) If $\mathbf{A}$ and $\mathbf{B}$ are permutation matrices, then $\mathbf{A} \otimes \mathbf{B}$ is a permutation matrix. (A permutation matrix is a square matrix with exactly one 1 in each row, one 1 in each column, and 0's elsewhere.) (g) If $\mathbf{A}$ and $\mathbf{B}$ are upper triangular, then $\mathbf{A} \otimes \mathbf{B}$ is upper triangular. (An upper triangular matrix is a square matrix whose elements below the diagonal are 0. I.e., if $\mathbf{A}$ is upper triangular, then $a_{ij} = 0$ if $i > j$.) (h) If $\mathbf{A}$ is upper triangular and $\mathbf{B}$ is not upper triangular, then $\mathbf{A} \otimes \mathbf{B}$ is upper triangular. (i) If $\mathbf{A}$ is not upper triangular and $\mathbf{B}$ is upper triangular, then $\mathbf{A} \otimes \mathbf{B}$ is upper triangular. (j) If $\mathbf{A}$ and $\mathbf{C}$ have the same dimensions, and $\mathbf{B}$ and $\mathbf{D}$ have the same dimensions, then $\mathbf{A} \otimes \mathbf{B} + \mathbf{C} \otimes \mathbf{D} = (\mathbf{A} + \mathbf{C}) \otimes (\mathbf{B} + \mathbf{D})$. (k) If $\mathbf{A}$ and $\mathbf{C}$ have the same dimensions, then $\mathbf{A} \otimes \mathbf{B} + \mathbf{C} \otimes \mathbf{B} = (\mathbf{A} + \mathbf{C}) \otimes \mathbf{B}$. (l) If $\mathbf{B}$ and $\mathbf{D}$ have the same dimensions, then $\mathbf{A} \otimes \mathbf{B} + \mathbf{A} \otimes \mathbf{D} = \mathbf{A} \otimes (\mathbf{B} + \mathbf{D})$.

Exercise 3.7.17. Prove (3.32a), (3.32b), and (3.32c).

Exercise 3.7.18. Take $\mathbf{C}$, $\mathbf{Y}$, and $\mathbf{D}$ to all be 2×2. Show (3.32d) explicitly.

Exercise 3.7.19. Suppose $\mathbf{A}$ is $a \times a$ and $\mathbf{B}$ is $b \times b$. (a) Show that (3.32e) for the trace of $\mathbf{A} \otimes \mathbf{B}$ holds. (b) Show that (3.32f) for the determinant of $\mathbf{A} \otimes \mathbf{B}$ holds. [Hint: Write $\mathbf{A} \otimes \mathbf{B} = (\mathbf{A} \otimes \mathbf{I}_b)(\mathbf{I}_a \otimes \mathbf{B})$. You can use the fact that the determinant of a product is the product of the determinants. For $|\mathbf{I}_a \otimes \mathbf{B}|$, permutate the rows and columns so it looks like $|\mathbf{B} \otimes \mathbf{I}_a|$.]

Exercise 3.7.20. Suppose the spectral decompositions of $\mathbf{A}$ and $\mathbf{B}$ are $\mathbf{A} = \mathbf{GLG}'$ and $\mathbf{B} = \mathbf{HKH}'$. Is the equation

$$\mathbf{A} \otimes \mathbf{B} = (\mathbf{G} \otimes \mathbf{H})(\mathbf{L} \otimes \mathbf{K})(\mathbf{G} \otimes \mathbf{H})' \tag{3.90}$$

the spectral decomposition of $\mathbf{A} \otimes \mathbf{B}$? If not, what may be wrong, and how can it be fixed?

Exercise 3.7.21. Suppose $\mathbf{W}$ is a $1 \times q$ vector with finite covariance matrix. Show that $q(\mathbf{c}) = \|\mathbf{W} - \mathbf{c}\|^2$ is minimized over $\mathbf{c} \in \mathbb{R}^q$ by $c = E[\mathbf{W}]$, and the minimum value is $q(E[\mathbf{W}]) = \text{trace}(Cov[\mathbf{W}])$. [Hint: Write

$$\begin{aligned}
q(\mathbf{c}) &= E[\|(\mathbf{W} - E[\mathbf{W}]) - (E[\mathbf{W}] - \mathbf{c})\|^2] \\
&= E[\|\mathbf{W} - E[\mathbf{W}]\|^2] + 2E[(\mathbf{W} - E[\mathbf{W}])(E[\mathbf{W}] - \mathbf{c})'] + E[\|E[\mathbf{W}] - \mathbf{c}\|^2] \quad (3.91) \\
&= E[\|\mathbf{W} - E[\mathbf{W}]\|^2] + E[\|E[\mathbf{W}] - \mathbf{c}\|^2]. \quad (3.92)
\end{aligned}$$

Show that the middle (cross-product) term in line (3.91) is zero ($E[\mathbf{W}]$ and $\mathbf{c}$ are constants), and argue that the second term in line (3.92) is uniquely minimized by $\mathbf{c} = E[\mathbf{W}]$. (No need to take derivatives.)]

Exercise 3.7.22. Verify the matrix multiplication in (3.48).

Exercise 3.7.23. Suppose $(\mathbf{X}, \mathbf{Y})$ is as in (3.53). (a) Show that (3.54) follows. [Be careful about the covariance, since $\text{row}(\mathbf{X}, \mathbf{Y}) \neq (\text{row}(\mathbf{X}), \text{row}(\mathbf{Y}))$ if $n > 1$.] (b) Apply Proposition 3.3 to (3.54) to obtain

$$\text{row}(\mathbf{Y}) \mid \text{row}(\mathbf{X}) = \text{row}(\mathbf{x}) \sim N_{nq}(\boldsymbol{\alpha}^* + \text{row}(\mathbf{x})\boldsymbol{\beta}^*, \boldsymbol{\Sigma}^*_{YY \cdot X}), \tag{3.93}$$

where

$$\boldsymbol{\alpha}^* = \text{row}(\boldsymbol{\mu}_Y) - \text{row}(\boldsymbol{\mu}_X)\boldsymbol{\beta}^*, \ \boldsymbol{\beta}^* = \mathbf{I}_n \otimes \boldsymbol{\beta}, \ \boldsymbol{\Sigma}^*_{YY \cdot X} = \mathbf{I}_n \otimes \boldsymbol{\Sigma}_{YY \cdot X}. \tag{3.94}$$

What are $\boldsymbol{\beta}$ and $\boldsymbol{\Sigma}_{YY \cdot X}$? (c) Use Proposition 3.2 to derive (3.56) from part (b).

Exercise 3.7.24. Suppose

$$(\mathbf{X}, \mathbf{Y}) \sim N_{p+q}\left((\boldsymbol{\mu}_X, \boldsymbol{\mu}_Y), \begin{pmatrix} \boldsymbol{\Sigma}_{XX} & \boldsymbol{\Sigma}_{XY} \\ \boldsymbol{\Sigma}_{YX} & \boldsymbol{\Sigma}_{YY} \end{pmatrix}\right), \tag{3.95}$$

as in (3.37), but with $\boldsymbol{\Sigma}_{XX}$ not invertible. We wish to find the conditional distribution of $\mathbf{Y}$ given $\mathbf{X} = \mathbf{x}$. Let r be the number of positive eigenvalues for $\boldsymbol{\Sigma}_{XX}$. As in (3.81), write the spectral decomposition

$$\boldsymbol{\Sigma}_{XX} = \boldsymbol{\Gamma}\boldsymbol{\Lambda}\boldsymbol{\Gamma}' = (\boldsymbol{\Gamma}_1, \boldsymbol{\Gamma}_2) \begin{pmatrix} \boldsymbol{\Lambda}_1 & \mathbf{0} \\ \mathbf{0} & \mathbf{0} \end{pmatrix} \begin{pmatrix} \boldsymbol{\Gamma}_1' \\ \boldsymbol{\Gamma}_2' \end{pmatrix} = \boldsymbol{\Gamma}_1\boldsymbol{\Lambda}_1\boldsymbol{\Gamma}_1', \tag{3.96}$$

where $\boldsymbol{\Lambda}_1$ is $r \times r$ with the positive eigenvalues on its diagonal. (a) Let $\mathbf{U} = \mathbf{X}\boldsymbol{\Gamma}$. Partition $\mathbf{U} = (\mathbf{U}_1, \mathbf{U}_2)$, where $\mathbf{U}_i = \mathbf{X}\boldsymbol{\Gamma}_i$. Find the distribution of $\mathbf{U}$. Show that $\mathbf{U}_2$ equals the constant $\mu_X \Gamma_2$, hence we can write $\mathbf{X}$ as a function of $\mathbf{U}_1$: $\mathbf{X} = (\mathbf{U}_1, \mu_X \Gamma_2)\Gamma'$. (b) Note that if $\mathbf{x}$ is in the space of $\mathbf{X}$, then part (a) shows that $\mathbf{x}\Gamma_2 = \mu_X \Gamma_2$. Argue that, for $\mathbf{x}$ in the space of $\mathbf{X}$,

$$[\mathbf{Y} \mid \mathbf{X} = \mathbf{x}] \ =^{\mathcal{D}} \ [Y \mid (\mathbf{U}_1, \mu_X \Gamma_2) = (\mathbf{x}\Gamma_1, \mathbf{x}\Gamma_2)] \ =^{\mathcal{D}} \ [\mathbf{Y} \mid \mathbf{U}_1 = \mathbf{x}\Gamma_1]. \tag{3.97}$$

(c) Show that the joint distribution of $(\mathbf{U}_1, \mathbf{Y})$ is

$$(\mathbf{U}_1, \mathbf{Y}) \sim N\left((\mu_X \Gamma_1, \mu_Y), \begin{pmatrix} \Lambda_1 & \Gamma_1' \Sigma_{XY} \\ \Sigma_{YX} \Gamma_1 & \Sigma_{YY} \end{pmatrix} \right), \tag{3.98}$$

and the conditional distribution of $\mathbf{Y}$ given $\mathbf{U}_1 = \mathbf{u}_1$ is

$$\mathbf{Y} \mid \mathbf{U}_1 = \mathbf{u}_1 \sim N(\alpha + \mathbf{u}_1 \beta^*, \Sigma_{YY \cdot U_1}). \tag{3.99}$$

(Is the covariance matrix of $\mathbf{U}_1$ invertible?) Give the parameters α, β^*, and $\Sigma_{YY \cdot U_1}$. (d) Use parts (b) and (c) to show that

$$\mathbf{Y} \mid \mathbf{X} = \mathbf{x} \sim N(\alpha + \mathbf{x}\beta, \Sigma_{YY \cdot U_1}), \tag{3.100}$$

where $\beta = \Gamma_1 \beta^*$. (e) The *Moore-Penrose inverse* of a matrix is a "pseudoinverse" in that it acts somewhat like an inverse for noninvertible matrices. For a symmetric matrix like Σ_{XX}, the Moore-Penrose inverse is given by

$$\Sigma_{XX}^{+} = \Gamma_1 \Lambda_1^{-1} \Gamma_1'. \tag{3.101}$$

Show that we can write the parameters in (3.100) as

$$\alpha = \mu_Y - \mu_X \beta, \ \beta = \Sigma_{XX}^{+} \Sigma_{XY}, \text{ and } \Sigma_{YY \cdot U_1} = \Sigma_{YY} - \Sigma_{YX} \Sigma_{XX}^{+} \Sigma_{XY}, \tag{3.102}$$

which look very much like the parameters in (3.40), (3.44), and (3.45) for the invertible case.

Exercise 3.7.25. Suppose (X, Y, Z) is multivariate normal with covariance matrix

$$(X, Y, Z) \ \sim \ N\left((0,0,0), \begin{pmatrix} 5 & 1 & 2 \\ 1 & 5 & 2 \\ 2 & 2 & 3 \end{pmatrix} \right) \tag{3.103}$$

(a) What is the correlation of X and Y? Consider the conditional distribution of $(X, Y) | Z = z$. (b) Give the conditional covariance matrix, $Cov[(X, Y) | Z = z]$. (c) The correlation from that matrix is the condition correlation of X and Y given $Z = z$, sometimes called the partial correlation. What is the conditional correlation in this case? (d) If the conditional correlation between two variables given a third variable is negative, is the marginal correlation between those two necessarily negative?

Exercise 3.7.26. Now suppose

$$(X, Y, Z) \ \sim \ N\left((0,0,0), \begin{pmatrix} 5 & 1 & c \\ 1 & 5 & 2 \\ c & 2 & 3 \end{pmatrix} \right). \tag{3.104}$$

Find c so that the conditional correlation between X and Y given $Z = z$ is 0 (so that X and Y are conditionally independent, because of their normality).

Exercise 3.7.27. Let $Y \mid X = x \sim N(0, x^2)$ and $X \sim N(2,1)$. (a) Find $E[Y]$ and $Var[Y]$. (b) Let $Z = Y/X$. What is the conditional distribution of $Z \mid X = x$? Is Z independent of X? What is the marginal distribution of Z? (c) What is the conditional distribution of $Y \mid |X| = r$?

Exercise 3.7.28. Suppose that conditionally, $(Y_1, Y_2) \mid X = x$ are iid $N(\alpha + \beta x, 10)$, and that marginally, $E[X] = Var[X] = 1$. (The X is not necessarily normal.) (a) Find $Var[Y_i]$, $Cov[Y_1, Y_2]$, and the (unconditional) correlation between Y_1 and Y_2. (b) What is the conditional distribution of $Y_1 + Y_2 \mid X = x$? Is $Y_1 + Y_2$ independent of X? (c) What is the conditional distribution of $Y_1 - Y_2 \mid X = x$? Is $Y_1 - Y_2$ independent of X?

Exercise 3.7.29. This question reverses the conditional distribution in a multivariate normal, without having to use Bayes' formula. Suppose conditionally $\mathbf{Y} \mid \mathbf{X} = \mathbf{x} \sim N(\boldsymbol{\alpha} + \mathbf{x}\boldsymbol{\beta}, \boldsymbol{\Sigma})$, and marginally $\mathbf{X} \sim N(\boldsymbol{\mu}_X, \boldsymbol{\Sigma}_{XX})$, where $\mathbf{Y}$ is $1 \times q$ and $\mathbf{X}$ is $1 \times p$. (a) Show that $(\mathbf{X}, \mathbf{Y})$ is multivariate normal, and find its mean vector, and show that

$$Cov[(\mathbf{X}\ \ \mathbf{Y})] = \begin{pmatrix} \boldsymbol{\Sigma}_{XX} & \boldsymbol{\Sigma}_{XX}\boldsymbol{\beta} \\ \boldsymbol{\beta}'\boldsymbol{\Sigma}_{XX} & \boldsymbol{\Sigma} + \boldsymbol{\beta}'\boldsymbol{\Sigma}_{XX}\boldsymbol{\beta} \end{pmatrix}. \tag{3.105}$$

[Hint: Show that $\mathbf{X}$ and $\mathbf{Y} - \boldsymbol{\alpha} - \mathbf{X}\boldsymbol{\beta}$ are independent normals, and find the $\mathbf{A}$ so that $(\mathbf{X}, \mathbf{Y}) = (\mathbf{X}, \mathbf{Y} - \boldsymbol{\alpha} - \mathbf{X}\boldsymbol{\beta})\mathbf{A} + (\mathbf{0}, \boldsymbol{\alpha})$.] (b) Show that the conditional distribution $\mathbf{X} \mid \mathbf{Y} = \mathbf{y}$ is multivariate normal with mean

$$E[\mathbf{X} \mid \mathbf{Y} = \mathbf{y}] = \boldsymbol{\mu}_X + (\mathbf{y} - \boldsymbol{\alpha} - \boldsymbol{\mu}_X\boldsymbol{\beta})(\boldsymbol{\Sigma} + \boldsymbol{\beta}'\boldsymbol{\Sigma}_{XX}\boldsymbol{\beta})^{-1}\boldsymbol{\beta}'\boldsymbol{\Sigma}_{XX}, \tag{3.106}$$

and

$$Cov[\mathbf{X} \mid \mathbf{Y} = \mathbf{y}] = \boldsymbol{\Sigma}_{XX} - \boldsymbol{\Sigma}_{XX}\boldsymbol{\beta}(\boldsymbol{\Sigma} + \boldsymbol{\beta}'\boldsymbol{\Sigma}_{XX}\boldsymbol{\beta})^{-1}\boldsymbol{\beta}'\boldsymbol{\Sigma}_{XX}. \tag{3.107}$$

(You can assume any covariance that needs to be invertible is invertible.)

Exercise 3.7.30 (Bayesian inference). A Bayesian approach to estimating the normal mean vector, when the covariance matrix is known, is to set

$$\mathbf{Y} \mid \boldsymbol{\mu} = \mathbf{m} \sim N_{1\times q}(\mathbf{m}, \boldsymbol{\Sigma}) \text{ and } \boldsymbol{\mu} \sim N_{1\times q}(\boldsymbol{\mu}_0, \boldsymbol{\Sigma}_0), \tag{3.108}$$

where $\boldsymbol{\Sigma}$, $\boldsymbol{\mu}_0$, and $\boldsymbol{\Sigma}_0$ are known. That is, the mean vector $\boldsymbol{\mu}$ is a random variable, with a multivariate normal prior. (a) Use Exercise 3.7.29 to show that the posterior distribution of $\boldsymbol{\mu}$, i.e., $\boldsymbol{\mu}$ given $\mathbf{Y} = \mathbf{y}$, is multivariate normal with

$$E[\boldsymbol{\mu} \mid \mathbf{Y} = \mathbf{y}] = \boldsymbol{\mu}_0 + (\mathbf{y} - \boldsymbol{\mu}_0)(\boldsymbol{\Sigma} + \boldsymbol{\Sigma}_0)^{-1}\boldsymbol{\Sigma}_0, \tag{3.109}$$

and

$$Cov[\boldsymbol{\mu} \mid \mathbf{Y} = \mathbf{y}] = \boldsymbol{\Sigma}_0 - \boldsymbol{\Sigma}_0(\boldsymbol{\Sigma} + \boldsymbol{\Sigma}_0)^{-1}\boldsymbol{\Sigma}_0. \tag{3.110}$$

[Hint: What are the $\boldsymbol{\alpha}$ and $\boldsymbol{\beta}$ in this case?] (b) [Extra credit] Show that the quantities found in part (a) can be written

$$E[\boldsymbol{\mu} \mid \mathbf{Y} = \mathbf{y}] = (\mathbf{y}\boldsymbol{\Sigma}^{-1} + \boldsymbol{\mu}_0\boldsymbol{\Sigma}_0^{-1})(\boldsymbol{\Sigma}^{-1} + \boldsymbol{\Sigma}_0^{-1})^{-1}, \tag{3.111}$$

and

$$Cov[\boldsymbol{\mu} \mid \mathbf{Y} = \mathbf{y}] = (\boldsymbol{\Sigma}^{-1} + \boldsymbol{\Sigma}_0^{-1})^{-1}. \tag{3.112}$$

Thus the posterior mean is a weighted average of the data $\mathbf{y}$ and the prior mean, with weights inversely proportional to their respective covariance matrices. (c) Show that the marginal distribution of $\mathbf{Y}$ is

$$\mathbf{Y} \sim N_{1\times q}(\boldsymbol{\mu}_0, \boldsymbol{\Sigma} + \boldsymbol{\Sigma}_0). \tag{3.113}$$

[Hint: See (3.105).][Note that the inverse of the posterior covariance is the sum of the inverses of the conditional covariance of $\mathbf{Y}$ and the prior covariance, while the marginal covariance of the $\mathbf{Y}$ is the sum of the conditional covariance of $\mathbf{Y}$ and the prior covariance.] (d) Replace $\mathbf{Y}$ with $\overline{\mathbf{Y}}$, the sample mean of n iid vectors, so that $\overline{\mathbf{Y}} \mid \boldsymbol{\mu} = \mathbf{m} \sim N(\mathbf{m}, \boldsymbol{\Sigma}/n)$. Keep the same prior on $\boldsymbol{\mu}$. Find the posterior distribution of $\boldsymbol{\mu}$ given the $\overline{\mathbf{Y}} = \overline{\mathbf{y}}$. (e) For the situation in part (d), consider the posterior distribution of $\sqrt{n}(\boldsymbol{\mu} - \overline{\mathbf{y}})$ given $\overline{\mathbf{Y}} = \overline{\mathbf{y}}$. What are the limits as $n \to \infty$ of the the posterior mean and covariance matrix?

Exercise 3.7.31 (Bayesian inference). Consider a matrix version of Exercise 3.7.30, i.e.,

$$\mathbf{Y} \mid \boldsymbol{\mu} = \mathbf{m} \sim N_{p\times q}(\mathbf{m}, \mathbf{K}^{-1} \otimes \boldsymbol{\Sigma}) \text{ and } \boldsymbol{\mu} \sim N_{p\times q}(\boldsymbol{\mu}_0, \mathbf{K}_0^{-1} \otimes \boldsymbol{\Sigma}), \tag{3.114}$$

where $\mathbf{K}$, $\boldsymbol{\Sigma}$, $\boldsymbol{\mu}_0$, and $\mathbf{K}_0$ are known, and the covariance matrices are invertible. [So if $\mathbf{Y}$ is a sample mean vector, $\mathbf{K}$ would be n, and, looking ahead to Chapter 6, if $\mathbf{Y}$ is $\widehat{\boldsymbol{\beta}}$ from multivariate regression, $\mathbf{K}$ would be $\mathbf{x}'\mathbf{x}$. See (6.4) with $\mathbf{z} = \mathbf{I}_q$.] Notice that the $\boldsymbol{\Sigma}$ is the same in the conditional distribution of $\mathbf{Y}$ and in the prior. Show that the posterior distribution of $\boldsymbol{\mu}$ is multivariate normal, with

$$E[\boldsymbol{\mu} \mid \mathbf{Y} = \mathbf{y}] = (\mathbf{K} + \mathbf{K}_0)^{-1}(\mathbf{K}\mathbf{y} + \mathbf{K}_0\boldsymbol{\mu}_0), \tag{3.115}$$

and

$$Cov[\boldsymbol{\mu} \mid \mathbf{Y} = \mathbf{y}] = (\mathbf{K} + \mathbf{K}_0)^{-1} \otimes \boldsymbol{\Sigma}. \tag{3.116}$$

[Hint: Use (3.111) and (3.112) on row($\mathbf{Y}$) and row($\boldsymbol{\mu}$), then use properties of Kronecker products, e.g., (3.32d) and Exercise 3.7.16 (l).]

Exercise 3.7.32. Suppose $\mathbf{X}$ is $n \times p$, $\mathbf{Y}$ is $n \times q$, and

$$\begin{pmatrix} \mathbf{X} & \mathbf{Y} \end{pmatrix} \sim N\left(\begin{pmatrix} \mathbf{M}_X & \mathbf{M}_Y \end{pmatrix}, \mathbf{I}_n \otimes \begin{pmatrix} \boldsymbol{\Sigma}_{XX} & \boldsymbol{\Sigma}_{XY} \\ \boldsymbol{\Sigma}_{YX} & \boldsymbol{\Sigma}_{YY} \end{pmatrix}\right). \tag{3.117}$$

Let

$$\mathbf{R} = \mathbf{Y} - \mathbf{X}\mathbf{C}' - \mathbf{D} \tag{3.118}$$

for some matrices $\mathbf{C}$ and $\mathbf{D}$. Instead of using least squares as in Section 3.4, here we try to find $\mathbf{C}$ and $\mathbf{D}$ so that the residuals have mean zero and are independent of $\mathbf{X}$. (a) What are the dimensions of $\mathbf{R}$, $\mathbf{C}$ and $\mathbf{D}$? (b) Show that $(\mathbf{X}, \mathbf{R})$ is an affine transformation of $(\mathbf{X}, \mathbf{Y})$. That is, find $\mathbf{A}$ and $\mathbf{B}$ so that

$$(\mathbf{X}, \mathbf{R}) = \mathbf{A} + (\mathbf{X}, \mathbf{Y})\mathbf{B}'. \tag{3.119}$$

(c) Find the distribution of $(\mathbf{X}, \mathbf{R})$. (d) What must $\mathbf{C}$ be in order for $\mathbf{X}$ and $\mathbf{R}$ to be independent? (You can assume $\boldsymbol{\Sigma}_{XX}$ is invertible.) (e) Using the $\mathbf{C}$ found in part (d), find $Cov[\mathbf{R}]$. (It should be $\mathbf{I}_n \otimes \boldsymbol{\Sigma}_{YY\cdot X}$.) (f) Sticking with the $\mathbf{C}$ from parts (d) and (e), find $\mathbf{D}$ so that $E[\mathbf{R}] = \mathbf{0}$. (g) Using the $\mathbf{C}$ and $\mathbf{D}$ from parts (d), (e), (f), what is the distribution of $\mathbf{R}$? The distribution of $\mathbf{R}'\mathbf{R}$?

Exercise 3.7.33. Let $\mathbf{Y} \sim N_{n\times p}(\mathbf{M}, \mathbf{I}_n \otimes \mathbf{\Sigma})$. Suppose $\mathbf{K}$ is an $n \times n$ symmetric idempotent matrix with trace$(\mathbf{K}) = k$, and that $\mathbf{KM} = \mathbf{0}$. Show that $\mathbf{Y}'\mathbf{KY}$ is Wishart, and give the parameters.

Exercise 3.7.34. Suppose $\mathbf{Y} \sim N(\mathbf{x}\boldsymbol{\beta}\mathbf{z}', \mathbf{I}_n \otimes \mathbf{\Sigma})$, where $\mathbf{x}$ is $n \times p$, and

$$\mathbf{z} = \begin{pmatrix} 1 & -1 & 1 \\ 1 & 0 & -2 \\ 1 & 1 & 1 \end{pmatrix}. \tag{3.120}$$

(a) Find $\mathbf{C}$ so that $E[\mathbf{YC}'] = \mathbf{x}\boldsymbol{\beta}$. (b) Assuming that $\mathbf{x}'\mathbf{x}$ is invertible, what is the distribution of $\mathbf{Q}_\mathbf{x}\mathbf{YC}'$, where $\mathbf{Q}_\mathbf{x} = \mathbf{I}_n - \mathbf{x}(\mathbf{x}'\mathbf{x})^{-1}\mathbf{x}'$? (Is $\mathbf{Q}_\mathbf{x}$ idempotent? Such matrices will appear again in equation 5.20.) (c) What is the distribution of $\mathbf{CY}'\mathbf{Q}_\mathbf{x}\mathbf{YC}'$?

Exercise 3.7.35. Here, $\mathbf{W} \sim \text{Wishart}_p(n, \mathbf{\Sigma})$. (a) Is $E[\text{trace}(\mathbf{W})] = n\,\text{trace}(\mathbf{\Sigma})$? (b) Are the diagonal elements of $\mathbf{W}$ independent? (c) Suppose $\mathbf{\Sigma} = \sigma^2\mathbf{I}_p$. What is the distribution of trace$(\mathbf{W})$?

Exercise 3.7.36. Suppose $\mathbf{W} \sim \text{Wishart}_{p+q}(\nu, \mathbf{\Sigma})$, where $\mathbf{W}$ and $\mathbf{\Sigma}$ are partitioned as

$$\mathbf{W} = \begin{pmatrix} \mathbf{W}_{11} & \mathbf{W}_{12} \\ \mathbf{W}_{21} & \mathbf{W}_{22} \end{pmatrix} \text{ and } \mathbf{\Sigma} = \begin{pmatrix} \mathbf{\Sigma}_{11} & \mathbf{\Sigma}_{12} \\ \mathbf{\Sigma}_{21} & \mathbf{\Sigma}_{22} \end{pmatrix} \tag{3.121}$$

with $\mathbf{W}_{11}$ and $\mathbf{\Sigma}_{11}$ being $p \times p$, etc. (a) What matrix $\mathbf{A}$ in (3.80) is used to show that $\mathbf{W}_{11} \sim \text{Wishart}_p(\nu, \mathbf{\Sigma}_{11})$? (b) Argue that if $\mathbf{\Sigma}_{12} = \mathbf{0}$, then $\mathbf{W}_{11}$ and $\mathbf{W}_{22}$ are independent.

Exercise 3.7.37. Suppose $\mathbf{Z} = (Z_1, Z_2) \sim N_{1\times 2}(\mathbf{0}, \mathbf{I}_2)$. Let (θ, R) be the polar coordinates, so that

$$Z_1 = R\cos(\theta) \text{ and } Z_2 = R\sin(\theta). \tag{3.122}$$

In order for the transformation to be one-to-one, remove $\mathbf{0}$ from the space of $\mathbf{Z}$. Then the space of (θ, R) is $[0, 2\pi) \times (0, \infty)$. The question is to derive the distribution of (θ, R). (a) Write down the density of $\mathbf{Z}$. (b) Show that the Jacobian of the transformation is r. (c) Find the density of (θ, R). What is the marginal distribution of θ? What is the marginal density of R? Are R and θ independent? (d) Find the distribution function $F_R(r)$ for R. (e) Find the inverse function of F_R. (f) Argue that if U_1 and U_2 are independent Uniform$(0,1)$ random variables, then

$$\sqrt{-2\log(U_2)} \times \begin{pmatrix} \cos(2\pi U_1) & \sin(2\pi U_1) \end{pmatrix} \sim N_{1\times 2}(0, \mathbf{I}_2). \tag{3.123}$$

Thus we can generate two random normals from two random uniforms. Equation (3.123) is called the **Box-Muller transformation** [Box and Muller, 1958]. [Hint: See Exercise 2.7.29.] (g) Find the pdf of $W = R^2$. What is the distribution of W? Does it check with (3.82)?

Chapter 4

Linear Models on Both Sides

This chapter presents some basic types of linear model. We start with the usual linear model, with just one $\mathbf{Y}$-variable. Multivariate regression extends the idea to several variables, placing the same model on each variable. We then introduce linear models that model the variables *within* the observations, basically reversing the roles of observations and variables. Finally, we introduce the **both-sides model**, which simultaneously models the observations and variables. Subsequent chapters present estimation and hypothesis testing for these models.

4.1 Linear regression

Section 3.4 presented conditional distributions in the multivariate normal. Interest was in the effect of one set of variables, $\mathbf{X}$, on another set, $\mathbf{Y}$. Conditional on $\mathbf{X} = \mathbf{x}$, the distribution of $\mathbf{Y}$ was normal with the mean being a linear function of $\mathbf{x}$, and the covariance being independent of $\mathbf{x}$. The normal linear regression model does not assume that the joint distribution of $(\mathbf{X}, \mathbf{Y})$ is normal, but only that given $\mathbf{x}$, $\mathbf{Y}$ is multivariate normal. Analysis is carried out considering $\mathbf{x}$ to be fixed. In fact, $\mathbf{x}$ need not be a realization of a random variable, but could be a quantity fixed by the researcher, such as the dose of a drug or the amount of fertilizer.

The multiple regression model uses the data matrix $(\mathbf{x}, \mathbf{Y})$, where $\mathbf{x}$ is $n \times p$ and is lower case to emphasize that those values are fixed, and $\mathbf{Y}$ is $n \times 1$. That is, there are p variables in $\mathbf{x}$ and a single variable in $\mathbf{Y}$. In Section 4.2, we allow $\mathbf{Y}$ to contain more than one variable.

The model is

$$\mathbf{Y} = \mathbf{x}\boldsymbol{\beta} + \mathbf{R}, \text{ where } \boldsymbol{\beta} \text{ is } p \times 1 \text{ and } \mathbf{R} \sim N_{n\times 1}(\mathbf{0}, \sigma_R^2 \mathbf{I}_n). \tag{4.1}$$

Compare this to (3.51). The variance σ_R^2 plays the role of $\sigma_{YY\cdot X}$. The model (4.1) assumes that the residuals R_i are iid $N(0, \sigma_R^2)$.

Some examples follow. There are thousands of books on linear regression and linear models. Scheffé [1999] is the classic theoretical reference, and Christensen [2011] provides a more modern treatment. A fine applied reference is Weisberg [2013].

Simple and multiple linear regression

One may wish to assess the relation between height and weight, or between cholesterol level and percentage of fat in the diet. A linear relation would be *cholesterol* $=$ $\alpha + \beta(fat) + residual$, so one would typically want both an intercept α and a slope β. Translating this model to (4.1), we would have $p = 2$, where the first column contains all 1's. That is, if $x_1, \ldots, x_n$ are the values of the explanatory variable (fat), the model would be

$$\begin{pmatrix} Y_1 \\ Y_2 \\ \vdots \\ Y_n \end{pmatrix} = \begin{pmatrix} 1 & x_1 \\ 1 & x_2 \\ \vdots & \vdots \\ 1 & x_n \end{pmatrix} \begin{pmatrix} \alpha \\ \beta \end{pmatrix} + \begin{pmatrix} R_1 \\ R_2 \\ \vdots \\ R_n \end{pmatrix}. \tag{4.2}$$

This model is called a **simple** linear regression because there is just one explanatory variable.

Multiple linear regression would add more explanatory variables, e.g., age, blood pressure, amount of exercise, etc., each one being represented by its own column in the $\mathbf{x}$ matrix. If $z_1, \ldots, z_n$ are the values of a second variable, we would have the model

$$\begin{pmatrix} Y_1 \\ Y_2 \\ \vdots \\ Y_n \end{pmatrix} = \begin{pmatrix} 1 & x_1 & z_1 \\ 1 & x_2 & z_2 \\ \vdots & \vdots & \vdots \\ 1 & x_n & z_n \end{pmatrix} \begin{pmatrix} \alpha \\ \beta \\ \gamma \end{pmatrix} + \begin{pmatrix} R_1 \\ R_2 \\ \vdots \\ R_n \end{pmatrix}. \tag{4.3}$$

Analysis of variance

In analysis of variance, observations are classified into different groups, and one wishes to compare the means of the groups. If there are three groups, with two observations in each group, the model could be

$$\begin{pmatrix} Y_1 \\ Y_2 \\ Y_3 \\ Y_4 \\ Y_5 \\ Y_6 \end{pmatrix} = \begin{pmatrix} 1 & 0 & 0 \\ 1 & 0 & 0 \\ 0 & 1 & 0 \\ 0 & 1 & 0 \\ 0 & 0 & 1 \\ 0 & 0 & 1 \end{pmatrix} \begin{pmatrix} \mu_1 \\ \mu_2 \\ \mu_3 \end{pmatrix} + \mathbf{R}. \tag{4.4}$$

Other design matrices $\mathbf{x}$ yield the same model (See Section 5.6), e.g., we could just as well write

$$\begin{pmatrix} Y_1 \\ Y_2 \\ Y_3 \\ Y_4 \\ Y_5 \\ Y_6 \end{pmatrix} = \begin{pmatrix} 1 & 2 & -1 \\ 1 & 2 & -1 \\ 1 & -1 & 2 \\ 1 & -1 & 2 \\ 1 & -1 & -1 \\ 1 & -1 & -1 \end{pmatrix} \begin{pmatrix} \mu \\ \alpha \\ \beta \end{pmatrix} + \mathbf{R}, \tag{4.5}$$

where μ is the grand mean, and α and β represent differences among the means. More complicated models arise when observations are classified in multiple ways, e.g., sex, age, and ethnicity.

Analysis of covariance

It may be that the main interest is in comparing the means of groups as in analysis of variance, but there are other variables that potentially affect the **Y**. For example, in a study comparing three drugs' effectiveness in treating leprosy, there were bacterial measurements before and after treatment. The **Y** is the "after" measurement, and one would expect the "before" measurement, in addition to the drugs, to affect the after measurement. Letting $\mathbf{x}_i$'s represent the before measurements, the model would be

$$\begin{pmatrix} Y_1 \\ Y_2 \\ Y_3 \\ Y_4 \\ Y_5 \\ Y_6 \end{pmatrix} = \begin{pmatrix} 1 & 0 & 0 & x_1 \\ 1 & 0 & 0 & x_2 \\ 0 & 1 & 0 & x_3 \\ 0 & 1 & 0 & x_4 \\ 0 & 0 & 1 & x_5 \\ 0 & 0 & 1 & x_6 \end{pmatrix} \begin{pmatrix} \mu_1 \\ \mu_2 \\ \mu_3 \\ \beta \end{pmatrix} + \mathbf{R}. \tag{4.6}$$

The actual experiment had ten observations in each group. See Section 6.4.4.

Polynomial and cyclic models

The "linear" in linear models refers to the linearity of the mean of **Y** in the parameter $\boldsymbol{\beta}$ for fixed values of **x**. Within the matrix **x**, there can be arbitrary nonlinear functions of variables. For example, in growth curves, one may be looking at Y_i's over time which grow as a quadratic in x_i, i.e., $E[Y_i] = \beta_0 + \beta_1 x_i + \beta_2 x_i^2$. Such a model is still considered a linear model because the β_j's come in linearly. The full model would be

$$\begin{pmatrix} Y_1 \\ Y_2 \\ \vdots \\ Y_n \end{pmatrix} = \begin{pmatrix} 1 & x_1 & x_1^2 \\ 1 & x_2 & x_2^2 \\ \vdots & \vdots & \vdots \\ 1 & x_n & x_n^2 \end{pmatrix} \begin{pmatrix} \beta_0 \\ \beta_1 \\ \beta_2 \end{pmatrix} + \begin{pmatrix} R_1 \\ R_2 \\ \vdots \\ R_n \end{pmatrix}. \tag{4.7}$$

Higher-order polynomials add on columns of x_i^3's, x_i^4's, etc.

Alternatively, the Y_i's might behave cyclically, such as temperature over the course of a year, or the circadian (daily) rhythms of animals. If the cycle is over 24 hours, and measurements are made at each hour, the model could be

$$\begin{pmatrix} Y_1 \\ Y_2 \\ \vdots \\ Y_{24} \end{pmatrix} = \begin{pmatrix} 1 & \cos(2\pi \cdot 1/24) & \sin(2\pi \cdot 1/24) \\ 1 & \cos(2\pi \cdot 2/24) & \sin(2\pi \cdot 2/24) \\ \vdots & \vdots & \vdots \\ 1 & \cos(2\pi \cdot 24/24) & \sin(2\pi \cdot 24/24) \end{pmatrix} \begin{pmatrix} \alpha \\ \gamma_1 \\ \gamma_2 \end{pmatrix} + \begin{pmatrix} R_1 \\ R_2 \\ \vdots \\ R_{24} \end{pmatrix}. \tag{4.8}$$

Based on the data, typical objectives in linear regression are to estimate $\boldsymbol{\beta}$, test whether certain components of $\boldsymbol{\beta}$ are 0, or predict future values of Y based on its x's. In Chapters 6 and 7, such formal inferences will be handled. In this chapter, we are concentrating on setting up the models.

4.2 Multivariate regression and analysis of variance

Consider $(\mathbf{x}, \mathbf{Y})$ to be a data matrix where $\mathbf{x}$ is again $n \times p$, but now $\mathbf{Y}$ is $n \times q$. The linear model analogous to the conditional model in (3.56) is

$$\mathbf{Y} = \mathbf{x}\boldsymbol{\beta} + \mathbf{R}, \quad \text{where } \boldsymbol{\beta} \text{ is } p \times q \text{ and } \mathbf{R} \sim N_{n\times q}(\mathbf{0}, \mathbf{I}_n \otimes \boldsymbol{\Sigma}_R). \tag{4.9}$$

This model looks very much like the linear regression model in (4.1), and it is. It is actually just a concatenation of q linear models, one for each variable (column) of $\mathbf{Y}$. Note that (4.9) places the *same* model on each variable, in the sense of using the same $\mathbf{x}$'s, but allows different coefficients represented by the different columns of $\boldsymbol{\beta}$. That is, (4.9) implies

$$\mathbf{Y}_1 = \mathbf{x}\boldsymbol{\beta}_1 + \mathbf{R}_1, \ldots, \mathbf{Y}_q = \mathbf{x}\boldsymbol{\beta}_q + \mathbf{R}_q, \tag{4.10}$$

where the subscript i indicates the ith column of the matrix.

The $\mathbf{x}$ matrix is the same as in the previous section, so rather than repeating the examples, just imagine them with extra columns of $\mathbf{Y}$ and $\boldsymbol{\beta}$, and prepend the word "multivariate" to the models, e.g., multivariate analysis of variance, multivariate polynomial regression, etc.

One might ask what the advantage is of doing all q regressions at once rather than doing q separate ones. Good question. The main reason is to gather strength from having several variables. For example, suppose one has an analysis of variance comparing drugs on a number of health-related variables. It may be that no single variable shows significant differences between drugs, but the variables together show strong differences. Using the overall model can also help deal with multiple comparisons, e.g., when one has many variables, there is a good chance at least one shows significance even when there is nothing going on.

These models are more compelling when they are expanded to model dependencies among the means of the variables, which is the subject of Section 4.3.

4.2.1 Examples of multivariate regression

Example: Grades

The data are the grades (in the data set grades), and sex (0=male, 1=female), of 107 students, a portion of which is below:

Obs i	Gender	HW	Labs	InClass	Midterms	Final	Total
1	0	30.47	0.00	0	60.38	52	43.52
2	1	37.72	20.56	75	69.84	62	59.34
3	1	65.56	77.33	75	68.81	42	63.18
4	0	65.50	75.83	100	58.88	56	64.04
5	1	72.36	65.83	25	74.93	60	65.92
⋮	⋮	⋮	⋮	⋮	⋮	⋮	⋮
105	1	93.18	97.78	100	94.75	92	94.64
106	1	97.54	99.17	100	91.23	96	94.69
107	1	94.17	97.50	100	94.64	96	95.67

(4.11)

Consider predicting the midterms and final exam scores from gender and the homework, labs, and inclass scores. The model is $\mathbf{Y} = \mathbf{x}\boldsymbol{\beta} + \mathbf{R}$, where $\mathbf{Y}$ is 107×2

(the midterms and finals), $\mathbf{x}$ is 107×5 (with gender, homework, labs, inclass, plus the first column of $\mathbf{1}_{107}$), and $\boldsymbol{\beta}$ is 5×2:

$$\boldsymbol{\beta} = \begin{pmatrix} \beta_{0M} & \beta_{0F} \\ \beta_{GM} & \beta_{GF} \\ \beta_{HM} & \beta_{HF} \\ \beta_{LM} & \beta_{LF} \\ \beta_{IM} & \beta_{IF} \end{pmatrix}. \tag{4.12}$$

Chapter 6 shows how to estimate the β_{ij}'s. In this case the estimates are

	Midterms	Final
Intercept	56.472	43.002
Gender	−3.044	−1.922
HW	0.244	0.305
Labs	0.052	0.005
InClass	0.048	0.076

(4.13)

Note that the largest slopes (not counting the intercepts) are the negative ones for gender, but to truly assess the sizes of the coefficients, we will need to find their standard errors, which we will do in Chapter 6.

Mouth sizes

Measurements were made on the size of mouths of 27 children at four ages: 8, 10, 12, and 14. The measurement is the distance from the "center of the pituitary to the **pteryomaxillary fissure**"[1] in millimeters. These data can be found in Potthoff and Roy [1964]. There are 11 girls (sex=1) and 16 boys (sex=0). See Table 4.1. Figure 4.1 contains a plot of the mouth sizes over time. These curves are generally increasing. There are some instances where the mouth sizes decrease over time. The measurements are between two defined locations in the mouth, and as people age, the mouth shape can change, so it is not that people mouths are really getting smaller. Note that generally the boys have bigger mouths than the girls, as they are generally bigger overall.

For the linear model, code $\mathbf{x}$ where the first column is 1=girl, 0=boy, and the second column is 0=girl, 1=boy:

$$\mathbf{Y} = \mathbf{x}\boldsymbol{\beta} + \mathbf{R} = \begin{pmatrix} \mathbf{1}_{11} & \mathbf{0}_{11} \\ \mathbf{0}_{16} & \mathbf{1}_{16} \end{pmatrix} \begin{pmatrix} \beta_{11} & \beta_{12} & \beta_{13} & \beta_{14} \\ \beta_{21} & \beta_{22} & \beta_{23} & \beta_{24} \end{pmatrix} + \mathbf{R}. \tag{4.14}$$

Here, $\mathbf{Y}$ and $\mathbf{R}$ are 27×4. So now the first row of $\boldsymbol{\beta}$ has the (population) means of the girls for the four ages, and the second row has the means for the boys. The sample means are

	Age8	Age10	Age12	Age14
Girls	21.18	22.23	23.09	24.09
Boys	22.88	23.81	25.72	27.47

(4.15)

The lower plot in Figure 4.1 shows the sample mean vectors. The boys' curve is higher than the girls', and the two are reasonably parallel, and linear.

[1] Actually, I believe it is the **pterygomaxillary fissure**. See Wikipedia [2014b] for an illustration and some references.

Obs i	Age8	Age10	Age12	Age14	Sex
1	21.0	20.0	21.5	23.0	1
2	21.0	21.5	24.0	25.5	1
3	20.5	24.0	24.5	26.0	1
4	23.5	24.5	25.0	26.5	1
5	21.5	23.0	22.5	23.5	1
6	20.0	21.0	21.0	22.5	1
7	21.5	22.5	23.0	25.0	1
8	23.0	23.0	23.5	24.0	1
9	20.0	21.0	22.0	21.5	1
10	16.5	19.0	19.0	19.5	1
11	24.5	25.0	28.0	28.0	1
12	26.0	25.0	29.0	31.0	0
13	21.5	22.5	23.0	26.5	0
14	23.0	22.5	24.0	27.5	0
15	25.5	27.5	26.5	27.0	0
16	20.0	23.5	22.5	26.0	0
17	24.5	25.5	27.0	28.5	0
18	22.0	22.0	24.5	26.5	0
19	24.0	21.5	24.5	25.5	0
20	23.0	20.5	31.0	26.0	0
21	27.5	28.0	31.0	31.5	0
22	23.0	23.0	23.5	25.0	0
23	21.5	23.5	24.0	28.0	0
24	17.0	24.5	26.0	29.5	0
25	22.5	25.5	25.5	26.0	0
26	23.0	24.5	26.0	30.0	0
27	22.0	21.5	23.5	25.0	0

Table 4.1: The mouth size data, from Potthoff and Roy [1964].

Histamine in dogs

Sixteen dogs were treated with drugs to see the effects on their blood histamine levels. The dogs were split into four groups: Two groups received the drug morphine, and two received the drug trimethaphan, both given intravenously. For one group within each pair of drug groups, the dogs had their supply of histamine depleted before treatment, while the other group had histamine intact. So this was a two-way analysis of variance model, the factors being "Drug" (morphine or trimethaphan) and "Depletion" (intact or depleted). These data are from a study by Morris and Zeppa [1963], analyzed also in Cole and Grizzle [1966]. See Table 4.2.

Each dog had four measurements: Histamine levels (in micrograms per milliliter of blood) before the inoculation, and then at 1, 3, and 5 minutes after. (The value "0.10" marked with an asterisk was actually missing. I filled it in arbitrarily.) Figure 4.2 has a plot of the 16 dogs' series of measurements. Most of the data is close to zero, so it is hard to distinguish many of the individuals.

The model is a two-way multivariate analysis of variance one: $\mathbf{Y} = \mathbf{x}\boldsymbol{\beta} + \mathbf{R}$, where $\boldsymbol{\beta}$ contains the mean effect (μ), two main effects (α and β), and interaction effect (γ)

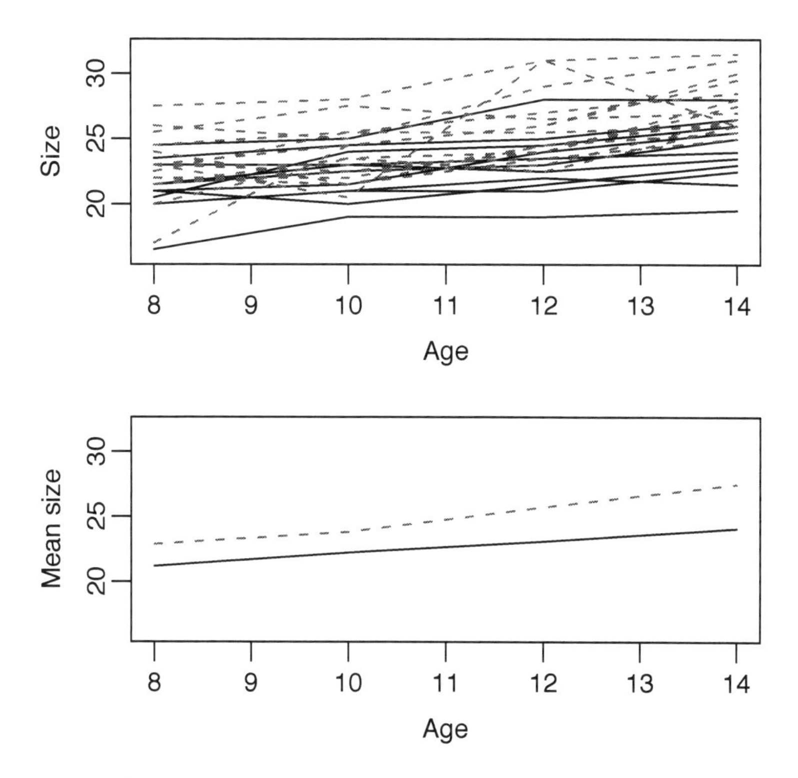

Figure 4.1: Mouth sizes over time. The boys are indicated by dashed lines, the girls by solid lines. The top plot has the individual graphs, the bottom the averages for the boys and girls.

for each time point:

$$\mathbf{Y} = \begin{pmatrix} \mathbf{1}_4 & -\mathbf{1}_4 & -\mathbf{1}_4 & \mathbf{1}_4 \\ \mathbf{1}_4 & -\mathbf{1}_4 & \mathbf{1}_4 & -\mathbf{1}_4 \\ \mathbf{1}_4 & \mathbf{1}_4 & -\mathbf{1}_4 & -\mathbf{1}_4 \\ \mathbf{1}_4 & \mathbf{1}_4 & \mathbf{1}_4 & \mathbf{1}_4 \end{pmatrix} \begin{pmatrix} \mu_0 & \mu_1 & \mu_3 & \mu_5 \\ \alpha_0 & \alpha_1 & \alpha_3 & \alpha_5 \\ \beta_0 & \beta_1 & \beta_3 & \beta_5 \\ \gamma_0 & \gamma_1 & \gamma_3 & \gamma_5 \end{pmatrix} + \mathbf{R}. \tag{4.16}$$

The estimate of $\boldsymbol{\beta}$ is

Effect	Before	After1	After3	After5
Mean	0.077	0.533	0.364	0.260
Drug	−0.003	0.212	0.201	0.140
Depletion	0.012	−0.449	−0.276	−0.169
Interaction	0.007	−0.213	−0.202	−0.144

(4.17)

See the middle plot in Figure 4.2 for the means of the groups, and the bottom plot for the effects, both plotted over time. Note that the mean and depletion effects are the largest, particularly at time point 2, after1.

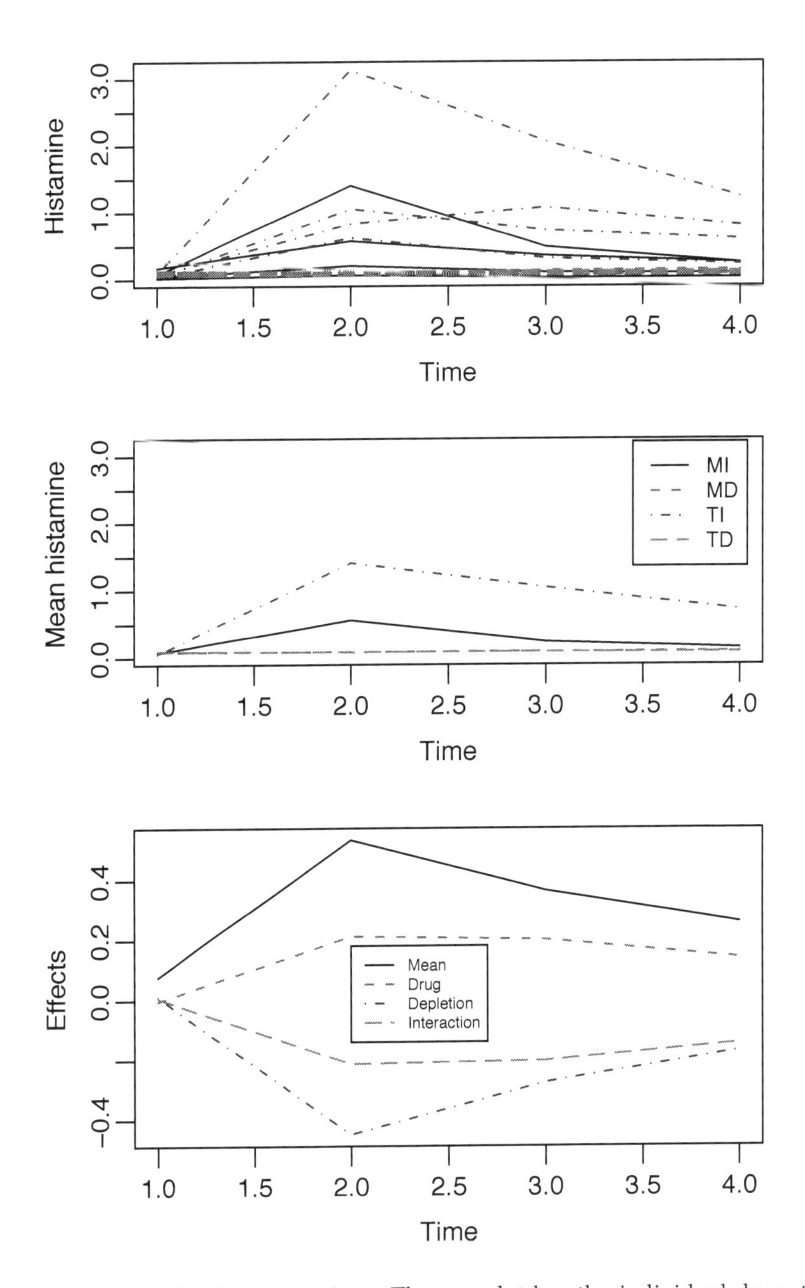

Figure 4.2: Plots of the dogs over time. The top plot has the individual dogs, the middle plot has the means of the groups, and the bottom plot has the effects in the analysis of variance. The groups: MI = morphine, intact; MD = morphine, depleted; TI = trimethaphan, intact; TD = trimethaphan, depleted.

Drug×Depletion	Obs i	Before	After1	After3	After5
Morphine	1	0.04	0.20	0.10	0.08
Intact	2	0.02	0.06	0.02	0.02
	3	0.07	1.40	0.48	0.24
	4	0.17	0.57	0.35	0.24
Morphine	5	0.10	0.09	0.13	0.14
Depleted	6	0.12	0.11	0.10	*0.10
	7	0.07	0.07	0.07	0.07
	8	0.05	0.07	0.06	0.07
Trimethaphan	9	0.03	0.62	0.31	0.22
Intact	10	0.03	1.05	0.73	0.60
	11	0.07	0.83	1.07	0.80
	12	0.09	3.13	2.06	1.23
Trimethaphan	13	0.10	0.09	0.09	0.08
Depleted	14	0.08	0.09	0.09	0.10
	15	0.13	0.10	0.12	0.12
	16	0.06	0.05	0.05	0.05

Table 4.2: The data on histamine levels in dogs. The value with the asterisk is missing, but for illustration purposes I filled it in. The dogs are classified according to the drug administered (morphine or trimethaphan), and whether the dog's histamine was artificially depleted.

4.3 Linear models on both sides

The regression and multivariate regression models in the previous sections model differences between the individuals: The rows of $\mathbf{x}$ are different for different individuals, but the same for each variable. Models on the variables switch the roles of variable and individual.

4.3.1 One individual

Start with just one individual, so that $\mathbf{Y} = (Y_1, \ldots, Y_q)$ is a $1 \times q$ row vector. A linear model on the variables is

$$\mathbf{Y} = \boldsymbol{\beta}\mathbf{z}' + \mathbf{R}, \text{ where } \boldsymbol{\beta} \text{ is } 1 \times l, \ \mathbf{R} \sim N_{1\times q}(\mathbf{0}, \boldsymbol{\Sigma}_R), \tag{4.18}$$

and $\mathbf{z}$ is a fixed $q \times l$ matrix. The model (4.18) looks like just a transpose of model (4.1), but (4.18) does not have iid residuals, because the observations are all on the same individual. Simple *repeated measures models* and *growth curve models* are special cases. (Simple because there is only one individual. Actual models would have more than one.)

A repeated measure model is used if the Y_j's represent replications of the same measurement. E.g., one may measure blood pressure of the same person several times, or take a sample of several leaves from the same tree. If no systematic differences are expected in the measurements, the model would have the same mean μ for each variable:

$$\mathbf{Y} = \mu(1, \ldots, 1) + \mathbf{R} = \mu\mathbf{1}_q' + \mathbf{R}. \tag{4.19}$$

It is common in this setting to assume $\mathbf{\Sigma}_R$ has the intraclass correlation structure, as in (1.61), i.e., the variances are all equal, and the covariances are all equal.

Growth curve models are used when the measurements are made over time, and growth (polynomial or otherwise) is expected. A quadratic model turns (4.7) on its side:

$$\mathbf{Y} = (\beta_0, \beta_1, \beta_2) \begin{pmatrix} 1 & 1 & \cdots & 1 \\ x_1 & x_2 & \cdots & x_q \\ x_1^2 & x_2^2 & \cdots & x_q^2 \end{pmatrix} + \mathbf{R}. \tag{4.20}$$

Similarly one can transpose cyclic models akin to (4.8).

4.3.2 IID observations

Now suppose we have a sample of n independent individuals, so that the $n \times q$ data matrix is distributed

$$\mathbf{Y} \sim N_{n\times q}(\mathbf{1}_n \otimes \boldsymbol{\mu}, \mathbf{I}_n \otimes \mathbf{\Sigma}_R), \tag{4.21}$$

which is the same as (3.28) with slightly different notation. Here, $\boldsymbol{\mu}$ is $1 \times q$, so the model says that the rows of $\mathbf{Y}$ are independent with the same mean $\boldsymbol{\mu}$ and covariance matrix $\mathbf{\Sigma}_R$. The simplest type of a repeated measure model assumes in addition that the elements of $\boldsymbol{\mu}$ are equal to μ, so that the linear model takes the mean in (4.21) and combines it with the mean in (4.19) to obtain

$$\mathbf{Y} = \mathbf{1}_n \mu \mathbf{1}_q' + \mathbf{R}, \quad \mathbf{R} \sim N_{n\times q}(\mathbf{0}, \mathbf{I}_n \otimes \mathbf{\Sigma}_R). \tag{4.22}$$

This model makes sense if one takes a random sample of n individuals, and makes repeated measurements from each. More generally, a growth curve model as in (4.20), but with n individuals measured, is

$$\mathbf{Y} = \mathbf{1}_n(\beta_0, \beta_1, \beta_2) \begin{pmatrix} 1 & \cdots & \cdots & 1 \\ z_1 & z_2 & \cdots & z_q \\ z_1^2 & z_2^2 & \cdots & z_q^2 \end{pmatrix} + \mathbf{R}. \tag{4.23}$$

Example: Births

The average births for each hour of the day for four different hospitals is given in Table 4.3. The data matrix $\mathbf{Y}$ here is 4×24, with the rows representing the hospitals and the columns the hours. Figure 4.3 plots the curves.

One might wish to fit sine waves (Figure 4.4) to the four hospitals' data, presuming one day reflects one complete cycle. The model is

$$\mathbf{Y} = \boldsymbol{\beta}\mathbf{z}' + \mathbf{R}, \tag{4.24}$$

where

$$\boldsymbol{\beta} = \begin{pmatrix} \beta_{10} & \beta_{11} & \beta_{12} \\ \beta_{20} & \beta_{21} & \beta_{22} \\ \beta_{30} & \beta_{31} & \beta_{32} \\ \beta_{40} & \beta_{41} & \beta_{42} \end{pmatrix} \tag{4.25}$$

	1	2	3	4	5	6	7	8
Hosp1	13.56	14.39	14.63	14.97	15.13	14.25	14.14	13.71
Hosp2	19.24	18.68	18.89	20.27	20.54	21.38	20.37	19.95
Hosp3	20.52	20.37	20.83	21.14	20.98	21.77	20.66	21.17
Hosp4	21.14	21.14	21.79	22.54	21.66	22.32	22.47	20.88

	9	10	11	12	13	14	15	16
Hosp1	14.93	14.21	13.89	13.60	12.81	13.27	13.15	12.29
Hosp2	20.62	20.86	20.15	19.54	19.52	18.89	18.41	17.55
Hosp3	21.21	21.68	20.37	20.49	19.70	18.36	18.87	17.32
Hosp4	22.14	21.86	22.38	20.71	20.54	20.66	20.32	19.36

	17	18	19	20	21	22	23	24
Hosp1	12.92	13.64	13.04	13.00	12.77	12.37	13.45	13.53
Hosp2	18.84	17.18	17.20	17.09	18.19	18.41	17.58	18.19
Hosp3	18.79	18.55	18.19	17.38	18.41	19.10	19.49	19.10
Hosp4	20.02	18.84	20.40	18.44	20.83	21.00	19.57	21.35

Table 4.3: The data on average number of births for each hour of the day for four hospitals.

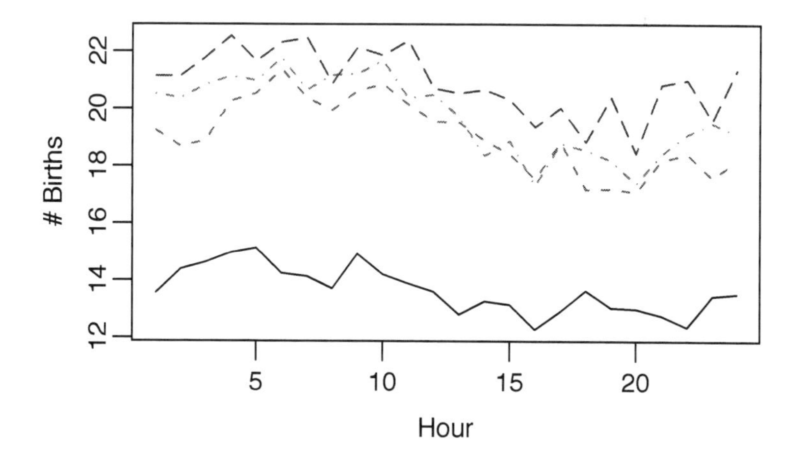

Figure 4.3: Plots of the four hospitals' average births for each hour, over twenty-four hours.

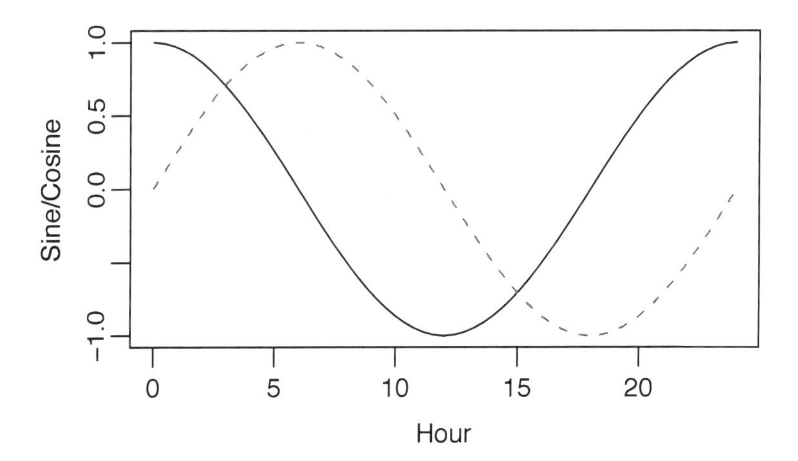

Figure 4.4: Sine and cosine waves, where one cycle spans twenty-four hours.

and

$$\mathbf{z}' = \begin{pmatrix} 1 & 1 & \cdots & 1 \\ \cos(1 \cdot 2\pi/24) & \cos(2 \cdot 2\pi/24) & \cdots & \cos(24 \cdot 2\pi/24) \\ \sin(1 \cdot 2\pi/24) & \sin(2 \cdot 2\pi/24) & \cdots & \sin(24 \cdot 2\pi/24) \end{pmatrix}, \tag{4.26}$$

the $\mathbf{z}$ here being the same as the $\mathbf{x}$ in (4.8).

The estimate of $\boldsymbol{\beta}$ is

	Mean	Cosine	Sine
Hosp1	13.65	0.03	0.93
Hosp2	19.06	−0.69	1.46
Hosp3	19.77	−0.22	1.70
Hosp4	20.93	−0.12	1.29

(4.27)

Then the "fits" are $\widehat{\mathbf{Y}} = \widehat{\boldsymbol{\beta}}\mathbf{z}'$, which is also 4×24. See Figure 4.5.

Now try the model with same curve for each hospital, $\mathbf{Y} = \mathbf{x}\boldsymbol{\beta}^*\mathbf{z}' + \mathbf{R}$, where $\mathbf{x} = \mathbf{1}_4$ (the star on the $\boldsymbol{\beta}^*$ is to distinguish it from the previous $\boldsymbol{\beta}$):

$$\mathbf{Y} = \mathbf{x}\boldsymbol{\beta}^*\mathbf{z}' + \mathbf{R} = \begin{pmatrix} 1 \\ 1 \\ 1 \\ 1 \end{pmatrix} \begin{pmatrix} \beta_0^* & \beta_1^* & \beta_2^* \end{pmatrix} \mathbf{z}' + \mathbf{R}. \tag{4.28}$$

The estimates of the coefficients are now $\widehat{\boldsymbol{\beta}}^* = (18.35, -0.25, 1.34)$, which is the average of the rows of $\widehat{\boldsymbol{\beta}}$. The fit is graphed as the thick line in Figure 4.5

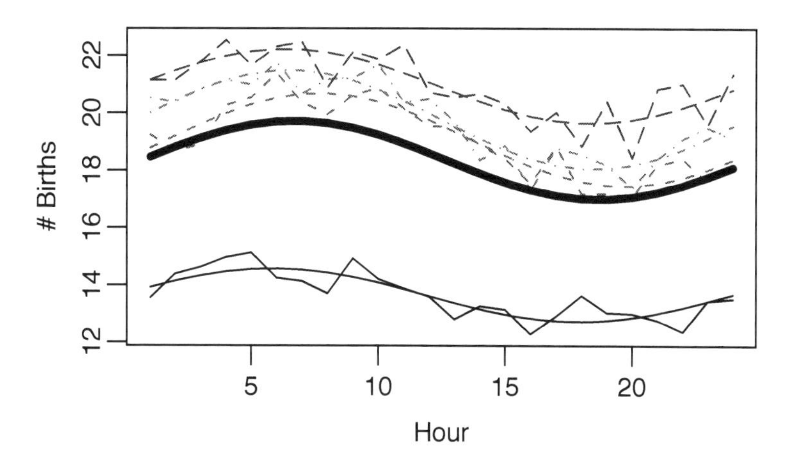

Figure 4.5: Plots of the four hospitals' births, with the fitted sign waves. The thick line fits one curve to all four hospitals.

4.3.3 The both-sides model

Note that the last two models, (4.20) and (4.23), have means with fixed matrices on both sides of the parameter. Generalizing, we have the model

$$\mathbf{Y} = \mathbf{x}\boldsymbol{\beta}\mathbf{z}' + \mathbf{R}, \tag{4.29}$$

where $\mathbf{x}$ is $n \times p$, $\boldsymbol{\beta}$ is $p \times l$, and $\mathbf{z}$ is $q \times l$. The $\mathbf{x}$ models differences between individuals, and the $\mathbf{z}$ models relationships between the variables. This formulation is by Potthoff and Roy [1964].

For example, consider the mouth size example in Section 4.2. A growth curve model seems reasonable, but one would not expect the iid model to hold. In particular, the mouths of the eleven girls would likely be smaller on average than those of the sixteen boys. An analysis of variance model, with two groups, models the differences between the individuals, while a growth curve models the relationship among the four time points. With $\mathbf{Y}$ being the 27×4 data matrix of measurements, the model is

$$\mathbf{Y} = \begin{pmatrix} \mathbf{1}_{11} & \mathbf{0}_{11} \\ \mathbf{0}_{16} & \mathbf{1}_{16} \end{pmatrix} \begin{pmatrix} \beta_{g0} & \beta_{g1} & \beta_{g2} \\ \beta_{b0} & \beta_{b1} & \beta_{b2} \end{pmatrix} \begin{pmatrix} 1 & 1 & 1 & 1 \\ 8 & 10 & 12 & 14 \\ 8^2 & 10^2 & 12^2 & 14^2 \end{pmatrix} + \mathbf{R}. \tag{4.30}$$

The "$\mathbf{0}_m$"'s are $m \times 1$ vectors of 0's. Thus $(\beta_{g0}, \beta_{g1}, \beta_{g2})$ contains the coefficients for the girls' growth curve, and $(\beta_{b0}, \beta_{b1}, \beta_{b2})$ the boys'. Some questions that can be addressed include

- Does the model fit, or are cubic terms necessary?
- Are the quadratic terms necessary (is $\beta_{g2} = \beta_{b2} = 0$)?
- Are the girls' and boys' curves the same (are $\beta_{gj} = \beta_{bj}$ for $j = 0, 1, 2$)?

- Are the girls' and boys' curves parallel (are $\beta_{g1} = \beta_{b1}$ and $\beta_{g2} = \beta_{b2}$, but maybe not $\beta_{g0} = \beta_{b0}$)?

See also Ware and Bowden [1977] for a circadian application and Zerbe and Jones [1980] for a time-series context. The model is often called the **generalized multivariate analysis of variance**, or **GMANOVA**, model. Extensions are many. For examples, see Gleser and Olkin [1970], Chinchilli and Elswick [1985], and the book by Kariya [1985].

4.4 Exercises

Exercise 4.4.1 (Prostaglandin). Below are data from Ware and Bowden [1977] taken at six four-hour intervals (labelled T1 to T6) over the course of a day for 10 individuals. The measurements are prostaglandin contents in their urine.

Person	T1	T2	T3	T4	T5	T6
1	146	280	285	215	218	161
2	140	265	289	231	188	69
3	288	281	271	227	272	150
4	121	150	101	139	99	103
5	116	132	150	125	100	86
6	143	172	175	222	180	126
7	174	276	317	306	139	120
8	177	313	237	135	257	152
9	294	193	306	204	207	148
10	76	151	333	144	135	99

(4.31)

(a) Write down the "$\mathbf{x}\boldsymbol{\beta}\mathbf{z}'$" part of the model that fits a separate sine wave to each person. (You don't have to calculate the estimates or anything. Just give the $\mathbf{x}$, $\boldsymbol{\beta}$ and $\mathbf{z}$ matrices.) (b) Do the same but for the model that fits one sine wave to all people.

Exercise 4.4.2 (Skulls). The data concern the sizes of Egyptian skulls over time, from Thomson and Randall-MacIver [1905]. There are 30 skulls from each of five time periods, so that $n = 150$ all together. There are four skull size measurements, all in millimeters: maximum length, basibregmatic height, basialveolar length, and nasal height. The model is multivariate regression, where $\mathbf{x}$ distinguishes between the time periods, and we do not use a $\mathbf{z}$. Use polynomials for the time periods (code them as 1, 2, 3, 4, 5), so that $\mathbf{x} = \mathbf{w} \otimes \mathbf{1}_{30}$. Find $\mathbf{w}$.

Exercise 4.4.3. Suppose $\mathbf{Y}_b$ and $\mathbf{Y}_a$ are $n \times 1$ with $n = 4$, and consider the model

$$(\mathbf{Y}_b\ \mathbf{Y}_a) \sim N(\mathbf{x}\boldsymbol{\beta}, \mathbf{I}_n \otimes \boldsymbol{\Sigma}), \tag{4.32}$$

where

$$\mathbf{x} = \begin{pmatrix} 1 & 1 & 1 \\ 1 & 1 & -1 \\ 1 & -1 & 1 \\ 1 & -1 & -1 \end{pmatrix}. \tag{4.33}$$

(a) What are the dimensions of $\boldsymbol{\beta}$ and $\boldsymbol{\Sigma}$? The conditional distribution of $\mathbf{Y}_a$ given $\mathbf{Y}_b = (4, 2, 6, 3)'$ is

$$\mathbf{Y}_a \mid \mathbf{Y}_b = (4, 2, 6, 3)' \sim N(\mathbf{x}^*\boldsymbol{\beta}^*, \mathbf{I}_n \otimes \boldsymbol{\Omega}) \tag{4.34}$$

for some fixed matrix $\mathbf{x}^*$, parameter matrix β^*, and covariance matrix $\mathbf{\Omega}$. (b) What are the dimensions of β^* and $\mathbf{\Omega}$? (c) What is $\mathbf{x}^*$? (d) What is the most precise description of the conditional model?

Exercise 4.4.4 (Caffeine). Henson et al. [1996] performed an experiment to see if caffeine has a negative effect on short-term visual memory. High school students were randomly chosen: 9 from eighth grade, 10 from tenth grade, and 9 from twelfth grade. Each person was tested once after having caffeinated Coke, and once after having decaffeinated Coke. After each drink, the person was given ten seconds to try to memorize twenty small, common objects, then allowed a minute to write down as many as could be remembered. The main question of interest is whether people remembered more objects after the Coke without caffeine than after the Coke with caffeine. The data are

Grade 8		Grade 10		Grade 12	
Without	With	Without	With	Without	With
5	6	6	3	7	7
9	8	9	11	8	6
6	5	4	4	9	6
8	9	7	6	11	7
7	6	6	8	5	5
6	6	7	6	9	4
8	6	6	8	9	7
6	8	9	8	11	8
6	7	10	7	10	9
		10	6		

(4.35)

"Grade" is the grade in school, and the "Without" and "With" entries are the numbers of items remembered after drinking Coke without or with caffeine. Consider the model

$$\mathbf{Y} = \mathbf{x}\beta\mathbf{z}' + \mathbf{R}, \tag{4.36}$$

where the $\mathbf{Y}$ is 28×2, the first column being the scores without caffeine, and the second being the scores with caffeine. The $\mathbf{x}$ is 28×3, being a polynomial (quadratic) matrix in the three grades. (a) The $\mathbf{z}$ has two columns. The first column of $\mathbf{z}$ represents the overall mean (of the number of objects a person remembers), and the second column represents the difference between the number of objects remembered with caffeine and without caffeine. Find $\mathbf{z}$. (b) What is the dimension of β? (c) What effects do the β_{ij}'s represent? (Choices: overall mean, overall linear effect of grade, overall quadratic effect of grade, overall difference in mean between caffeinated and decaffeinated coke, linear effect of grade in the difference between caffeinated and decaffeinated coke, quadratic effect of grade in the difference between caffeinated and decaffeinated coke, interaction of linear and quadratic effects of grade.)

Exercise 4.4.5 (Histamine in dogs). In Table 4.2, we have the data for the model,

$$\mathbf{Y} = \mathbf{x}\beta\mathbf{z}' + \mathbf{R}, \tag{4.37}$$

where $\mathbf{x}$ ($n \times 4$) describes a balanced two-way analysis of variance, as in (4.16). The columns represent, respectively, the overall mean, the drug effect, the depletion effect, and the drug $\times$ depletion interaction. For the $\mathbf{z}$, the first column is the effect of the "before" measurement (at time 0), and the last three columns represent polynomial effects (constant, linear, and quadratic) for just the three "after" time points (times 1, 3, 5). (a) What is $\mathbf{z}$? (b) Which of the β_{ij}'s represents each of the following effects? (i) Overall drug effect for the after measurements, (ii) overall drug effect for the before measurement, (iii) average after measurement, (iv) drug $\times$ depletion interaction for the before measurement, (v) linear effect in after time points for the drug effect.

Exercise 4.4.6 (Leprosy). Below are data on leprosy patients, found in Snedecor and Cochran [1989]. There were 30 patients, randomly allocated to three groups of 10. The first group received drug A, the second drug D, and the third group received a placebo. Each person had their bacterial count taken before and after receiving the treatment.

Drug A		Drug D		Placebo	
Before	After	Before	After	Before	After
11	6	6	0	16	13
8	0	6	2	13	10
5	2	7	3	11	18
14	8	8	1	9	5
19	11	18	18	21	23
6	4	8	4	16	12
10	13	19	14	12	5
6	1	8	9	12	16
11	8	5	1	7	1
3	0	15	9	12	20

(4.38)

(a) Consider the model $\mathbf{Y} = \mathbf{x}\boldsymbol{\beta} + \mathbf{R}$ for the multivariate analysis of variance with three groups and two variables (so that $\mathbf{Y}$ is 30×2), where $\mathbf{R} \sim N_{30\times 2}(\mathbf{0}, \mathbf{I}_{30} \otimes \boldsymbol{\Sigma}_R)$. The $\mathbf{x}$ has vectors for the overall mean, the contrast between the drugs and the placebo, and the contrast between drug A and drug D. Because there are ten people in each group, $\mathbf{x}$ can be written as $\mathbf{w} \otimes \mathbf{1}_{10}$. Find $\mathbf{w}$. (b) Because the before measurements were taken before any treatment, the means for the three groups on that variable should be the same. Describe that constraint in terms of the $\boldsymbol{\beta}$. (c) With $\mathbf{Y} = (\mathbf{Y}_b \;\; \mathbf{Y}_a)$, find the model for the conditional distribution

$$\mathbf{Y}_a \mid \mathbf{Y}_b = \mathbf{y}_b \;\sim\; N(\mathbf{x}^*\boldsymbol{\beta}^*, \mathbf{I}_n \otimes \boldsymbol{\Omega}). \tag{4.39}$$

Give the $\mathbf{x}^*$ in terms of $\mathbf{x}$ and $\mathbf{y}_b$, and give $\boldsymbol{\Omega}$ in terms of the elements of $\boldsymbol{\Sigma}_R$. (Hint: Write down what it would be with $E[\mathbf{Y}] = (\boldsymbol{\mu}_b \;\; \boldsymbol{\mu}_a)$ using the conditional formula, then see what you get when $\boldsymbol{\mu}_b = \mathbf{x}\boldsymbol{\beta}_b$ and $\boldsymbol{\mu}_a = \mathbf{x}\boldsymbol{\beta}_a$.)

Exercise 4.4.7 (Parity). Johnson and Wichern [2007] present data (in their Exercise 6.17) on an experiment. Each of 32 subjects was given several sets of pairs of integers, and had to say whether the two numbers had the same parity (i.e., both odd or both even), or different parity. So $(1,3)$ have the same parity, while $(4,5)$ have different parity. Some of the integer pairs were given numerically, like $(2,4)$, and some were written out, i.e., $(two, four)$. The time it took to decide the parity for each pair was measured. Each person had a little two-way analysis of variance, where the two factors are parity, with levels *different* and *same*, and format, with levels *word* and *numeric*. The measurements were the median time for each parity/format combination for that person. Person i then had observation vector $\mathbf{y}_i = (y_{i1}, y_{i2}, y_{i3}, y_{i4})$, which in the ANOVA could be arranged as

	Format	
Parity	Word	Numeric
Different	y_{i1}	y_{i2}
Same	y_{i3}	y_{i4}

(4.40)

The model is of the form

$$\mathbf{Y} = \mathbf{x}\boldsymbol{\beta}\mathbf{z}' + \mathbf{R}. \tag{4.41}$$

(a) What are $\mathbf{x}$ and $\mathbf{z}$ for the model where each person has a possibly different ANOVA, and each ANOVA has effects for overall mean, parity effect, format effect, and parity/format interaction? How many, and which, elements of $\boldsymbol{\beta}$ must be set to zero to model no-interaction? (b) What are $\mathbf{x}$ and $\mathbf{z}$ for the model where each person has the same mean vector, and that vector represents the ANOVA with effects for overall mean, parity effect, format effect, and parity/format interaction? How many, and which, elements of $\boldsymbol{\beta}$ must be set to zero to model no-interaction?

Exercise 4.4.8 (Sine waves). Let θ be an angle running from 0 to 2π, so that a sine/-cosine wave with one cycle has the form

$$g(\theta) = A + B\ \cos(\theta + C) \tag{4.42}$$

for parameters A, B, and C. Suppose we observe the wave at the q equally-spaced points

$$\theta_j = \frac{2\pi}{q}\ j,\ j = 1, \ldots, q, \tag{4.43}$$

plus error, so that the model is

$$Y_j = g(\theta_j) + R_j = A + B\ \cos\left(\frac{2\pi}{q}\ j + C\right) + R_j,\ \ j = 1, \ldots, q, \tag{4.44}$$

where the R_j are the residuals. (a) Is the model linear in the parameters A, B, C? Why or why not? (b) Show that the model can be rewritten as

$$Y_j = \beta_1 + \beta_2\ \cos\left(\frac{2\pi}{q}\ j\right) + \beta_3\ \sin\left(\frac{2\pi}{q}\ j\right) + R_j,\ \ j = 1, \ldots, q, \tag{4.45}$$

and give the β_k's in terms of A, B, C. [Hint: What is $\cos(a+b)$?] (c) Write this model as a linear model, $\mathbf{Y} = \boldsymbol{\beta}\mathbf{z}' + \mathbf{R}$, where $\mathbf{Y}$ is $1 \times q$. What is the $\mathbf{z}$? (d) Waves with $m \geq 1$ cycles can be added to the model by including cosine and sine terms with θ replaced by $m\theta$:

$$\cos\left(\frac{2\pi m}{q}\ j\right),\ \ \sin\left(\frac{2\pi m}{q}\ j\right). \tag{4.46}$$

If $q = 6$, then with the constant term, we can fit in the cosine and sign terms for the wave with $m = 1$ cycle, and the cosine and sine terms for the wave with $m = 2$ cycles. The $\mathbf{x}$ cannot have more than 6 columns (or else it won't be invertible). Find the cosine and sine terms for $m = 3$. What do you notice? Which one should you put in the model?

Chapter 5

Linear Models: Least Squares and Projections

In this chapter, we briefly review linear subspaces and projections onto them. Most of the chapter is abstract, in the sense of not necessarily tied to statistics. The main result we need for the rest of the book is the least squares estimate given in Theorem 5.2. Further results can be found in Chapter 1 of Rao [1973], an excellent compendium of facts on linear subspaces and matrices that are useful in statistics.

5.1 Linear subspaces

We start with the space $\mathbb{R}^N$. The elements $\mathbf{v} \in \mathbb{R}^N$ may be considered row vectors, or column vectors, or matrices, or any other configuration. We will generically call them "vectors." This space could represent vectors for individuals, in which case $N = q$, the number of variables, or it could represent vectors for variables, so $N = n$, the number of individuals, or it could represent the entire data matrix, so that $N = nq$. A linear subspace is a subset of $\mathbb{R}^N$ closed under addition and multiplication by a scalar. Because we will deal with Euclidean space, everyone knows what addition and multiplication mean. Here is the definition.

Definition 5.1. *A subset $\mathcal{W} \subset \mathbb{R}^N$ is a* ***linear subspace*** *of $\mathbb{R}^N$ if*

$$\mathbf{u}, \mathbf{v} \in \mathcal{W} \Longrightarrow \mathbf{u} + \mathbf{v} \in \mathcal{W}, \quad \text{and} \tag{5.1}$$

$$c \in \mathbb{R}, \mathbf{u} \in \mathcal{W} \Longrightarrow c\mathbf{u} \in \mathcal{W}. \tag{5.2}$$

We often shorten "linear subspace" to "subspace," or even "space." Note that $\mathbb{R}^N$ is itself a linear (sub)space, as is the set $\{\mathbf{0}\}$. Because c in (5.2) can be 0, any subspace must contain $\mathbf{0}$. Any line through $\mathbf{0}$, or plane through $\mathbf{0}$, is a subspace. One convenient representation of subspaces is the set of linear combinations of some elements:

Definition 5.2. *The* ***span*** *of the set of vectors $\{\mathbf{d}_1, \ldots, \mathbf{d}_K\} \subset \mathbb{R}^N$ is*

$$\text{span}\{\mathbf{d}_1, \ldots, \mathbf{d}_K\} = \{\gamma_1 \mathbf{d}_1 + \cdots + \gamma_K \mathbf{d}_K \mid \boldsymbol{\gamma} = (\gamma_1, \ldots, \gamma_K) \in \mathbb{R}^K\}. \tag{5.3}$$

By convention, the span of the empty set is just $\{\mathbf{0}\}$. It is not hard to show that any span is a linear subspace. Some examples: For $K = 2$, span$\{(1, 1)\}$ is the set of vectors of the form (a, a), that is, the equiangular line through $\mathbf{0}$. For $K = 3$, $\text{span}\{(1,0,0),(0,1,0)\}$ is the set of vectors of the form $(a, b, 0)$, which is the x/y plane, considering the axes to be x, y, z.

We will usually write the span in matrix form. Let $\mathbf{D}$ be the $N \times K$ matrix with columns $\mathbf{d}_1, \ldots, \mathbf{d}_K$. We have the following representations of subspace $\mathcal{W}$:

$$\begin{aligned} \mathcal{W} &= \text{span}\{\mathbf{d}_1, \ldots, \mathbf{d}_K\} \\ &= \text{span}\{\text{columns of } \mathbf{D}\} \\ &= \{\mathbf{D}\gamma \mid \gamma \in \mathbb{R}^K \ (\gamma \text{ is } K \times 1)\} \\ &= \text{span}\{\text{rows of } \mathbf{D}'\} \\ &= \{\gamma\mathbf{D}' \mid \gamma \in \mathbb{R}^K \ (\gamma \text{ is } 1 \times K)\}. \end{aligned} \tag{5.4}$$

Not only is any span a subspace, but any subspace is a span of some vectors. In fact, any subspace of $\mathbb{R}^N$ can be written as a span of at most N vectors, although not necessarily in a unique way. For example, for $N = 3$,

$$\begin{aligned} \text{span}\{(1,0,0),(0,1,0)\} &= \text{span}\{(1,0,0),(0,1,0),(1,1,0)\} \\ &= \text{span}\{(1,0,0),(1,1,0)\} \\ &= \text{span}\{(2,0,0),(0,-7,0),(33,2,0)\}. \end{aligned} \tag{5.5}$$

Any invertible transformation of the vectors yields the same span, as in the next lemma. See Exercise 5.8.4 for the proof.

Lemma 5.1. *Suppose $\mathcal{W}$ is the span of the columns of the $N \times K$ matrix $\mathbf{D}$ as in (5.4), and $\mathbf{A}$ is an invertible $K \times K$ matrix. Then $\mathcal{W}$ is also the span of the columns of $\mathbf{DA}$, i.e.,*

$$\text{span}\{\textit{columns of } \mathbf{D}\} = \text{span}\{\textit{columns of } \mathbf{DA}\}. \tag{5.6}$$

Note that the space in (5.5) can be a span of two or three vectors, or a span of any number more than three as well. It cannot be written as a span of only one vector. In the two sets of three vectors, there is a redundancy, that is, one of the vectors can be written as a linear combination of the other two: $(1,1,0) = (1,0,0) + (0,1,0)$ and $(2,0,0) = (\frac{4}{33\times 7}(0,-7,0) + \frac{2}{33} \times (33,2,0)$. Such sets are called **linearly dependent**. We first define the opposite.

Definition 5.3. *The vectors $\mathbf{d}_1, \ldots, \mathbf{d}_K$ in $\mathbb{R}^N$ are **linearly independent** if*

$$\gamma_1\mathbf{d}_1 + \cdots + \gamma_K\mathbf{d}_K = \mathbf{0} \implies \gamma_1 = \cdots = \gamma_K = 0. \tag{5.7}$$

The vectors are **linearly dependent** if and only if they are not linearly independent. Equivalently, the vectors are linearly dependent if and only if one of them can be written as a linear combination of the others. That is, they are linearly dependent iff there is a $\mathbf{d}_i$ and set of coefficients γ_j such that

$$\mathbf{d}_i = \gamma_1\mathbf{d}_1 + \cdots + \gamma_{i-1}\mathbf{d}_{i-1} + \gamma_{i+1}\mathbf{d}_{i+1} + \ldots + \gamma_K\mathbf{d}_K. \tag{5.8}$$

In (5.5), the sets with three vectors are linearly dependent, and those with two vectors are linearly independent. To see that latter fact for $\{(1,0,0),(1,1,0)\}$, suppose that $\gamma_1(1,0,0) + \gamma_2(1,1,0) = (0,0,0)$. Then

$$\gamma_1 + \gamma_2 = 0 \text{ and } \gamma_2 = 0 \implies \gamma_1 = \gamma_2 = 0, \tag{5.9}$$

which verifies (5.7).

If a set of vectors is linearly dependent, then one can remove one of the redundant vectors (5.8), and still have the same span. A basis is a set of vectors that has the same span but no dependencies.

Definition 5.4. *The set of vectors* $\{\mathbf{d}_1, \ldots, \mathbf{d}_K\}$ *is a* ***basis*** *for the subspace* $\mathcal{W}$ *if the vectors are linearly independent and* $\mathcal{W} = \text{span}\{\mathbf{d}_1, \ldots, \mathbf{d}_K\}$.

Although a (nontrivial) subspace has many bases, each basis has the same number of elements, which is the dimension. See Exercise 5.8.47.

Definition 5.5. *The* ***dimension*** *of a subspace is the number of vectors in any of its bases.*

5.2 Projections

In linear models, the mean of the data matrix is presumed to lie in a linear subspace, and an aspect of fitting the model is to find the point in the subspace closest to the data. This closest point is called the projection. Before we get to the formal definition, we need to define orthogonality. Recall from Section 1.5 that two column vectors $\mathbf{v}$ and $\mathbf{w}$ are orthogonal if $\mathbf{v}'\mathbf{w} = 0$ (or $\mathbf{v}\mathbf{w}' = 0$ if they are row vectors).

Definition 5.6. *The vector* $\mathbf{v} \in \mathbb{R}^N$ *is orthogonal to the subspace* $\mathcal{W} \subset \mathbb{R}^N$ *if* $\mathbf{v}$ *is orthogonal to* $\mathbf{w}$ *for all* $\mathbf{w} \in \mathcal{W}$. *Also, subspace* $\mathcal{V} \subset \mathbb{R}^N$ *is orthogonal to* $\mathcal{W}$ *if* $\mathbf{v}$ *and* $\mathbf{w}$ *are orthogonal for all* $\mathbf{v} \in \mathcal{V}$ *and* $\mathbf{w} \in \mathcal{W}$.

Geometrically, two objects are orthogonal if they are perpendicular. For example, in $\mathbb{R}^3$, the z-axis is orthogonal to the x/y-plane. Exercise 5.8.6 is to prove the next result.

Lemma 5.2. *Suppose* $\mathcal{W} = \text{span}\{\mathbf{d}_1, \ldots, \mathbf{d}_K\}$. *Then* $\mathbf{v}$ *is orthogonal to* $\mathcal{W}$ *if and only if* $\mathbf{v}$ *is orthogonal to each* $\mathbf{d}_j$.

Definition 5.7. *The* ***orthogonal projection*** *of* $\mathbf{v}$ *onto* $\mathcal{W}$ *is the* $\widehat{\mathbf{v}}$ *that satisfies*

$$\widehat{\mathbf{v}} \in \mathcal{W} \;\; \textit{and} \;\; \mathbf{v} - \widehat{\mathbf{v}} \;\; \textit{is orthogonal to} \;\; \mathcal{W}. \tag{5.10}$$

In statistical parlance, the projection $\widehat{\mathbf{v}}$ is the **fit** and $\mathbf{v} - \widehat{\mathbf{v}}$ is the **residual**. Because of the orthogonality, we have the decomposition of squared norms,

$$\|\mathbf{v}\|^2 = \|\widehat{\mathbf{v}}\|^2 + \|\mathbf{v} - \widehat{\mathbf{v}}\|^2, \tag{5.11}$$

which is Pythagoras' Theorem. In a regression setting, the left-hand side is the *total sum-of-squares*, and the right-hand side is the *regression sum-of-squares* ($\|\widehat{\mathbf{v}}\|^2$) plus the *residual sum-of-squares*, although usually the sample mean of the v_i's is subtracted from $\mathbf{v}$ and $\widehat{\mathbf{v}}$. See Section 7.2.5.

Because the only projections we will be dealing with directly are orthogonal projections, we will drop the "orthogonal" modifier. So, e.g., the matrix $\mathbf{P}$ in part (e) of the theorem below is actually the *orthogonal* projection matrix. Exercises 5.8.8, 5.8.48 and 5.8.49 prove the following.

Theorem 5.1 (Projection). *Suppose* $\mathbf{v} \in \mathbb{R}^K$, $\mathcal{W}$ *is a subspace of* $\mathbb{R}^K$, *and* $\widehat{\mathbf{v}}$ *is the projection of* $\mathbf{v}$ *onto* $\mathcal{W}$. *Then*

(a) *The projection is unique: If $\widehat{\mathbf{v}}_1$ and $\widehat{\mathbf{v}}_2$ are both in $\mathcal{W}$, and $\mathbf{v} - \widehat{\mathbf{v}}_1$ and $\mathbf{v} - \widehat{\mathbf{v}}_2$ are both orthogonal to $\mathcal{W}$, then $\widehat{\mathbf{v}}_1 = \widehat{\mathbf{v}}_2$.*

(b) *If $\mathbf{v} \in \mathcal{W}$, then $\widehat{\mathbf{v}} = \mathbf{v}$.*

(c) *If $\mathbf{v}$ is orthogonal to $\mathcal{W}$, then $\widehat{\mathbf{v}} = \mathbf{0}$.*

(d) *The projection $\widehat{\mathbf{v}}$ uniquely minimizes the Euclidean distance between $\mathbf{v}$ and $\mathcal{W}$, that is,*

$$\|\mathbf{v} - \widehat{\mathbf{v}}\|^2 < \|\mathbf{v} - \mathbf{w}\|^2 \text{ for all } \mathbf{w} \in \mathcal{W}, \mathbf{w} \neq \widehat{\mathbf{v}}. \tag{5.12}$$

(e) *There exists a unique $N \times N$ matrix $\mathbf{P}$, called the **projection matrix**, such that*

$$\mathbf{vP} = \widehat{\mathbf{v}} \text{ for all } \mathbf{v} \in \mathbb{R}^N. \tag{5.13}$$

The final equation is assuming $\mathbf{v}$ is a row vector. If it is a column vector, then we have $\mathbf{Pv} = \widehat{\mathbf{v}}$. We present an explicit formula for the projection matrix in (5.20).

5.3 Least squares

In this section, we explicitly find the projection of $\mathbf{v}$ $(1 \times N)$ onto $\mathcal{W}$. Suppose $\mathbf{d}_1, \ldots, \mathbf{d}_K$, the columns of the $N \times K$ matrix $\mathbf{D}$, form a basis for $\mathcal{W}$, so that the final expression in (5.4) holds. Consider the linear model

$$\mathbf{V} = \boldsymbol{\gamma}\mathbf{D}' + \mathbf{R}, \tag{5.14}$$

where $\boldsymbol{\gamma}$ is $1 \times K$. Our objective is to find a vector $\widehat{\boldsymbol{\gamma}}$ so that $\mathbf{v} \approx \widehat{\boldsymbol{\gamma}}\mathbf{D}'$. One approach is least squares.

Definition 5.8. *A **least squares** estimate of $\boldsymbol{\gamma}$ in the equation (5.14) is any $\widehat{\boldsymbol{\gamma}}$ such that*

$$\|\mathbf{v} - \widehat{\boldsymbol{\gamma}}\mathbf{D}'\|^2 = \min_{\boldsymbol{\gamma} \in \mathbb{R}^K} \|\mathbf{v} - \boldsymbol{\gamma}\mathbf{D}'\|^2. \tag{5.15}$$

Part (d) of Theorem 5.1 implies that a least squares estimate of $\boldsymbol{\gamma}$ is any $\widehat{\boldsymbol{\gamma}}$ for which $\widehat{\boldsymbol{\gamma}}\mathbf{D}'$ is the projection of $\mathbf{v}$ onto the subspace $\mathcal{W}$. Thus $\mathbf{v} - \widehat{\boldsymbol{\gamma}}\mathbf{D}'$ is orthogonal to $\mathcal{W}$, and by Lemma 5.2, is orthogonal to each $\mathbf{d}_j$. These orthogonality conditions result in the **normal equations**:

$$(\mathbf{v} - \widehat{\boldsymbol{\gamma}}\mathbf{D}')\mathbf{d}_j = 0 \text{ for each } j = 1, \ldots, K. \tag{5.16}$$

We then have

$$\begin{aligned}
(5.16) &\Longrightarrow (\mathbf{v} - \widehat{\boldsymbol{\gamma}}\mathbf{D}')\mathbf{D} = \mathbf{0} \\
&\Longrightarrow \widehat{\boldsymbol{\gamma}}\mathbf{D}'\mathbf{D} = \mathbf{vD} && (5.17) \\
&\Longrightarrow \widehat{\boldsymbol{\gamma}} = \mathbf{vD}(\mathbf{D}'\mathbf{D})^{-1}, && (5.18)
\end{aligned}$$

where the final equation holds if $\mathbf{D}'\mathbf{D}$ is invertible, which occurs if and only if the columns of $\mathbf{D}$ constitute a basis of $\mathcal{W}$. See Exercise 5.8.46. (If $\mathbf{D}'\mathbf{D}$ is not invertible, then we can substitute its Moore-Penrose inverse for $(\mathbf{D}'\mathbf{D})^{-1}$ in (5.18) to obtain one of the solutions to (5.17). See Exercise 3.7.24.) Summarizing:

Theorem 5.2 (Least squares). *Any solution $\widehat{\gamma}$ to the least squares equation (5.15) satisfies the normal equations (5.17). The solution is unique if and only if $\mathbf{D}'\mathbf{D}$ is invertible, in which case (5.18) holds.*

If $\mathbf{D}'\mathbf{D}$ is invertible, the projection of $\mathbf{v}$ onto $\mathcal{W}$ can be written

$$\widehat{\mathbf{v}} = \widehat{\gamma}\mathbf{D}' = \mathbf{v}\mathbf{D}(\mathbf{D}'\mathbf{D})^{-1}\mathbf{D}' \equiv \mathbf{v}\mathbf{P}_{\mathbf{D}}, \tag{5.19}$$

where

$$\mathbf{P}_{\mathbf{D}} = \mathbf{D}(\mathbf{D}'\mathbf{D})^{-1}\mathbf{D}' \tag{5.20}$$

is the projection matrix for $\mathcal{W}$ as in part (e) of Theorem 5.1. The residual vector is then

$$\mathbf{v} - \widehat{\mathbf{v}} = \mathbf{v} - \mathbf{v}\mathbf{P}_{\mathbf{D}} = \mathbf{v}(\mathbf{I}_N - \mathbf{P}_{\mathbf{D}}) = \mathbf{v}\mathbf{Q}_{\mathbf{D}}, \tag{5.21}$$

where

$$\mathbf{Q}_{\mathbf{D}} = \mathbf{I}_N - \mathbf{P}_{\mathbf{D}}. \tag{5.22}$$

The minimum value in (5.15) can be written

$$\|\mathbf{v} - \widehat{\mathbf{v}}\|^2 = \mathbf{v}\mathbf{Q}_{\mathbf{D}}\mathbf{v}'. \tag{5.23}$$

Exercises 5.8.12 to 5.8.14 prove the following properties of projection matrices.

Proposition 5.1 (Projection matrices). *Suppose $\mathbf{P}_{\mathbf{D}}$ is defined as in (5.20), where $\mathbf{D}'\mathbf{D}$ is invertible. Then the following hold.*

(a) $\mathbf{P}_{\mathbf{D}}$ is symmetric and idempotent, with $trace(\mathbf{P}_{\mathbf{D}}) = K$, the dimension of $\mathcal{W}$;

(b) Any symmetric idempotent matrix is a projection matrix for some subspace $\mathcal{W}$;

(c) $\mathbf{Q}_{\mathbf{D}} = \mathbf{I}_N - \mathbf{P}_{\mathbf{D}}$ is also a projection matrix, and is orthogonal to $\mathbf{P}_{\mathbf{D}}$ in the sense that $\mathbf{P}_{\mathbf{D}}\mathbf{Q}_{\mathbf{D}} = \mathbf{Q}_{\mathbf{D}}\mathbf{P}_{\mathbf{D}} = \mathbf{0}$;

(d) $\mathbf{P}_{\mathbf{D}}\mathbf{D} = \mathbf{D}$ and $\mathbf{Q}_{\mathbf{D}}\mathbf{D} = \mathbf{0}$.

The matrix $\mathbf{Q}_{\mathbf{D}}$ is the projection matrix onto the **orthogonal complement** of $\mathcal{W}$, where the orthogonal complement contains all vectors in $\mathbb{R}^N$ that are orthogonal to $\mathcal{W}$.

5.4 Best linear unbiased estimators

One might wonder whether the least squares estimate is a good one. We add moment conditions on $\mathbf{R}$ to the model in (5.14):

$$\mathbf{V} = \gamma\mathbf{D}' + \mathbf{R}, \text{ where } E[\mathbf{R}] = \mathbf{0} \text{ and } Cov[\mathbf{R}] = \mathbf{\Omega}. \tag{5.24}$$

We also assume that $\mathbf{D}'\mathbf{D}$ is invertible. The goal in this section is to find the "best linear unbiased estimator" of γ.

An estimator $\widehat{\gamma}$ is linear in $\mathbf{v}$ if

$$\widehat{\gamma} = \mathbf{v}\mathbf{L} \tag{5.25}$$

for some $N \times K$ constant matrix $\mathbf{L}$. Because (5.24) implies that $E[\mathbf{V}] = \gamma\mathbf{D}'$, the estimator (5.25) is unbiased if and only if

$$E[\widehat{\gamma}] = \gamma\mathbf{D}'\mathbf{L} = \gamma \text{ for all } \gamma \in \mathbb{R}^K \iff \mathbf{D}'\mathbf{L} = \mathbf{I}_K. \tag{5.26}$$

(See Exercise 5.8.17.) From (5.18), the least squares estimator is linear,

$$\widehat{\gamma}_{LS} = \mathbf{v}\mathbf{L}_{LS} \text{ with } \mathbf{L}_{LS} = \mathbf{D}(\mathbf{D}'\mathbf{D})^{-1}, \tag{5.27}$$

and because $\mathbf{D}'\mathbf{L}_{LS} = \mathbf{D}'\mathbf{D}(\mathbf{D}'\mathbf{D})^{-1} = \mathbf{I}_K$ as in (5.26), the estimator is unbiased.

We are interested in finding the linear unbiased estimator with the smallest covariance matrix. In general, $Cov[\mathbf{VL}] = \mathbf{L}'\mathbf{\Omega}\mathbf{L}$, but we will initially assume that $\mathbf{\Omega}$ is the identity matrix, so that

$$\mathbf{\Omega} = \mathbf{I}_N \Longrightarrow Cov[\mathbf{VL}] = \mathbf{L}'\mathbf{L} \text{ and } Cov[\mathbf{VL}_{LS}] = \mathbf{L}_{LS}'\mathbf{L}_{LS} = (\mathbf{D}'\mathbf{D})^{-1}. \tag{5.28}$$

The main result is next.

Theorem 5.3. ***Gauss-Markov Theorem.*** *In model (5.24) with $\mathbf{\Omega} = \mathbf{I}_N$, the least squares estimator $\widehat{\gamma}_{LS}$ is the* ***best linear unbiased estimator (BLUE)*** *of γ in the sense that it is linear and unbiased, and for any other linear unbiased estimator $\widehat{\gamma} = \mathbf{vL}$,*

$$Cov[\widehat{\gamma}] - Cov[\widehat{\gamma}_{LS}] \text{ is nonnegative definite.} \tag{5.29}$$

Proof. Let $\mathbf{M} = \mathbf{L} - \mathbf{L}_{LS}$. Then by (5.26) and assumption of unbiasedness,

$$\mathbf{I}_K = \mathbf{D}'\mathbf{L} = \mathbf{D}'(\mathbf{L}_{LS} + \mathbf{M}) = \mathbf{D}'\mathbf{L}_{LS} + \mathbf{D}'\mathbf{M} = \mathbf{I}_K + \mathbf{D}'\mathbf{M}, \tag{5.30}$$

hence $\mathbf{D}'\mathbf{M} = \mathbf{0}$. Then

$$\begin{aligned} Cov[\mathbf{VL}] &= (\mathbf{L}_{LS} + \mathbf{M})'(\mathbf{L}_{LS} + \mathbf{M}) \\ &= \mathbf{L}_{LS}'\mathbf{L}_{LS} + \mathbf{M}'\mathbf{M} + \mathbf{L}_{LS}'\mathbf{M} + \mathbf{M}'\mathbf{L}_{LS} \\ &= Cov[\mathbf{VL}_{LS}] + \mathbf{M}'\mathbf{M}, \end{aligned} \tag{5.31}$$

since

$$\mathbf{L}_{LS}'\mathbf{M} = (\mathbf{D}'\mathbf{D})^{-1}\mathbf{D}'\mathbf{M} = \mathbf{0}. \tag{5.32}$$

The matrix $\mathbf{M}'\mathbf{M}$ is nonnegative definite, hence we have (5.29). □

The conclusion in the theorem means that for any linear combination $\gamma\mathbf{a}$, where $\mathbf{a}$ is $K \times 1$,

$$Var[\widehat{\gamma}\mathbf{a}] \geq Var[\widehat{\gamma}_{LS}\mathbf{a}], \tag{5.33}$$

so that the least squares estimate has the lowest variance of any linear unbiased estimator. In particular, it is best for each component of γ_i.

Theorem 5.3 holds also if $Cov[\mathbf{R}] = \sigma^2\mathbf{I}_N$ for scalar $\sigma^2 > 0$ in (5.24), but not necessarily for $Cov[\mathbf{R}] = \mathbf{\Omega}$ for arbitrary positive definite $\mathbf{\Omega}$. But it is easy to rewrite the model to find the BLUE. Let

$$\mathbf{V}^* = \mathbf{V}\mathbf{\Omega}^{-1/2}, \ \mathbf{D}^* = \mathbf{\Omega}^{-1/2}\mathbf{D}, \ \mathbf{R}^* = \mathbf{R}\mathbf{\Omega}^{-1/2}. \tag{5.34}$$

Then we have the situation in (5.24), where $\mathbf{V}^* = \gamma\mathbf{D}^{*'} + \mathbf{R}^*$ and $Cov[\mathbf{R}^*] = \mathbf{I}_N$. Thus the least squares estimate in this new model is BLUE, where here

$$\begin{aligned} \widehat{\gamma}_{WLS} &= \mathbf{v}^*\mathbf{D}^*(\mathbf{D}^{*\prime}\mathbf{D}^*)^{-1} \\ &= \mathbf{v}\mathbf{\Omega}^{-1}\mathbf{D}(\mathbf{D}'\mathbf{\Omega}^{-1}\mathbf{D})^{-1}. \end{aligned} \tag{5.35}$$

The "WLS" in the subscript of the estimator stands for "weighted least squares," because this estimator minimizes the weighted objective function

$$(\mathbf{v} - \gamma\mathbf{D}')\boldsymbol{\Omega}^{-1}(\mathbf{v} - \gamma\mathbf{D}')'. \tag{5.36}$$

See Exercise 5.8.18. The covariance matrix for the BLUE in (5.35) is by (5.28)

$$\begin{aligned} Cov[\widehat{\gamma}_{WLS}] &= (\mathbf{D}^{*\prime}\mathbf{D}^*)^{-1} \\ &= (\mathbf{D}'\boldsymbol{\Omega}^{-1}\mathbf{D})^{-1}. \end{aligned} \tag{5.37}$$

5.5 Least squares in the both-sides model

We now specialize to the both-sides model as defined in (4.29), i.e.,

$$\mathbf{Y} = \mathbf{x}\boldsymbol{\beta}\mathbf{z}' + \mathbf{R}, \;\; \mathbf{R} \sim N(\mathbf{0}, \mathbf{I}_n \otimes \boldsymbol{\Sigma}_R), \tag{5.38}$$

where $\mathbf{Y}$ is $n \times q$, $\mathbf{x}$ is $n \times p$, $\boldsymbol{\beta}$ is $p \times l$, and $\mathbf{z}$ is $q \times l$. We assume $\mathbf{x}'\mathbf{x}$, $\mathbf{z}'\mathbf{z}$, and $\boldsymbol{\Sigma}_R$ are invertible. In order to find the best linear unbiased estimator of $\boldsymbol{\beta}$, we make the following identifications between (5.24) and (5.38):

$$\mathbf{V} = \text{row}(\mathbf{Y}), \;\; \gamma = \text{row}(\boldsymbol{\beta}), \;\; \mathbf{D} = \mathbf{x} \otimes \mathbf{z}, \;\; \boldsymbol{\Omega} = \mathbf{I}_n \otimes \boldsymbol{\Sigma}_R. \tag{5.39}$$

Also, in (5.24), $\mathbf{R}$ is $1 \times N$, while in (5.38), it is $n \times q$.

Theorem 5.3 shows that the BLUE of $\boldsymbol{\beta}$, assuming $\boldsymbol{\Sigma}_R$ is known, is the weighted least squares estimate in (5.35). Using Proposition 3.2, we translate the formula to

$$\begin{aligned} \widehat{\text{row}(\boldsymbol{\beta})}_{WLS} &= \mathbf{v}\boldsymbol{\Omega}^{-1}\mathbf{D}(\mathbf{D}'\boldsymbol{\Omega}^{-1}\mathbf{D})^{-1} \\ &= \text{row}(\mathbf{y})(\mathbf{I}_n \otimes \boldsymbol{\Sigma}_R)^{-1}(\mathbf{x} \otimes \mathbf{z})((\mathbf{x} \otimes \mathbf{z})'(\mathbf{I}_n \otimes \boldsymbol{\Sigma}_R)^{-1}(\mathbf{x} \otimes \mathbf{z}))^{-1} \\ &= \text{row}(\mathbf{y})(\mathbf{x} \otimes \boldsymbol{\Sigma}_R^{-1}\mathbf{z})(\mathbf{x}'\mathbf{x} \otimes \mathbf{z}'\boldsymbol{\Sigma}_R^{-1}\mathbf{z})^{-1} \\ &= \text{row}(\mathbf{y})(\mathbf{x}(\mathbf{x}'\mathbf{x})^{-1} \otimes \boldsymbol{\Sigma}_R^{-1}\mathbf{z}(\mathbf{z}'\boldsymbol{\Sigma}_R^{-1}\mathbf{z})^{-1}). \end{aligned} \tag{5.40}$$

Thus, unrowing,

$$\widehat{\boldsymbol{\beta}}_{WLS} = (\mathbf{x}'\mathbf{x})^{-1}\mathbf{x}'\mathbf{y}\boldsymbol{\Sigma}_R^{-1}\mathbf{z}(\mathbf{z}'\boldsymbol{\Sigma}_R^{-1}\mathbf{z})^{-1}. \tag{5.41}$$

By (5.37),

$$\begin{aligned} Cov[\widehat{\boldsymbol{\beta}}_{WLS}] &= ((\mathbf{x} \otimes \mathbf{z})'(\mathbf{I}_n \otimes \boldsymbol{\Sigma}_R)^{-1}(\mathbf{x} \otimes \mathbf{z}))^{-1} \\ &= (\mathbf{x}'\mathbf{x})^{-1} \otimes (\mathbf{z}'\boldsymbol{\Sigma}_R^{-1}\mathbf{z})^{-1}. \end{aligned} \tag{5.42}$$

Because $\boldsymbol{\Sigma}_R$ is typically not known, its presence in the estimator of $\boldsymbol{\beta}$ can be problematic. Thus we often instead use the unweighted least squares estimator, which is as in (5.41) but with $\mathbf{I}_q$ instead of $\boldsymbol{\Sigma}_R$:

$$\widehat{\boldsymbol{\beta}}_{LS} = (\mathbf{x}'\mathbf{x})^{-1}\mathbf{x}'\mathbf{Y}\mathbf{z}(\mathbf{z}'\mathbf{z})^{-1}. \tag{5.43}$$

This estimator is not necessarily BLUE. By (3.34), its covariance is

$$\begin{aligned} Cov[\widehat{\boldsymbol{\beta}}_{LS}] &= (\mathbf{D}'\mathbf{D})^{-1}\mathbf{D}'\boldsymbol{\Omega}\mathbf{D}(\mathbf{D}'\mathbf{D})^{-1} \\ &= ((\mathbf{x}'\mathbf{x})^{-1}\mathbf{x}')((\mathbf{x}'\mathbf{x})^{-1}\mathbf{x}')' \otimes (\mathbf{z}(\mathbf{z}'\mathbf{z})^{-1})'\boldsymbol{\Sigma}_R\mathbf{z}(\mathbf{z}'\mathbf{z})^{-1} \\ &= \mathbf{C}_{\mathbf{x}} \otimes \boldsymbol{\Sigma}_{\mathbf{z}}, \end{aligned} \tag{5.44}$$

where we define

$$\mathbf{C_x} = (\mathbf{x}'\mathbf{x})^{-1} \text{and } \boldsymbol{\Sigma}_\mathbf{z} = (\mathbf{z}'\mathbf{z})^{-1}\mathbf{z}'\boldsymbol{\Sigma}_R\mathbf{z}(\mathbf{z}'\mathbf{z})^{-1}. \tag{5.45}$$

But in the multivariate regression case (4.9), where $\mathbf{z} = \mathbf{I}_q$, or in fact whenever $\mathbf{z}$ is square (hence invertible, since $\mathbf{z}'\mathbf{z}$ is assumed invertible), the dependence on $\boldsymbol{\Sigma}_R$ disappears:

$$\begin{aligned}\mathbf{z} \text{ is } q \times q \implies \widehat{\boldsymbol{\beta}}_{WLS} &= (\mathbf{x}'\mathbf{x})^{-1}\mathbf{x}'\mathbf{y}\boldsymbol{\Sigma}_R^{-1}\mathbf{z}(\boldsymbol{\Sigma}_R^{-1}\mathbf{z})^{-1}(\mathbf{z}')^{-1} \\ &= (\mathbf{x}'\mathbf{x})^{-1}\mathbf{x}'\mathbf{y}(\mathbf{z}')^{-1} \\ &= \widehat{\boldsymbol{\beta}}_{LS}. \end{aligned} \tag{5.46}$$

Chapter 6 treats estimation of $\boldsymbol{\Sigma}_R$ and standard errors of the $\widehat{\beta}_{ij}$'s for the least squares estimator. Consideration of the weighted least squares estimator appears in Section 9.3.1 on maximum likelihood estimates.

5.6 What is a linear model?

We have been working with linear models for a while, so perhaps it is time to formally define them. Basically, a linear model for $\mathbf{Y}$ is one for which the mean of $\mathbf{Y}$ lies in a given linear subspace. A model itself is a set of distributions. The linear model does not describe the entire distribution, thus the actual distribution, e.g., multivariate normal with a particular covariance structure, needs to be specified as well. In the both-sides model (6.1), with the identifications in (5.39), we have

$$E[\text{row}(\mathbf{Y})] = \text{row}(\boldsymbol{\beta})(\mathbf{x} \otimes \mathbf{z})'. \tag{5.47}$$

Letting $\boldsymbol{\beta}$ range over all the $p \times l$ matrices, we have the restriction

$$E[\text{row}(\mathbf{Y})] \in \mathcal{W} \equiv \{\text{row}(\boldsymbol{\beta})(\mathbf{x} \otimes \mathbf{z})' \mid \text{row}(\boldsymbol{\beta}) \in \mathbb{R}^{pl}\}. \tag{5.48}$$

Is $\mathcal{W}$ a linear subspace? Indeed, as in Definition 5.2, it is the span of the columns of $\mathbf{x} \otimes \mathbf{z}$, the columns being $\mathbf{x}_i \otimes \mathbf{z}_j$, where $\mathbf{x}_i$ and $\mathbf{z}_j$ are the columns of $\mathbf{x}$ and $\mathbf{z}$, respectively. That is,

$$\text{row}(\boldsymbol{\beta})(\mathbf{x} \otimes \mathbf{z})' = \sum_{i=1}^{p}\sum_{j=1}^{l} \beta_{ij}(\mathbf{x}_i \otimes \mathbf{z}_j)'. \tag{5.49}$$

The linear model is then the set of distributions

$$\mathcal{M} = \{N(\mathbf{M}, \mathbf{I}_n \otimes \boldsymbol{\Sigma}_R) \mid \mathbf{M} \in \mathcal{W} \text{ and } \boldsymbol{\Sigma}_R \in \mathcal{S}_q^+\}, \tag{5.50}$$

denoting

$$\mathcal{S}_q^+ = \text{The set of } q \times q \text{ positive definite symmetric matrices.} \tag{5.51}$$

Other linear models can have different distributional assumptions, e.g., covariance restrictions, but do have to have the mean lie in a linear subspace.

There are many different parametrizations of a given linear model, for the same reason that there are many different bases for the mean space $\mathcal{W}$. For example, it may not be obvious, but

$$\mathbf{x} = \begin{pmatrix} 1 & 0 \\ 0 & 1 \end{pmatrix}, \ \mathbf{z} = \begin{pmatrix} 1 & 1 & 1 \\ 1 & 2 & 4 \\ 1 & 3 & 9 \end{pmatrix} \tag{5.52}$$

and

$$\mathbf{x}^* = \begin{pmatrix} 1 & -1 \\ 1 & 1 \end{pmatrix}, \ \mathbf{z}^* = \begin{pmatrix} 1 & -1 & 1 \\ 1 & 0 & -2 \\ 1 & 1 & 1 \end{pmatrix} \tag{5.53}$$

lead to exactly the same model, though different interpretations of the parameters. In fact, with $\mathbf{x}$ being $n \times p$ and $\mathbf{z}$ being $q \times l$,

$$\mathbf{x}^* = \mathbf{x}\mathbf{A} \ \text{ and } \ \mathbf{z}^* = \mathbf{z}\mathbf{B} \tag{5.54}$$

yields the same model as long as $\mathbf{A}$ $(p \times p)$ and $\mathbf{B}$ $(l \times l)$ are invertible:

$$\mathbf{x}\beta\mathbf{z}' = \mathbf{x}^*\beta^*\mathbf{z}^{*\prime} \ \text{ with } \ \beta^* = \mathbf{A}^{-1}\beta(\mathbf{B}')^{-1}. \tag{5.55}$$

The representation in (5.53) has the advantage that the columns of the $\mathbf{x}^* \otimes \mathbf{z}^*$ are orthogonal, which makes it easy to find the least squares estimates as the $\mathbf{D}'\mathbf{D}$ matrix is diagonal, hence easy to invert. Note the $\mathbf{z}$ is the matrix for a quadratic. The $\mathbf{z}^*$ is the corresponding set of orthogonal polynomials, as discussed in Section 5.7.2.

5.7 Gram-Schmidt orthogonalization

We have seen polynomial models in (4.7), (4.23), and (4.30). Note that, especially in the latter case, one can have a design matrix ($\mathbf{x}$ or $\mathbf{z}$) whose entries have widely varying magnitudes, as well as highly correlated column vectors, which can lead to numerical difficulties in calculation. Orthogonalizing the vectors, without changing their span, can help both numerically and for interpretation. **Gram-Schmidt orthogonalization** is a well-known constructive approach. It is based on the following lemma.

Lemma 5.3. *Suppose* $(\mathbf{D}_1, \mathbf{D}_2)$ *is* $N \times K$*, where* $\mathbf{D}_1$ *is* $N \times K_1$*,* $\mathbf{D}_2$ *is* $N \times K_2$*, and* $\mathcal{W}$ *is the span of the combined columns:*

$$\mathcal{W} = \text{span}\{\textit{columns of } (\mathbf{D}_1, \mathbf{D}_2)\}. \tag{5.56}$$

Suppose $\mathbf{D}_1'\mathbf{D}_1$ *is invertible, and let*

$$\mathbf{D}_{2\cdot 1} = \mathbf{Q}_{\mathbf{D}_1}\mathbf{D}_2, \tag{5.57}$$

for $\mathbf{Q}_{\mathbf{D}_1}$ *defined in (5.22) and (5.20). Then the columns of* $\mathbf{D}_1$ *and* $\mathbf{D}_{2\cdot 1}$ *are orthogonal,*

$$\mathbf{D}_{2\cdot 1}'\mathbf{D}_1 = \mathbf{0}, \tag{5.58}$$

and

$$\mathcal{W} = \text{span}\{\textit{columns of } (\mathbf{D}_1, \mathbf{D}_{2\cdot 1})\}. \tag{5.59}$$

Proof. $\mathbf{D}_{2\cdot 1}$ is the residual matrix for the least squares model $\mathbf{D}_2 = \mathbf{D}_1\beta + \mathbf{R}$, i.e., $\mathbf{D}_1$ is the $\mathbf{x}$ and $\mathbf{D}_2$ is the $\mathbf{Y}$ in the multivariate regression model (4.9). Equation (5.58) then follows from part (d) of Proposition 5.1: $\mathbf{D}_{2\cdot 1}'\mathbf{D}_1 = \mathbf{D}_2'\mathbf{Q}_{\mathbf{D}_1}\mathbf{D}_1 = \mathbf{0}$. For (5.59),

$$\begin{pmatrix} \mathbf{D}_1 & \mathbf{D}_{2\cdot 1} \end{pmatrix} = \begin{pmatrix} \mathbf{D}_1 & \mathbf{D}_2 \end{pmatrix} \begin{pmatrix} \mathbf{I}_{K_1} & -(\mathbf{D}_1'\mathbf{D}_1)^{-1}\mathbf{D}_1'\mathbf{D}_2 \\ \mathbf{0} & \mathbf{I}_{K_2} \end{pmatrix}. \tag{5.60}$$

The final matrix is invertible, hence by Lemma 5.1, the spans of the columns of $(\mathbf{D}_1, \mathbf{D}_2)$ and $(\mathbf{D}_1, \mathbf{D}_{2\cdot 1})$ are the same. □

Now let $\mathbf{d}_1, \ldots \mathbf{d}_K$ be the columns of the $N \times K$ matrix $\mathbf{D}$, and $\mathcal{W}$ their span. The Gram-Schmidt process starts by applying Lemma 5.3 with $\mathbf{D}_1 = \mathbf{d}_1$ and $\mathbf{D}_2 = (\mathbf{d}_2, \ldots, \mathbf{d}_K)$. Write the resulting columns of $\mathbf{D}_{2\cdot 1}$ as the vectors

$$\mathbf{d}_{2\cdot 1}, \cdots, \mathbf{d}_{K\cdot 1}, \text{ where } \mathbf{d}_{j\cdot 1} = \mathbf{d}_j - \frac{\mathbf{d}_j'\mathbf{d}_1}{\|\mathbf{d}_1\|^2}\mathbf{d}_1. \tag{5.61}$$

In other words, $\mathbf{d}_{j\cdot 1}$ is the residual of the projection of $\mathbf{d}_j$ onto span$\{\mathbf{d}_1\}$. Thus $\mathbf{d}_1$ is orthogonal to all the $\mathbf{d}_{j\cdot 1}$'s in (5.61), and $\mathcal{W} = \text{span}\{\mathbf{d}_1, \mathbf{d}_{2\cdot 1}, \ldots, \mathbf{d}_{K\cdot 1}\}$ by (5.59).

Second step is to apply the lemma again, this time with $\mathbf{D}_1 = \mathbf{d}_{2\cdot 1}$, and $\mathbf{D}_2 = (\mathbf{d}_{3\cdot 1}, \ldots, \mathbf{d}_{K\cdot 1})$, leaving aside the $\mathbf{d}_1$ for the time being. Now write the columns of the new $\mathbf{D}_{2\cdot 1}$ dotting out the "1" and "2":

$$\mathbf{d}_{3\cdot 12}, \cdots, \mathbf{d}_{K\cdot 12}, \text{ where } \mathbf{d}_{j\cdot 12} = \mathbf{d}_{j\cdot 1} - \frac{\mathbf{d}_{j\cdot 1}'\mathbf{d}_{2\cdot 1}}{\|\mathbf{d}_{2\cdot 1}\|^2}\mathbf{d}_{2\cdot 1}. \tag{5.62}$$

Now $\mathbf{d}_{2\cdot 1}$, as well as $\mathbf{d}_1$, are orthogonal to the vectors in (5.62), and

$$\mathcal{W} = \text{span}\{\mathbf{d}_1, \mathbf{d}_{2\cdot 1}, \mathbf{d}_{3\cdot 12}, \ldots, \mathbf{d}_{K\cdot 12}\}. \tag{5.63}$$

We continue until we have the set of vectors

$$\mathbf{d}_1, \mathbf{d}_{2\cdot 1}, \mathbf{d}_{3\cdot 12}, \ldots, \mathbf{d}_{K\cdot\{1:(K-1)\}}, \tag{5.64}$$

which are mutually orthogonal and span $\mathcal{W}$. Here, we are using the R-based notation

$$\{a:b\} = \{a, a+1, \ldots, b\} \text{ for integers } a < b. \tag{5.65}$$

It is possible that one or more of the vectors we use for $\mathbf{D}_1$ will be zero. In such cases, we just leave the vectors in $\mathbf{D}_2$ alone, i.e., $\mathbf{D}_{2\cdot 1} = \mathbf{D}_2$, because the projection of any vector on the space $\{\mathbf{0}\}$ is $\mathbf{0}$, hence the residual equals the original vector. We can describe the entire resulting process iteratively, for $i = 1, \ldots, K-1$ and $j = i+1, \ldots, K$, as setting

$$\mathbf{d}_{j\cdot\{1:i\}} = \mathbf{d}_{j\cdot\{1:(i-1)\}} - \gamma_{ij}\,\mathbf{d}_{i\cdot\{1:(i-1)\}}, \tag{5.66}$$

where

$$\gamma_{ij} = \frac{\mathbf{d}_{j\cdot\{1:(i-1)\}}'\mathbf{d}_{i\cdot\{1:(i-1)\}}}{\|\mathbf{d}_{i\cdot\{1:(i-1)\}}\|^2} \tag{5.67}$$

if its denominator is nonzero. Otherwise, set $\gamma_{ij} = 0$, although any value will do.

Optionally, one can multiply any of these vectors by a nonzero constant, e.g., so that it has a norm of one, or for esthetics, so that the entries are small integers. Any zero vectors left in the set can be eliminated without affecting the span.

Note that by the stepwise nature of the algorithm, we have that the spans of the first k vectors from each set are equal, that is,

$$\begin{aligned} \text{span}\{\mathbf{d}_1, \mathbf{d}_2\} &= \text{span}\{\mathbf{d}_1, \mathbf{d}_{2\cdot 1}\} \\ \text{span}\{\mathbf{d}_1, \mathbf{d}_2, \mathbf{d}_3\} &= \text{span}\{\mathbf{d}_1, \mathbf{d}_{2\cdot 1}, \mathbf{d}_{3\cdot 12}\} \\ &\vdots \\ \text{span}\{\mathbf{d}_1, \mathbf{d}_2, \ldots, \mathbf{d}_K\} &= \text{span}\{\mathbf{d}_1, \mathbf{d}_{2\cdot 1}, \mathbf{d}_{3\cdot 12}, \ldots, \mathbf{d}_{K\cdot\{1:(K-1)\}}\}. \end{aligned} \tag{5.68}$$

The next section derives some important matrix decompositions based on the Gram-Schmidt orthogonalization. Section 5.7.2 applies the orthogonalization to polynomials.

5.7.1 The QR and Cholesky decompositions

We can write the Gram-Schmidt process in matrix form. The first step is

$$\begin{pmatrix} \mathbf{d}_1 & \mathbf{d}_{2\cdot 1} & \cdots & \mathbf{d}_{K\cdot 1} \end{pmatrix} = \mathbf{D} \begin{pmatrix} 1 & -\gamma_{12} & \cdots & -\gamma_{1K} \\ 0 & 1 & \cdots & 0 \\ \vdots & \vdots & \ddots & \vdots \\ 0 & 0 & \cdots & 1 \end{pmatrix}. \tag{5.69}$$

The γ_{ij}'s are defined in (5.67). Next,

$$\begin{aligned} &\begin{pmatrix} \mathbf{d}_1 & \mathbf{d}_{2\cdot 1} & \mathbf{d}_{3\cdot 12} & \cdots & \mathbf{d}_{K\cdot 12} \end{pmatrix} \\ &= \begin{pmatrix} \mathbf{d}_1 & \mathbf{d}_{2\cdot 1} & \cdots & \mathbf{d}_{K\cdot 1} \end{pmatrix} \begin{pmatrix} 1 & 0 & 0 & \cdots & 0 \\ 0 & 1 & -\gamma_{23} & \cdots & -\gamma_{2K} \\ 0 & 0 & 1 & \cdots & 0 \\ \vdots & \vdots & \vdots & \ddots & \vdots \\ 0 & 0 & 0 & \cdots & 1 \end{pmatrix}. \end{aligned} \tag{5.70}$$

We continue, so that the final result is

$$\mathbf{D}^* \equiv \begin{pmatrix} \mathbf{d}_1 & \mathbf{d}_{2\cdot 1} & \mathbf{d}_{3\cdot 12} \cdots & \mathbf{d}_{K\cdot\{1:(K-1)\}} \end{pmatrix} = \mathbf{D}\mathbf{B}^{(1)}\mathbf{B}^{(2)} \cdots \mathbf{B}^{(K-1)}, \tag{5.71}$$

where $\mathbf{B}^{(k)}$ is the identity except for the elements kj, $j > k$:

$$\mathbf{B}^{(k)}_{ij} = \begin{cases} 1 & \text{if } i = j \\ -\gamma_{kj} & \text{if } j > k = i \\ 0 & \text{otherwise.} \end{cases} \tag{5.72}$$

These matrices are **upper unitriangular**, meaning they are upper triangular (i.e., all elements below the diagonal are zero), and all diagonal elements are one. We will use the notation

$$\mathcal{T}^1_q = \{\mathbf{T} \mid \mathbf{T} \text{ is } q \times q, t_{ii} = 1 \text{ for all } i, t_{ij} = 0 \text{ for } i > j\}. \tag{5.73}$$

Such matrices form an algebraic group. A group of matrices is a set $\mathcal{G}$ of $N \times N$ invertible matrices $\mathbf{g}$ that is closed under multiplication and inverse:

$$\mathbf{g}_1, \mathbf{g}_2 \in \mathcal{G} \;\Rightarrow\; \mathbf{g}_1\mathbf{g}_2 \in \mathcal{G}, \tag{5.74}$$

$$\mathbf{g} \in \mathcal{G} \;\Rightarrow\; \mathbf{g}^{-1} \in \mathcal{G}. \tag{5.75}$$

Thus we can write

$$\mathbf{D} = \mathbf{D}^*\mathbf{B}^{-1}, \text{ where } \mathbf{B} = \mathbf{B}^{(1)} \cdots \mathbf{B}^{(K-1)}. \tag{5.76}$$

Exercise 5.8.30 shows that

$$\mathbf{B}^{-1} = \begin{pmatrix} 1 & \gamma_{12} & \gamma_{13} & \cdots & \gamma_{1K} \\ 0 & 1 & \gamma_{23} & \cdots & \gamma_{2K} \\ 0 & 0 & 1 & \cdots & \gamma_{3K} \\ \vdots & \vdots & \vdots & \ddots & \vdots \\ 0 & 0 & 0 & \cdots & 1 \end{pmatrix}. \tag{5.77}$$

Now suppose the columns of $\mathbf{D}$ are linearly independent, which means that all the columns of $\mathbf{D}^*$ are nonzero. (See Exercise 5.8.32.) Then we can divide each column of $\mathbf{D}^*$ by its norm, so that the resulting vectors are orthonormal:

$$\mathbf{q}_i = \frac{\mathbf{d}_{i\cdot\{1:(i-1)\}}}{\|\mathbf{d}_{i\cdot\{1:(i-1)\}}\|}, \quad \mathbf{Q} = \begin{pmatrix} \mathbf{q}_1 & \cdots & \mathbf{q}_K \end{pmatrix} = \mathbf{D}^*\mathbf{\Delta}^{-1}, \tag{5.78}$$

where $\mathbf{\Delta}$ is the diagonal matrix with the norms on the diagonal. Letting $\mathbf{R} = \mathbf{\Delta}\mathbf{B}^{-1}$, we have that

$$\mathbf{D} = \mathbf{Q}\mathbf{R}, \tag{5.79}$$

where $\mathbf{R}$ is upper triangular with positive diagonal elements, the $\mathbf{\Delta}_{ii}$'s. The set of such matrices $\mathbf{R}$ is also group, denoted by

$$\mathcal{T}_q^+ = \{\mathbf{T} \mid \mathbf{T} \text{ is } q \times q, t_{ii} > 0 \text{ for all } i, t_{ij} = 0 \text{ for } i > j\}. \tag{5.80}$$

Hence we have the next result. The uniqueness for $N = K$ is shown in Exercise 5.8.37.

Theorem 5.4 (QR-decomposition). *Suppose the $N \times K$ matrix $\mathbf{D}$ has linearly independent columns (hence $K \leq N$). Then there is a unique decomposition $\mathbf{D} = \mathbf{Q}\mathbf{R}$, where $\mathbf{Q}$, $N \times K$, has orthonormal columns and $\mathbf{R} \in \mathcal{T}_K^+$.*

Gram-Schmidt also has useful implications for the matrix $\mathbf{S} = \mathbf{D}'\mathbf{D}$. From (5.60) we have

$$\begin{aligned} \mathbf{S} &= \begin{pmatrix} \mathbf{I}_{K_1} & \mathbf{0} \\ \mathbf{D}_2'\mathbf{D}_1(\mathbf{D}_1'\mathbf{D}_1)^{-1} & \mathbf{I}_{K_2} \end{pmatrix} \begin{pmatrix} \mathbf{D}_1'\mathbf{D}_1 & \mathbf{0} \\ \mathbf{0} & \mathbf{D}_{2\cdot1}'\mathbf{D}_{2\cdot1} \end{pmatrix} \begin{pmatrix} \mathbf{I}_{K_1} & (\mathbf{D}_1'\mathbf{D}_1)^{-1}\mathbf{D}_1'\mathbf{D}_2 \\ \mathbf{0} & \mathbf{I}_{K_2} \end{pmatrix} \\ &= \begin{pmatrix} \mathbf{I}_{K_1} & \mathbf{0} \\ \mathbf{S}_{21}\mathbf{S}_{11}^{-1} & \mathbf{I}_{K_2} \end{pmatrix} \begin{pmatrix} \mathbf{S}_{11} & \mathbf{0} \\ \mathbf{0} & \mathbf{S}_{22\cdot1} \end{pmatrix} \begin{pmatrix} \mathbf{I}_{K_1} & \mathbf{S}_{11}^{-1}\mathbf{S}_{12} \\ \mathbf{0} & \mathbf{I}_{K_2} \end{pmatrix}, \end{aligned} \tag{5.81}$$

where $\mathbf{S}_{22\cdot 1} = \mathbf{S}_{22} - \mathbf{S}_{21}\mathbf{S}_{11}^{-1}\mathbf{S}_{12}$ as in (3.49). See Exercise 5.8.38. Then using steps as in Gram-Schmidt, we have

$$\begin{aligned}\mathbf{S} &= (\mathbf{B}^{-1})' \begin{pmatrix} S_{11} & 0 & 0 & \cdots & 0 \\ 0 & S_{22\cdot 1} & 0 & \cdots & 0 \\ 0 & 0 & S_{33\cdot 12} & \cdots & 0 \\ \vdots & \vdots & \vdots & \ddots & \vdots \\ 0 & 0 & 0 & \cdots & S_{KK\cdot\{1:(K-1)\}} \end{pmatrix} \mathbf{B}^{-1} \\ &= \mathbf{R}'\mathbf{R}, \end{aligned} \tag{5.82}$$

because the inner matrix is $\mathbf{\Delta}^2$. Also, note that

$$\gamma_{ij} = S_{ij\cdot\{1:(i-1)\}}/S_{ii\cdot\{1:(i-1)\}} \text{ for } j > i, \tag{5.83}$$

and $\mathbf{R}$ is given by

$$R_{ij} = \begin{cases} \sqrt{S_{ii\cdot\{1:(i-1)\}}} & \text{if } j = i, \\ S_{ij\cdot\{1:(i-1)\}}/\sqrt{S_{ii\cdot\{1:(i-1)\}}} & \text{if } j > i, \\ 0 & \text{if } j < i. \end{cases} \tag{5.84}$$

Exercise 5.8.43 shows this decomposition works for any positive definite symmetric matrix. It is then called the Cholesky decomposition:

Theorem 5.5 (Cholesky decomposition). *If $\mathbf{S} \in \mathcal{S}_q^+$ (5.51), then there exists a unique $\mathbf{R} \in \mathcal{T}_q^+$ such that $\mathbf{S} = \mathbf{R}'\mathbf{R}$.*

Note that this decomposition yields a particular square root of $\mathbf{S}$.

5.7.2 Orthogonal polynomials

Turn to polynomials. We will illustrate with the example on mouth sizes in (4.30). Here $K = 4$, and we will consider the cubic model, so that the vectors are

$$\begin{pmatrix} \mathbf{d}_1 & \mathbf{d}_2 & \mathbf{d}_3 & \mathbf{d}_4 \end{pmatrix} = \begin{pmatrix} 1 & 8 & 8^2 & 8^3 \\ 1 & 10 & 10^2 & 10^3 \\ 1 & 12 & 12^2 & 12^3 \\ 1 & 14 & 14^2 & 14^3 \end{pmatrix}. \tag{5.85}$$

Note that the ages (values 8, 10, 12, 14) are equally spaced. Thus we can just as well code the ages as (0,1,2,3), so that we actually start with

$$\begin{pmatrix} \mathbf{d}_1 & \mathbf{d}_2 & \mathbf{d}_3 & \mathbf{d}_4 \end{pmatrix} = \begin{pmatrix} 1 & 0 & 0 & 0 \\ 1 & 1 & 1 & 1 \\ 1 & 2 & 4 & 8 \\ 1 & 3 & 9 & 27 \end{pmatrix}. \tag{5.86}$$

Dotting $\mathbf{d}_1$ out of vector $\mathbf{w}$ is equivalent to subtracting the mean of the elements of $\mathbf{w}$ for each element. Hence

$$\mathbf{d}_{2\cdot 1} = \begin{pmatrix} -3/2 \\ -1/2 \\ 1/2 \\ 3/2 \end{pmatrix} \rightarrow \begin{pmatrix} -3 \\ -1 \\ 1 \\ 3 \end{pmatrix}, \mathbf{d}_{3\cdot 1} = \begin{pmatrix} -7/2 \\ -5/2 \\ 1/2 \\ 11/2 \end{pmatrix} \rightarrow \begin{pmatrix} -7 \\ -5 \\ 1 \\ 11 \end{pmatrix}, \mathbf{d}_{4\cdot 1} = \begin{pmatrix} -9 \\ -8 \\ -1 \\ 18 \end{pmatrix}. \tag{5.87}$$

We multiplied the first two vectors in (5.87) by 2 for simplicity. Next, we dot $\mathbf{d}_{2\cdot 1}$ out of the last two vectors. So for $\mathbf{d}_{3\cdot 1}$, we have

$$\mathbf{d}_{3\cdot 12} = \begin{pmatrix} -7 \\ -5 \\ 1 \\ 11 \end{pmatrix} - \frac{(-7,-5,1,11)(-3,-1,1,3)'}{\|(-3,-1,1,3)\|^2} \begin{pmatrix} -3 \\ -1 \\ 1 \\ 3 \end{pmatrix} = \begin{pmatrix} 2 \\ -2 \\ -2 \\ 2 \end{pmatrix} \rightarrow \begin{pmatrix} 1 \\ -1 \\ -1 \\ 1 \end{pmatrix} \tag{5.88}$$

and, similarly, $\mathbf{d}_{4\cdot 12} = (4.2, -3.6, -5.4, 4.8)' \rightarrow (7, -6, -9, 8)'$. Finally, we dot $\mathbf{d}_{3\cdot 12}$ out of $\mathbf{d}_{4\cdot 12}$ to obtain $\mathbf{d}_{4\cdot 123} = (-1, 3, -3, 1)'$. Then our final orthogonal polynomial matrix is the very nice

$$\begin{pmatrix} \mathbf{d}_1 & \mathbf{d}_{2\cdot 1} & \mathbf{d}_{3\cdot 12} & \mathbf{d}_{4\cdot 123} \end{pmatrix} = \begin{pmatrix} 1 & -3 & 1 & -1 \\ 1 & -1 & -1 & 3 \\ 1 & 1 & -1 & -3 \\ 1 & 3 & 1 & 1 \end{pmatrix}. \tag{5.89}$$

Some older statistics books (e.g., Snedecor and Cochran [1989]) contain tables of orthogonal polynomials for small K, and statistical packages will calculate them for you. In R, the function is poly.

A key advantage to using orthogonal polynomials over the original polynomial vectors is that, by virtue of the sequence in (5.68), one can estimate the parameters for models of all degrees at once. For example, consider the mean of the girls' mouth sizes in (4.15) as the $\mathbf{V}$, and the matrix in (5.89) as the $\mathbf{D}$, in the model (5.24):

$$(21.18, 22.23, 23.09, 24.09) \approx (\gamma_1, \gamma_2, \gamma_3, \gamma_4) \begin{pmatrix} 1 & 1 & 1 & 1 \\ -3 & -1 & 1 & 3 \\ 1 & -1 & -1 & 1 \\ -1 & 3 & -3 & 1 \end{pmatrix}. \tag{5.90}$$

Because $\mathbf{D}'\mathbf{D}$ is diagonal, the least squares estimates of the coefficients are found via

$$\widehat{\gamma}_j = \frac{\mathbf{v}\mathbf{d}_{j\cdot\{1:(j-1)\}}}{\|\mathbf{d}_{j\cdot\{1:(j-1)\}}\|^2}, \tag{5.91}$$

which here yields

$$\widehat{\gamma} = (22.6475, 0.4795, -0.0125, 0.0165). \tag{5.92}$$

These are the coefficients for the cubic model. The coefficients for the quadratic model set $\widehat{\gamma}_4 = 0$, but the other three are as for the cubic. Likewise, the linear model has $\widehat{\gamma}$ equalling $(22.6475, 0.4795, 0, 0)$, and the constant model has $(22.6475, 0, 0, 0)$.

In contrast, if one uses the original vectors in either (5.85) or (5.86), one has to recalculate the coefficients separately for each model. Using (5.86), we have the following estimates:

Model	$\widehat{\gamma}_1^*$	$\widehat{\gamma}_2^*$	$\widehat{\gamma}_3^*$	$\widehat{\gamma}_4^*$
Cubic	21.1800	1.2550	−0.2600	0.0550
Quadratic	21.1965	0.9965	−0.0125	0
Linear	21.2090	0.9590	0	0
Constant	22.6475	0	0	0

(5.93)

Note that the nonzero values in each column are not equal.

5.8 Exercises

Exercise 5.8.1. Show that the span in (5.3) is indeed a linear subspace.

Exercise 5.8.2. Verify that the four spans given in (5.5) are the same.

Exercise 5.8.3. Show that for matrices $\mathbf{C}$ ($N \times J$) and $\mathbf{D}$ ($N \times K$),

$$\text{span}\{\text{columns of } \mathbf{D}\} \subset \text{span}\{\text{columns of } \mathbf{C}\} \Rightarrow \mathbf{D} = \mathbf{CA}, \tag{5.94}$$

for some $J \times K$ matrix $\mathbf{A}$. [Hint: Each column of $\mathbf{D}$ must be a linear combination of the columns of $\mathbf{C}$.]

Exercise 5.8.4. Here, $\mathbf{D}$ is an $N \times K$ matrix, and $\mathbf{A}$ is $K \times L$. (a) Show that

$$\text{span}\{\text{columns of } \mathbf{DA}\} \subset \text{span}\{\text{columns of } \mathbf{D}\}. \tag{5.95}$$

[Hint: Any vector in the left-hand space equals $\mathbf{DA}\boldsymbol{\gamma}$ for some $L \times 1$ vector $\boldsymbol{\gamma}$. For what vector $\boldsymbol{\gamma}^*$ is $\mathbf{DA}\boldsymbol{\gamma} = \mathbf{D}\boldsymbol{\gamma}^*$?] (b) Prove Lemma 5.1. [Use part (a) twice, once for $\mathbf{A}$ and once for $\mathbf{A}^{-1}$.] (c) Show that if the columns of $\mathbf{D}$ are linearly independent, and $\mathbf{A}$ is $K \times K$ and invertible, then the columns of $\mathbf{DA}$ are linearly independent. [Hint: Suppose the columns of $\mathbf{DA}$ are linearly dependent, so that for some $\boldsymbol{\gamma} \neq \mathbf{0}$, $\mathbf{DA}\boldsymbol{\gamma} = \mathbf{0}$. Then there is a $\boldsymbol{\gamma}^* \neq \mathbf{0}$ with $\mathbf{D}\boldsymbol{\gamma}^* = \mathbf{0}$. What is it?]

Exercise 5.8.5. Let $\mathbf{d}_1, \ldots, \mathbf{d}_K$ be vectors in $\mathbb{R}^N$. (a) Suppose (5.8) holds. Show that the vectors are linearly dependent. [That is, find γ_j's, not all zero, so that $\sum \gamma_i \mathbf{d}_i = \mathbf{0}$.] (b) Suppose the vectors are linearly dependent. Find an index i and constants γ_j so that (5.8) holds.

Exercise 5.8.6. Prove Lemma 5.2.

Exercise 5.8.7. Suppose the set of $M \times 1$ vectors $\mathbf{g}_1, \ldots, \mathbf{g}_K$ are nonzero and mutually orthogonal. Show that they are linearly independent. [Hint: Suppose they are linearly dependent, and let $\mathbf{g}_i$ be the vector on the left-hand side in (5.8). Then take $\mathbf{g}_i'$ times each side of the equation, to arrive at a contradiction.]

Exercise 5.8.8. Prove part (a) of Theorem 5.1. [Hint: Show that the difference of $\mathbf{v} - \widehat{\mathbf{v}}_1$ and $\mathbf{v} - \widehat{\mathbf{v}}_2$ is orthogonal to $\mathcal{W}$, as well as in $\mathcal{W}$. Then show that such a vector must be zero.] (b) Prove part (b) of Theorem 5.1. (c) Prove part (c) of Theorem 5.1. (d) Prove part (d) of Theorem 5.1. [Hint: Start by writing $\|\mathbf{v} - \mathbf{w}\|^2 = \|(\mathbf{v} - \widehat{\mathbf{v}}) - (\mathbf{w} - \widehat{\mathbf{v}})\|^2$, then expand. Explain why $\mathbf{v} - \widehat{\mathbf{v}}$ and $\mathbf{w} - \widehat{\mathbf{v}}$ are orthogonal.]

Exercise 5.8.9. Suppose $\mathbf{D}_1$ is $N \times K_1$, $\mathbf{D}_2$ is $N \times K_2$, and $\mathbf{D}_1'\mathbf{D}_2 = \mathbf{0}$. Let $\mathcal{V}_i =$ span{columns of $\mathbf{D}_i$} for $i = 1, 2$. (a) Show that $\mathcal{V}_1$ and $\mathcal{V}_2$ are orthogonal spaces. (b) Define $\mathcal{W} =$ span{columns of $(\mathbf{D}_1, \mathbf{D}_2)$}. (The subspace $\mathcal{W}$ is called the **direct sum** of the subspaces $\mathcal{V}_1$ and $\mathcal{V}_2$.) Show that $\mathbf{u}$ is orthogonal to $\mathcal{V}_1$ and $\mathcal{V}_2$ if and only if it is orthogonal to $\mathcal{W}$. (c) For $\mathbf{v} \in \mathbb{R}^N$, let $\widehat{\mathbf{v}}_i$ be the projection onto $\mathcal{V}_i$, $i = 1, 2$. Show that $\widehat{\mathbf{v}}_1 + \widehat{\mathbf{v}}_2$ is the projection of $\mathbf{v}$ onto $\mathcal{W}$. [Hint: Show that $\mathbf{v} - \widehat{\mathbf{v}}_1 - \widehat{\mathbf{v}}_2$ is orthogonal to $\mathcal{V}_1$ and $\mathcal{V}_2$, then use part (b).]

Exercise 5.8.10. Derive the normal equations (5.16) by differentiating $\|\mathbf{v} - \boldsymbol{\gamma}\mathbf{D}'\|^2$ with respect to the γ_i's.

Exercise 5.8.11. Suppose $\mathbf{A}$ and $\mathbf{B}$ are both $n \times q$ matrices. Exercise 3.7.6 showed that $\text{trace}(\mathbf{AB}') = \text{trace}(\mathbf{B}'\mathbf{A})$. Show further that

$$\text{trace}(\mathbf{AB}') = \text{trace}(\mathbf{B}'\mathbf{A}) = \text{row}(\mathbf{A})\,\text{row}(\mathbf{B})'. \tag{5.96}$$

Exercise 5.8.12. This exercise proves part (a) of Proposition 5.1. Suppose $\mathcal{W} =$ span{columns of $\mathbf{D}$}, where $\mathbf{D}$ is $N \times K$ and $\mathbf{D}'\mathbf{D}$ is invertible. (a) Show that the projection matrix $\mathbf{P}_\mathbf{D} = \mathbf{D}(\mathbf{D}'\mathbf{D})^{-1}\mathbf{D}'$ as in (5.20) is symmetric and idempotent. (b) Show that $\text{trace}(\mathbf{P}_\mathbf{D}) = K$. [Use Exercise 5.8.11.]

Exercise 5.8.13. This exercise proves part (b) of Proposition 5.1. Suppose $\mathbf{P}$ is a symmetric and idempotent $N \times N$ matrix. Find a set of linearly independent vectors $\mathbf{d}_1, \ldots, \mathbf{d}_K$, where $K = \text{trace}(\mathbf{P})$, so that span$\{\mathbf{d}_1, \ldots, \mathbf{d}_K\}$ has projection matrix $\mathbf{P}$. [Hint: Write $\mathbf{P} = \boldsymbol{\Gamma}_1\boldsymbol{\Gamma}_1'$ where $\boldsymbol{\Gamma}_1$ has orthonormal columns, as in Lemma 3.1. Show that $\mathbf{P}$ is the projection matrix onto the span of the columns of the $\boldsymbol{\Gamma}_1$, and use Exercise 5.8.7 to show that those columns are a basis. What is $\mathbf{D}$, then?]

Exercise 5.8.14. (a) Prove part (c) of Proposition 5.1. (b) Prove part (d) of Proposition 5.1. (c) Prove (5.23).

Exercise 5.8.15. Consider the projection of $\mathbf{v} \in \mathbb{R}^N$ onto span$\{\mathbf{1}_N'\}$. (a) Find the projection. (b) Find the residual. What does it contain? (c) Find the projection matrix $\mathbf{P}$. What is $\mathbf{Q} = \mathbf{I}_N - \mathbf{P}$? Have we seen it before?

Exercise 5.8.16. Suppose $\mathbf{D}$ is $N \times K$ and $\mathbf{D}^*$ is $N \times L$, where span{columns of $\mathbf{D}$} $\subset$ span{columns of $\mathbf{D}^*$}. (a) Show that $\mathbf{P}_\mathbf{D}\mathbf{P}_{\mathbf{D}^*} = \mathbf{P}_{\mathbf{D}^*}\mathbf{P}_\mathbf{D} = \mathbf{P}_\mathbf{D}$. [Hint: What is $\mathbf{P}_{\mathbf{D}^*}\mathbf{D}$?] (b) Show that $\mathbf{Q}_\mathbf{D}\mathbf{Q}_{\mathbf{D}^*} = \mathbf{Q}_{\mathbf{D}^*}\mathbf{Q}_\mathbf{D} = \mathbf{Q}_{\mathbf{D}^*}$.

Exercise 5.8.17. Suppose $\mathbf{A}$ is a $K \times K$ matrix such that $\boldsymbol{\gamma}\mathbf{A} = \boldsymbol{\gamma}$ for all $1 \times K$ vectors $\boldsymbol{\gamma}$. Show that $\mathbf{A} = \mathbf{I}_K$.

Exercise 5.8.18. Show that the quadratic form in (5.36) equals $\|\mathbf{v}^* - \boldsymbol{\gamma}\mathbf{D}^{*\prime}\|^2$ for $\mathbf{v}^*$ and $\mathbf{D}^*$ in (5.34). Conclude that the weighted least squares estimator (5.35) does minimize (5.34).

Exercise 5.8.19. Show that the weighted least squares estimate (5.35) is the same as the regular least squares estimate (5.18) when in (5.24) the covariance $\boldsymbol{\Omega} = \sigma^2\mathbf{I}_N$ for some $\sigma^2 > 0$.

Exercise 5.8.20. Verify the steps in (5.40), (5.41), and (5.42), detailing which parts of Proposition 3.2 are used at each step.

Exercise 5.8.21. Verify (5.44).

Exercise 5.8.22. Verify (5.60).

Exercise 5.8.23. Show that for any $K_1 \times K_2$ matrix $\mathbf{A}$,

$$\begin{pmatrix} \mathbf{I}_{K_1} & \mathbf{A} \\ \mathbf{0} & \mathbf{I}_{K_2} \end{pmatrix}^{-1} = \begin{pmatrix} \mathbf{I}_{K_1} & -\mathbf{A} \\ \mathbf{0} & \mathbf{I}_{K_2} \end{pmatrix}. \tag{5.97}$$

Exercise 5.8.24. Suppose $\mathbf{D} = (\mathbf{D}_1, \mathbf{D}_2)$, where $\mathbf{D}_1$ is $N \times K_1$, $\mathbf{D}_2$ is $N \times K_2$, and $\mathbf{D}_1'\mathbf{D}_2 = \mathbf{0}$. Use Exercise 5.8.9 to show that $\mathbf{P}_\mathbf{D} = \mathbf{P}_{\mathbf{D}_1} + \mathbf{P}_{\mathbf{D}_2}$.

Exercise 5.8.25. Suppose $\mathbf{D} = (\mathbf{D}_1, \mathbf{D}_2)$, where $\mathbf{D}_1$ is $N \times K_1$, $\mathbf{D}_2$ is $N \times K_2$ (but $\mathbf{D}_1'\mathbf{D}_2 \neq \mathbf{0}$, maybe). (a) Use Exercise 5.8.24 and Lemma 5.3 to show that $\mathbf{P}_\mathbf{D} = \mathbf{P}_{\mathbf{D}_1} + \mathbf{P}_{\mathbf{D}_{2\cdot 1}}$, where $\mathbf{D}_{2\cdot 1} = \mathbf{Q}_{\mathbf{D}_1}\mathbf{D}_2$ as in (5.57). (b) Show that $\mathbf{Q}_{\mathbf{D}_1} = \mathbf{Q}_\mathbf{D} + \mathbf{P}_{\mathbf{D}_{2\cdot 1}}$.

Exercise 5.8.26. Show that the equation for $\mathbf{d}_{j\cdot 1}$ in (5.61) does follow from the derivation of $\mathbf{D}_{2\cdot 1}$.

Exercise 5.8.27. Give an argument for why the set of equations in (5.68) follows from the Gram-Schmidt algorithm.

Exercise 5.8.28. Given that a subspace is a span of a set of vectors, explain how one would obtain an orthogonal basis for the space.

Exercise 5.8.29. Let $\mathbf{Z}_1$ be a $N \times K$ matrix with linearly independent columns. (a) How would you find a $N \times (N-K)$ matrix $\mathbf{Z}_2$ so that $(\mathbf{Z}_1, \mathbf{Z}_2)$ is an invertible $N \times N$ matrix, and $\mathbf{Z}_1'\mathbf{Z}_2 = \mathbf{0}$ (i.e., the columns of $\mathbf{Z}_1$ are orthogonal to those of $\mathbf{Z}_2$). [Hint: Start by using Lemma 5.3 with $\mathbf{D}_1 = \mathbf{Z}_1$ and $\mathbf{D}_2 = \mathbf{I}_N$. (What is the span of the columns of $(\mathbf{Z}_1, \mathbf{I}_N)$?) Then use either Gram-Schmidt or Exercise 5.8.13 on $\mathbf{D}_{2\cdot 1}$ to find a set of vectors to use as the $\mathbf{Z}_2$. Do you recognize $\mathbf{D}_{2\cdot 1}$?] (b) Suppose the columns of $\mathbf{Z}_1$ are orthonormal. How would you modify the $\mathbf{Z}_2$ obtained in part (a) so that $(\mathbf{Z}_1, \mathbf{Z}_2)$ is an orthogonal matrix?

Exercise 5.8.30. Consider the matrix $\mathbf{B}^{(k)}$ defined in (5.72). (a) Use Exercise 5.8.23 to show that the inverse of $\mathbf{B}^{(k)}$ is of the same form, but with the $-b_{kj}$'s changed to b_{kj}'s. That is, the inverse is the $K \times K$ matrix $\mathbf{C}^{(k)}$, where

$$\mathbf{C}_{ij}^{(k)} = \begin{cases} 1 & \text{if } i = j \\ b_{kj} & \text{if } j > k = i \\ 0 & \text{otherwise.} \end{cases} \tag{5.98}$$

Thus $\mathbf{C}$ is the inverse of the $\mathbf{B}$ in (5.76), where $\mathbf{C} = \mathbf{C}^{(K-1)} \cdots \mathbf{C}^{(1)}$. (b) Show that $\mathbf{C}$ is unitriangular, where the b_{ij}'s are in the upper triangular part, i.e, $\mathbf{C}_{ij} = b_{ij}$ for $j > i$, as in (5.77). (c) The $\mathbf{R}$ in (5.79) is then $\mathbf{\Delta C}$, where $\mathbf{\Delta}$ is the diagonal matrix with diagonal elements being the norms of the columns of $\mathbf{D}^*$. Show that $\mathbf{R}$ is given by

$$\mathbf{R}_{ij} = \begin{cases} \|\mathbf{d}_{i\cdot\{1:(i-1)\}}\| & \text{if } j = i \\ \mathbf{d}_{j\cdot\{1:(i-1)\}}'\mathbf{d}_{i\cdot\{1:(i-1)\}} / \|\mathbf{d}_{i\cdot\{1:(i-1)\}}\| & \text{if } j > i \\ 0 & \text{if } j < i. \end{cases} \tag{5.99}$$

Exercise 5.8.31. Verify (5.83).

Exercise 5.8.32. Suppose $\mathbf{d}_1, \ldots, \mathbf{d}_K$ are vectors in $\mathbb{R}^N$, and $\mathbf{d}_1^*, \ldots, \mathbf{d}_K^*$ are the corresponding orthogonal vectors resulting from the Gram-Schmidt algorithm, i.e., $\mathbf{d}_1^* = \mathbf{d}_1$, and for $i > 1$, $\mathbf{d}_i^* = \mathbf{d}_{i \cdot \{1:(i-1)\}}$ in (5.66). (a) Show that the $\mathbf{d}_1^*, \ldots, \mathbf{d}_K^*$ are linearly independent if and only if they are all nonzero. Why? [Hint: Recall Exercise 5.8.7.] (b) Show that $\mathbf{d}_1, \ldots, \mathbf{d}_K$ are linearly independent if and only if all the $\mathbf{d}_j^*$ are nonzero.

Exercise 5.8.33. Suppose $\mathbf{D}$ is $N \times K$, with linearly independent columns, and $\mathbf{D} = \mathbf{QR}$ is its QR decomposition. Show that span$\{$columns of $\mathbf{D}\}$ = span$\{$columns of $\mathbf{Q}\}$.

Exercise 5.8.34. Suppose $\mathbf{D}$ is an $N \times N$ matrix whose columns are linearly independent. Show that $\mathbf{D}$ is invertible. [Hint: Use the QR decomposition in Theorem 5.4. What kind of a matrix is the $\mathbf{Q}$ here? Is it invertible?]

Exercise 5.8.35. (a) Show that span$\{$columns of $\mathbf{Q}\} = \mathbb{R}^N$ if $\mathbf{Q}$ is an $N \times N$ orthogonal matrix. (b) Suppose the $N \times 1$ vectors $\mathbf{d}_1, \ldots, \mathbf{d}_N$ are linearly independent, and $\mathcal{W} = \text{span}\{\mathbf{d}_1, \ldots, \mathbf{d}_N\}$. Show that $\mathcal{W} = \mathbb{R}^N$. [Hint: Use Theorem 5.4, Lemma 5.1, and part (a).]

Exercise 5.8.36. Show that if $\mathbf{d}_1, \ldots, \mathbf{d}_K$ are vectors in $\mathbb{R}^N$ with $K > N$, that the $\mathbf{d}_i$'s are linearly dependent. (This fact should make sense, since there cannot be more axes than there are dimensions in Euclidean space.) [Hint: Use Exercise 5.8.35 on the first N vectors, then show how $\mathbf{d}_{N+1}$ is a linear combination of them.]

Exercise 5.8.37. Show that the QR decomposition in Theorem 5.4 is unique when $N = K$. That is, suppose $\mathbf{Q}_1$ and $\mathbf{Q}_2$ are $K \times K$ orthogonal matrices, and $\mathbf{R}_1$ and $\mathbf{R}_2$ are $K \times K$ upper triangular matrices with positive diagonals, and $\mathbf{Q}_1\mathbf{R}_1 = \mathbf{Q}_2\mathbf{R}_2$. Show that $\mathbf{Q}_1 = \mathbf{Q}_2$ and $\mathbf{R}_1 = \mathbf{R}_2$. [Hint: Show that $\mathbf{Q} \equiv \mathbf{Q}_2'\mathbf{Q}_1 = \mathbf{R}_2\mathbf{R}_1^{-1} \equiv \mathbf{R}$, so that the orthogonal matrix $\mathbf{Q}$ equals the upper triangular matrix $\mathbf{R}$ with positive diagonals. Show that therefore $\mathbf{Q} = \mathbf{R} = \mathbf{I}_K$.] [Extra credit: Show the uniqueness when $M > K$.]

Exercise 5.8.38. Verify (5.81). [Exercise 5.8.23 helps.] In particular: (a) Argue that the $\mathbf{0}$'s in the middle matrix on the left-hand side of (5.81) are correct. (b) Show $\mathbf{S}_{22\cdot 1} = \mathbf{D}_{2\cdot 1}'\mathbf{D}_{2\cdot 1}$.

Exercise 5.8.39. Suppose

$$\mathbf{S} = \begin{pmatrix} \mathbf{S}_{11} & \mathbf{S}_{12} \\ \mathbf{S}_{21} & \mathbf{S}_{22} \end{pmatrix}, \tag{5.100}$$

where $\mathbf{S}_{11}$ is $K_1 \times K_1$ and $\mathbf{S}_{22}$ is $K_2 \times K_2$, and $\mathbf{S}_{11}$ is invertible. (a) Show that

$$|\mathbf{S}| = |\mathbf{S}_{11}|\,|\mathbf{S}_{22\cdot 1}|. \tag{5.101}$$

[Hint: Use (5.81).] (b) Show that

$$\mathbf{S}^{-1} = \begin{pmatrix} \mathbf{S}_{11}^{-1} + \mathbf{S}_{11}^{-1}\mathbf{S}_{12}\mathbf{S}_{22\cdot 1}^{-1}\mathbf{S}_{21}\mathbf{S}_{11}^{-1} & -\mathbf{S}_{11}^{-1}\mathbf{S}_{12}\mathbf{S}_{22\cdot 1}^{-1} \\ -\mathbf{S}_{22\cdot 1}^{-1}\mathbf{S}_{21}\mathbf{S}_{11}^{-1} & \mathbf{S}_{22\cdot 1}^{-1} \end{pmatrix}. \tag{5.102}$$

[Hint: Use (5.81) and (5.97).] (c) Use part (b) to show that

$$[\mathbf{S}^{-1}]_{22} = \mathbf{S}_{22\cdot 1}^{-1}, \tag{5.103}$$

where $[\mathbf{S}^{-1}]_{22}$ is the lower-right $K_2 \times K_2$ block of $\mathbf{S}^{-1}$. Under what condition on $\mathbf{S}_{12}$ is $[\mathbf{S}^{-1}]_{22} = \mathbf{S}_{22}^{-1}$?

Exercise 5.8.40. For $\mathbf{S} \in \mathcal{S}_K^+$, show that

$$|\mathbf{S}| = S_{11}S_{22\cdot 1}S_{33\cdot 12}\cdots S_{KK\cdot\{1:(K-1)\}}. \tag{5.104}$$

[Hint: Use (5.82). What is the determinant of a unitriangular matrix?]

Exercise 5.8.41. Consider the multivariate regression model, as in (5.38) with $\mathbf{z} = \mathbf{I}_q$. Partition $\mathbf{x}$ and $\boldsymbol{\beta}$:

$$\mathbf{Y} = \mathbf{x}\boldsymbol{\beta} + \mathbf{R} = (\mathbf{x}_1 \ \ \mathbf{x}_2)\begin{pmatrix}\boldsymbol{\beta}_1 \\ \boldsymbol{\beta}_2\end{pmatrix} + \mathbf{R}, \tag{5.105}$$

where $\mathbf{x}_1$ is $n \times p_1$ and $\mathbf{x}_2$ is $n \times p_2$. Let $\mathbf{x}_{2\cdot 1} = \mathbf{Q}_{\mathbf{x}_1}\mathbf{x}_2$. (a) Show that

$$[\mathbf{C}_\mathbf{x}]_{22} = (\mathbf{x}_{2\cdot 1}'\mathbf{x}_{2\cdot 1})^{-1}, \tag{5.106}$$

where $[\mathbf{C}_\mathbf{x}]_{22}$ is the lower-right $p_2 \times p_2$ part of $\mathbf{C}_\mathbf{x}$ in (5.45). (See 5.103.) (b) Show that $\widehat{\boldsymbol{\beta}}_{LS2}$, the lower $p_2 \times q$ part of $\widehat{\boldsymbol{\beta}}_{LS}$ in (5.43), satisfies

$$\widehat{\boldsymbol{\beta}}_{LS2} = (\mathbf{x}_{2\cdot 1}'\mathbf{x}_{2\cdot 1})^{-1}\mathbf{x}_{2\cdot 1}'\mathbf{Y}. \tag{5.107}$$

Exercise 5.8.42. Consider the model (5.38) where $\mathbf{z}$ is the $q \times l$ matrix

$$\mathbf{z} = \begin{pmatrix}\mathbf{I}_l \\ \mathbf{0}\end{pmatrix}, \tag{5.108}$$

with $l < q$, and partition $\boldsymbol{\Sigma}_\mathbf{R}$ as

$$\boldsymbol{\Sigma}_R = \begin{pmatrix}\boldsymbol{\Sigma}_{11} & \boldsymbol{\Sigma}_{12} \\ \boldsymbol{\Sigma}_{21} & \boldsymbol{\Sigma}_{22}\end{pmatrix}, \ \ \boldsymbol{\Sigma}_{11} \text{ is } l \times l, \ \ \boldsymbol{\Sigma}_{22} \text{ is } (q-l) \times (q-l). \tag{5.109}$$

Show that in (5.45), $\boldsymbol{\Sigma}_\mathbf{z} = \boldsymbol{\Sigma}_{11}$, and use (5.103) to show that in (5.42),

$$(\mathbf{z}'\boldsymbol{\Sigma}_R^{-1}\mathbf{z})^{-1} = \boldsymbol{\Sigma}_{11\cdot 2}. \tag{5.110}$$

Then find an explicit equation for $Cov[\widehat{\boldsymbol{\beta}}_{LS}] - Cov[\widehat{\boldsymbol{\beta}}_{WLS}]$ from(5.42) and (5.44). Which estimator has a better covariance? [We already know that Gauss-Markov answers this last question.]

Exercise 5.8.43. Suppose $\mathbf{S} \in \mathcal{S}_K^+$. Prove Theorem 5.5, i.e., show that we can write $\mathbf{S} = \mathbf{R}'\mathbf{R}$, where $\mathbf{R}$ is upper triangular with positive diagonal elements. [Hint: Use the spectral decomposition $\mathbf{S} = \mathbf{G}\mathbf{L}\mathbf{G}'$ from (1.33). Then let $\mathbf{D} = \mathbf{L}^{1/2}\mathbf{G}'$ in (5.79). Are the columns of this $\mathbf{D}$ linearly independent?]

Exercise 5.8.44. Show that if $\mathbf{W} = \mathbf{R}'\mathbf{R}$ is the Cholesky decomposition of $\mathbf{W}$ ($K \times K$), then

$$|\mathbf{W}| = \prod_{j=1}^{K} r_{jj}^2. \tag{5.111}$$

Exercise 5.8.45. Show that the Cholesky decomposition in Theorem 5.5 is unique. That is, if $\mathbf{R}_1$ and $\mathbf{R}_2$ are $K \times K$ upper triangular matrices with positive diagonals, that $\mathbf{R}_1'\mathbf{R}_1 = \mathbf{R}_2'\mathbf{R}_2$ implies that $\mathbf{R}_1 = \mathbf{R}_2$. [Hint: Let $\mathbf{R} = \mathbf{R}_1\mathbf{R}_2^{-1}$, and show that $\mathbf{R}'\mathbf{R} = \mathbf{I}_K$. Then show that this $\mathbf{R}$ must be $\mathbf{I}_K$, just as in Exercise 5.8.37.]

Exercise 5.8.46. Show that the $N \times K$ matrix $\mathbf{D}$ has linearly independent columns if and only if $\mathbf{D}'\mathbf{D}$ is invertible. [Hint: If $\mathbf{D}$ has linearly independent columns, then $\mathbf{D}'\mathbf{D} = \mathbf{R}'\mathbf{R}$ as Theorem 5.5, and $\mathbf{R}$ is invertible. If the columns are linearly dependent, there is a $\boldsymbol{\gamma} \neq \mathbf{0}$ with $\mathbf{D}'\mathbf{D}\boldsymbol{\gamma} = \mathbf{0}$. Why does that equation imply $\mathbf{D}'\mathbf{D}$ has no inverse?]

Exercise 5.8.47. Suppose $\mathbf{D}$ is $N \times K$ and $\mathbf{C}$ is $N \times J$, $K > J$, and both matrices have linearly independent columns. Furthermore, suppose

$$\text{span}\{\text{columns of } \mathbf{D}\} = \text{span}\{\text{columns of } \mathbf{C}\}. \tag{5.112}$$

Thus this space has two bases with differing numbers of elements. (a) Let $\mathbf{A}$ be the $J \times K$ matrix such that $\mathbf{D} = \mathbf{CA}$, guaranteed by Exercise 5.8.3. Show that the columns of $\mathbf{A}$ are linearly independent. [Hint: Note that $\mathbf{D}\boldsymbol{\gamma} \neq \mathbf{0}$ for any $K \times 1$ vector $\boldsymbol{\gamma} \neq \mathbf{0}$. Hence $\mathbf{A}\boldsymbol{\gamma} \neq \mathbf{0}$ for any $\boldsymbol{\gamma} \neq \mathbf{0}$.] (b) Use Exercise 5.8.36 to show that such an $\mathbf{A}$ cannot exist. (c) What do you conclude?

Exercise 5.8.48. This exercise is to show that any linear subspace $\mathcal{W}$ in $\mathbb{R}^N$ has a basis. If $\mathcal{W} = \{\mathbf{0}\}$, the basis is the empty set. So you can assume $\mathcal{W}$ has more than just the zero vector. (a) Suppose $\mathbf{d}_1, \ldots, \mathbf{d}_J$ are linearly independent vectors in $\mathbb{R}^N$. Show that $\mathbf{d} \in \mathbb{R}^N$ but $\mathbf{d} \notin \text{span}\{\mathbf{d}_1, \ldots, \mathbf{d}_J\}$ implies that $\mathbf{d}_1, \ldots, \mathbf{d}_J, \mathbf{d}$ are linearly independent. [Hint: If they are not linearly independent, then some linear combination of them equals zero. The coefficient of $\mathbf{d}$ in that linear combination must be nonzero. (Why?) Thus $\mathbf{d}$ must be in the span of the others.] (b) Take $\mathbf{d}_1 \in \mathcal{W}$, $\mathbf{d}_1 \neq \mathbf{0}$. [I guess we are assuming the Axiom of Choice.] If $\text{span}\{\mathbf{d}_1\} = \mathcal{W}$, then we have the basis. If not, there must be a $\mathbf{d}_2 \in \mathcal{W} - \text{span}\{\mathbf{d}_1\}$. If $\text{span}\{\mathbf{d}_1, \mathbf{d}_2\} = \mathcal{W}$, we are done. Explain how to continue. (Also, explain why part (a) is important here.) How do you know this process stops? (c) Argue that any linear subspace has a corresponding projection matrix.

Exercise 5.8.49. Suppose $\mathbf{P}$ and $\mathbf{P}^*$ are projection matrices for the linear subspace $\mathcal{W} \subset \mathbb{R}^N$. Show that $\mathbf{P} = \mathbf{P}^*$, i.e., the projection matrix is unique to the subspace. [Hint: Because the projection of any vector is unique, $\mathbf{Pv} = \mathbf{P}^*\mathbf{v}$ for all $\mathbf{v}$. Consider $\mathbf{v}$ being each of the columns of $\mathbf{I}_N$.]

Exercise 5.8.50. Find the orthogonal polynomial matrix (up to cubic) for the four time points 1, 2, 4, 5.

Exercise 5.8.51 (Skulls). For the model on skull measurements described in Exercise 4.4.2, replace the polynomial matrix $\mathbf{w}$ with that for orthogonal polynomials.

Exercise 5.8.52 (Caffeine). In Exercise 4.4.4, the $\mathbf{x}$ is a quadratic polynomial matrix in grade (8, 10, 12). Replace it with the orthogonal polynomial matrix (also 28×3), where the first column is all ones, the second is the linear vector $(-\mathbf{1}_9', \mathbf{0}_{10}', \mathbf{1}_9')'$, and the third is the quadratic vector $(\mathbf{1}_9', -c\mathbf{1}_{10}', \mathbf{1}_9')'$ for some c. Find c.

Exercise 5.8.53 (Leprosy). Consider again the model for the leprosy data in Exercise 4.4.6. An alternate expression for $\mathbf{x}$ is $\mathbf{w}^* \otimes \mathbf{1}_{10}$, where the first column of $\mathbf{w}^*$ represents the overall mean, the second tells whether the treatment is one of the drugs, and the third whether the treatment is Drug A, so that

$$\mathbf{w}^* = \begin{pmatrix} 1 & 1 & 1 \\ 1 & 1 & 0 \\ 1 & 0 & 0 \end{pmatrix}. \tag{5.113}$$

Use Gram-Schmidt to orthogonalize the columns of $\mathbf{w}^*$. How does this matrix differ from $\mathbf{w}$? How does the model using $\mathbf{w}^*$ differ from that using $\mathbf{w}$?

Chapter 6

Both-Sides Models: Distribution of the Least Squares Estimator

6.1 Distribution of $\widehat{\beta}$

The both-sides model as defined in (4.29) is

$$\mathbf{Y} = \mathbf{x}\boldsymbol{\beta}\mathbf{z}' + \mathbf{R}, \quad \mathbf{R} \sim N(\mathbf{0}, \mathbf{I}_n \otimes \boldsymbol{\Sigma}_R), \tag{6.1}$$

where $\mathbf{Y}$ is $n \times q$, $\mathbf{x}$ is $n \times p$, $\boldsymbol{\beta}$ is $p \times l$, and $\mathbf{z}$ is $q \times l$. Assuming that $\mathbf{x}'\mathbf{x}$ and $\mathbf{z}'\mathbf{z}$ are invertible, the least squares estimate of $\boldsymbol{\beta}$ is given in (5.43) to be

$$\widehat{\boldsymbol{\beta}} = (\mathbf{x}'\mathbf{x})^{-1}\mathbf{x}'\mathbf{Y}\mathbf{z}(\mathbf{z}'\mathbf{z})^{-1}. \tag{6.2}$$

From (5.27) we have seen that $\widehat{\boldsymbol{\beta}}$ is unbiased, and (5.44) gives the covariance matrix. Because in (6.1), $\mathbf{Y}$ is multivariate normal, and $\widehat{\boldsymbol{\beta}}$ is a linear function of $\mathbf{Y}$, $\widehat{\boldsymbol{\beta}}$ is multivariate normal. Thus we have that its distribution is

$$\widehat{\boldsymbol{\beta}} \sim N_{p\times l}(\boldsymbol{\beta}, \mathbf{C}_\mathbf{x} \otimes \boldsymbol{\Sigma}_\mathbf{z}), \tag{6.3}$$

where from (5.45),

$$\mathbf{C}_\mathbf{x} = (\mathbf{x}'\mathbf{x})^{-1} \text{ and } \boldsymbol{\Sigma}_\mathbf{z} = (\mathbf{z}'\mathbf{z})^{-1}\mathbf{z}'\boldsymbol{\Sigma}_R\mathbf{z}(\mathbf{z}'\mathbf{z})^{-1}. \tag{6.4}$$

In order to find confidence intervals and hypothesis tests for the coefficients, we need to estimate $\boldsymbol{\Sigma}_\mathbf{z}$.

6.2 Estimating the covariance

6.2.1 Multivariate regression

We start with the multivariate regression model, i.e., $\mathbf{z} = \mathbf{I}_q$:

$$\mathbf{Y} = \mathbf{x}\boldsymbol{\beta} + \mathbf{R}, \quad \mathbf{R} \sim N(\mathbf{0}, \mathbf{I}_n \otimes \boldsymbol{\Sigma}_R). \tag{6.5}$$

As in (5.19), the fit of the model to the data is given by

$$\widehat{\mathbf{Y}} = \mathbf{x}\widehat{\boldsymbol{\beta}} = \mathbf{x}(\mathbf{x}'\mathbf{x})^{-1}\mathbf{x}'\mathbf{Y} = \mathbf{P}_{\mathbf{x}}\mathbf{Y}, \tag{6.6}$$

where $\mathbf{P}_{\mathbf{x}} = \mathbf{x}(\mathbf{x}'\mathbf{x})^{-1}\mathbf{x}'$, the projection matrix onto the span of the columns of $\mathbf{x}$, as in (5.20). The residual vector is then

$$\widehat{\mathbf{R}} = \mathbf{Y} - \widehat{\mathbf{Y}} = \mathbf{Y} - \mathbf{P}_{\mathbf{x}}\mathbf{Y} = \mathbf{Q}_{\mathbf{x}}\mathbf{Y}, \tag{6.7}$$

where $\mathbf{Q}_{\mathbf{x}} = \mathbf{I}_n - \mathbf{P}_{\mathbf{x}}$ as in (5.22).

The joint distribution of $\widehat{\mathbf{Y}}$ and $\widehat{\mathbf{R}}$ is multivariate normal because the collection is a linear transformation of $\mathbf{Y}$. The means are straightforward:

$$E[\widehat{\mathbf{Y}}] = \mathbf{x}E[\widehat{\boldsymbol{\beta}}] = \mathbf{x}\boldsymbol{\beta} \tag{6.8}$$

because $\widehat{\boldsymbol{\beta}}$ is unbiased from (6.3), hence

$$E[\widehat{\mathbf{R}}] = E[\mathbf{Y}] - E[\widehat{\mathbf{Y}}] = \mathbf{x}\boldsymbol{\beta} - \mathbf{x}\boldsymbol{\beta} = \mathbf{0}. \tag{6.9}$$

For the joint covariance of the fit and residual, write

$$\begin{pmatrix} \widehat{\mathbf{Y}} \\ \widehat{\mathbf{R}} \end{pmatrix} = \begin{pmatrix} \mathbf{P}_{\mathbf{x}} \\ \mathbf{Q}_{\mathbf{x}} \end{pmatrix} \mathbf{Y}. \tag{6.10}$$

Using Proposition 3.1 on the covariance in (6.1), we have

$$\begin{aligned} Cov\left[\begin{pmatrix} \widehat{\mathbf{Y}} \\ \widehat{\mathbf{R}} \end{pmatrix}\right] &= \begin{pmatrix} \mathbf{P}_{\mathbf{x}}\mathbf{P}_{\mathbf{x}}' & \mathbf{P}_{\mathbf{x}}\mathbf{Q}_{\mathbf{x}}' \\ \mathbf{Q}_{\mathbf{x}}\mathbf{P}_{\mathbf{x}}' & \mathbf{Q}_{\mathbf{x}}\mathbf{Q}_{\mathbf{x}}' \end{pmatrix} \otimes \boldsymbol{\Sigma}_R \\ &= \begin{pmatrix} \mathbf{P}_{\mathbf{x}} & \mathbf{0} \\ \mathbf{0} & \mathbf{Q}_{\mathbf{x}} \end{pmatrix} \otimes \boldsymbol{\Sigma}_R, \end{aligned} \tag{6.11}$$

where we use Proposition 5.1, parts (a) and (c), on the projection matrices. Note that the fit and residual are independent because they are uncorrelated and jointly multivariate normal. Furthermore, we can write $\widehat{\boldsymbol{\beta}}$ as a function of the fit,

$$\widehat{\boldsymbol{\beta}} = (\mathbf{x}'\mathbf{x})^{-1}\mathbf{x}'\mathbf{Y} = (\mathbf{x}'\mathbf{x})^{-1}\mathbf{x}'\mathbf{P}_{\mathbf{x}}\mathbf{Y} = (\mathbf{x}'\mathbf{x})^{-1}\mathbf{x}'\widehat{\mathbf{Y}}, \tag{6.12}$$

because $\mathbf{x}'\mathbf{P}_{\mathbf{x}} = \mathbf{x}'$. Thus $\widehat{\boldsymbol{\beta}}$ and $\widehat{\mathbf{R}}$ are independent.

From (6.9) and (6.11), we have that

$$\widehat{\mathbf{R}} = \mathbf{Q}_{\mathbf{x}}\mathbf{Y} \sim N(\mathbf{0}, \mathbf{Q}_{\mathbf{x}} \otimes \boldsymbol{\Sigma}_R). \tag{6.13}$$

Because $\mathbf{Q}_{\mathbf{x}}$ is idempotent, Corollary 3.1 shows that

$$\widehat{\mathbf{R}}'\widehat{\mathbf{R}} = \mathbf{Y}'\mathbf{Q}_{\mathbf{x}}\mathbf{Y} \sim \text{Wishart}_q(n-p, \boldsymbol{\Sigma}_R), \tag{6.14}$$

where by Proposition 5.1 (a), $\text{trace}(\mathbf{Q}_{\mathbf{x}}) = \text{trace}(\mathbf{I}_n - \mathbf{P}_{\mathbf{x}}) = n - p$. We collect these results.

Theorem 6.1. *In the model (6.5), $\widehat{\boldsymbol{\beta}}$ in (6.12) and $\mathbf{Y}'\mathbf{Q}_{\mathbf{x}}\mathbf{Y}$ are independent, with distributions given in (6.3) and (6.14), respectively.*

The Wishart result in (6.14) implies that an unbiased estimate of $\boldsymbol{\Sigma}_R$ is

$$\widehat{\boldsymbol{\Sigma}}_R = \frac{1}{n-p}\mathbf{Y}'\mathbf{Q}_{\mathbf{x}}\mathbf{Y}. \tag{6.15}$$

6.2.2 Both-sides model

Now consider the general both-sides model (6.1). We start by using just the $\mathbf{z}$ part of the estimate (6.2), that is, define the $n \times l$ matrix

$$\mathbf{Y_z} = \mathbf{Yz}(\mathbf{z}'\mathbf{z})^{-1}. \tag{6.16}$$

Exercise 6.6.2 shows that

$$\mathbf{Y_z} = \mathbf{x}\boldsymbol{\beta} + \mathbf{R_z}, \quad \mathbf{R_z} \sim N(\mathbf{0}, \mathbf{I}_n \otimes \boldsymbol{\Sigma}_\mathbf{z}) \tag{6.17}$$

for $\boldsymbol{\Sigma}_\mathbf{z}$ in (6.4). Notice that $\mathbf{Y_z}$ follows a multivariate regression model (6.5). Theorem 6.1 can be applied to show that $\widehat{\boldsymbol{\beta}}$ and $\mathbf{Y_z}'\mathbf{Q_x}\mathbf{Y_z}$ are independent, with

$$\mathbf{Y_z}'\mathbf{Q_x}\mathbf{Y_z} \sim \text{Wishart}_l(n-p, \boldsymbol{\Sigma}_\mathbf{z}). \tag{6.18}$$

Also, as in (6.15),

$$\widehat{\boldsymbol{\Sigma}}_\mathbf{z} = \frac{1}{n-p}\mathbf{Y_z}'\mathbf{Q_x}\mathbf{Y_z} \tag{6.19}$$

is an unbiased estimator of $\boldsymbol{\Sigma}_\mathbf{z}$. Thus an unbiased estimator of the covariance of $\widehat{\boldsymbol{\beta}}$ is

$$\widehat{Cov}[\widehat{\boldsymbol{\beta}}] = \mathbf{C_x} \otimes \widehat{\boldsymbol{\Sigma}}_\mathbf{z}. \tag{6.20}$$

6.3 Standard errors and t-statistics

In order to find confidence intervals and perform hypothesis tests of the individual β_{ij}'s, we need to find the standard errors of the estimates. From (6.3), the individual coefficients satisfy

$$\widehat{\beta}_{ij} \sim N(\beta_{ij}, C_{\mathbf{x}ii}\sigma_{\mathbf{z}jj}), \tag{6.21}$$

where $C_{\mathbf{x}ii}$ is the ith diagonal element of $\mathbf{C_x}$, and $\sigma_{\mathbf{z}jj}$ is the jth diagonal element of $\boldsymbol{\Sigma}_\mathbf{z}$. Let $\widehat{\sigma}_{\mathbf{z}jj}$ be the jth diagonal of $\widehat{\boldsymbol{\Sigma}}_\mathbf{z}$, and define the (estimated) standard error by

$$se(\widehat{\beta}_{ij}) = \sqrt{C_{\mathbf{x}ii}\widehat{\sigma}_{\mathbf{z}jj}}. \tag{6.22}$$

Inference is based on the t-statistic

$$T = \frac{\widehat{\beta}_{ij} - \beta_{ij}}{se(\widehat{\beta}_{ij})}. \tag{6.23}$$

We now need the Student's t distribution:

Definition 6.1. *If $Z \sim N(0,1)$ and $U \sim \chi^2_\nu$, and Z and U are independent, then*

$$T \equiv \frac{Z}{\sqrt{U/\nu}} \tag{6.24}$$

has the Student's t distribution on ν degrees of freedom, written $T \sim t_\nu$.

To see that T in (6.23) has a Student's t distribution, note that since $(n-p)\widehat{\mathbf{\Sigma}}_\mathbf{z}$ is Wishart as in (6.18), by the marginals discussion in Section 3.6,

$$U \equiv (n-p)\widehat{\sigma}_{\mathbf{z}jj}/\sigma_{\mathbf{z}jj} \sim \chi^2_{n-p}. \tag{6.25}$$

Also, from (6.21),

$$Z = (\widehat{\beta}_{ij} - \beta_{ij})/\sqrt{\mathbf{C}_{\mathbf{x}ii}\sigma_{\mathbf{z}jj}} \sim N(0,1). \tag{6.26}$$

Theorem 6.1 guarantees that U and Z are independent, hence Definition 6.1 can be used to show that T in (6.23) has the distribution of that in (6.24) with $\nu = n-p$. See Exercise 6.6.3.

Thus a $100 \times (1-\alpha)\%$ confidence interval for β_{ij} is

$$\widehat{\beta}_{ij} \pm t_{n-p,\alpha/2}\, se(\widehat{\beta}_{ij}), \tag{6.27}$$

where $t_{n-p,\alpha/2}$ is the cutoff point for which $P[|T| > t_{n-p,\alpha/2}] = \alpha$ for $T \sim t_{n-p}$. A level α test of the null hypothesis $H_0 : \beta_{ij} = 0$ rejects the null if

$$\frac{|\widehat{\beta}_{ij}|}{se(\widehat{\beta}_{ij})} > t_{n-p,\alpha/2}. \tag{6.28}$$

6.4 Examples

6.4.1 Mouth sizes

Section 4.2.1 contains data on the size of mouths of 11 girls (sex=1) and 16 boys (sex=0) at four ages, so $n = 27$ and $q = 4$. The model where the $\mathbf{x}$ matrix compares the boys and girls, and the $\mathbf{z}$ matrix specifies orthogonal polynomial growth curves (Section 5.7.2), is

$$\begin{aligned} \mathbf{Y} &= \mathbf{x}\beta\mathbf{z}' + \mathbf{R} \\ &= \begin{pmatrix} \mathbf{1}_{11} & \mathbf{1}_{11} \\ \mathbf{1}_{16} & \mathbf{0}_{16} \end{pmatrix} \begin{pmatrix} \beta_0 & \beta_1 & \beta_2 & \beta_3 \\ \delta_0 & \delta_1 & \delta_2 & \delta_3 \end{pmatrix} \begin{pmatrix} 1 & 1 & 1 & 1 \\ -3 & -1 & 1 & 3 \\ 1 & -1 & -1 & 1 \\ -1 & 3 & -3 & 1 \end{pmatrix} + \mathbf{R}, \end{aligned} \tag{6.29}$$

Compare this model to that in (4.14). Here the β_j's are the boys' coefficients, and the $(\beta_j + \delta_j)$'s are the girls' coefficients, hence the δ_j's are the girls' minus the boys'.

In this case, $\mathbf{z}$ is square, hence invertible, so that we have

$$\begin{aligned} \mathbf{z} \text{ is } q \times q &\Longrightarrow \mathbf{z}(\mathbf{z}'\mathbf{z})^{-1} = (\mathbf{z}')^{-1} \\ &\Longrightarrow \mathbf{Y_z} = \mathbf{Yz}(\mathbf{z}'\mathbf{z})^{-1} = \mathbf{Y}(\mathbf{z}')^{-1} \end{aligned} \tag{6.30}$$

for $\mathbf{Y_z}$ from (6.16). The rows of the $\mathbf{Y}$ contain the individuals' original mouth-size measurements over time. By contrast, the rows of $\mathbf{Y_z}$ contain the coefficients for the individuals' growth curves. The following table exhibits a few of the rows of $\mathbf{Y_z}$:

$$\mathbf{Y_z}: \quad \begin{array}{ccccc} \text{Observation} & \text{Constant} & \text{Linear} & \text{Quadratic} & \text{Cubic} \\ \hline 1 & 21.375 & 0.375 & 0.625 & -0.125 \\ 2 & 23.000 & 0.800 & 0.250 & -0.150 \\ & & \cdots & & \\ 27 & 23.000 & 0.550 & 0.500 & -0.150 \end{array} \tag{6.31}$$

The first element "21.375" of $\mathbf{Y_z}$ is the average mouth size for the first observation (a girl), the next element "0.375" is the linear coefficient for her growth curve, and the next two elements are her quadratic and cubic terms, respectively.

The data are in the 27×5 R matrix mouths, where the first four columns contain the mouth size measurements, and the fifth is the indicator for sex. We first create the $\mathbf{x}$, $\mathbf{y}$, and $\mathbf{z}$ matrices, then find $\widehat{\boldsymbol{\beta}}$:

```
x <- cbind(1,mouths[,5])
y <- mouths[,1:4]
z <- cbind(c(1,1,1,1),c(-3,-1,1,3),c(1,-1,-1,1),c(-1,3,-3,1))
yz <- y%*%solve(t(z))
cx <- solve(t(x)%*%x)
betahat <- cx%*%t(x)%*%yz
```

In R, solve(a) finds the inverse of the matrix a, and the symbol for matrix multiplication is "%*%." Note that if $\mathbf{z}$ is not square, then the $\mathbf{Y_z}$ is calculated via

```
yz <- y%*%z%*%solve(t(z)%*%z)
```

The estimate $\widehat{\boldsymbol{\beta}}$ is

$$\begin{array}{lcccc} & \text{Intercept} & \text{Linear} & \text{Quadratic} & \text{Cubic} \\ \hline \text{Boys} & 24.969 & 0.784 & 0.203 & -0.056 \\ \text{Girls} - \text{Boys} & -2.321 & -0.305 & -0.214 & 0.072 \end{array} \tag{6.32}$$

Next we find the estimated covariance $\widehat{\boldsymbol{\Sigma}}_{\mathbf{z}}$ in (6.19):

```
resid <- yz-x%*%beta
sigmaz <- t(resid)%*%resid/(27-2) # n=27, p=2
```

We now have

$$\mathbf{C_x} = \begin{pmatrix} 0.0625 & -0.0625 \\ -0.0625 & 0.1534 \end{pmatrix}, \tag{6.33}$$

and

$$\widehat{\boldsymbol{\Sigma}}_{\mathbf{z}} = \begin{pmatrix} 3.7791 & 0.0681 & -0.0421 & -0.1555 \\ 0.0681 & 0.1183 & -0.0502 & 0.0091 \\ -0.0421 & -0.0502 & 0.2604 & -0.0057 \\ -0.1555 & 0.0091 & -0.0057 & 0.1258 \end{pmatrix}. \tag{6.34}$$

By (6.22), the standard errors of the $\widehat{\beta}_{ij}$'s are estimated by multiplying the ith diagonal of $\mathbf{C_x}$ and j^{th} diagonal of $\widehat{\boldsymbol{\Sigma}}_{\mathbf{z}}$, then taking the square root. We can obtain the matrix of standard errors using the command

```
se <- sqrt(outer(diag(cx),diag(sigmaz),"*"))
```

The t-statistics then divide the estimates by their standard errors, betahat/se:

Standard Errors

	Intercept	Linear	Quadratic	Cubic
Boys	0.4860	0.0860	0.1276	0.0887
Girls−Boys	0.7614	0.1347	0.1999	0.1389

(6.35)

t-statistics

	Intercept	Linear	Quadratic	Cubic
Boys	51.376	9.121	1.592	−0.634
Girls−Boys	−3.048	−2.262	−1.073	0.519

It looks like the quadratic and cubic terms are unnecessary, so that straight lines for each sex fit well. It is clear that the linear term for boys is necessary, and the intercepts for the boys and girls are different (the p-value for $t = 3.048$ with 25 df is 0.005). The p-value for the girls−boys slope is 0.033, which is borderline significant.

The function bothsidesmodel in Section A.2.1 organizes these calculations, which can be invoked using bothsidesmodel(x,y,z), or simply bothsidesmodel(x,y) for the multivariate regression model.

6.4.2 Using linear regression routines

In the previous section, we used explicit matrix calculations. It turns out that the estimates and standard errors can be found using any software that will perform multiple linear regression. In R, the lm function will accept matrices $\mathbf{Y}$ and $\mathbf{x}$, and calculate one multiple regression for each column of $\mathbf{Y}$, each time using the same $\mathbf{x}$. It is invoked by lm(y~x−1). The function by default adds the $\mathbf{1}_n$ vector to the model, but that addition can be prevented by using the "−1" after the $\mathbf{x}$.

For the both-sides model, we need two steps. The first finds the $\mathbf{Y_z} = \mathbf{Y}\mathbf{z}(\mathbf{z}'\mathbf{z})^{-1}$. Note that this matrix looks like the estimate of $\boldsymbol{\beta}$ in the multivariate regression model in (6.6), but with $\mathbf{z}$ in place of $\mathbf{x}$, and the $\mathbf{z}$ parts on the wrong side. Thus we need to transpose the $\mathbf{Y}$, using lm(t(y)~z−1). The $\mathbf{Y_z}$ is then in the coefficients component of the regression, again transposed. So to find $\mathbf{Y_z}$, we use

```
yz <- t(lm(t(y)~z-1)$coef)
```

Step two is to put the $\mathbf{Y_z}$ into the function for the regression (6.17):

```
regr <- lm(yz~x-1)
summary(regr)
```

The summary function prints out the estimates, standard errors, and t-statistics, as well as other quantities, though not in a very compact form. The covariance matrix of the $\widehat{\boldsymbol{\beta}}$ is found using the vcov function, though it gives $\widehat{\boldsymbol{\Sigma}}_\mathbf{z} \otimes \mathbf{C_x}$ rather than the $\mathbf{C_x} \otimes \widehat{\boldsymbol{\Sigma}}_\mathbf{z}$ in (6.20). The function reverse.kronecker in Section A.2.2 will turn the former into the latter:

```
covbeta <- reverse.kronecker(vcov(regr),ncol(z),ncol(z))
```

The function takes the Kronecker product, as well as the dimensions of the left-hand matrix $\widehat{\boldsymbol{\Sigma}}_\mathbf{z}$, which in this case is 4×4.

6.4.3 Leprosy data

Recall the Leprosy example described in Exercise 4.4.6. There are three groups of ten observations each, corresponding to the three treatments of drug A, drug D, and placebo. The people were randomly assigned to the treatments. Each observation consists of a bacterial count before treatment, and another bacterial count after treatment. The $\mathbf{x}$ matrix has columns representing the overall mean, the drug vs. placebo effect, and the drug A vs. drug D effect. Thus the multivariate analysis of variance model, with the before and after counts as the two $\mathbf{Y}$ variables, is

$$\begin{aligned}\mathbf{Y} = \begin{pmatrix} \mathbf{Y}_b & \mathbf{Y}_a \end{pmatrix} &= \mathbf{x}\boldsymbol{\beta} + \mathbf{R} \\ &= \left[\begin{pmatrix} 1 & 1 & 1 \\ 1 & 1 & -1 \\ 1 & -2 & 0 \end{pmatrix} \otimes \mathbf{1}_{10} \right] \begin{pmatrix} \mu_b & \mu_a \\ \alpha_b & \alpha_a \\ \beta_b & \beta_a \end{pmatrix} + \mathbf{R}, \end{aligned} \tag{6.36}$$

where

$$Cov[\mathbf{R}] = \mathbf{I}_{30} \otimes \mathbf{\Sigma}_R = \mathbf{I}_{30} \otimes \begin{pmatrix} \sigma_{bb} & \sigma_{ba} \\ \sigma_{ab} & \sigma_{aa} \end{pmatrix}. \tag{6.37}$$

The R matrix leprosy is 30×3, with the first two columns consisting of the before and after bacterial counts, and the final column indicating the group. Because this is a multivariate regression problem, i.e., $\mathbf{z} = \mathbf{I}_q$, we need to use the linear regression function in Section 6.4.2 just once:

```
y <- leprosy[,1:2]
x <- kronecker(cbind(1,c(1,1,-2),c(1,-1,0)),rep(1,10))
summary(lm(y~x-1))
```

The results for $\widehat{\boldsymbol{\beta}}$:

	Estimates		Standard errors		t-statistics	
	Before	After	Before	After	Before	After
Mean	10.733	7.900	0.856	1.108	12.544	7.127
Drug effect	−1.083	−2.200	0.605	0.784	−1.791	−2.807
Drugs A vs. D	−0.350	−0.400	1.048	1.357	−0.334	−0.295

(6.38)

The degrees of freedom for the t-statistics are $n - p = 30 - 3 = 27$. The main interest is the effect of the treatments on the after-treatment bacterial counts. The t for the drug effect on the after counts (α_a) is -2.807, which yields a p-value of 0.009, suggesting that indeed the drugs are effective. The t for the drug A vs. drug D after-effect (β_a) is only about $-.3$, which is not significant at all. Interestingly, the before counts for the drug effect (drugs versus placebo) show a p-value of 0.08, which is almost significant. What this means is that the randomization, by chance, assigned less healthy people to the placebo group than to the drug groups. Thus one may wonder whether the positive effect we see for the drugs is due to the favorable randomization. This question we take up in the next section.

6.4.4 Covariates: Leprosy data

A covariate is a variable that is of (possibly) secondary importance, but is recorded because it might help adjust some estimates in order to make them more precise. In the leprosy example above, the before-treatment bacterial counts constitute the covariate.

These measurements are indications of health of the subjects before treatment, the higher the less healthy.

Note that the before counts should not be affected by the treatments, because they were measured before the treatment, and the people were randomly assigned to treatments. The implication for the model (6.37) is that the before parameters indicating drug effects, α_b and β_b, are zero in the population. That is,

$$\beta = \begin{pmatrix} \mu_b & \mu_a \\ 0 & \alpha_a \\ 0 & \beta_a \end{pmatrix}. \tag{6.39}$$

To take into account the before counts, we condition on them, that is, consider the conditional distribution of the after counts ($\mathbf{Y}_a$) given the before counts ($\mathbf{Y}_b$). Using equation 3.56 (with $\mathbf{Y}$ there being $\mathbf{Y}_a$ here, and $\mathbf{X}$ there being $\mathbf{Y}_b$ here), we obtain

$$\mathbf{Y}_a \mid \mathbf{Y}_b = \mathbf{y}_b \sim N_{30\times 1}(\boldsymbol{\alpha} + \mathbf{y}_b\eta, \mathbf{I}_{30} \times \sigma_{aa\cdot b}), \tag{6.40}$$

where $\eta = \sigma_{ab}/\sigma_{bb}$, and recalling (3.40),

$$\begin{aligned} \boldsymbol{\alpha} + \mathbf{y}_b\eta &= E[\mathbf{Y}_a] - E[\mathbf{Y}_b]\eta + \mathbf{y}_b\eta \\ &= \mathbf{x}\begin{pmatrix} \mu_a \\ \alpha_a \\ \beta_a \end{pmatrix} - \mathbf{x}\begin{pmatrix} \mu_b \\ 0 \\ 0 \end{pmatrix}\eta + \mathbf{y}_b\eta \\ &= \mathbf{x}\begin{pmatrix} \mu^* \\ \alpha_a \\ \beta_a \end{pmatrix} + \mathbf{y}_b\eta \\ &= (\mathbf{x}\ \mathbf{y}_b)\begin{pmatrix} \mu^* \\ \alpha_a \\ \beta_a \\ \eta \end{pmatrix}, \end{aligned} \tag{6.41}$$

where $\mu^* = \mu_a - \mu_b\eta$. Thus conditionally we have another linear model, this one with one $\mathbf{Y}$ vector and four $\mathbf{X}$ variables, instead of two $\mathbf{Y}$'s and three $\mathbf{X}$'s:

$$\mathbf{Y}_a = (\mathbf{x}\ \mathbf{y}_b)\begin{pmatrix} \mu^* \\ \alpha_a \\ \beta_a \\ \eta \end{pmatrix} + \mathbf{R}^*, \text{ where } \mathbf{R}^* \mid \mathbf{Y}_b = \mathbf{y}_b \sim N(\mathbf{0}, \mathbf{I}_n \otimes \sigma_{aa\cdot b}). \tag{6.42}$$

A key point here is that the parameters of interest, α_a and β_a, are still estimable in this model. The μ_a and μ_b are not estimable in this conditional model, though if we estimate μ_b by $\overline{\mathbf{y}}_b$, we can estimate μ_a by $\widehat{\mu}^* + \overline{\mathbf{y}}_b\widehat{\eta}$.

The calculations proceed as for any (multivariate) regression model, as in Theorem 6.1 but with $(\mathbf{x}, \mathbf{y}_b)$ in place of $\mathbf{x}$, and $\mathbf{y}_a$ in place of $\mathbf{y}$. The first column of leprosy has the before bacterial counts ($\mathbf{y}_b$), and the second column has the after counts ($\mathbf{y}_a$). The following code finds the estimates:

```
ya <- leprosy[,2]
xyb <- cbind(x,leprosy[,1])
summary(lm(ya~xyb-1))
```

Note that we have lost a degree of freedom, because there are now 4 columns in $(\mathbf{x}, \mathbf{y}_b)$.

The next tables have the comparisons of the original estimates and the covariate-adjusted estimates for α_a and β_a:

Original			
	Estimate	se	t
α_a	−2.200	0.784	−2.807
β_a	−0.400	1.357	−0.295

Covariate-adjusted			
	Estimate	se	t
α_a	−1.131	0.547	−2.067
β_a	−0.054	0.898	−0.061

(6.43)

The covariates helped the precision of the estimates, lowering the standard errors by about 30%. Also, the p-value for the after-treatment drug effect is 0.049, still significant, but just barely. Thus using the covariate presented a more defensible, though not necessarily more publishable, result.

6.4.5 Histamine in dogs

Turn to the example in Section 4.2.1 where sixteen dogs were split into four groups, and each had four histamine levels measured over the course of time.

In the model (4.16), the $\mathbf{x}$ part of the model is for a 2×2 ANOVA with interaction. We add a matrix $\mathbf{z}$ to obtain a both-sides model. The $\mathbf{z}$ we will take has a separate mean for the "before" measurements, because these are taken before any treatment, and a quadratic model for the three "after" time points:

$$\begin{aligned} \mathbf{Y} &= \mathbf{x}\boldsymbol{\beta}\mathbf{z}' + \mathbf{R} \\ &= \left[\begin{pmatrix} 1 & -1 & -1 & 1 \\ 1 & -1 & 1 & -1 \\ 1 & 1 & -1 & -1 \\ 1 & 1 & 1 & 1 \end{pmatrix} \otimes \mathbf{1}_4 \right] \begin{pmatrix} \mu_b & \mu_0 & \mu_1 & \mu_2 \\ \alpha_b & \alpha_0 & \alpha_1 & \alpha_2 \\ \beta_b & \beta_0 & \beta_1 & \beta_2 \\ \gamma_b & \gamma_0 & \gamma_1 & \gamma_2 \end{pmatrix} \\ &\qquad \times \begin{pmatrix} 1 & 0 & 0 & 0 \\ 0 & 1 & 1 & 1 \\ 0 & -1 & 0 & 1 \\ 0 & 1 & -2 & 1 \end{pmatrix} + \mathbf{R}. \end{aligned} \tag{6.44}$$

The μ's are for the overall mean of the groups, the α's for the drug effects, the β's for the depletion effect, and the γ's for the interactions. The "b" subscript indicates the before means, and the 0, 1, and 2 subscripts indicate the constant, linear, and quadratic terms of the growth curves. We first set up the design matrices, then use the two steps from Section 6.4.2, where the $\mathbf{Y}$ is the R matrix histamine.

```
x <- kronecker(cbind(1,c(-1,-1,1,1),c(-1,1,-1,1),c(1,-1,-1,1)),rep(1,4))
z <- cbind(c(1,0,0,0),c(0,1,1,1),c(0,-1,0,1),c(0,1,-2,1))
yz <- t(lm(t(histamine)~z-1)$coef)
summary(lm(yz~x-1))
```

The coefficients' estimates, standard errors, and t-statistics are below.

Estimates

	Before	Intercept	Linear	Quadratic
Mean	0.0769	0.3858	−0.1366	0.0107
Drug	−0.0031	0.1842	−0.0359	−0.0082
Depletion	0.0119	−0.2979	0.1403	−0.0111
Interaction	0.0069	−0.1863	0.0347	0.0078

Standard Errors

	Before	Intercept	Linear	Quadratic
Mean	0.0106	0.1020	0.0607	0.0099
Drug	0.0106	0.1020	0.0607	0.0099
Depletion	0.0106	0.1020	0.0607	0.0099
Interaction	0.0106	0.1020	0.0607	0.0099

(6.45)

t-statistics

	Before	Intercept	Linear	Quadratic
Mean	7.252	3.156	−2.248	1.086
Drug	−0.295	1.805	−0.592	−0.833
Depletion	1.120	−2.920	2.310	−1.128
Interaction	0.649	−1.826	0.571	0.791

Here $n = 16$ and $p = 4$, so the degrees of freedom in the t-statistics are 12. It looks like the quadratic terms are not needed, and that the basic assumption that the treatment effects for the before measurements is 0 is reasonable. The statistically significant effects are the intercept and linear effects for the mean and depletion effects, as well as possibly the intercept terms for the drug and interaction effects, with p-values of 0.08 and 0.07, respectively. See Figure 4.2 (bottom plot) for a plot of these effects. Chapter 7 deals with testing blocks of β_{ij}'s equal to zero, which may be more appropriate for these data. Exercise 6.6.13 considers using the before measurements as covariates.

6.5 Submodels of the both-sides model

As with any linear regression, we often wish to consider submodels of the main model, that is, set some of the β_{ij}'s to zero. For example, in the model (6.29) for the mouth sizes, the quadratic and cubic parts of the growth curves appear unnecessary, hence we could imagine setting $\beta_2 = \beta_3 = \delta_2 = \delta_3 = 0$. The model can then be written

$$\mathbf{Y} = \begin{pmatrix} \mathbf{1}_{11} & \mathbf{1}_{11} \\ \mathbf{1}_{16} & \mathbf{0}_{16} \end{pmatrix} \begin{pmatrix} \beta_0 & \beta_1 \\ \delta_0 & \delta_1 \end{pmatrix} \begin{pmatrix} 1 & 1 & 1 & 1 \\ -3 & -1 & 1 & 3 \end{pmatrix} + \mathbf{R}. \tag{6.46}$$

This is again a both-sides model. The $\mathbf{x}$ is the same as before, but the last two columns of the original $\mathbf{z}$ have been dropped, as well as the corresponding last two columns of $\boldsymbol{\beta}$. The calculations proceed as before. (In this case, because the columns of $\mathbf{z}$ are orthogonal, redoing the calculations is unnecessary, because they yield the same ones we saw in (6.32) through (6.35), dropping the quadratic and cubic quantities.)

In the histamine in dogs example of Section 6.4.5, it would be reasonable to try the model without the drug and interaction effects, which involves dropping the second and fourth columns of the $\mathbf{x}$, and the second and fourth rows of the $\boldsymbol{\beta}$. Furthermore, we might wish to drop the quadratic terms from the $\mathbf{z}$. In either case, we are left with a straightforward both-sides model.

In the above submodels, the nonzero β_{ij}'s formed a rectangle. Other submodels do not leave a rectangle, hence the model cannot be defined simply by dropping columns from the $\mathbf{x}$ and or $\mathbf{z}$. For example, in the mouth size data, consider the model where the two growth curves are parallel straight lines. Then $\delta_1 = 0$ in (6.46), so that

$$\boldsymbol{\beta} = \begin{pmatrix} \beta_0 & \beta_1 \\ \delta_0 & 0 \end{pmatrix}. \tag{6.47}$$

The usual formula (6.2) will not give us the least squares estimate of $\boldsymbol{\beta}$, even if we set the $\widehat{\delta}_1 = 0$. Instead, we first find $\mathbf{Y_z} = \mathbf{Yz}(\mathbf{z}'\mathbf{z})^{-1}$ as in (6.16) and (6.17), then calculate a separate regression for each column, where for the second column we use only the first column of $\mathbf{x}$, so that

$$\mathbf{Y}_{\mathbf{z}1} = \begin{pmatrix} \mathbf{1}_{11} & \mathbf{1}_{11} \\ \mathbf{1}_{16} & \mathbf{0}_{16} \end{pmatrix} \begin{pmatrix} \beta_0 \\ \delta_0 \end{pmatrix} + \mathbf{R}_{\mathbf{z}1} \text{ and } \mathbf{Y}_{\mathbf{z}2} = \begin{pmatrix} \mathbf{1}_{11} \\ \mathbf{1}_{16} \end{pmatrix} \beta_1 + \mathbf{R}_{\mathbf{z}2}. \tag{6.48}$$

Then to find the estimates, use the x and y from Section 6.4.1, and

```
x <- cbind(1,mouths[,5])
y <- mouths[,1:4]
z <- cbind(1,c(-3,-1,1,3))
yz <- t(lm(t(y)~z-1)$coef)
regr1 <- lm(yz[,1]~x-1)
regr2 <- lm(yz[,2]~x[,1]-1) # or lm(yz[,2]~1)
beta <- matrix(0,ncol=2,nrow=2) # blank beta
beta[,1] <- regr1$coef # fill in the first column
beta[1,2] <- regr2$coef #fill in the second column, first row
```

The result is

$$\widehat{\boldsymbol{\beta}} = \begin{pmatrix} 24.969 & 0.660 \\ -2.321 & 0 \end{pmatrix}. \tag{6.49}$$

The standard errors can also be found in the regression structures, using the vcov function, which extracts the covariance matrix of the coefficients. We just need to take the square roots of the diagonals, then find the t-statistics:

```
tt <- se <- matrix(0,ncol=2,nrow=2)
se[,1] <- sqrt(diag(vcov(regr1)))
se[1,2] <- sqrt(diag(vcov(regr2)))
tt[,1] <- beta[,1]/se[,1]
tt[1,2] <- beta[1,2]/se[1,2]
```

The answers are

	Standard errors		t-statistics	
	Constant	Linear	Constant	Linear
Boys	0.4860	0.0713	51.38	9.27
Girls−Boys	0.7614		−3.05	

(6.50)

The results for the first column of coefficients are the same as in (6.35), but the estimate and standard error of β_1 have changed slightly. In either case, the t is strongly significant, which just means that the slopes are not zero. Note that the degrees of freedom for the t are different for the two columns. For the first column of coefficients they are $n-2=25$, and for the second column they are $n-1=26$.

In general, we consider submodels of the both-sides model that set a subset of the β_{ij}'s to zero. To specify such restrictions on the $p \times l$ matrix $\boldsymbol{\beta}$, we define a "pattern" $\mathbf{P}$, which is also a $p \times l$ matrix, but with just 0's and 1's as entries indicating which elements are allowed to be nonzero. That is, for given pattern $\mathbf{P}$, the set of such $\boldsymbol{\beta}$'s is defined to be

$$\mathcal{B}(\mathbf{P}) = \{\boldsymbol{\beta} \in \mathbb{R}^{p\times l} \mid \beta_{ij} = 0 \text{ if } p_{ij} = 0, \beta_{ij} \in \mathbb{R} \text{ if } p_{ij} = 1\}. \tag{6.51}$$

Then given the pattern $\mathbf{P}$, the model is

$$\mathbf{Y} = \mathbf{x}\boldsymbol{\beta}\mathbf{z}' + \mathbf{R}, \quad \boldsymbol{\beta} \in \mathcal{B}(\mathbf{P}), \tag{6.52}$$

where $\mathbf{R} \sim N(\mathbf{0}, \mathbf{I}_n \otimes \boldsymbol{\Sigma}_R)$ as usual. For example, the model (6.46) can be obtained from the big model in (6.29) with

$$\mathbf{P} = \begin{pmatrix} 1 & 1 & 0 & 0 \\ 1 & 1 & 0 & 0 \end{pmatrix}. \tag{6.53}$$

Other examples, paralleling the $\boldsymbol{\beta}$'s in (6.41) and (6.47), respectively, are

$$\mathbf{P} = \begin{pmatrix} 1 & 1 \\ 0 & 1 \\ 0 & 1 \end{pmatrix} \text{ and } \mathbf{P} = \begin{pmatrix} 1 & 1 \\ 1 & 0 \end{pmatrix}. \tag{6.54}$$

The model (6.52) can be fit as above, where each column of $\mathbf{Y_z}$ is fit using some submatrix of $\mathbf{x}$. The jth column uses

```
lm(yz[,j]~x[,pattern[,j]==1]−1)
```

where here pattern has the pattern. This approach is fine for obtaining the coefficients' estimates and standard errors, but it takes some extra work to find an unbiased estimator of $\boldsymbol{\Sigma}_\mathbf{z}$ and the entire $Cov[\widehat{\boldsymbol{\beta}}]$. See Exercises 6.6.7 and 6.6.8. Or, just use the bothsidesmodel function in Section A.2.1, including the pattern in the list of arguments. So the model in (6.47) can be fit using

```
pattern <− cbind(c(1,1),c(1,0))
bothsidesmodel(x,y,z,pattern)
```

6.6 Exercises

Exercise 6.6.1. Verify the calculations in (6.11).

Exercise 6.6.2. Show (6.17).

Exercise 6.6.3. Use (6.25) and (6.26) in (6.24) to show that the T in (6.23) has the Student's t distribution..

Exercise 6.6.4. Prove (6.30).

Exercise 6.6.5. Verify the equations in (6.41).

Exercise 6.6.6 (Bayesian inference). This exercise extends the Bayesian results in Exercises 3.7.30 and 3.7.31 to the $\boldsymbol{\beta}$ in multivariate regression. We start with the estimator $\widehat{\boldsymbol{\beta}}$ in (6.3), where the $\mathbf{z} = \mathbf{I}_q$, hence $\boldsymbol{\Sigma}_\mathbf{z} = \boldsymbol{\Sigma}_R$. The model is then

$$\widehat{\boldsymbol{\beta}} \mid \boldsymbol{\beta} = \mathbf{b} \sim N_{p\times q}(\mathbf{b}, (\mathbf{x}'\mathbf{x})^{-1} \otimes \boldsymbol{\Sigma}_R) \text{ and } \boldsymbol{\beta} \sim N_{p\times q}(\boldsymbol{\beta}_0, \mathbf{K}_0^{-1} \otimes \boldsymbol{\Sigma}_R), \tag{6.55}$$

where $\boldsymbol{\Sigma}_R$, $\boldsymbol{\beta}_0$, and $\mathbf{K}_0$ are known. Note that the $\boldsymbol{\Sigma}_R$ matrix appears in the prior, which makes the posterior tractable. (a) Show that the posterior distribution of $\boldsymbol{\beta}$ is multivariate normal, with

$$E[\boldsymbol{\beta} \mid \widehat{\boldsymbol{\beta}} = \widehat{\mathbf{b}}] = (\mathbf{x}'\mathbf{x} + \mathbf{K}_0)^{-1}((\mathbf{x}'\mathbf{x})\widehat{\mathbf{b}} + \mathbf{K}_0\boldsymbol{\beta}_0), \tag{6.56}$$

and

$$Cov[\boldsymbol{\beta} \mid \widehat{\boldsymbol{\beta}} = \widehat{\mathbf{b}}] = (\mathbf{x}'\mathbf{x} + \mathbf{K}_0)^{-1} \otimes \boldsymbol{\Sigma}_R. \tag{6.57}$$

[Hint: Same hint as in Exercise 3.7.31.] (b) Set the prior parameters $\boldsymbol{\beta}_0 = \mathbf{0}$ and $\mathbf{K}_0 = k_0\mathbf{I}_p$ for some $k_0 > 0$. Show that

$$E[\boldsymbol{\beta} \mid \widehat{\boldsymbol{\beta}} = \widehat{\mathbf{b}}] = (\mathbf{x}'\mathbf{x} + k_0\mathbf{I}_p)^{-1}\mathbf{x}'\mathbf{y}. \tag{6.58}$$

This conditional mean is the **ridge regression** estimator of $\boldsymbol{\beta}$. See Hoerl and Kennard [1970]. This estimator can be better than the least squares estimator (a little biased, but much less variable) when $\mathbf{x}'\mathbf{x}$ is nearly singular, that is, one or more of its eigenvalues are close to zero.

Exercise 6.6.7. Consider the submodel (6.52) in the multivariate regression case, i.e., $\mathbf{z} = \mathbf{I}_q$. Write the model for the jth column of $\mathbf{Y}$ as $\mathbf{Y}_j = \mathbf{x}_{(j)}\boldsymbol{\beta}_j + \mathbf{R}_j$, where $\boldsymbol{\beta}_j$ contains just the nonzero coefficients in the jth column of $\boldsymbol{\beta}$, and $\mathbf{x}_{(j)}$ has the corresponding columns of $\mathbf{x}$. For example, suppose $\boldsymbol{\beta}$ is 4×3,

$$\mathbf{P} = \begin{pmatrix} 1 & 1 & 0 \\ 1 & 0 & 1 \\ 1 & 0 & 1 \\ 0 & 1 & 0 \end{pmatrix} \text{ and } \mathbf{x} = (\mathbf{x}_1 \ \mathbf{x}_2 \ \mathbf{x}_3 \ \mathbf{x}_4). \tag{6.59}$$

Then

$$\begin{aligned} \mathbf{x}_{(1)} &= (\mathbf{x}_1 \ \mathbf{x}_2 \ \mathbf{x}_3) \text{ and } \boldsymbol{\beta}_1 = \begin{pmatrix} \beta_{11} \\ \beta_{21} \\ \beta_{31} \end{pmatrix}; \\ \mathbf{x}_{(2)} &= (\mathbf{x}_1 \ \mathbf{x}_4) \text{ and } \boldsymbol{\beta}_2 = \begin{pmatrix} \beta_{12} \\ \beta_{42} \end{pmatrix}; \\ \mathbf{x}_{(3)} &= (\mathbf{x}_2 \ \mathbf{x}_3) \text{ and } \boldsymbol{\beta}_3 = \begin{pmatrix} \beta_{23} \\ \beta_{33} \end{pmatrix}. \end{aligned} \tag{6.60}$$

With $\widehat{\boldsymbol{\beta}}_j = (\mathbf{x}'_{(j)}\mathbf{x}_{(j)})^{-1}\mathbf{x}'_{(j)}\mathbf{y}_j$, show that

$$Cov[\widehat{\boldsymbol{\beta}}_j, \widehat{\boldsymbol{\beta}}_k] = \sigma_{jk}\,(\mathbf{x}'_{(j)}\mathbf{x}_{(j)})^{-1}\mathbf{x}'_{(j)}\mathbf{x}_{(k)}(\mathbf{x}'_{(k)}\mathbf{x}_{(k)})^{-1}, \tag{6.61}$$

where σ_{jk} is the $(jk)^{th}$ element of $\boldsymbol{\Sigma}_R$. [Hint: Recall (2.107).]

Exercise 6.6.8. Continue with the setup in Exercise 6.6.7. The residual for the jth column of $\mathbf{y}_j$ is $\widehat{\mathbf{R}}_j = \mathbf{Q}_{\mathbf{x}_{(j)}}\mathbf{y}_j$ as in (6.6) and (6.7). (a) Show that

$$Cov[\widehat{\mathbf{R}}_j, \widehat{\mathbf{R}}_k] = \sigma_{jk}\, \mathbf{Q}_{\mathbf{x}_{(j)}} \mathbf{Q}_{\mathbf{x}_{(k)}}. \tag{6.62}$$

(b) Show that consequently

$$E[\widehat{\mathbf{R}}_j' \widehat{\mathbf{R}}_k] = \sigma_{jk}\ \text{trace}(\mathbf{Q}_{\mathbf{x}_{(j)}} \mathbf{Q}_{\mathbf{x}_{(k)}}). \tag{6.63}$$

(c) What does the $(jk)^{th}$ element of the matrix $\widehat{\mathbf{R}}'\widehat{\mathbf{R}}$ have to be divided by in order to obtain an unbiased estimator of $\mathbf{\Sigma}_R$? To answer the question, let p_j be the number of components of $\boldsymbol{\beta}_j$. Show that

$$\text{trace}(\mathbf{Q}_{\mathbf{x}_{(j)}} \mathbf{Q}_{\mathbf{x}_{(k)}}) = n - p_j - p_k + \text{trace}((\mathbf{x}_{(j)}'\mathbf{x}_{(j)})^{-1}\mathbf{x}_{(j)}'\mathbf{x}_{(k)}(\mathbf{x}_{(k)}'\mathbf{x}_{(k)})^{-1}\mathbf{x}_{(k)}'\mathbf{x}_{(j)}). \tag{6.64}$$

[Hint: Write the $\mathbf{Q}$'s as $\mathbf{I}_n - \mathbf{P}$'s, then multiply out the matrices.]

Exercise 6.6.9 (Prostaglandin). Continue with the data described in Exercise 4.4.1. The data are in the R matrix prostaglandin. Consider the both-sides model (6.1), where the ten people have the same mean, so that $\mathbf{x} = \mathbf{1}_{10}$, and $\mathbf{z}$ contains the cosine and sine vectors for $m = 1, 2$ and 3, as in Exercise 4.4.8. (Thus $\mathbf{z}$ is 6×6.) (a) What is $\mathbf{z}$? (b) Are the columns of $\mathbf{z}$ orthogonal? What are the squared norms of the columns? (c) Find $\widehat{\boldsymbol{\beta}}$. (d) Find $\widehat{\mathbf{\Sigma}}_z$. (e) Find the (estimated) standard errors of the $\widehat{\beta}_j$'s. (f) Find the t-statistics for the β_j's. (g) Based on the t-statistics, which model appears most appropriate? Choose from the constant model; the one-cycle model (just m=1); the model with one cycle and two cycles; the model with one, two and three cycles.

Exercise 6.6.10 (Skulls). This question continues with the data described in Exercise 4.4.2. The data are in the R matrix skulls, obtained from `http://lib.stat.cmu.edu/DASL/Datafiles/EgyptianSkulls.html` at DASL Project [1996]. The $\mathbf{Y} \sim N(\mathbf{x}\boldsymbol{\beta}, \mathbf{I}_n \otimes \mathbf{\Sigma}_R)$, where the $\mathbf{x}$ represents the orthogonal polynomials over time periods (from Exercise 5.8.51). (a) Find $\widehat{\boldsymbol{\beta}}$. (b) Find $(\mathbf{x}'\mathbf{x})^{-1}$. (c) Find $\widehat{\mathbf{\Sigma}}_R$. What are the degrees of freedom? (d) Find the standard errors of the $\widehat{\beta}_{ij}$'s. (e) Which of the $\widehat{\beta}_{ij}$'s have t-statistic larger than 2 in absolute value? (Ignore the first row, since those are the overall means.) (f) Explain what the parameters with $|t| > 2$ are measuring. (g) There is a significant linear trend for which measurements? (h) There is a significant quadratic trend for which measurements?

Exercise 6.6.11 (Caffeine). Consider the caffeine data (in the R matrix caffeine) and the model from Exercise 5.8.52. (a) Fit the model, and find the relevant estimates. (b) Find the t-statistics for the $\widehat{\beta}_{ij}$'s. (c) What do you conclude? (Choose as many conclusions as appropriate from the following: On average the students do about the same with or without caffeine; on average the students do significantly better without caffeine; on average the students do significantly better with caffeine; the older students do about the same as the younger ones on average; the older students do significantly better than the younger ones on average; the older students do significantly worse than the younger ones on average; the deleterious effects of caffeine are not significantly different for the older students than for the younger; the deleterious effects of caffeine are significantly greater for the older students than for the younger; the deleterious effects of caffeine are significantly greater for the younger students than for the older; the quadratic effects are not significant.)

Exercise 6.6.12 (Grades). Consider the grades data in (4.11). Let $\mathbf{Y}$ be the 107×5 matrix consisting of the variables homework, labs, inclass, midterms, and final. The $\mathbf{x}$ matrix indicates gender. Let the first column of $\mathbf{x}$ be $\mathbf{1}_n$. There are 70 women and 37 men in the class, so let the second column have 0.37 for the women and -0.70 for the men. (That way, the columns of $\mathbf{x}$ are orthogonal.) For the $\mathbf{z}$, we want the overall mean score; a contrast between the exams (midterms and final) and other scores (homework, labs, inclass); a contrast between (homework, labs) and inclass; a contrast between homework and labs; and a contrast between midterms and final. Thus

$$\mathbf{z}' = \begin{pmatrix} 1 & 1 & 1 & 1 & 1 \\ 2 & 2 & 2 & -3 & -3 \\ 1 & 1 & -2 & 0 & 0 \\ 1 & -1 & 0 & 0 & 0 \\ 0 & 0 & 0 & 1 & -1 \end{pmatrix}. \tag{6.65}$$

Let

$$\boldsymbol{\beta} = \begin{pmatrix} \beta_1 & \beta_2 & \beta_3 & \beta_4 & \beta_5 \\ \delta_1 & \delta_2 & \delta_3 & \delta_4 & \delta_5 \end{pmatrix}. \tag{6.66}$$

(a) Briefly describe what each of the parameters represents. (b) Find $\widehat{\boldsymbol{\beta}}$. (c) Find the standard errors of the $\widehat{\beta}_{ij}$'s. (d) Which of the parameters have $|t|$ over 2? (e) Based on the results in part (d), discuss whether there is any difference between the grade profiles of the men and women.

Exercise 6.6.13 (Histamine). Continue with the histamine in dogs example in Section 6.4.5. Redo the analysis, but using the before measurements (the first column of $\mathbf{Y}$) as a covariate. (a) Find the standard errors of the $(\alpha_i, \beta_i, \gamma_i), i = 0, 1, 2$, in the covariate model, and compare them to the corresponding standard errors in (6.45). (b) Which of those effects are statistically significant? Is the overall conclusion different than in Section 6.4.5?

Chapter 7

Both-Sides Models: Hypothesis Tests on $\boldsymbol{\beta}$

Once again the model for this section is the both-sides model, $\mathbf{Y} = \mathbf{x}\boldsymbol{\beta}\mathbf{z}' + \mathbf{R}$, which we transform to a multivariate regression by shifting the $\mathbf{z}$ to the $\mathbf{Y}$ as in (6.16) and (6.17). That is, $\mathbf{Y_z} = \mathbf{Yz}(\mathbf{z}'\mathbf{z})^{-1}$, and

$$\mathbf{Y_z} = \mathbf{x}\boldsymbol{\beta} + \mathbf{R_z}, \;\; \mathbf{R_z} \sim N(\mathbf{0}, \mathbf{I}_n \otimes \boldsymbol{\Sigma}_\mathbf{z}), \tag{7.1}$$

where $\mathbf{Y_z}$ is $n \times l$, $\mathbf{x}$ is $n \times p$, and $\boldsymbol{\beta}$ is $p \times l$.

Testing a single $\beta_{ij} = 0$ is easy using the t-test as in Section 6.3. It is often informative to test a set of β_{ij}'s is 0, e.g., a row from $\boldsymbol{\beta}$, or a column, or a block, or some other configuration. In Section 7.1, we present a general test statistic and its χ^2 approximation for testing any set of parameters equals zero. Section 7.2 refines the test statistic when the set of β_{ij}'s of interest is a block. Likelihood ratio tests are covered in Section 9.4. Section 7.5 covers Mallows' C_p statistic, which yields an approach to choosing among several models, rather than just two models as in traditional hypothesis tests. Some likelihood-based selection procedures are presented in Section 9.5.

7.1 Approximate χ^2 test

Start by placing the parameters of interest in the $1 \times K$ vector $\boldsymbol{\theta}$. We assume we have a vector of estimates $\widehat{\boldsymbol{\theta}}$ such that

$$\widehat{\boldsymbol{\theta}} \sim N(\boldsymbol{\theta}, \boldsymbol{\Omega}), \tag{7.2}$$

and we wish to test

$$H_0 : \boldsymbol{\theta} = \mathbf{0}. \tag{7.3}$$

We could test whether the vector equals a fixed nonzero vector, but then we can subtract that hypothesized value from $\boldsymbol{\theta}$ and return to the case (7.3). Assuming $\boldsymbol{\Omega}$ is invertible, we have that under H_0,

$$\widehat{\boldsymbol{\theta}}\boldsymbol{\Omega}^{-1/2} \sim N(\mathbf{0}, \mathbf{I}_K), \tag{7.4}$$

hence

$$\widehat{\boldsymbol{\theta}}\boldsymbol{\Omega}^{-1}\widehat{\boldsymbol{\theta}}' \sim \chi^2_K. \tag{7.5}$$

Typically, $\mathbf{\Omega}$ will have to be estimated, in which case we use

$$T^2 \equiv \widehat{\boldsymbol{\theta}}\widehat{\mathbf{\Omega}}^{-1}\widehat{\boldsymbol{\theta}}', \tag{7.6}$$

where under appropriate conditions (e.g., large sample size relative to K), under H_0,

$$T^2 \approx \chi^2_K. \tag{7.7}$$

7.1.1 Example: Mouth sizes

In the mouth size example in Section 6.4.1, consider testing the fit of the model specifying parallel straight lines for the boys and girls, with possibly differing intercepts. Then in the model (6.29), only β_0, β_1, and δ_0 would be nonzero, so we would be testing whether the other five are zero. Place those in the vector $\boldsymbol{\theta}$:

$$\boldsymbol{\theta} = (\beta_2, \beta_3, \delta_1, \delta_2, \delta_3). \tag{7.8}$$

The estimate is

$$\widehat{\boldsymbol{\theta}} = (0.203, -0.056, -0.305, -0.214, 0.072). \tag{7.9}$$

To find the estimated covariance matrix $\widehat{\mathbf{\Omega}}$, we need to pick off the relevant elements of the matrix $\mathbf{C_x} \otimes \widehat{\mathbf{\Sigma}}_z$ using the values in (6.34). In terms of $\text{row}(\boldsymbol{\beta})$, we are interested in elements 3, 4, 6, 7, and 8. Continuing the R work from Section 6.4.1, we have

```
omegahat <- kronecker(cx,sigmaz)[c(3,4,6,7,8),c(3,4,6,7,8)]
```

so that

$$\widehat{\mathbf{\Omega}} = \begin{pmatrix} 0.01628 & -0.00036 & 0.00314 & -0.01628 & 0.00036 \\ -0.00036 & 0.00786 & -0.00057 & 0.00036 & -0.00786 \\ 0.00314 & -0.00057 & 0.01815 & -0.00770 & 0.00139 \\ -0.01628 & 0.00036 & -0.00770 & 0.03995 & -0.00088 \\ 0.00036 & -0.00786 & 0.00139 & -0.00088 & 0.01930 \end{pmatrix}. \tag{7.10}$$

The statistic (7.7) is

$$T^2 = \widehat{\boldsymbol{\theta}}\widehat{\mathbf{\Omega}}^{-1}\widehat{\boldsymbol{\theta}}' = 10.305. \tag{7.11}$$

The degrees of freedom $K = 5$, which yields an approximate p-value of 0.067, borderline significant. Judging from the individual t-statistics, the $\widehat{\beta}_{22}$ element, indicating a difference in slopes, may be the reason for the almost-significance.

7.2 Testing blocks of $\boldsymbol{\beta}$ are zero

In this section, we focus on blocks in the both-sides model, for which we can find a better approximation to the distribution of T^2, or in some cases the exact distribution. A block $\boldsymbol{\beta}^*$ of the $p \times l$ matrix $\boldsymbol{\beta}$ is a $p^* \times l^*$ rectangular (though not necessarily contiguous) submatrix of $\boldsymbol{\beta}$. For example, if $\boldsymbol{\beta}$ is 5×4, we might have $\boldsymbol{\beta}^*$ be the 3×2 submatrix

$$\boldsymbol{\beta}^* = \begin{pmatrix} \beta_{11} & \beta_{13} \\ \beta_{41} & \beta_{43} \\ \beta_{51} & \beta_{53} \end{pmatrix}, \tag{7.12}$$

which uses the rows 1, 4, and 5, and the columns 1 and 3. Formally, a block is defined using a subset of the row indices for $\boldsymbol{\beta}$, $\mathcal{I} \subset \{1, 2, \ldots, p\}$, and a subset of the column indices, $\mathcal{J} \subset \{1, 2, \ldots, l\}$. Then $\boldsymbol{\beta}^*$ is the $\#\mathcal{I} \times \#\mathcal{J}$ matrix containing elements β_{ij}, for $i \in \mathcal{I}, j \in \mathcal{J}$.

Consider testing the hypothesis

$$H_0 : \boldsymbol{\beta}^* = \mathbf{0}. \tag{7.13}$$

Using (6.20) for the covariance, the corresponding estimate of $\boldsymbol{\beta}^*$ has distribution

$$\widehat{\boldsymbol{\beta}}^* \sim N_{p^* \times l^*}(\boldsymbol{\beta}^*, \mathbf{C}_{\mathbf{x}}^* \otimes \boldsymbol{\Sigma}_{\mathbf{z}}^*), \tag{7.14}$$

where $\mathbf{C}_{\mathbf{x}}^*$ and $\boldsymbol{\Sigma}_{\mathbf{z}}^*$ are the appropriate $p^* \times p^*$ and $l^* \times l^*$ submatrices of, respectively, $\mathbf{C}_x$ and $\boldsymbol{\Sigma}_{\mathbf{z}}$. In the example (7.12),

$$\mathbf{C}_{\mathbf{x}}^* = \begin{pmatrix} C_{\mathbf{x}11} & C_{\mathbf{x}14} & C_{\mathbf{x}15} \\ C_{\mathbf{x}41} & C_{\mathbf{x}44} & C_{\mathbf{x}45} \\ C_{\mathbf{x}51} & C_{\mathbf{x}54} & C_{\mathbf{x}55} \end{pmatrix} \text{ and } \boldsymbol{\Sigma}_{\mathbf{z}}^* = \begin{pmatrix} \sigma_{\mathbf{z}11} & \sigma_{\mathbf{z}13} \\ \sigma_{\mathbf{z}31} & \sigma_{\mathbf{z}33} \end{pmatrix}. \tag{7.15}$$

Also, letting $\widehat{\boldsymbol{\Sigma}}_{\mathbf{z}}^*$ be the corresponding submatrix of (6.19), we have

$$\mathbf{W} \equiv \nu\, \widehat{\boldsymbol{\Sigma}}_{\mathbf{z}}^* \sim \text{Wishart}_{l^*}(\nu, \boldsymbol{\Sigma}_{\mathbf{z}}^*),\ \nu = n - p. \tag{7.16}$$

We take

$$\boldsymbol{\theta} = \text{row}(\boldsymbol{\beta}^*) \text{ and } \boldsymbol{\Omega} = \mathbf{C}_{\mathbf{x}}^* \otimes \boldsymbol{\Sigma}_{\mathbf{z}}^*, \tag{7.17}$$

and $\widehat{\boldsymbol{\theta}}, \widehat{\boldsymbol{\Omega}}$ as the obvious estimates. Then using (3.32c) and (3.32d), we have that

$$\begin{aligned} T^2 &= \text{row}(\widehat{\boldsymbol{\beta}}^*)(\mathbf{C}_{\mathbf{x}}^* \otimes \widehat{\boldsymbol{\Sigma}}_{\mathbf{z}}^*)^{-1}\, \text{row}(\widehat{\boldsymbol{\beta}}^*)' \\ &= \text{row}(\widehat{\boldsymbol{\beta}}^*)(\mathbf{C}_{\mathbf{x}}^{*-1} \otimes \widehat{\boldsymbol{\Sigma}}_{\mathbf{z}}^{*-1})\, \text{row}(\widehat{\boldsymbol{\beta}}^*)' \\ &= \text{row}(\mathbf{C}_{\mathbf{x}}^{*-1} \widehat{\boldsymbol{\beta}}^* \widehat{\boldsymbol{\Sigma}}_{\mathbf{z}}^{*-1})\, \text{row}(\widehat{\boldsymbol{\beta}}^*)' \\ &= \text{trace}(\widehat{\boldsymbol{\beta}}^{*\prime} \mathbf{C}_{\mathbf{x}}^{*-1} \widehat{\boldsymbol{\beta}}^* \widehat{\boldsymbol{\Sigma}}_{\mathbf{z}}^{*-1}), \end{aligned} \tag{7.18}$$

where the final equation uses (5.96).

To clean up the notation a bit, we write

$$T^2 = \nu\, \text{trace}(\mathbf{W}^{-1}\mathbf{B}), \tag{7.19}$$

where $\mathbf{W}$ is given in (7.16), and

$$\mathbf{B} = \widehat{\boldsymbol{\beta}}^{*\prime} \mathbf{C}_{\mathbf{x}}^{*-1} \widehat{\boldsymbol{\beta}}^*. \tag{7.20}$$

Thus by Theorem 6.1, $\mathbf{B}$ and $\mathbf{W}$ are independent, and by (7.14) and (7.16), under H_0,

$$\mathbf{B} \sim \text{Wishart}_{l^*}(p^*, \boldsymbol{\Sigma}_{\mathbf{z}}^*) \text{ and } \mathbf{W} \sim \text{Wishart}_{l^*}(\nu, \boldsymbol{\Sigma}_{\mathbf{z}}^*). \tag{7.21}$$

In multivariate analysis of variance, we usually call $\mathbf{B}$ the "between-group" sum of squares and cross-products matrix, and $\mathbf{W}$ the "within-group" matrix. See Section 7.2.5. The test based on T^2 in (7.19) is called the **Lawley-Hotelling trace test**, where the statistic is usually defined to be T^2/ν.

We could use the chi-squared approximation (7.7) with $K = p^* l^*$, but we do a little better with the approximation

$$F \equiv \frac{\nu - l^* + 1}{\nu p^* l^*} T^2 \approx F_{p^* l^*, \nu - l^* + 1}. \tag{7.22}$$

Recall the following definition of $F_{\mu,\nu}$.

Definition 7.1. *If $B \sim \chi^2_\mu$ and $W \sim \chi^2_\nu$, and B and W are independent, then*

$$F = \frac{\nu}{\mu} \frac{B}{W}, \tag{7.23}$$

an F distribution with degrees of freedom μ and ν.

When $p^* = 1$ or $l^* = 1$, so that we are testing elements within a single row or column, the distribution in (7.22) is exact. In fact, when we are testing just one β_{ij} ($p^* = l^* = 1$), the T^2 is the square of the usual t-statistic, hence distributed $F_{1,\nu}$. In other cases, at least the mean of the test statistic matches that of the F. The rest of this section verifies these statements.

7.2.1 Just one column: F test

Suppose $l^* = 1$, so that B and W are independent scalars, and from (7.21),

$$B \sim \sigma^{*2}_{\mathbf{z}} \chi^2_{p^*} \text{ and } W \sim \sigma^{*2}_{\mathbf{z}} \chi^2_\nu. \tag{7.24}$$

Then $T^2 = \nu B/W$, and the constant multiplying T^2 in (7.22) is simply $1/p^*$, so that

$$F = \frac{\nu}{p^*} \frac{B}{W} \sim F_{p^*,\nu}. \tag{7.25}$$

This is the classical problem in multiple (univariate) regression, and this test is the regular F test.

7.2.2 Just one row: Hotelling's T^2

Now $p^* = 1$, so that $\widehat{\boldsymbol{\beta}}^*$ is $1 \times l^*$ and $C^*_{\mathbf{x}}$ is a scalar. Thus we can write

$$\mathbf{B} = \mathbf{Z}'\mathbf{Z}, \quad \mathbf{Z} \sim N_{1\times l^*}(\mathbf{0}, \boldsymbol{\Sigma}^*_z), \tag{7.26}$$

where

$$\mathbf{Z} = \widehat{\boldsymbol{\beta}}^* / \sqrt{C^*_{\mathbf{x}}}. \tag{7.27}$$

From (7.19), T^2 can be variously written

$$T^2 = \nu \operatorname{trace}(\mathbf{W}^{-1}\mathbf{Z}'\mathbf{Z}) = \nu\, \mathbf{Z}\mathbf{W}^{-1}\mathbf{Z}' = \frac{\widehat{\boldsymbol{\beta}}^* \boldsymbol{\Sigma}^{*-1}_{\mathbf{z}} \widehat{\boldsymbol{\beta}}^{*\prime}}{C^*_{\mathbf{x}}}. \tag{7.28}$$

In this situation, the statistic is called **Hotelling's T^2**. The next proposition shows that the distribution of the F version of Hotelling's T^2 in (7.22) is exact, setting $p^* = 1$. The proof of the proposition is in Section 8.4.

Proposition 7.1. *Suppose* $\mathbf{W}$ *and* $\mathbf{Z}$ *are independent,* $\mathbf{W} \sim \text{Wishart}_{l^*}(\nu, \boldsymbol{\Sigma})$ *and* $\mathbf{Z} \sim N_{1\times l^*}(\mathbf{0}, \boldsymbol{\Sigma})$*, where* $\nu \geq l^*$ *and* $\boldsymbol{\Sigma}$ *is invertible. Then*

$$\frac{\nu - l^* + 1}{l^*}\, \mathbf{Z}\mathbf{W}^{-1}\mathbf{Z}' \sim F_{l^*,\nu-l^*+1}. \tag{7.29}$$

7.2.3 General blocks

In this section we verify that the expected values of the two sides of (7.22) are the same. It is not hard to show that $E[\chi^2_\mu] = \mu$ and $E[1/\chi^2_\nu] = 1/(\nu-2)$ if $\nu > 2$. Thus by the definition of F in (7.23), independence yields

$$E[F_{\mu,\nu}] = \frac{\nu}{\nu - 2} \tag{7.30}$$

if $\nu > 2$. Otherwise, the expected value is $+\infty$. For T^2 in (7.19) and (7.21), again by independence of $\mathbf{B}$ and $\mathbf{W}$,

$$E[T^2] = \nu\, \text{trace}(E[\mathbf{W}^{-1}]\; E[\mathbf{B}]) = \nu p^*\, \text{trace}(E[\mathbf{W}^{-1}]\; \boldsymbol{\Sigma}^*_{\mathbf{z}}), \tag{7.31}$$

because $E[\mathbf{B}] = p^*\boldsymbol{\Sigma}^*_{\mathbf{z}}$ by (3.74). To finish, we need the following lemma, which extends the results on $E[1/\chi^2_\nu]$.

Lemma 7.1. *If* $\mathbf{W} \sim \text{Wishart}_{l^*}(\nu, \boldsymbol{\Sigma})$, $\nu > l^* + 1$, *and* $\boldsymbol{\Sigma}$ *is invertible,*

$$E[\mathbf{W}^{-1}] = \frac{1}{\nu - l^* - 1}\, \boldsymbol{\Sigma}^{-1}. \tag{7.32}$$

The proof is in Section 8.3. Continuing from (7.31),

$$E[T^2] = \frac{\nu p^*}{\nu - l^* - 1}\, \text{trace}(\boldsymbol{\Sigma}^{*-1}_{\mathbf{z}}\; \boldsymbol{\Sigma}^*_{\mathbf{z}}) = \frac{\nu p^* l^*}{\nu - l^* - 1}. \tag{7.33}$$

Using (7.33) and (7.30) on (7.22), we have

$$\frac{\nu - l^* + 1}{\nu p^* l^*}\, E[T^2] = \frac{\nu - l^* + 1}{\nu - l^* - 1} = E[F_{p^*l^*,\nu-l^*+1}]. \tag{7.34}$$

7.2.4 Additional test statistics

In addition to the Lawley-Hotelling trace statistic T^2 (7.19), other popular test statistics for testing blocks based on $\mathbf{W}$ and $\mathbf{B}$ in (7.19) include the following.

Wilks' Λ

The statistic is based on the likelihood ratio statistic (see Section 9.4.1), and is defined as

$$\Lambda = \frac{|\mathbf{W}|}{|\mathbf{W} + \mathbf{B}|}. \tag{7.35}$$

Its distribution under the null hypothesis has the **Wilks' Λ distribution**, which is a generalization of the beta distribution.

Definition 7.2 (Wilks' Λ). *Suppose* $\mathbf{W}$ *and* $\mathbf{B}$ *are independent Wishart's, with distributions as in (7.21). Then* Λ *has the* **Wilks' $\mathbf{\Lambda}$** *distribution with dimension* l^* *and degrees of freedom* (p^*, ν), *written*

$$\Lambda \sim \text{Wilks}_{l^*}(p^*, \nu). \tag{7.36}$$

Bartlett [1954] has a number of approximations for multivariate statistics, including one for Wilks' Λ:

$$-\left(\nu - \frac{l^* - p^* + 1}{2}\right) \log(\Lambda) \approx \chi^2_{p^* l^*}. \tag{7.37}$$

Pillai trace

This one is the locally most powerful invariant test. (Don't worry about what that means exactly, but it has relatively good power if in the alternative the $\boldsymbol{\beta}^*$ is not far from 0.) The statistic is

$$\text{trace}((\mathbf{W} + \mathbf{B})^{-1}\mathbf{B}). \tag{7.38}$$

Asymptotically, as $\nu \to \infty$,

$$\nu\, \text{trace}((\mathbf{W} + \mathbf{B})^{-1}\mathbf{B}) \to \chi^2_{l^* p^*}, \tag{7.39}$$

which is the same limit as for the Lawley-Hotelling $\mathbf{T}^2$.

Roy's maximum root

This test is based on the largest root, i.e., largest eigenvalue of $(\mathbf{W} + \mathbf{B})^{-1}\mathbf{B}$.

If $p^* = 1$ or $l^* = 1$, these statistics are all equivalent to T^2. In general, Lawley-Hotelling, Pillai, and Wilks' have similar operating characteristics. Each of these four tests is admissible in the sense that there is no other test of the same level that always has better power. See Anderson [2003] for discussions of these statistics, including some asymptotic approximations and tables.

7.2.5 The between and within matrices

In this section, we explain the "between" and "within" appellations given to the matrices $\mathbf{B}$ and $\mathbf{W}$, respectively. To keep the notation clean, we consider the multivariate regression setup ($\mathbf{z} = \mathbf{I}_q$) in Exercise 5.8.41, i.e.,

$$\mathbf{Y} = \mathbf{x}\boldsymbol{\beta} + \mathbf{R} = (\mathbf{x}_1 \;\; \mathbf{x}_2) \begin{pmatrix} \boldsymbol{\beta}_1 \\ \boldsymbol{\beta}_2 \end{pmatrix} + \mathbf{R}, \tag{7.40}$$

where $\mathbf{x}_i$ is $n \times p_i$ and $\boldsymbol{\beta}_i$ is $p_i \times q$, $i = 1, 2$. We wish to test

$$H_0 : \boldsymbol{\beta}_2 = 0 \;\; \textit{versus} \;\; H_A : \boldsymbol{\beta}_2 \neq 0. \tag{7.41}$$

Thus $\boldsymbol{\beta}_2$ is playing the role of $\boldsymbol{\beta}^*$ in (7.13). The $\mathbf{W}$ in (7.16) can be written

$$\mathbf{W} = \mathbf{Y}'\mathbf{Q}_{(\mathbf{x}_1, \mathbf{x}_2)}\mathbf{Y}, \tag{7.42}$$

the residual sum of squares and crossproduct matrix (residual SSCP). From (5.106) and (5.107), where $\mathbf{x}_{2\cdot 1} = \mathbf{Q}_{\mathbf{x}_1}\mathbf{x}_2$, we have that

$$\widehat{\boldsymbol{\beta}}^* = \widehat{\boldsymbol{\beta}}_2 = (\mathbf{x}_{2\cdot 1}'\mathbf{x}_{2\cdot 1})^{-1}\mathbf{x}_{2\cdot 1}'\mathbf{Y} \text{ and } \mathbf{C}_{\mathbf{x}}^* = (\mathbf{x}_{2\cdot 1}'\mathbf{x}_{2\cdot 1})^{-1}. \tag{7.43}$$

Thus for $\mathbf{B}$ in (7.20), we find that

$$\begin{aligned}\mathbf{B} &= \mathbf{Y}'\mathbf{x}_{2\cdot 1}(\mathbf{x}_{2\cdot 1}'\mathbf{x}_{2\cdot 1})^{-1}\mathbf{x}_{2\cdot 1}'\mathbf{Y} \\ &= \mathbf{Y}'\mathbf{P}_{\mathbf{x}_{2\cdot 1}}\mathbf{Y}.\end{aligned} \tag{7.44}$$

By Exercise 5.8.25, $\mathbf{Q}_{\mathbf{x}_1} = \mathbf{Q}_{(\mathbf{x}_1,\mathbf{x}_2)} + \mathbf{P}_{\mathbf{x}_{2\cdot 1}}$, hence

$$\mathbf{W} + \mathbf{B} = \mathbf{Y}'\mathbf{Q}_{\mathbf{x}_1}\mathbf{Y}, \tag{7.45}$$

which is the residual SSCP for the smaller model $\mathbf{Y} = \mathbf{x}_1\boldsymbol{\beta}_1 + \mathbf{R}$. This sum is called the **total SSCP**.

Now specialize to the one-way multivariate analysis of variance with K groups. Suppose there are n_k observations in group k, and the rows of $\mathbf{Y}$ are arranged so that those in group 1 come first, then group 2, etc. We wish to test whether the groups all have the same mean. Here, $\mathbf{x}_1 = \mathbf{1}_n$, and $\mathbf{x}_2$ is any $n \times (K-1)$ matrix that distinguishes the groups. E.g.,

$$\mathbf{x}_2 = \begin{pmatrix} \mathbf{1}_{n_1} & \mathbf{0} & \cdots & \mathbf{0} \\ \mathbf{0} & \mathbf{1}_{n_2} & \cdots & \mathbf{0} \\ & & \vdots & \\ \mathbf{0} & \mathbf{0} & \cdots & \mathbf{1}_{n_{K-1}} \\ \mathbf{0} & \mathbf{0} & \cdots & \mathbf{0} \end{pmatrix}. \tag{7.46}$$

Since the big model fits a separate mean to each group, the residual SSCP sums the sample SSCP's *within* each group, i.e,

$$\mathbf{W} = \sum_{k=1}^{K} \sum_{i \in \text{Group } k} (\mathbf{Y}_i - \overline{\mathbf{Y}}_k)'(\mathbf{Y}_i - \overline{\mathbf{Y}}_k), \tag{7.47}$$

where $\mathbf{Y}_i$ is the ith row of $\mathbf{Y}$, and $\overline{\mathbf{Y}}_k$ is the average of the observations in group k. The total SSCP is the residual SSCP for the model with just the $\mathbf{1}_n$ vector as the $\mathbf{x}$, hence is the usual deviation SSCP:

$$\mathbf{B} + \mathbf{W} = \sum_{i=1}^{n} (\mathbf{Y}_i - \overline{\mathbf{Y}})'(\mathbf{Y}_i - \overline{\mathbf{Y}}), \tag{7.48}$$

where $\overline{\mathbf{Y}}$ is the overall average of the $\mathbf{Y}_i$'s. It can be shown (Exercise 7.6.6) that

$$\begin{aligned}\mathbf{B} &= \sum_{k=1}^{K} \sum_{i \in \text{Group } k} (\overline{\mathbf{Y}}_k - \overline{\mathbf{Y}})'(\overline{\mathbf{Y}}_k - \overline{\mathbf{Y}}) \\ &= \sum_{k=1}^{K} n_k(\overline{\mathbf{Y}}_k - \overline{\mathbf{Y}})'(\overline{\mathbf{Y}}_k - \overline{\mathbf{Y}}).\end{aligned} \tag{7.49}$$

The final summation measures the differences *between* the group means. Thus

$$\text{Total SSCP} = \text{Between SSCP} + \text{Within SSCP}. \tag{7.50}$$

7.3 Examples

In this section we further analyze the mouth size and histamine data. The book Hand and Taylor [1987] contains a number of other nice examples.

7.3.1 Mouth sizes

We will use the model from (6.29) with orthogonal polynomials for the four age variables and the second row of the $\boldsymbol{\beta}$ representing the differences between the girls and boys:

$$\begin{aligned} \mathbf{Y} &= \mathbf{x}\boldsymbol{\beta}\mathbf{z}' + \mathbf{R} \\ &= \begin{pmatrix} \mathbf{1}_{11} & \mathbf{1}_{11} \\ \mathbf{1}_{16} & \mathbf{0}_{16} \end{pmatrix} \begin{pmatrix} \beta_0 & \beta_1 & \beta_2 & \beta_3 \\ \delta_0 & \delta_1 & \delta_2 & \delta_3 \end{pmatrix} \begin{pmatrix} 1 & 1 & 1 & 1 \\ -3 & -1 & 1 & 3 \\ 1 & -1 & -1 & 1 \\ -1 & 3 & -3 & 1 \end{pmatrix} + \mathbf{R}. \end{aligned} \tag{7.51}$$

See Section 6.4.1 for calculation of estimates of the parameters.

We start by testing equality of the boys' and girls' curves. Consider the last row of $\boldsymbol{\beta}$:

$$H_0 : (\delta_0, \delta_1, \delta_2, \delta_3) = (0,0,0,0). \tag{7.52}$$

The estimate is

$$(\widehat{\delta}_0, \widehat{\delta}_1, \widehat{\delta}_2, \widehat{\delta}_3) = (-2.321, -0.305, -0.214, 0.072). \tag{7.53}$$

Because $p^* = 1$, the T^2 is Hotelling's T^2 from Section 7.2.2, where $l^* = l = 4$ and $\nu = n - p = 27 - 2 = 25$. Here $C_{\mathbf{x}}^* = C_{x22} = 0.1534$ and $\widehat{\boldsymbol{\Sigma}}_{\mathbf{z}}^* = \widehat{\boldsymbol{\Sigma}}_{\mathbf{z}}$ from (6.34). We calculate $T^2 = 16.5075$, using

```
t2 <− betahat[2,]%*%solve(sigmaz,betahat[2,])/cx[2,2]
```

By (7.22), under the null,

$$\frac{\nu - l^* + 1}{\nu l^*} T^2 \sim F_{l^*, \nu - l^* + 1};\ \frac{22}{100}\, 16.5075 = 3.632, \tag{7.54}$$

which, compared to a $F_{4,22}$, has p-value 0.02. So we reject H_0, showing there is a difference in the sexes.

Next, consider testing that the two curves are actually linear, that is, the quadratic and cubic terms are 0 for both curves:

$$H_0 : \begin{pmatrix} \beta_2 & \beta_3 \\ \delta_2 & \delta_3 \end{pmatrix} = \mathbf{0}. \tag{7.55}$$

Now $p^* = l^* = 2$, $\mathbf{C}_{\mathbf{x}}^* = \mathbf{C}_{\mathbf{x}}$, and $\widehat{\boldsymbol{\Sigma}}_{\mathbf{z}}^*$ is the lower right 2×2 submatrix of $\widehat{\boldsymbol{\Sigma}}_{\mathbf{z}}$. Calculating:

```
sigmazstar <- sigmaz[3:4,3:4]
betastar <- betahat[,3:4]
b <- t(betastar)%*%solve(cx)%*%betastar # Note that solve(cx) = t(x)%*%x here
t2 <- tr(solve(sigmazstar)%*%b)
```

The function tr is a simple function that finds the trace of a square matrix defined by

```
tr <- function(x) sum(diag(x))
```

This $T^2 = 2.9032$, and the F form in (7.22) is $(24/100) \times T^2 = 0.697$, which is not at all significant for an $F_{4,24}$.

The other tests in Section 7.2.4 are also easy to implement, where here $\mathbf{W} = 25\,\widehat{\mathbf{\Sigma}}_{\mathbf{z}}^{*}$. Wilks' Λ (7.35) is

```
w <- 25*sigmazstar
lambda <- det(w)/det(b+w)
```

The $\Lambda = 0.8959$. For the large-sample approximation (7.37), the factor is 24.5, and the statistic is $-24.5\log(\Lambda) = 2.693$, which is not significant for a χ^2_4. Pillai's trace test statistic (7.38) is

```
tr(solve(b+w)%*%b)
```

which equals 0.1041, and the statistic (7.39) is 2.604, similar to Wilks' Λ. The final one is Roy's maximum root test. The eigenvalues are found using

```
eigen(solve(b+w)%*%b)$values
```

being 0.1036 and 0.0005. Thus the statistic here is 0.1036. Anderson [2003] has tables and other information about these tests. For this situation, $(\nu + p^*)/p^*$ times the statistic, which is $(27/2) \times 0.1036 = 1.40$, has 0.05 cutoff point of 5.75. The function bothsidesmodel.hotelling in Section A.2.5 will perform the T^2 and Wilks' tests.

The conclusion is that we need not worry about the quadratic or cubic terms. Just for fun, go back to the original model, and test the equality of the boys' and girls' curves presuming the quadratic and cubic terms are 0. The $\boldsymbol{\beta}^* = (\delta_0, \delta_1)$, so that $p^* = 1$ and $l^* = 2$. Then $\widehat{\boldsymbol{\beta}}^* = (-2.321, -0.305)$, Hotelling's $T^2 = 13.1417$, and the $F = [(25-2+1)/(25\times 2)] \times 13.1417 = 6.308$. Compared to an $F_{2,24}$, the p-value is 0.006. Note that this is quite a bit smaller than the p-value before (0.02), since we have narrowed the focus by eliminating insignificant terms from the statistic.

Our conclusion that the boys' and girls' curves are linear but different appears reasonable given the Figure 4.1.

7.3.2 Histamine in dogs

Consider again the two-way multivariate analysis of variance model $\mathbf{Y} = \mathbf{x}\boldsymbol{\beta} + \mathbf{R}$ from (4.16), where

$$\boldsymbol{\beta} = \begin{pmatrix} \mu_b & \mu_1 & \mu_2 & \mu_3 \\ \alpha_b & \alpha_1 & \alpha_2 & \alpha_3 \\ \beta_b & \beta_1 & \beta_2 & \beta_3 \\ \gamma_b & \gamma_1 & \gamma_2 & \gamma_3 \end{pmatrix}. \tag{7.56}$$

Recall that the μ's are for the overall mean of the groups, the α's for the drug effects, the β's for the depletion effect, and the γ's for the interactions. The "b" subscript indicates the before means, and the 1, 2, 3's represent the means for the three after time points.

We do an overall test of equality of the four groups based on the three after time points. Thus

$$H_0 : \begin{pmatrix} \alpha_1 & \alpha_2 & \alpha_3 \\ \beta_1 & \beta_2 & \beta_3 \\ \gamma_1 & \gamma_2 & \gamma_3 \end{pmatrix} = \mathbf{0}. \tag{7.57}$$

Section 6.4.5 contains the initial calculations. Here we will use the both-sides model Hotelling T^2 function:

```
x <− kronecker(cbind(1,c(−1,−1,1,1),c(−1,1,−1,1),c(1,−1,−1,1)),rep(1,4))
z <− cbind(c(1,0,0,0),c(0,1,1,1),c(0,−1,0,1),c(0,1,−2,1))
bothsidesmodel.hotelling(x,histamine,z,2:4,2:4)
```

The last two arguments to the bothsidesmodel.hotelling function give the indices of the rows and columns that define the block set to zero. In this case, both are $(2,3,4)$. We find that $T^2 = 41.5661$ and $F = 3.849$, which has degrees of freedom (9,10). The p-value is 0.024, which does indicate a difference in groups.

7.4 Testing linear restrictions

Instead of testing that some of the β_{ij}'s are 0, one often wishes to test equalities among them, or other linear restrictions. For example, consider the one-way multivariate analysis of variance with three groups, n_k observations in group k, and $q = 2$ variables, written as

$$\mathbf{Y} = \begin{pmatrix} \mathbf{1}_{n_1} & \mathbf{0} & \mathbf{0} \\ \mathbf{0} & \mathbf{1}_{n_2} & \mathbf{0} \\ \mathbf{0} & \mathbf{0} & \mathbf{1}_{n_3} \end{pmatrix} \begin{pmatrix} \mu_{11} & \mu_{12} \\ \mu_{21} & \mu_{22} \\ \mu_{31} & \mu_{32} \end{pmatrix} + \mathbf{R}. \tag{7.58}$$

The hypothesis that the groups have the same means is

$$H_0 : \mu_{11} = \mu_{21} = \mu_{31} \text{ and } \mu_{12} = \mu_{22} = \mu_{32}. \tag{7.59}$$

That hypothesis can be expressed in matrix form as

$$\begin{pmatrix} 1 & -1 & 0 \\ 1 & 0 & -1 \end{pmatrix} \begin{pmatrix} \mu_{11} & \mu_{12} \\ \mu_{21} & \mu_{22} \\ \mu_{31} & \mu_{32} \end{pmatrix} = \begin{pmatrix} 0 & 0 \\ 0 & 0 \end{pmatrix}. \tag{7.60}$$

Or, if only the second column of $\mathbf{Y}$ is of interest, then one might wish to test

$$H_0 : \mu_{12} = \mu_{22} = \mu_{32}, \tag{7.61}$$

which in matrix form can be expressed as

$$\begin{pmatrix} 1 & -1 & 0 \\ 1 & 0 & -1 \end{pmatrix} \begin{pmatrix} \mu_{11} & \mu_{12} \\ \mu_{21} & \mu_{22} \\ \mu_{31} & \mu_{32} \end{pmatrix} \begin{pmatrix} 0 \\ 1 \end{pmatrix} = \begin{pmatrix} 0 \\ 0 \end{pmatrix}. \tag{7.62}$$

Turning to the both-sides model, such hypotheses can be written as

$$H_0 : \mathbf{C}\boldsymbol{\beta}\mathbf{D}' = \mathbf{0}, \tag{7.63}$$

where $\mathbf{C}$ ($p^* \times p$) and $\mathbf{D}$ ($l^* \times l$) are fixed matrices that express the desired restrictions. To test the hypothesis, we use

$$\mathbf{C}\widehat{\boldsymbol{\beta}}\mathbf{D}' \sim N(\mathbf{C}\boldsymbol{\beta}\mathbf{D}', \mathbf{C}\mathbf{C}_{\mathbf{x}}\mathbf{C}' \otimes \mathbf{D}\boldsymbol{\Sigma}_{\mathbf{z}}\mathbf{D}'), \tag{7.64}$$

and

$$\mathbf{D}\widehat{\boldsymbol{\Sigma}}_{\mathbf{z}}\mathbf{D}' \sim \frac{1}{n-p} \text{ Wishart}(n-p, \mathbf{D}\boldsymbol{\Sigma}_{\mathbf{z}}\mathbf{D}'). \tag{7.65}$$

Then, assuming the appropriate matrices are invertible, we set

$$\mathbf{B} = \mathbf{D}\widehat{\boldsymbol{\beta}}'\mathbf{C}'(\mathbf{C}\mathbf{C}_{\mathbf{x}}\mathbf{C}')^{-1}\mathbf{C}\widehat{\boldsymbol{\beta}}\mathbf{D}',\ \nu = n-p,\ \mathbf{W} = \nu\mathbf{D}\widehat{\boldsymbol{\Sigma}}_{\mathbf{z}}\mathbf{D}', \tag{7.66}$$

which puts us back at the distributions in (7.21). Thus T^2 or any of the other test statistics can be used as above. In fact, the hypothesis $\boldsymbol{\beta}^*$ in (7.13) and (7.12) can be written as $\boldsymbol{\beta}^* = \mathbf{C}\boldsymbol{\beta}\mathbf{D}'$ for $\mathbf{C}$ and $\mathbf{D}$ with 0's and 1's in the right places.

7.5 Model selection: Mallows' C_p

Hypothesis testing is a popular method for deciding between two nested models, but often in linear models, and in any statistical analysis, there are many models up for consideration. For example, in a linear model with a $p \times l$ matrix $\boldsymbol{\beta}$ of parameters, there are 2^{pl} models attainable by setting subsets of the β_{ij}'s to 0. There is also an infinite number of submodels obtained by setting linear (and nonlinear) restrictions among the parameters. One approach to choosing among many models is to minimize some criterion that measures the efficacy of a model. In linear models, some function of the residual sum of squares and cross products matrix is an obvious choice. The drawback is that, typically, the more parameters in the model, the lower the residual sum of squares, hence the best model always ends up being the one with all the parameters, i.e., the entire $\boldsymbol{\beta}$ in the multivariate linear model. Thus one often assesses a penalty depending on the number of parameters in the model, the larger the number of parameters, the higher the penalty, so that there is some balance between the residual sum of squares and number of parameters.

In this section we present **Mallows' C_p** criterion, [Mallows, 1973]. Section 9.5 exhibits the Akaike information criterion (AIC) and the Bayes information criterion (BIC), which are based on the likelihood. Mallows' C_p is motivated by prediction. We develop this idea first for the multivariate regression model,

$$\mathbf{Y} = \mathbf{x}\boldsymbol{\beta} + \mathbf{R}, \text{ where } \mathbf{R} \sim N_{n\times q}(\mathbf{0}, \mathbf{I}_n \otimes \boldsymbol{\Sigma}_R) \tag{7.67}$$

and $\boldsymbol{\beta}$ is $p \times q$, then use the technique in Section 6.2.2 to extend it to the general both-sides model.

The observed $\mathbf{Y}$ is dependent on the value of other variables represented by $\mathbf{x}$. Imagine using the observed data to predict a new variable $\mathbf{Y}^{New}$ based on its $\mathbf{x}^{New}$. For example, an insurance company may have data on $\mathbf{Y}$, the payouts the company has made to a number of people, and $\mathbf{x}$, the basic data (age, sex, overall health, etc.) on these same people. But the company is really wondering whether to insure new people, whose $\mathbf{x}^{New}$ they know, but $\mathbf{Y}^{New}$ they cannot observe yet but wish to predict. The prediction, $\widehat{\mathbf{Y}}^*$, is a function of $\mathbf{x}^{New}$ and the observed $\mathbf{Y}$. A good predictor has $\widehat{\mathbf{Y}}^*$ close to $\mathbf{Y}^{New}$.

We act as if we wish to predict observations $\mathbf{Y}^{New}$ that have the same explanatory $\mathbf{x}$ as the data, so that

$$\mathbf{Y}^{New} = \mathbf{x}\boldsymbol{\beta} + \mathbf{R}^{New}, \tag{7.68}$$

where $\mathbf{R}^{New}$ has the same distribution as $\mathbf{R}$, and $\mathbf{Y}^{New}$ and $\mathbf{Y}$ are independent. It is perfectly reasonable to want to predict $\mathbf{Y}^{New}$'s for different $\mathbf{x}$ than in the data, but the analysis is a little easier if they are the same, plus it is a good starting point.

We consider the submodels of (7.67) from Section 6.5, which are defined by patterns $\mathbf{P}$ of 0's and 1's as in (6.51). The $\mathbf{P}$ has the same dimensions as the $\boldsymbol{\beta}$, where a "0" in the $\mathbf{P}$ indicates that the corresponding element in $\boldsymbol{\beta}$ is set to 0, and a "1" indicates the corresponding element is unrestricted. Then the submodel is given by restricting $\boldsymbol{\beta}$ to the set $\mathcal{B}(\mathbf{P})$ as in (6.52). The idea is to use the fit from the submodel to predict $\mathbf{Y}^{New}$. Hasegawa [1986] and Gupta and Kabe [2000] study this criterion for a slightly different set of submodels.

If we let $\widehat{\boldsymbol{\beta}}^*$ denote the least squares estimate of $\boldsymbol{\beta}$ for the submodel, then the predictor using this submodel is

$$\widehat{\mathbf{Y}}^* = \mathbf{x}\widehat{\boldsymbol{\beta}}^*. \tag{7.69}$$

We assess the predictor using a scaled version of the expected sum of squares, specifically

$$EPredSS^* = E[\text{trace}(\boldsymbol{\Sigma}_R^{-1}(\mathbf{Y}^{New} - \widehat{\mathbf{Y}}^*)'(\mathbf{Y}^{New} - \widehat{\mathbf{Y}}^*))]. \tag{7.70}$$

The covariance is there to scale the variables, so that, for example, the criterion will not be dependent upon the units in which the $\mathbf{Y}$'s are recorded. We take the expected value because the $\mathbf{Y}^{New}$ is not observed. A natural first try at estimating the $EPredSS^*$ would be to substitute $\widehat{\boldsymbol{\Sigma}}_R$ for $\boldsymbol{\Sigma}_R$, and the observed $\mathbf{Y}$ for $\mathbf{Y}^{New}$, though the resulting estimator would likely be an underestimate, because the fit $\widehat{\mathbf{Y}}^*$ is tailored to the $\mathbf{Y}$.

There are two steps to deriving an estimate of the prediction error. Step 1 finds the bias in the residual sum of squares. Specifically, we show that

$$EPredSS^* - EResidSS^* = 2\#\{\beta_{ij}^* \neq 0\} = 2\sum_{i,j} p_{ij}, \tag{7.71}$$

where

$$EResidSS^* = E[\text{trace}(\boldsymbol{\Sigma}_R^{-1}(\mathbf{Y} - \widehat{\mathbf{Y}}^*)'(\mathbf{Y} - \widehat{\mathbf{Y}}^*))]. \tag{7.72}$$

Step 2 finds an unbiased estimator of $EResidSS^*$, which is

$$\widehat{EResidSS}^* = \frac{n-p-q-1}{n-p}\,\text{trace}(\widehat{\boldsymbol{\Sigma}}_R^{-1}(\mathbf{Y} - \widehat{\mathbf{Y}}^*)'(\mathbf{Y} - \widehat{\mathbf{Y}}^*)) + q(q+1), \tag{7.73}$$

where

$$\widehat{\boldsymbol{\Sigma}}_R = \frac{1}{n-p}\mathbf{Y}'\mathbf{Q}_\mathbf{x}\mathbf{Y}, \tag{7.74}$$

the estimated covariance under the model (7.68) with no restrictions. Thus an unbiased estimator of $EPredSS^*$ adds the results in (7.71) and (7.73), which yields Mallow's C_p,

$$C_p^* = \frac{n-p-q-1}{n-p}\,\text{trace}(\widehat{\boldsymbol{\Sigma}}_R^{-1}(\mathbf{Y} - \widehat{\mathbf{Y}}^*)'(\mathbf{Y} - \widehat{\mathbf{Y}}^*)) + 2d^*, \tag{7.75}$$

where d^* is the number of free parameters in the submodel, which parameters include the nonzero β^*_{ij}'s and the $q(q+1)/2$ parameters in the $\mathbf{\Sigma}_R$:

$$d^* = \#\{\beta^*_{ij} \neq 0\} + \frac{q(q+1)}{2}. \tag{7.76}$$

It is perfectly fine to drop the $q(q+1)/2$ part from the d^*, since it will be the same for all submodels. The verification of these results is in Section 7.5.2.

For the both-sides model, we find $\mathbf{Y}_z = \mathbf{Y}\mathbf{z}(\mathbf{z}'\mathbf{z})^{-1}$ as in Section 6.2.2, so that the model is

$$\mathbf{Y}_\mathbf{z} = \mathbf{x}\boldsymbol{\beta} + \mathbf{R}_\mathbf{z}, \text{ where } \mathbf{R}_\mathbf{z} \sim N_{n\times l}(\mathbf{0}, \mathbf{I}_n \otimes \mathbf{\Sigma}_\mathbf{z}), \tag{7.77}$$

where now $\boldsymbol{\beta}$ is $p \times l$. Mallow's C_p in this case adds a few $\mathbf{z}$'s in the subscripts, and substitutes l for q:

$$C^*_p = \frac{n-p-l-1}{n-p}\,\widehat{\mathbf{\Sigma}}_\mathbf{z}^{-1}\,(\mathbf{Y}_\mathbf{z} - \widehat{\mathbf{Y}}^*_\mathbf{z})'(\mathbf{Y}_\mathbf{z} - \widehat{\mathbf{Y}}^*_\mathbf{z}) + 2d^*, \tag{7.78}$$

where $d^* = \#\{\text{nonzero } \beta^*_{ij}\text{'s}\} + l(l+1)/2$.

7.5.1 Example: Mouth sizes

Return to the both-sides models (6.29) for the mouth size data. Here, the coefficient matrix is

$$\boldsymbol{\beta} = \left(\begin{array}{lcccc} & \text{Constant} & \text{Linear} & \text{Quadratic} & \text{Cubic} \\ \text{Boys} & \beta_0 & \beta_1 & \beta_2 & \beta_3 \\ \text{Girls}-\text{Boys} & \delta_0 & \delta_1 & \delta_2 & \delta_3 \end{array} \right). \tag{7.79}$$

We are interested in finding a good submodel. There are eight β_{ij}'s, so in principle there are 2^8 possible submodels obtained by setting a subset of parameters to zero. But there are certain monotonicities that should not be violated without good reason. First, if a difference parameter (δ_j) is in the model, so is the corresponding overall boys parameter (β_j). Next, if a monomial parameter of degree j is in the model, so are the monomials of lesser degree. That is, if δ_2, the quadratic parameter for the difference, is in, so are δ_0 and δ_1, the constant and linear parameters. There are 15 such models. We describe the models with the pair of integers (b, d), indicating the number of nonzero parameters for the boys' and girls'−boys' effects, respectively. Thus the degree of the boys' polynomial is $b+1$ if $b > 0$. The monotonicities require that

$$0 \le d \le b \le 4, \tag{7.80}$$

and the models are

$$M_{bd} \Rightarrow \beta_j = 0 \text{ if } j \ge b, \;\; \delta_j = 0 \text{ if } j \ge d. \tag{7.81}$$

Some examples:

$$\begin{array}{cc|cc} \text{Model} & \text{Pattern} & \text{Model} & \text{Pattern} \\ \hline M_{10} & \begin{pmatrix} 1000 \\ 0000 \end{pmatrix} & M_{32} & \begin{pmatrix} 1110 \\ 1100 \end{pmatrix} \\ M_{21} & \begin{pmatrix} 1100 \\ 1000 \end{pmatrix} & M_{40} & \begin{pmatrix} 1111 \\ 0000 \end{pmatrix} \\ M_{22} & \begin{pmatrix} 1100 \\ 1100 \end{pmatrix} & M_{44} & \begin{pmatrix} 1111 \\ 1111 \end{pmatrix} \end{array} \tag{7.82}$$

We use the x, y, z from Section 6.4.1, and the function bothsidesmodel discussed in Section 6.5. For example, to find the C_p for model M_{21}, we use the pattern given in (7.82), then call the function:

```
pattern <- rbind(c(1,1,0,0),c(1,0,0,0))
bothsidesmodel(x,y,z,pattern)
```

The function returns, among other quantities, the components ResidSS, Dim and Cp, which for this model are 88.47, 13, and 114.47, respectively.

To find the C_p's for all fifteen models, we loop over b and d:

```
cps <- NULL
for(b in (0:4)) for(d in (0:b)) {
        pattern <- matrix(0,ncol=4,nrow=2)
        if(b>0) pattern[1,1:b] <- 1
        if(d>0) pattern[2,1:d] <- 1
        bsm <- bothsidesmodel(x,y,z,pattern)
        b0 <- c(b,d,bsm$ResidSS,bsm$Dim,bsm$Cp)
        cps <- rbind(cps,b0)
}
```

Here are the results:

(7.83)

b	d	$ResidSS^*$	d^*	C_p^*
0	0	3584.72	10	3604.72
1	0	189.12	11	211.12
1	1	183.06	12	207.06
2	0	94.54	12	118.54
2	1	88.47	13	114.47
2	2	82.55	14	$*$110.55
3	0	93.34	13	119.34
3	1	87.28	14	115.28
3	2	81.35	15	111.35
3	3	80.36	16	112.36
4	0	93.21	14	121.21
4	1	87.14	15	117.14
4	2	81.22	16	113.22
4	3	80.23	17	114.23
4	4	80.00	18	116.00

Note that in general, the larger the model (in terms of d^*, the number of free parameters), the smaller the $ResidSS^*$ but the larger the penalty. The C_p statistic aims to balance the fit and complexity. The model with the lowest C_p is the $(2,2)$ model, which fits separate linear growth curves to the boys and girls. We arrived at this model in Section 7.3.1 as well. The $(3,2)$ model is essentially as good, but is a little more complicated, allowing the intercepts and slopes to differ for the two groups, but adding a common quadratic term. Generally, one looks for the model with the smallest C_p, but if several models have approximately the same low C_p, one might choose the simplest.

7.5.2 Mallows' C_p verification

We show that Mallows' C_p in (7.75) is an unbiased estimator of $EPredSS^*$ in (7.70) for a given submodel of the multivariate regression model (7.67) specified by the pattern matrix $\mathbf{P}$. In both $EPredSS^*$ and $EResidSS^*$ of (7.72), the fit $\widehat{\mathbf{Y}}^*$ to the submodel appears. To prove their difference is as in (7.71), it is easier to use the rowed version of the models as in Section 6.5. Since $\mathbf{z} = \mathbf{I}_q$, the big model can be written

$$\mathbf{V} = \boldsymbol{\gamma}\mathbf{D}' + \text{row}(\mathbf{R}), \tag{7.84}$$

where

$$\mathbf{V} = \text{row}(\mathbf{Y}),\ \ \boldsymbol{\gamma} = \text{row}(\boldsymbol{\beta}),\ \ \mathbf{D} = \mathbf{x}\otimes\mathbf{I}_q,\ \ \text{and}\ \ Cov[\mathbf{V}] = \boldsymbol{\Omega} = \mathbf{I}_n\otimes\boldsymbol{\Sigma}_R. \tag{7.85}$$

Similarly, $\mathbf{V}^{New} = \text{row}(\mathbf{Y}^{New})$, which has the same distribution as $\mathbf{V}$, and is independent of $\mathbf{V}$. The submodel we write as

$$\mathbf{V}^* = \boldsymbol{\gamma}^*\mathbf{D}^{*\prime} + \text{row}(\mathbf{R}), \tag{7.86}$$

where $\boldsymbol{\gamma}^*$ contains the nonzero elements of $\boldsymbol{\gamma}$ as indicated by row$(\mathbf{P})$. $\mathbf{D}^*$ then is $\mathbf{D}$ with just the corresponding columns. The prediction of $\mathbf{V}^{New}$ given by the submodel is then

$$\text{row}(\widehat{\mathbf{Y}}^*) = \widehat{\mathbf{V}}^* = \mathbf{V}\mathbf{P}_{\mathbf{D}^*}, \tag{7.87}$$

where $\mathbf{P}_{\mathbf{D}^*}$ is the projection matrix for $\mathbf{D}^*$. The quantities of interest are now given by

$$\begin{aligned} EPredSS^* &= \text{trace}(\boldsymbol{\Omega}^{-1}E[(\mathbf{V}^{New} - \widehat{\mathbf{V}}^*)'(\mathbf{V}^{New} - \widehat{\mathbf{V}}^*)]),\ \text{and} \\ EResidSS^* &= \text{trace}(\boldsymbol{\Omega}^{-1}E[(\mathbf{V} - \widehat{\mathbf{V}}^*)'(\mathbf{V} - \widehat{\mathbf{V}}^*)]). \end{aligned} \tag{7.88}$$

For any row vector $\mathbf{U}$,

$$E[\mathbf{U}'\mathbf{U}] = E[\mathbf{U}]'E[\mathbf{U}] + Cov[\mathbf{U}]. \tag{7.89}$$

Apply this formula to $\mathbf{U} = \mathbf{V}^{New} - \widehat{\mathbf{V}}^*$ and $\mathbf{U} = \mathbf{V} - \widehat{\mathbf{V}}^*$. Note that since $\mathbf{V}$ and $\mathbf{V}^{New}$ have the same distributions, they have the same means, hence the terms based on the $E[\mathbf{U}]$'s cancel when we subtract. Thus,

$$EPredSS^* - EResidSS^* = \text{trace}(\boldsymbol{\Omega}^{-1}(Cov[\mathbf{V}^{New} - \widehat{\mathbf{V}}^*] - Cov[\mathbf{V} - \widehat{\mathbf{V}}^*])). \tag{7.90}$$

Because $\mathbf{V}^{New}$ and $\widehat{\mathbf{V}}^*$ are independent, their covariances sum:

$$\begin{aligned} Cov[\mathbf{V}^{New} - \widehat{\mathbf{V}}^*] &= Cov[\mathbf{V}^{New}] + Cov[\widehat{\mathbf{V}}^*] \\ &= \boldsymbol{\Omega} + Cov[\mathbf{V}\mathbf{P}_{\mathbf{D}^*}] \\ &= \boldsymbol{\Omega} + \mathbf{P}_{\mathbf{D}^*}\boldsymbol{\Omega}\mathbf{P}_{\mathbf{D}^*}. \end{aligned} \tag{7.91}$$

The second covariance is that of the residual, i.e.,

$$\begin{aligned} Cov[\mathbf{V} - \widehat{\mathbf{V}}^*] &= Cov[\mathbf{V}(\mathbf{I}_{nq} - \mathbf{P}_{\mathbf{D}^*})] \\ &= (\mathbf{I}_{nq} - \mathbf{P}_{\mathbf{D}^*})\mathbf{\Omega}(\mathbf{I}_{nq} - \mathbf{P}_{\mathbf{D}^*}) \\ &= \mathbf{\Omega} - \mathbf{P}_{\mathbf{D}^*}\mathbf{\Omega} - \mathbf{\Omega}\mathbf{P}_{\mathbf{D}^*} + \mathbf{P}_{\mathbf{D}^*}\mathbf{\Omega}\mathbf{P}_{\mathbf{D}^*}. \end{aligned} \tag{7.92}$$

Thus the difference is

$$\begin{aligned} EPredSS^* - EResidSS^* &= \text{trace}(\mathbf{\Omega}^{-1}(\mathbf{P}_{\mathbf{D}^*}\mathbf{\Omega} + \mathbf{\Omega}\mathbf{P}_{\mathbf{D}^*})) \\ &\quad - 2\,\text{trace}(\mathbf{P}_{\mathbf{D}^*}) \\ &= 2\,\#\{\text{columns in } \mathbf{D}^*\} = 2\sum_{i,j} p_{ij} \end{aligned} \tag{7.93}$$

as in (7.71), since $\sum_{i,j} p_{ij}$ equals the number of nonzero β_{ij}'s, which equals the number of elements in γ^*.

To show the estimator in (7.73) is unbiased for $EResidSS^*$, we first write

$$\begin{aligned} \mathbf{Y} - \widehat{\mathbf{Y}}^* &= \mathbf{Q}_\mathbf{x}\mathbf{Y} + \mathbf{P}_\mathbf{x}\mathbf{Y} - \mathbf{x}\widehat{\boldsymbol{\beta}}^*\mathbf{z}' \\ &= \mathbf{Q}_\mathbf{x}\mathbf{Y} + \mathbf{x}\widehat{\boldsymbol{\beta}} - \mathbf{x}\widehat{\boldsymbol{\beta}}^*, \end{aligned} \tag{7.94}$$

where we emphasize that $\widehat{\boldsymbol{\beta}}$ is the estimator from the big model, and $\widehat{\boldsymbol{\beta}}^*$ is the estimator from the submodel. Because $\mathbf{Q}_\mathbf{x}\mathbf{x} = \mathbf{0}$,

$$(\mathbf{Y} - \widehat{\mathbf{Y}}^*)'(\mathbf{Y} - \widehat{\mathbf{Y}}^*) = \mathbf{W} + \mathbf{B}^*, \tag{7.95}$$

where

$$\mathbf{W} = \mathbf{Y}'\mathbf{Q}_\mathbf{x}\mathbf{Y} \text{ and } \mathbf{B}^* = (\widehat{\boldsymbol{\beta}} - \widehat{\boldsymbol{\beta}}^*)'\mathbf{x}'\mathbf{x}(\widehat{\boldsymbol{\beta}} - \widehat{\boldsymbol{\beta}}^*). \tag{7.96}$$

Recalling that this $\mathbf{W} \sim \text{Wishart}_q(n-p, \mathbf{\Sigma}_R)$, we have

$$\begin{aligned} EResidSS^* &= \text{trace}(\mathbf{\Sigma}_R^{-1}\, E[\mathbf{W} + \mathbf{B}^*]) \\ &= \text{trace}(\mathbf{\Sigma}_R^{-1}\, ((n-p)\mathbf{\Sigma}_R + E[\mathbf{B}^*])) \\ &= (n-p)q + \text{trace}(\mathbf{\Sigma}_R^{-1}E[\mathbf{B}^*]). \end{aligned} \tag{7.97}$$

For the estimator (7.73), where (7.74) shows that $\widehat{\mathbf{\Sigma}}_R = \mathbf{W}/(n-p)$, we use (7.95) again, so that

$$\begin{aligned} \widehat{EResidSS}^* &= \frac{n-p-q-1}{n-p}\,\text{trace}((\mathbf{W}/(n-p))^{-1}\,(\mathbf{W} + \mathbf{B}^*)) + q(q+1) \\ &= (n-p-q-1)(q + \text{trace}(\mathbf{W}^{-1}\mathbf{B}^*)) + q(q+1). \end{aligned} \tag{7.98}$$

To find the expected value of $\widehat{EResidSS}^*$, we note that $\mathbf{W}$ and $\mathbf{B}^*$ are independent, because they depend on the residual and fit, respectively. Also, Lemma 7.1 shows that $E[\mathbf{W}^{-1}] = \mathbf{\Sigma}_R^{-1}/(n-p-q-1)$, hence

$$\begin{aligned} E[\widehat{EResidSS}^*] &= (n-p-q-1)(q + \text{trace}(E[\mathbf{W}^{-1}]E[\mathbf{B}^*])) + q(q+1) \\ &= (n-p-q-1)(q + \text{trace}(\mathbf{\Sigma}_R^{-1}E[\mathbf{B}^*]/(n-p-q-1))) + q(q+1) \\ &= (n-p)q + \text{trace}(\mathbf{\Sigma}_R^{-1}E[\mathbf{B}^*]) \\ &= EResidSS^* \end{aligned} \tag{7.99}$$

as in (7.97).

7.6 Exercises

Exercise 7.6.1. Verify the equations (7.4) and (7.5).

Exercise 7.6.2. Show that when $p^* = l^* = 1$, the T^2 in (7.18) equals the square of the t-statistic in (6.23), assuming $\beta_{ij} = 0$ there.

Exercise 7.6.3. Verify the Wishart distribution result for $\mathbf{B}$ given in (7.21) when $\boldsymbol{\beta}^* = 0$.

Exercise 7.6.4. Verify the equalities in (7.28).

Exercise 7.6.5. Verify the first equality in (7.44).

Exercise 7.6.6. Verify the two equalities in (7.49).

Exercise 7.6.7. If $A \sim Gamma(\alpha, \lambda)$ and $B \sim Gamma(\beta, \lambda)$, and A and B are independent, then $U = A/(A+B)$ is distributed $Beta(\alpha, \beta)$. Show that when $l^* = 1$, Wilks' Λ (Definition 7.2) is $Beta(\alpha, \beta)$, and give the parameters in terms of p^* and ν. [Hint: See Exercise 3.7.9 for the gamma distribution, whose pdf is given in (3.83). Also, the beta pdf is found in Exercise 2.7.13, in equation (2.96), though you do not need it here.]

Exercise 7.6.8. Suppose $F \sim F_{\mu,\nu}$ as in Definition 7.1. Show that $U = \nu/(\mu F + \nu)$ is $Beta(\alpha, \beta)$ from Exercise 7.6.7, and give the parameters in terms of μ and ν.

Exercise 7.6.9. Show that $E[1/\chi^2_\nu] = 1/(\nu - 2)$ if $\nu \geq 2$, as used in (7.30). The pdf of the chi-square is given in (3.82).

Exercise 7.6.10. Verify the distribution results in (7.64) and (7.65).

Exercise 7.6.11. Find the matrices $\mathbf{C}$ and $\mathbf{D}$ so that the hypothesis in (7.63) is the same as that in (7.13), where $\boldsymbol{\beta}$ is 5×4.

Exercise 7.6.12. Verify the steps in (7.93).

Exercise 7.6.13. Verify (7.95).

Exercise 7.6.14 (Mouth sizes). In the mouth size data in Section 7.1.1, there are $n_G = 11$ girls and $n_B = 16$ boys, and $q = 4$ measurements on each. Thus $\mathbf{Y}$ is 27×4. Assume that

$$\mathbf{Y} \sim N(\mathbf{x}\boldsymbol{\beta}, \mathbf{I}_n \otimes \boldsymbol{\Sigma}_R), \tag{7.100}$$

where this time

$$\mathbf{x} = \begin{pmatrix} \mathbf{1}_{11} & \mathbf{0}_{11} \\ \mathbf{0}_{16} & \mathbf{1}_{16} \end{pmatrix}, \tag{7.101}$$

and

$$\boldsymbol{\beta} = \begin{pmatrix} \boldsymbol{\mu}_G \\ \boldsymbol{\mu}_B \end{pmatrix} = \begin{pmatrix} \mu_{G1} & \mu_{G2} & \mu_{G3} & \mu_{G4} \\ \mu_{B1} & \mu_{B2} & \mu_{B3} & \mu_{B4} \end{pmatrix}. \tag{7.102}$$

The sample means for the two groups are $\widehat{\boldsymbol{\mu}}_G$ and $\widehat{\boldsymbol{\mu}}_B$. Consider testing

$$H_0 : \boldsymbol{\mu}_G = \boldsymbol{\mu}_B. \tag{7.103}$$

(a) What is the constant c so that

$$\mathbf{Z} = \frac{\widehat{\boldsymbol{\mu}}_G - \widehat{\boldsymbol{\mu}}_B}{c} \sim N(\mathbf{0}, \boldsymbol{\Sigma}_R)? \tag{7.104}$$

(b) The unbiased estimate of $\mathbf{\Sigma}_R$ is $\widehat{\mathbf{\Sigma}}_R$. Then

$$\mathbf{W} = \nu\, \widehat{\mathbf{\Sigma}}_R \sim \text{Wishart}(\nu, \mathbf{\Sigma}_R). \tag{7.105}$$

What is ν? (c) What is the value of Hotelling's T^2? (d) Find the constant d and the degrees of freedom a, b so that $F = d\ T^2 \sim F_{a,b}$. (e) What is the value of F? What is the resulting p-value? (f) What do you conclude? (g) Compare these results to those in Section 7.3.1, equations (7.54) and below.

Exercise 7.6.15 (Skulls). This question continues Exercise 6.6.10 on Egyptian skulls. (a) Consider testing that there is no difference among the five time periods for all four measurements at once. What are p^*, l^*, ν and T^2 in (7.22) for this hypothesis? What is the F and its degrees of freedom? What is the p-value? What do you conclude? (b) Now consider testing whether there is a nonlinear effect on skull size over time, that is, test whether the last three rows of the $\boldsymbol{\beta}$ matrix are all zero. What are l^*, ν, the F-statistic obtained from T^2, the degrees of freedom, and the p-value? What do you conclude? (c) Finally, consider testing whether there is a linear effect on skull size over time assuming there is no nonlinear effect. Find the F-statistic obtained from T^2. What do you conclude?

Exercise 7.6.16 (Prostaglandin). This question continues Exercise 6.6.9 on prostaglandin levels over time. The model is $\mathbf{Y} = \mathbf{1}_{10}\boldsymbol{\beta}\mathbf{z}' + \mathbf{R}$, where $\boldsymbol{\beta}$ is 1×6, and the ith row of $\mathbf{z}$ is

$$(1,\ \cos(\theta_i),\ \sin(\theta_i),\ \cos(2\theta_i),\ \sin(2\theta_i),\ \cos(3\theta_i)) \tag{7.106}$$

for $\theta_i = 2\pi i/6$, $i = 1,\ldots,6$. (a) In Exercise 4.4.8, the one-cycle wave is given by the equation $A + B\ \cos(\theta + C)$. The null hypothesis that the model does not include that wave is expressed by setting $B = 0$. What does this hypothesis translate to in terms of the β_{ij}'s? (b) Test whether the one-cycle wave is in the model. What is p^*? [It is the same for all these tests.] What is l^*? (c) Test whether the two-cycle wave is in the model. What is l^*? (d) Test whether the three-cycle wave is in the model. What is l^*? (e) Test whether just the one-cycle wave needs to be in the model. (I.e., test whether the two- and three-cycle waves have zero coefficients.) (f) Using the results from parts (b) through (e), choose the best model among the models with (1) no waves; (2) just the one-cycle wave; (3) just the one- and two-cycle waves; (4) the one-, two-, and three-cycle waves. (g) Use Mallows' C_p to choose among the four models listed in part (f).

Exercise 7.6.17 (Histamine in dogs). Consider the model for the histamine in dogs example in (4.16), i.e.,

$$\mathbf{Y} = \mathbf{x}\boldsymbol{\beta} + \mathbf{R} = \left(\begin{pmatrix} 1 & -1 & -1 & 1 \\ 1 & -1 & 1 & -1 \\ 1 & 1 & -1 & -1 \\ 1 & 1 & 1 & 1 \end{pmatrix} \otimes \mathbf{1}_4 \right) \begin{pmatrix} \mu_0 & \mu_1 & \mu_2 & \mu_3 \\ \alpha_0 & \alpha_1 & \alpha_2 & \alpha_3 \\ \beta_0 & \beta_1 & \beta_2 & \beta_3 \\ \gamma_0 & \gamma_1 & \gamma_2 & \gamma_3 \end{pmatrix} + \mathbf{R}. \tag{7.107}$$

For the two null hypotheses described in parts (a) and (b) below, specify which parameters are set to zero, then find p^*, l^*, ν, the T^2 and its F version, the degrees of freedom for the F, the p-value, and whether you accept or reject. Interpret the finding in terms of the groups and variables. (a) The four groups have equal means (for all four time points). Compare the results to that for the hypothesis in (7.57). (b) The four groups have equal before means. (c) Now consider testing the null hypothesis that the after means are equal, but using the before measurements as a covariate. (So we assume that $\alpha_b = \beta_b = \gamma_b = 0$.) What are the dimensions of the resulting $\mathbf{Y}_a$ and the $\mathbf{x}$ matrix augmented with the covariate? What are p^*, l^*, ν, and the degrees of freedom in the F for testing the null hypothesis? (d) Find the T^2, the F statistic, and the p-value for testing the hypothesis using the covariate. What do you conclude? How does this result compare to that without the covariates?

Exercise 7.6.18 (Histamine, cont.). Continue the previous question, using as a starting point the model with the before measurements as the covariate, so that

$$\mathbf{Y}^* = \mathbf{x}^* \begin{pmatrix} \mu_1^* & \mu_2^* & \mu_3^* \\ \alpha_1 & \alpha_2 & \alpha_3 \\ \beta_1 & \beta_2 & \beta_3 \\ \gamma_1 & \gamma_2 & \gamma_3 \\ \delta_1 & \delta_2 & \delta_3 \end{pmatrix} \mathbf{z}' + \mathbf{R}^*, \tag{7.108}$$

where $\mathbf{Y}^*$ has just the after measurements, $\mathbf{x}^*$ is the $\mathbf{x}$ in (7.107) augmented with the before measurements, and $\mathbf{z}$ represents orthogonal polynomials for the after time points,

$$\mathbf{z} = \begin{pmatrix} 1 & -1 & 1 \\ 1 & 0 & -2 \\ 1 & 1 & 1 \end{pmatrix}. \tag{7.109}$$

Now consider the equivalent model resulting from multiplying both sides of the equation on the right by $(\mathbf{z}')^{-1}$. (a) Find the estimates and standard errors for the quadratic terms, $(\mu_3^*, \alpha_3, \beta_3, \gamma_3)$. Test the null hypothesis that $(\mu_3^*, \alpha_3, \beta_3, \gamma_3) = (0, 0, 0, 0)$. What is ν? What is the p-value? Do you reject this null? (The answer should be no.) (b) Now starting with the null model from part (a) (i.e., the quadratic terms are zero), use the vector of quadratic terms as the covariate. Find the estimates and standard errors of the relevant parameters, i.e.,

$$\begin{pmatrix} \mu_1^* & \mu_2^* \\ \alpha_1 & \alpha_2 \\ \beta_1 & \beta_2 \\ \gamma_1 & \gamma_2 \end{pmatrix}. \tag{7.110}$$

(c) Use Hotelling's T^2 to test the interaction terms are zero, i.e., that $(\gamma_1, \gamma_2) = (0, 0)$. (What are l^* and ν?) Also, do the t-tests for the individual parameters. What do you conclude?

Exercise 7.6.19 (Caffeine). This question uses the data on the effects of caffeine on memory described in Exercise 4.4.4. The model is as in (4.36), with $\mathbf{x}$ as described there, and

$$\mathbf{z} = \begin{pmatrix} 1 & -1 \\ 1 & 1 \end{pmatrix}. \tag{7.111}$$

The goal of this problem is to use Mallows' C_p to find a good model, choosing among the constant, linear and quadratic models for $\mathbf{x}$, and the "overall mean" and "overall mean + difference models" for the scores. Thus there are six models. (a) For each of the 6 models, find the p^*, l^*, residual sum of squares, penalty, and C_p values. (b) Which model is best in terms of C_p? (c) Find the estimate of $\boldsymbol{\beta}^*$ for the best model. (d) What do you conclude?

Exercise 7.6.20 (Skulls). Continue with the skulls data from Exercise 7.6.15. Find Mallows' C_p for the 2^4 models wherein each measurement (column of $\mathbf{Y}$) has either a constant model or a linear model.

Chapter 8

Some Technical Results

This chapter contains a number of results useful for linear models and other models, including the densities of the multivariate normal and Wishart. We collect them here so as not to interrupt the flow of the narrative.

8.1 The Cauchy-Schwarz inequality

Lemma 8.1. ***Cauchy-Schwarz inequality.*** *Suppose* $\mathbf{y}$ *and* $\mathbf{d}$ *are* $1 \times K$ *vectors. Then*

$$(\mathbf{y}\mathbf{d}')^2 \leq \|\mathbf{y}\|^2 \|\mathbf{d}\|^2, \tag{8.1}$$

with equality if and only if $\mathbf{d}$ *is zero, or*

$$\mathbf{y} = \widehat{\gamma}\mathbf{d} \tag{8.2}$$

for some constant $\widehat{\gamma}$.

Proof. If $\mathbf{d}$ is zero, the result is immediate. Suppose $\mathbf{d} \neq \mathbf{0}$, and let $\widehat{\mathbf{y}}$ be the projection of $\mathbf{y}$ onto span$\{\mathbf{d}\}$. (See Definitions 5.2 and 5.7.) Then by least squares, Theorem 5.2 (with $\mathbf{D} = \mathbf{d}'$), $\widehat{\mathbf{y}} = \widehat{\gamma}\mathbf{d}$, where here $\widehat{\gamma} = \mathbf{y}\mathbf{d}'/\|\mathbf{d}\|^2$. The sum-of-squares decomposition in (5.11) implies that

$$\|\mathbf{y}\|^2 - \|\widehat{\mathbf{y}}\|^2 = \|\mathbf{y} - \widehat{\mathbf{y}}\|^2 \geq 0, \tag{8.3}$$

which yields

$$\|\mathbf{y}\|^2 \geq \|\widehat{\mathbf{y}}\|^2 = \frac{(\mathbf{y}\mathbf{d}')^2}{\|\mathbf{d}\|^2}, \tag{8.4}$$

from which (8.1) follows. Equality in (8.1) holds if and only if $\mathbf{y} = \widehat{\mathbf{y}}$, which holds if and only if (8.2). □

If U and V are random variables, with $E[|UV|] < \infty$, then the Cauchy-Schwarz inequality becomes

$$E[UV]^2 \leq E[U^2]E[V^2], \tag{8.5}$$

with equality if and only if V is zero with probability one, or

$$U = bV \tag{8.6}$$

for constant $b = E[UV]/E[V^2]$. See Exercise 8.8.2. The next result is well-known in statistics.

Corollary 8.1 (Correlation inequality). *Suppose Y and X are random variables with finite positive variances. Then*

$$-1 \leq Corr[Y, X] \leq 1, \tag{8.7}$$

with equality if and only if, for some constants a and b,

$$Y = a + bX. \tag{8.8}$$

Proof. Apply (8.5) with $U = Y - E[Y]$ and $V = X - E[X]$ to obtain

$$Cov[Y, X]^2 \leq Var[Y]Var[X], \tag{8.9}$$

from which (8.7) follows. Then (8.8) follows from (8.6), with $b = Cov[Y, X]/Var[X]$ and $a = E[Y] - bE[X]$, so that (8.8) is the least squares fit of X to Y. □

This inequality for the sample correlation coefficient of $n \times 1$ vectors $\mathbf{x}$ and $\mathbf{y}$ follows either by using Lemma 8.1 on $\mathbf{H}_n\mathbf{y}$ and $\mathbf{H}_n\mathbf{x}$, where $\mathbf{H}_n$ is the centering matrix (1.12), or by using Corollary 8.1 with (X, Y) having the empirical distribution given by $\mathbf{x}$ and $\mathbf{y}$, i.e.,

$$P[(X, Y) = (x, y)] = \frac{1}{n}\,\#\{i \mid (x_i, y_i) = (x, y)\}. \tag{8.10}$$

The next result also follows from Cauchy-Schwarz. It will be useful for Hotelling's T^2 in Section 8.4.1, and for canonical correlations in Section 13.3.

Corollary 8.2. *Suppose $\mathbf{y}$ and $\mathbf{d}$ are $1 \times K$ vectors, and $\|\mathbf{y}\| = 1$. Then*

$$(\mathbf{y}\mathbf{d}')^2 \leq \|\mathbf{d}\|^2, \tag{8.11}$$

with equality if and only if $\mathbf{d}$ is zero, or $\mathbf{d}$ is nonzero and

$$\mathbf{y} = \pm\frac{\mathbf{d}}{\|\mathbf{d}\|}. \tag{8.12}$$

8.2 Conditioning in a Wishart

We start with $\mathbf{W} \sim \text{Wishart}_{p+q}(\nu, \mathbf{\Sigma})$ as in Definition 3.6, where $\mathbf{\Sigma}$ is partitioned

$$\mathbf{\Sigma} = \begin{pmatrix} \mathbf{\Sigma}_{XX} & \mathbf{\Sigma}_{XY} \\ \mathbf{\Sigma}_{YX} & \mathbf{\Sigma}_{YY} \end{pmatrix}, \tag{8.13}$$

$\mathbf{\Sigma}_{XX}$ is $p \times p$, $\mathbf{\Sigma}_{YY}$ is $q \times q$, and $\mathbf{W}$ is partitioned similarly. We are mainly interested in the distribution of

$$\mathbf{W}_{YY\cdot X} = \mathbf{W}_{YY} - \mathbf{W}_{YX}\mathbf{W}_{XX}^{-1}\mathbf{W}_{XY} \tag{8.14}$$

(see Equation 3.49), but some additional results will easily come along for the ride.

Proposition 8.1. *Consider the situation above, where $\mathbf{\Sigma}_{XX}$ is invertible and $\nu \geq p$. Then*

$$(\mathbf{W}_{XX}, \mathbf{W}_{XY}) \text{ is independent of } \mathbf{W}_{YY\cdot X}, \tag{8.15}$$

$$\mathbf{W}_{YY\cdot X} \sim \text{Wishart}_q(\nu - p, \mathbf{\Sigma}_{YY\cdot X}), \tag{8.16}$$

and

$$\mathbf{W}_{XY} \mid \mathbf{W}_{XX} = \mathbf{w}_{xx} \sim N(\mathbf{w}_{xx}\mathbf{\Sigma}_{XX}^{-1}\mathbf{\Sigma}_{XY}, \mathbf{w}_{xx} \otimes \mathbf{\Sigma}_{YY\cdot X}), \tag{8.17}$$

$$\mathbf{W}_{XX} \sim \text{Wishart}_p(\nu, \mathbf{\Sigma}_{XX}). \tag{8.18}$$

Proof. The final equation is just the marginal of a Wishart, as in Section 3.6. By Definition 3.6 of the Wishart,

$$\mathbf{W} =^{\mathcal{D}} (\mathbf{X}\ \mathbf{Y})'(\mathbf{X}\ \mathbf{Y}), \text{ where } (\mathbf{X}\ \mathbf{Y}) \sim N(\mathbf{0}, \mathbf{I}_n \otimes \mathbf{\Sigma}), \tag{8.19}$$

$\mathbf{X}$ is $n \times p$, and $\mathbf{Y}$ is $n \times q$. Conditioning as in (3.56), we have

$$\mathbf{Y} \mid \mathbf{X} = \mathbf{x} \sim N_{n\times q}(\mathbf{x}\boldsymbol{\beta}, \mathbf{I}_n \otimes \mathbf{\Sigma}_{YY\cdot X}),\ \boldsymbol{\beta} = \mathbf{\Sigma}_{XX}^{-1}\mathbf{\Sigma}_{XY}. \tag{8.20}$$

(The $\boldsymbol{\alpha} = \mathbf{0}$ because the means of $\mathbf{X}$ and $\mathbf{Y}$ are zero.) Note that (8.20) is the both-sides model (6.1), with $\mathbf{z} = \mathbf{I}_q$ and $\mathbf{\Sigma}_R = \mathbf{\Sigma}_{YY\cdot X}$. Thus by Theorem 6.1 and the plug-in property (2.62) of conditional distributions, $\widehat{\boldsymbol{\beta}} = (\mathbf{X}'\mathbf{X})^{-1}\mathbf{X}'\mathbf{Y}$ and $\mathbf{Y}'\mathbf{Q}_\mathbf{X}\mathbf{Y}$ are conditionally independent given $\mathbf{X} = \mathbf{x}$,

$$\widehat{\boldsymbol{\beta}} \mid \mathbf{X} = \mathbf{x} \sim N(\boldsymbol{\beta}, (\mathbf{x}'\mathbf{x})^{-1} \otimes \mathbf{\Sigma}_{YY\cdot X}), \tag{8.21}$$

and

$$[\mathbf{Y}'\mathbf{Q}_\mathbf{X}\mathbf{Y} \mid \mathbf{X} = \mathbf{x}] =^{\mathcal{D}} [\mathbf{Y}'\mathbf{Q}_\mathbf{x}\mathbf{Y} \mid \mathbf{X} = \mathbf{x}] \sim \text{Wishart}_q(n - p, \mathbf{\Sigma}_{YY\cdot X}). \tag{8.22}$$

The conditional distribution in (8.22) does not depend on $\mathbf{x}$, hence $\mathbf{Y}'\mathbf{Q}_\mathbf{X}\mathbf{Y}$ is (unconditionally) independent of the pair $(\mathbf{X}, \widehat{\boldsymbol{\beta}})$, as in (2.65) and therebelow, hence

$$\mathbf{Y}'\mathbf{Q}_\mathbf{X}\mathbf{Y} \text{ is independent of } (\mathbf{X}'\mathbf{X}, \mathbf{X}'\mathbf{Y}). \tag{8.23}$$

Property (2.66) implies that

$$\widehat{\boldsymbol{\beta}} \mid \mathbf{X}'\mathbf{X} = \mathbf{x}'\mathbf{x} \sim N(\boldsymbol{\beta}, (\mathbf{x}'\mathbf{x})^{-1} \otimes \mathbf{\Sigma}_{YY\cdot X}), \tag{8.24}$$

hence

$$\mathbf{X}'\mathbf{Y} = (\mathbf{X}'\mathbf{X})\widehat{\boldsymbol{\beta}} \mid \mathbf{X}'\mathbf{X} = \mathbf{x}'\mathbf{x} \sim N((\mathbf{x}'\mathbf{x})\boldsymbol{\beta}, (\mathbf{x}'\mathbf{x}) \otimes \mathbf{\Sigma}_{YY\cdot X}). \tag{8.25}$$

Translating to $\mathbf{W}$ using (8.19), noting that $\mathbf{Y}'\mathbf{Q}_\mathbf{X}\mathbf{Y} = \mathbf{W}_{YY\cdot X}$, we have that (8.23) is (8.15), (8.22) is (8.16), and (8.25) is (8.17). □

8.3 Expected value of the inverse Wishart

We first prove Lemma 7.1 for

$$\mathbf{U} \sim \text{Wishart}_{l^*}(\nu, \mathbf{I}_{l^*}). \tag{8.26}$$

For any $l^* \times l^*$ orthogonal matrix $\mathbf{\Gamma}$, $\mathbf{\Gamma U \Gamma}'$ has the same distribution as $\mathbf{U}$, hence in particular

$$E[\mathbf{U}^{-1}] = E[(\mathbf{\Gamma U \Gamma}')^{-1}] = \mathbf{\Gamma} E[\mathbf{U}^{-1}]\mathbf{\Gamma}'. \tag{8.27}$$

Exercise 8.8.6 shows that any symmetric $l^* \times l^*$ matrix $\mathbf{A}$ for which $\mathbf{A} = \mathbf{\Gamma A \Gamma}'$ for all orthogonal $\mathbf{\Gamma}$ must be of the form $a_{11}\mathbf{I}_{l^*}$. Thus

$$E[\mathbf{U}^{-1}] = E[(\mathbf{U}^{-1})_{11}]\ \mathbf{I}_{l^*}. \tag{8.28}$$

Using (5.103),

$$E[(\mathbf{U}^{-1})_{11}] = E\left[\frac{1}{U_{11\cdot\{2:l^*\}}}\right] = E\left[\frac{1}{\chi^2_{\nu-l^*+1}}\right] = \frac{1}{\nu - l^* - 1}. \tag{8.29}$$

Equations (8.28) and (8.29) show that

$$E[\mathbf{U}^{-1}] = \frac{1}{\nu - l^* - 1}\ \mathbf{I}_{l^*}. \tag{8.30}$$

Next take $\mathbf{W} \sim \text{Wishart}_{l^*}(\nu, \mathbf{\Sigma})$, with $\mathbf{\Sigma}$ invertible. Then, $\mathbf{W} =^{\mathcal{D}} \mathbf{\Sigma}^{1/2}\mathbf{U}\mathbf{\Sigma}^{1/2}$, and

$$\begin{aligned} E[\mathbf{W}^{-1}] &= E[(\mathbf{\Sigma}^{1/2}\mathbf{U}\mathbf{\Sigma}^{1/2})^{-1}] \\ &= \mathbf{\Sigma}^{-1/2}\ \frac{1}{\nu - l^* - 1}\ \mathbf{I}_{l^*}\ \mathbf{\Sigma}^{-1/2} \\ &= \frac{1}{\nu - l^* - 1}\ \mathbf{\Sigma}^{-1}, \end{aligned} \tag{8.31}$$

verifying (7.32).

8.4 Distribution of Hotelling's T^2

Here we prove Proposition 7.1. Exercise 8.8.5 shows that we can assume $\mathbf{\Sigma} = \mathbf{I}_{l^*}$ in the proof, which we do. Divide and multiply the $\mathbf{Z}\mathbf{W}^{-1}\mathbf{Z}'$ by $\|\mathbf{Z}\|^2$:

$$\mathbf{Z}\mathbf{W}^{-1}\mathbf{Z}' = \frac{\mathbf{Z}\mathbf{W}^{-1}\mathbf{Z}'}{\|\mathbf{Z}\|^2}\ \|\mathbf{Z}\|^2. \tag{8.32}$$

Because $\mathbf{Z}$ is a vector of l^* independent standard normals,

$$\|\mathbf{Z}\|^2 \sim \chi^2_{l^*}. \tag{8.33}$$

Consider the distribution of the ratio in (8.32) conditional on $\mathbf{Z} = \mathbf{z}$. Because $\mathbf{Z}$ and $\mathbf{W}$ are independent, we can use the plugin formula (2.63), to obtain

$$\left[\frac{\mathbf{Z}\mathbf{W}^{-1}\mathbf{Z}'}{\|\mathbf{Z}\|^2}\ \Big|\ \mathbf{Z} = \mathbf{z}\right] =^{\mathcal{D}} \frac{\mathbf{z}\mathbf{W}^{-1}\mathbf{z}'}{\|\mathbf{z}\|^2} = \mathbf{g}_1\mathbf{W}^{-1}\mathbf{g}_1', \tag{8.34}$$

where $\mathbf{g}_1 = \mathbf{z}/\|\mathbf{z}\|$. Note that on the right-hand side we have the unconditional distribution for $\mathbf{W}$. Let $\mathbf{G}$ be any $l^* \times l^*$ orthogonal matrix with $\mathbf{g}_1$ as its first row. (Exercise 5.8.29 guarantees there is one.) Then

$$\mathbf{g}_1\mathbf{W}^{-1}\mathbf{g}_1' = \mathbf{e}_1\mathbf{G}\mathbf{W}^{-1}\mathbf{G}'\mathbf{e}_1',\ \mathbf{e}_1 = (1, 0, \ldots, 0). \tag{8.35}$$

Because the covariance parameter in the Wishart distribution for $\mathbf{W}$ is $\mathbf{I}_{l^*}$, $\mathbf{U} \equiv \mathbf{G}\mathbf{W}\mathbf{G}' \sim \text{Wishart}_{l^*}(\nu, \mathbf{I}_{l^*})$. But

$$\mathbf{e}_1\mathbf{G}\mathbf{W}^{-1}\mathbf{G}'\mathbf{e}_1' = \mathbf{e}_1\mathbf{U}^{-1}\mathbf{e}_1' = [\mathbf{U}^{-1}]_{11} = U_{11\cdot\{2:l^*\}}^{-1} \tag{8.36}$$

by (5.103).

Note that the distribution of $\mathbf{U}$, hence $[\mathbf{U}^{-1}]_{11}$, does not depend on $\mathbf{z}$, which means that

$$\frac{\mathbf{Z}\mathbf{W}^{-1}\mathbf{Z}'}{\|\mathbf{Z}\|^2} \text{ is independent of } \mathbf{Z}. \tag{8.37}$$

Furthermore, by (8.16), where $p = l^* - 1$ and $q = 1$,

$$U_{11\cdot\{2:l^*\}} \sim \text{Wishart}_1(\nu - l^* + 1, 1) \equiv \chi^2_{\nu - l^* + 1}. \tag{8.38}$$

Now (8.32) can be expressed as

$$\mathbf{Z}\mathbf{W}^{-1}\mathbf{Z}' = \frac{\|\mathbf{Z}\|^2}{U_{11\cdot\{2:l^*\}}} =^{\mathcal{D}} \frac{\chi^2_{l^*}}{\chi^2_{\nu - l^* + 1}}, \tag{8.39}$$

where the two χ^2's are independent. Then (7.29) follows from Definition 7.1 for the F.

8.4.1 A motivation for Hotelling's T^2

Hotelling's T^2 test can be motivated using the projection pursuit idea. We have that $\mathbf{Z} \sim N_{l^*}(\boldsymbol{\mu}, \boldsymbol{\Sigma})$ and $\mathbf{W} \sim \text{Wishart}_{l^*}(\nu, \boldsymbol{\Sigma})$, independent, and wish to test $H_0 : \boldsymbol{\mu} = \mathbf{0}$. Let $\mathbf{a} \neq \mathbf{0}$ be a $1 \times l^*$ vector of constants. Then under H_0,

$$\mathbf{Z}\mathbf{a}' \sim N(0, \mathbf{a}\boldsymbol{\Sigma}\mathbf{a}') \text{ and } \mathbf{a}\mathbf{W}\mathbf{a}' \sim \text{Wishart}_1(\nu, \mathbf{a}\boldsymbol{\Sigma}\mathbf{a}') = (\mathbf{a}\boldsymbol{\Sigma}\mathbf{a}')\,\chi^2_\nu. \tag{8.40}$$

Now we are basically in the univariate t (6.23) case, i.e.,

$$T_{\mathbf{a}} = \frac{\mathbf{Z}\mathbf{a}'}{\sqrt{\mathbf{a}\mathbf{W}\mathbf{a}'/\nu}} \sim t_\nu, \tag{8.41}$$

or, since $t_\nu^2 = F_{1,\nu}$,

$$T_{\mathbf{a}}^2 = \frac{(\mathbf{Z}\mathbf{a}')^2}{\mathbf{a}\mathbf{W}\mathbf{a}'/\nu} \sim F_{1,\nu}. \tag{8.42}$$

For any $\mathbf{a}$, we can do a regular F test. The projection pursuit approach is to find the $\mathbf{a}$ that gives the most significant result. That is, we wish to find

$$T^2 = \max_{\mathbf{a} \neq \mathbf{0}} T_{\mathbf{a}}^2. \tag{8.43}$$

To find the best $\mathbf{a}$, first simplify the denominator by setting

$$\mathbf{b} = \mathbf{a}\mathbf{W}^{1/2}, \text{ so that } \mathbf{a} = \mathbf{b}\mathbf{W}^{-1/2}. \tag{8.44}$$

Then

$$T^2 = \max_{\mathbf{b} \neq \mathbf{0}} \nu \frac{(\mathbf{V}\mathbf{b}')^2}{\mathbf{b}\mathbf{b}'}, \text{ where } \mathbf{V} = \mathbf{Z}\mathbf{W}^{-1/2}. \tag{8.45}$$

Letting $\mathbf{g} = \mathbf{b}/\|\mathbf{b}\|$, so that $\|\mathbf{g}\| = 1$, Corollary 8.2 of Cauchy-Schwarz shows that (see Exercise 8.8.9)

$$T^2 = \nu \max_{\mathbf{g} \,|\, \|\mathbf{g}\|=1} (\mathbf{V}\mathbf{g}')^2 = \nu \mathbf{V}\mathbf{V}' = \nu \mathbf{Z}\mathbf{W}^{-1}\mathbf{Z}', \tag{8.46}$$

which is indeed Hotelling's T^2 of (7.28), a multivariate generalization of Student's t^2. Even though $T^2_{\mathbf{a}}$ has an $F_{1,\nu}$ distribution, the T^2 does not have that distribution, because it maximizes over many $F_{1,\nu}$'s.

8.5 Density of the multivariate normal

Except for when using likelihood methods in Chapter 9, we do not need the density of the multivariate normal, nor of the Wishart, for our main purposes, but present them here because of their intrinsic interest. We start with the multivariate normal, with positive definite covariance matrix.

Lemma 8.2. *Suppose* $\mathbf{Y} \sim N_{1\times N}(\boldsymbol{\mu}, \boldsymbol{\Omega})$*, where* $\boldsymbol{\Omega}$ *is positive definite. Then the pdf of* $\mathbf{Y}$ *is*

$$f(\mathbf{y} \,|\, \boldsymbol{\mu}, \boldsymbol{\Omega}) = \frac{1}{(2\pi)^{N/2}} \frac{1}{|\boldsymbol{\Omega}|^{1/2}} e^{-\frac{1}{2}(\mathbf{y}-\boldsymbol{\mu})\boldsymbol{\Omega}^{-1}(\mathbf{y}-\boldsymbol{\mu})'}. \tag{8.47}$$

Proof. Recall that a multivariate normal vector is an affine transform of a vector of independent standard normals,

$$\mathbf{Y} = \mathbf{Z}\mathbf{A}' + \mu, \; \mathbf{Z} \sim N_{1\times N}(\mathbf{0}, \mathbf{I}_N), \;\; \mathbf{A}\mathbf{A}' = \boldsymbol{\Omega}. \tag{8.48}$$

We will take $\mathbf{A}$ to be $N \times N$, so that $\boldsymbol{\Omega}$ being positive definite implies that $\mathbf{A}$ is invertible. Then

$$\mathbf{Z} = (\mathbf{Y} - \boldsymbol{\mu})(\mathbf{A}')^{-1}, \tag{8.49}$$

and the Jacobian is

$$|\frac{\partial \mathbf{z}}{\partial \mathbf{y}}| \equiv \left| \begin{pmatrix} \partial z_1/\partial y_1 & \partial z_1/\partial y_2 & \cdots & \partial z_1/\partial y_N \\ \partial z_2/\partial y_1 & \partial z_2/\partial y_2 & \cdots & \partial z_2/\partial y_N \\ \vdots & \vdots & \ddots & \vdots \\ \partial z_N/\partial y_1 & \partial z_N/\partial y_2 & \cdots & \partial z_N/\partial y_N \end{pmatrix} \right| = |(\mathbf{A}')^{-1}|. \tag{8.50}$$

The density of $\mathbf{Z}$ is

$$f(\mathbf{z} \,|\, \mathbf{0}, \mathbf{I}_N) = \frac{1}{(2\pi)^{N/2}} e^{-\frac{1}{2}(z_1^2 + \cdots + z_N^2)} = \frac{1}{(2\pi)^{N/2}} e^{-\frac{1}{2}\mathbf{z}\mathbf{z}'}, \tag{8.51}$$

so that

$$\begin{aligned} f(\mathbf{y} \,|\, \boldsymbol{\mu}, \boldsymbol{\Omega}) &= \frac{1}{(2\pi)^{N/2}} \text{abs}|(\mathbf{A}')^{-1}| \, e^{-\frac{1}{2}(\mathbf{y}-\boldsymbol{\mu})(\mathbf{A}')^{-1}\mathbf{A}^{-1}(\mathbf{y}-\mu)'} \\ &= \frac{1}{(2\pi)^{N/2}} |\mathbf{A}\mathbf{A}'|^{-1/2} e^{-\frac{1}{2}(\mathbf{y}-\boldsymbol{\mu})(\mathbf{A}\mathbf{A}')^{-1}(\mathbf{y}-\mu)'}, \end{aligned} \tag{8.52}$$

from which (8.47) follows. □

When $\mathbf{Y}$ can be written as a matrix with a Kronecker product for its covariance matrix, as is often the case for us, the pdf can be compactified.

Corollary 8.3. *Suppose* $\mathbf{Y} \sim N_{n\times q}(\mathbf{M}, \mathbf{C} \otimes \mathbf{\Sigma})$, *where* $\mathbf{C}$ $(n \times n)$ *and* $\mathbf{\Sigma}$ $(q \times q)$ *are positive definite. Then*

$$f(\mathbf{y} \mid \mathbf{M}, \mathbf{C}, \mathbf{\Sigma}) = \frac{1}{(2\pi)^{nq/2}} \frac{1}{|\mathbf{C}|^{q/2}|\mathbf{\Sigma}|^{n/2}} e^{-\frac{1}{2}\operatorname{trace}(\mathbf{C}^{-1}(\mathbf{y}-\mathbf{M})\mathbf{\Sigma}^{-1}(\mathbf{y}-\mathbf{M})')}. \tag{8.53}$$

See Exercise 8.8.15 for the proof.

8.6 The QR decomposition for the multivariate normal

Here we discuss the distributions of the $\mathbf{Q}$ and $\mathbf{R}$ matrices in the QR decomposition of a multivariate normal matrix. From the distribution of the upper triangular $\mathbf{R}$ we obtain **Bartlett's decomposition** [Bartlett, 1939], useful for randomly generating Wisharts, as well as for deriving the Wishart density in Section 8.7. Also, we see that $\mathbf{Q}$ has a certain uniform distribution, which provides a method for generating random orthogonal matrices from random normals. The results are found in Olkin and Roy [1954], and this presentation is close to that of Kshirsagar [1959]. (Old school, indeed!)

We start with the data matrix

$$\mathbf{Z} \sim N_{\nu\times q}(\mathbf{0}, \mathbf{I}_\nu \otimes \mathbf{I}_q), \tag{8.54}$$

a matrix of independent $N(0,1)$'s, where $\nu \geq q$, and consider the QR decomposition (Theorem 5.4)

$$\mathbf{Z} = \mathbf{QR}. \tag{8.55}$$

We find the distribution of the $\mathbf{R}$. Let

$$\mathbf{S} \equiv \mathbf{Z}'\mathbf{Z} = \mathbf{R}'\mathbf{R} \sim \text{Wishart}_q(\nu, \mathbf{I}_q). \tag{8.56}$$

Apply Proposition 8.1 with $\mathbf{S}_{XX}$ being the single element S_{11}. Because $\mathbf{\Sigma} = \mathbf{I}_q$, $(S_{11}, \mathbf{S}_{1\{2:q\}})$ is independent of $\mathbf{S}_{\{2:q\}\{2:q\}\cdot 1}$,

$$\begin{aligned} S_{11} &\sim \text{Wishart}_1(\nu, \mathbf{I}_1) = \chi^2_\nu, \\ \mathbf{S}_{1\{2:q\}} \mid S_{11} = s_{11} &\sim N_{1\times(q-1)}(\mathbf{0}, s_{11}\mathbf{I}_{q-1}), \\ \text{and } \mathbf{S}_{\{2:q\}\{2:q\}\cdot 1} &\sim \text{Wishart}_{q-1}(\nu - 1, \mathbf{I}_{q-1}). \end{aligned} \tag{8.57}$$

Note that $\mathbf{S}_{1\{2:q\}}/\sqrt{S_{11}}$, conditional on S_{11}, is $N(\mathbf{0}, \mathbf{I}_{q-1})$, in particular, is independent of S_{11}. Thus the three quantities S_{11}, $\mathbf{S}_{1\{2:q\}}/\sqrt{S_{11}}$, and $\mathbf{S}_{\{2:q\}\{2:q\}\cdot 1}$ are mutually independent. Equation (5.84) shows that

$$\begin{aligned} R_{11} = \sqrt{S_{11}} &\sim \sqrt{\chi^2_\nu} \\ \text{and } \begin{pmatrix} R_{12} & \cdots & R_{1q} \end{pmatrix} = \mathbf{S}_{1\{2:q\}}/\sqrt{S_{11}} &\sim N_{1\times(q-1)}(\mathbf{0}, \mathbf{I}_{q-1}). \end{aligned} \tag{8.58}$$

Next, work on the first component of $S_{22\cdot 1}$ of $\mathbf{S}_{\{2:q\}\{2:q\}\cdot 1}$. We find that

$$\begin{aligned} R_{22} = \sqrt{S_{22\cdot 1}} &\sim \sqrt{\chi^2_{\nu-1}} \\ \text{and } \begin{pmatrix} R_{23} & \cdots & R_{2q} \end{pmatrix} = \mathbf{S}_{2\{3:q\}\cdot 1}/\sqrt{S_{22\cdot 1}} &\sim N_{1\times(q-2)}(\mathbf{0}, \mathbf{I}_{q-2}), \end{aligned} \tag{8.59}$$

both independent of each other, and of $\mathbf{S}_{\{3:q\}\{3:q\}\cdot 12}$. We continue, obtaining the following result.

Lemma 8.3 (Bartlett's decomposition). *Suppose* $\mathbf{S} \sim \text{Wishart}_q(\nu, \mathbf{I}_q)$, *where* $\nu \geq q$, *and let its Cholesky decomposition be* $\mathbf{S} = \mathbf{R}'\mathbf{R}$. *Then the elements of* $\mathbf{R}$ *are mutually independent, where*

$$R_{ii}^2 \sim \chi^2_{\nu-i+1}, i = 1, \ldots, q, \text{ and } R_{ij} \sim N(0,1), 1 \leq i < j \leq q. \tag{8.60}$$

Next, suppose $\mathbf{Y} \sim N_{\nu \times q}(\mathbf{0}, \mathbf{I}_\nu \otimes \mathbf{\Sigma})$, where $\mathbf{\Sigma}$ is invertible. Let $\mathbf{A}$ be the matrix such that

$$\mathbf{\Sigma} = \mathbf{A}'\mathbf{A}, \text{ where } \mathbf{A} \in \mathcal{T}_q^+ \text{ of (5.80)}, \tag{8.61}$$

i.e., $\mathbf{A}'\mathbf{A}$ is the Cholesky decomposition of $\mathbf{\Sigma}$. Thus we can take $\mathbf{Y} = \mathbf{ZA}$ for $\mathbf{Z}$ in (8.54). Now by (8.55), $\mathbf{Y} = \mathbf{QV}$, where $\mathbf{V} \equiv \mathbf{RA}$ is also in $\mathcal{T}_q^+$, and $\mathbf{Q}$ still has orthonormal columns. By the uniqueness of the QR decomposition, $\mathbf{QV}$ is the QR decomposition for $\mathbf{Y}$. Then

$$\mathbf{Y}'\mathbf{Y} = \mathbf{V}'\mathbf{V} \sim \text{Wishart}(\mathbf{0}, \mathbf{\Sigma}). \tag{8.62}$$

We call the distribution of $\mathbf{V}$ the **Half-Wishart**$_q(\nu, \mathbf{A})$.

To generate a random $\mathbf{W} \sim \text{Wishart}_q(\nu, \mathbf{\Sigma})$ matrix, one can first generate $q(q-1)/2$ $N(0,1)$'s, and q χ^2's, all independently, then set the R_{ij}'s as in (8.60), then calculate $\mathbf{V} = \mathbf{RA}$, and $\mathbf{W} = \mathbf{V}'\mathbf{V}$. If ν is large, this process is more efficient than generating the νq normals in $\mathbf{Z}$ or $\mathbf{Y}$. The next section derives the density of the Half-Wishart, then that of the Wishart itself.

We end this section by completing description of the joint distribution of $(\mathbf{Q}, \mathbf{V})$. Exercise 3.7.37 handled the case $\mathbf{Z} \sim N_{2\times 1}(\mathbf{0}, \mathbf{I}_2)$.

Lemma 8.4. *Suppose* $\mathbf{Y} = \mathbf{QV}$ *as above. Then*

(i) $\mathbf{Q}$ *and* $\mathbf{V}$ *are independent;*

(ii) The distribution of $\mathbf{Q}$ *does not depend on* $\mathbf{\Sigma}$*;*

(iii) The distribution of $\mathbf{Q}$ *is invariant under orthogonal transforms: If* $\mathbf{\Gamma} \in \mathcal{O}_n$, *the group of* $n \times n$ *orthogonal matrices (see (5.75)), then*

$$\mathbf{Q} =^{\mathcal{D}} \mathbf{\Gamma Q}. \tag{8.63}$$

Proof. From above, we see that $\mathbf{Z}$ and $\mathbf{Y} = \mathbf{ZA}$ have the same $\mathbf{Q}$. The distribution of $\mathbf{Z}$ does not depend on $\mathbf{\Sigma}$, hence neither does the distribution of $\mathbf{Q}$, proving part (ii). For part (iii), consider $\mathbf{\Gamma Y}$, which has the same distribution as $\mathbf{Y}$. We have $\mathbf{\Gamma Y} = (\mathbf{\Gamma Q})\mathbf{V}$. Since $\mathbf{\Gamma Q}$ also has orthonormal columns, the uniqueness of the QR decomposition implies that $\mathbf{\Gamma Q}$ is the "$\mathbf{Q}$" for $\mathbf{\Gamma Y}$. Thus $\mathbf{Q}$ and $\mathbf{\Gamma Q}$ have the same distribution.

Proving the independence result of part (i) takes some extra machinery from mathematical statistics. See, e.g., Lehmann and Casella [2003]. Rather than providing all the details, we outline how one can go about the proof. First, $\mathbf{V}$ can be shown to be a complete sufficient statistic for the model $\mathbf{Y} \sim N(\mathbf{0}, \mathbf{I}_\nu \otimes \mathbf{\Sigma})$. Basu's Lemma says that any statistic whose distribution does not depend on the parameter, in this case $\mathbf{\Sigma}$, is independent of the complete sufficient statistic. Thus by part (ii), $\mathbf{Q}$ is independent of $\mathbf{V}$. □

If $n = q$, the $\mathbf{Q}$ is an orthogonal matrix, and its distribution has the **Haar** probability measure, or uniform distribution, over $\mathcal{O}_\nu$. It is the only probability distribution that does have the above invariance property, although proving the fact is beyond this book. See Halmos [1950]. Thus one can generate a random $q \times q$ orthogonal matrix by first generating an $q \times q$ matrix of independent $N(0,1)$'s, then performing Gram-Schmidt orthogonalization on the columns, normalizing the results so that the columns have norm 1.

8.7 Density of the Wishart

We derive the density of the Half-Wishart, then the Wishart. We need to be careful with constants, and find two Jacobians. Some details are found in Exercises 8.8.16 to 8.8.21.

We start by writing down the density of $\mathbf{R} \sim \text{Half-Wishart}_q(\nu, \mathbf{I}_q)$, assuming $n \geq q$, as in (8.60). The density of U (> 0), where $U^2 \sim \chi^2_k$, is

$$f_k(u) = \frac{1}{\Gamma(k/2)2^{(k/2)-1}}\, u^{k-1} e^{-\frac{1}{2}u^2}. \tag{8.64}$$

Thus the density for $\mathbf{R}$, with the q χ^2's and $q(q-1)/2$ $N(0,1)$'s (all independent), is

$$f_{\mathbf{R}}(\mathbf{r}) = \frac{1}{c(\nu,q)}\, r_{11}^{\nu-1} r_{22}^{\nu-2} \cdots r_{qq}^{n-q}\, e^{-\frac{1}{2}\,\text{trace}(\mathbf{r}'\mathbf{r})}, \tag{8.65}$$

where

$$c(\nu,q) = \pi^{q(q-1)/4}\, 2^{(\nu q/2)-q} \prod_{j=1}^{q} \Gamma\left(\frac{\nu - j + 1}{2}\right). \tag{8.66}$$

For $\mathbf{V} \sim \text{Half-Wishart}_q(\nu, \mathbf{\Sigma})$, where $\mathbf{\Sigma}$ is invertible, we set $\mathbf{V} = \mathbf{RA}$, where $\mathbf{A}'\mathbf{A}$ is the Cholesky decomposition of $\mathbf{\Sigma}$ in (8.61). The Jacobian J is given by

$$\frac{1}{J} = \left|\frac{\partial \mathbf{v}}{\partial \mathbf{r}}\right| = a_{11} a_{22}^2 \cdots a_{qq}^q. \tag{8.67}$$

Thus, since $v_{jj} = a_{jj} r_{jj}$, the density of $\mathbf{V}$ is

$$\begin{aligned} f_{\mathbf{V}}(\mathbf{v} \mid \mathbf{\Sigma}) &= \frac{1}{c(\nu,q)} \frac{v_{11}^{\nu-1} v_{22}^{\nu-2} \cdots v_{qq}^{\nu-q}}{a_{11}^{\nu-1} a_{22}^{\nu-2} \cdots a_{qq}^{\nu-q}}\, e^{-\frac{1}{2}\,\text{trace}((\mathbf{A}')^{-1}\mathbf{v}'\mathbf{v}\mathbf{A}^{-1})}\, \frac{1}{a_{11} a_{22}^2 \cdots a_{qq}^q} \\ &= \frac{1}{c(\nu,q)} \frac{1}{|\mathbf{\Sigma}|^{\nu/2}}\, v_{11}^{\nu-1} v_{22}^{\nu-2} \cdots v_{qq}^{\nu-q}\, e^{-\frac{1}{2}\,\text{trace}(\mathbf{\Sigma}^{-1}\mathbf{v}'\mathbf{v})}, \end{aligned} \tag{8.68}$$

since $|\mathbf{\Sigma}| = \prod a_{ii}^2$. See Exercise 5.8.44.

Finally, suppose $\mathbf{W} \sim \text{Wishart}_q(\nu, \mathbf{\Sigma})$, so that we can take $\mathbf{W} = \mathbf{V}'\mathbf{V}$. The Jacobian is

$$\frac{1}{J^*} = \left|\frac{\partial \mathbf{w}}{\partial \mathbf{v}}\right| = 2^q v_{11}^q v_{22}^{q-1} \cdots v_{qq}. \tag{8.69}$$

Thus from (8.68),

$$f_{\mathbf{W}}(\mathbf{w} \mid \mathbf{\Sigma}) = \frac{1}{2^q} \frac{1}{c(\nu, q)} \frac{1}{|\mathbf{\Sigma}|^{\nu/2}} \frac{v_{11}^{\nu-1} v_{22}^{\nu-2} \cdots v_{qq}^{\nu-q}}{v_{11}^{q} v_{22}^{q-1} \cdots v_{qq}} e^{-\frac{1}{2}\,\mathrm{trace}(\mathbf{\Sigma}^{-1}\mathbf{w})}$$
$$= \frac{1}{d(\nu,q)} \frac{1}{|\mathbf{\Sigma}|^{\nu/2}} |\mathbf{w}|^{(\nu-q-1)/2} e^{-\frac{1}{2}\,\mathrm{trace}(\mathbf{\Sigma}^{-1}\mathbf{w})}, \tag{8.70}$$

where

$$d(\nu, q) = \pi^{q(q-1)/4} 2^{\nu q/2} \prod_{j=1}^{q} \Gamma\left(\frac{\nu - j + 1}{2}\right). \tag{8.71}$$

8.8 Exercises

Exercise 8.8.1. Suppose $\mathbf{y}$ is $1 \times K$, $\mathbf{D}$ is $N \times K$, and $\mathbf{D}'\mathbf{D}$ is invertible. Let $\widehat{\mathbf{y}}$ be the projection of $\mathbf{y}$ onto the span of the rows of $\mathbf{D}'$, so that $\widehat{\mathbf{y}} = \widehat{\gamma}\mathbf{D}'$, where $\widehat{\gamma} = \mathbf{y}\mathbf{D}(\mathbf{D}'\mathbf{D})^{-1}$ is the least squares estimate as in (5.18). Show that

$$\|\widehat{\mathbf{y}}\|^2 = \mathbf{y}\mathbf{D}(\mathbf{D}'\mathbf{D})^{-1}\mathbf{D}'\mathbf{y}'. \tag{8.72}$$

(Notice the projection matrix, from (5.20).) Show that in the case $\mathbf{D} = \mathbf{d}'$, i.e., $N = 1$, (8.72) yields the equality in (8.4).

Exercise 8.8.2. Prove the Cauchy-Schwarz inequality for random variables U and V given in (8.5) and (8.6), assuming that V is nonzero with probability one. [Hint: Use least squares, by finding b to minimize $E[(U - bV)^2]$.]

Exercise 8.8.3. Prove Corollary 8.2. [Hint: Show that (8.11) follows from (8.1), and that (8.2) implies that $\widehat{\gamma} = \pm 1/\|\mathbf{d}\|$, using the fact that $\|\mathbf{y}\| = 1$.]

Exercise 8.8.4. For $\mathbf{W}$ in (8.19), verify that $\mathbf{X}'\mathbf{X} = \mathbf{W}_{XX}$, $\mathbf{X}'\mathbf{Y} = \mathbf{W}_{XY}$, and $\mathbf{Y}'\mathbf{Q_X}\mathbf{Y} = \mathbf{W}_{YY\cdot X}$, where $\mathbf{Q_X} = \mathbf{I}_n - \mathbf{X}(\mathbf{X}'\mathbf{X})^{-1}\mathbf{X}'$.

Exercise 8.8.5. Suppose $\mathbf{Z} \sim N_{1\times l^*}(\mathbf{0}, \mathbf{\Sigma})$ and $\mathbf{W} \sim \mathrm{Wishart}_{l^*}(\nu, \mathbf{\Sigma})$ are as in Proposition 7.1. (a) Show that for any $l^* \times l^*$ invertible matrix $\mathbf{A}$,

$$\mathbf{Z}\mathbf{W}^{-1}\mathbf{Z}' = (\mathbf{Z}\mathbf{A})(\mathbf{A}'\mathbf{W}\mathbf{A})^{-1}(\mathbf{Z}\mathbf{A})'. \tag{8.73}$$

(b) For what $\mathbf{A}$ do we have $\mathbf{Z}\mathbf{A} \sim N_{1\times l^*}(\mathbf{0}, \mathbf{I}_{l^*})$ and $\mathbf{A}'\mathbf{W}\mathbf{A} \sim \mathrm{Wishart}_{l^*}(\nu, \mathbf{I}_{l^*})$?

Exercise 8.8.6. Let $\mathbf{A}$ be a $q \times q$ symmetric matrix, and for $q \times q$ orthogonal matrix $\mathbf{\Gamma}$, contemplate the equality

$$\mathbf{A} = \mathbf{\Gamma}\mathbf{A}\mathbf{\Gamma}'. \tag{8.74}$$

(a) Suppose (8.74) holds for all permutation matrices $\mathbf{\Gamma}$ (matrices with one "1" in each row, and one "1" in each column, and zeroes elsewhere). Show that all the diagonals of $\mathbf{A}$ must be equal (i.e., $a_{11} = a_{22} = \cdots = a_{qq}$), and that all off-diagonals must be equal (i.e., $a_{ij} = a_{kl}$ if $i \neq j$ and $k \neq l$). [Hint: You can use the permutation matrix that switches the first two rows and first two columns,

$$\mathbf{\Gamma} = \begin{pmatrix} 0 & 1 & \mathbf{0} \\ 1 & 0 & \mathbf{0} \\ \mathbf{0} & \mathbf{0} & \mathbf{I}_{q-2} \end{pmatrix}, \tag{8.75}$$

to show that $a_{11} = a_{22}$ and $a_{1i} = a_{2i}$ for $i = 3, \ldots, q$. Similar equalities can be obtained by switching other pairs of rows and columns.] (b) Suppose (8.74) holds for all $\mathbf{\Gamma}$ that are diagonal, with each diagonal element being either $+1$ or -1. (They needn't all be the same sign.) Show that all off-diagonals must be 0. (c) Suppose (8.74) holds for all orthogonal $\mathbf{\Gamma}$. Show that $\mathbf{A}$ must be of the form $a_{11}\mathbf{I}_q$. [Hint: Use parts (a) and (b).]

Exercise 8.8.7. Verify the three equalities in (8.36).

Exercise 8.8.8. Show that $t_\nu^2 = F_{1,\nu}$.

Exercise 8.8.9. Find the $\mathbf{g}$ in (8.46) that maximizes $(\mathbf{V}\mathbf{g}')^2$, and show that the maximum is indeed $\mathbf{V}\mathbf{V}'$. (Use Corollary 8.2.) What is a maximizing $\mathbf{a}$ in (8.43)?

Exercise 8.8.10. Suppose that $\mathbf{z} = \mathbf{y}\mathbf{B}$, where $\mathbf{z}$ and $\mathbf{y}$ are $1 \times N$, and $\mathbf{B}$ is $N \times N$ and invertible. Show that $|\partial\mathbf{z}/\partial\mathbf{y}| = |\mathbf{B}|$. (Recall (8.50).)

Exercise 8.8.11. Show that for $N \times N$ matrix $\mathbf{A}$, $\text{abs}|(\mathbf{A}')^{-1}| = |\mathbf{A}\mathbf{A}'|^{-1/2}$.

Exercise 8.8.12. Let

$$(\mathbf{X}, \mathbf{Y}) \sim N_{p+q}\left((\mathbf{0}, \mathbf{0}), \begin{pmatrix} \mathbf{\Sigma}_{XX} & 0 \\ 0 & \mathbf{\Sigma}_{YY} \end{pmatrix}\right), \tag{8.76}$$

where $\mathbf{X}$ is $1 \times p$, $\mathbf{Y}$ is $1 \times q$, $Cov[\mathbf{X}] = \mathbf{\Sigma}_{XX}$, and $Cov[\mathbf{Y}] = \mathbf{\Sigma}_{YY}$. By writing out the density of $(\mathbf{X}, \mathbf{Y})$, show that $\mathbf{X}$ and $\mathbf{Y}$ are independent. (Assume the covariances are invertible.)

Exercise 8.8.13. Take $(\mathbf{X}, \mathbf{Y})$ as in (8.76). Show that $\mathbf{X}$ and $\mathbf{Y}$ are independent by using moment generating functions. Do you need that the covariances are invertible?

Exercise 8.8.14. With

$$(X, Y) \sim N_{1\times 2}\left((\mu_X, \mu_Y), \begin{pmatrix} \sigma_{XX} & \sigma_{XY} \\ \sigma_{YX} & \sigma_{YY} \end{pmatrix}\right), \tag{8.77}$$

derive the conditional distribution $Y \mid X = x$ explicitly using densities, assuming the covariance matrix is invertible. That is, show that $f_{Y|X}(y|x) = f(x, y)/f_X(x)$.

Exercise 8.8.15. Prove Corollary 8.3. [Hint: Make the identifications $\mathbf{y} \to \text{row}(\mathbf{y})$, $\boldsymbol{\mu} \to \text{row}(\mathbf{M})$, and $\mathbf{\Omega} \to \mathbf{C} \otimes \mathbf{\Sigma}$ in (8.47). Use (3.32f) for the determinant term in the density. For the term in the exponent, use (3.32d) to help show that

$$\begin{aligned}(\text{row}(\mathbf{y}) - \text{row}(\mathbf{M}))&(\mathbf{C}^{-1} \otimes \mathbf{\Sigma}^{-1})(\text{row}(\mathbf{y}) - \text{row}(\mathbf{M}))' \\ &= \text{trace}(\mathbf{C}^{-1}(\mathbf{y} - \mathbf{M})\mathbf{\Sigma}^{-1}(\mathbf{y} - \mathbf{M})').]\end{aligned} \tag{8.78}$$

Exercise 8.8.16. Show that $U = \sqrt{X}$, where $X \sim \chi_k^2$, has density as in (8.64).

Exercise 8.8.17. Verify (8.65) and (8.66). [Hint: Collect the constants as in (8.64), along with the constants from the normals (the R_{ij}'s, $j > i$). The trace in the exponent collects all the r_{ij}^2.]

Exercise 8.8.18. Verify (8.67). [Hint: Vectorize the matrices by row, leaving out the structural zeroes, i.e., for $q = 3$, $\mathbf{v} \to (v_{11}, v_{12}, v_{13}, v_{22}, v_{23}, v_{33})$. Then the matrix of derivatives will be lower triangular.]

Exercise 8.8.19. Verify (8.68). In particular, show that

$$\text{trace}((\mathbf{A}')^{-1}\mathbf{v}'\mathbf{v}\mathbf{A}^{-1}) = \text{trace}(\boldsymbol{\Sigma}^{-1}\mathbf{v}'\mathbf{v}) \tag{8.79}$$

and $\prod a_{jj} = |\boldsymbol{\Sigma}|^{1/2}$. [Recall (5.111).]

Exercise 8.8.20. Verify (8.69). [Hint: Vectorize the matrices as in Exercise 8.8.18, where for $\mathbf{w}$ just take the elements in the upper triangular part.]

Exercise 8.8.21. Verify (8.70) and (8.71).

Exercise 8.8.22. Suppose $\mathbf{V} \sim \text{Half-Wishart}_q(\nu, \boldsymbol{\Sigma})$ as in (8.62), where $\boldsymbol{\Sigma}$ is positive definite and $\nu \geq p$. Show that the diagonals V_{jj} are independent, and

$$V_{ii}^2 \sim \sigma_{jj \cdot \{1:(j-1)\}} \chi^2_{\nu-j+1}. \tag{8.80}$$

[Hint: Show that with $\mathbf{V} = \mathbf{RA}$ as in (8.60) and (8.61), $V_{jj} = a_{jj}R_{jj}$. Apply (5.84) to the $\mathbf{A}$ and $\boldsymbol{\Sigma}$.]

Exercise 8.8.23. Verify the Jacobian in (8.69). [Hint: Vectorize the $\mathbf{V}$ as in Exercise 8.8.18, and the $\mathbf{S}$ similarly, recognizing that by symmetry, one needs only the elements s_{ij} with $j \geq i$.]

Exercise 8.8.24. For a covariance matrix $\boldsymbol{\Sigma}$, $|\boldsymbol{\Sigma}|$ is called the population **generalized variance**. It is an overall measure of spread. Suppose $\mathbf{W} \sim \text{Wishart}_q(\nu, \boldsymbol{\Sigma})$, where $\boldsymbol{\Sigma}$ is positive definite and $\nu \geq q$. Show that

$$\widehat{|\boldsymbol{\Sigma}|} = \frac{1}{\nu(\nu-1)\cdots(\nu-q+1)} |\mathbf{W}| \tag{8.81}$$

is an unbiased estimate of the generalized variance. [Hint: Find the Cholesky decomposition $\mathbf{W} = \mathbf{V}'\mathbf{V}$, then use (8.80) and (5.104).]

Exercise 8.8.25 (Bayesian inference). Consider Bayesian inference for the covariance matrix. It turns out that the conjugate prior is an inverse Wishart on the covariance matrix, which means $\boldsymbol{\Sigma}^{-1}$ has a Wishart prior. Specifically, let

$$\boldsymbol{\Psi} = \boldsymbol{\Sigma}^{-1} \text{ and } \nu_0 \boldsymbol{\Psi}_0 = \boldsymbol{\Sigma}_0^{-1}, \tag{8.82}$$

where $\boldsymbol{\Sigma}_0$ is the prior guess of $\boldsymbol{\Sigma}$, and ν_0 is the "prior sample size." (The larger the ν_0, the more weight is placed on the prior vs. the data.) Then the model in terms of the inverse covariance parameter matrices is

$$\begin{aligned} \mathbf{W} \mid \boldsymbol{\Psi} = \boldsymbol{\psi} &\sim \text{Wishart}_q(\nu, \boldsymbol{\psi}^{-1}) \\ \boldsymbol{\Psi} &\sim \text{Wishart}_q(\nu_0, \boldsymbol{\Psi}_0), \end{aligned} \tag{8.83}$$

where $\nu \geq q$, $\nu_0 \geq q$, and $\boldsymbol{\Psi}_0$ is positive definite, so that $\boldsymbol{\Psi}$, hence $\boldsymbol{\Sigma}$, is invertible with probability one. Note that the prior mean for $\boldsymbol{\Psi}$ is $\boldsymbol{\Sigma}_0^{-1}$. (a) Show that the joint density of $(\mathbf{W}, \boldsymbol{\Psi})$ is

$$f_{\mathbf{W}\mid\boldsymbol{\Psi}}(\mathbf{w} \mid \boldsymbol{\psi}) f_{\boldsymbol{\Psi}}(\boldsymbol{\psi}) = c(\mathbf{w}) |\boldsymbol{\psi}|^{(\nu+\nu_0-q-1)/2} e^{-\frac{1}{2}\text{trace}((\mathbf{w}+\boldsymbol{\Psi}_0^{-1})\boldsymbol{\psi})}, \tag{8.84}$$

where $c(\mathbf{w})$ is some constant that does not depend on ψ, though it does depend on $\mathbf{\Psi}_0$ and ν_0. (b) Without doing any calculations, show that the posterior distribution of $\mathbf{\Psi}$ is

$$\mathbf{\Psi} \mid \mathbf{W} = \mathbf{w} \sim \text{Wishart}_q(\nu + \nu_0, (\mathbf{w} + \mathbf{\Psi}_0^{-1})^{-1}). \tag{8.85}$$

[Hint: Dividing the joint density in (8.84) by the marginal density of $\mathbf{W}$, $f_{\mathbf{W}}(\mathbf{w})$, yields the posterior density just like the joint density, but with a different constant, say, $c^*(\mathbf{w})$. With ψ as the variable, the density is a Wishart one, with given parameters.] (c) Letting $\mathbf{S} = \mathbf{W}/\nu$ be the sample covariance matrix, show that the posterior mean of $\mathbf{\Sigma}$ is

$$E[\mathbf{\Sigma} \mid \mathbf{W} = \mathbf{w}] = \frac{1}{\nu + \nu_0 - q - 1}(\nu \mathbf{S} + \nu_0 \mathbf{\Sigma}_0), \tag{8.86}$$

close to a weighted average of the prior guess and observed covariance matrices. [Hint: Use Lemma 7.1 on $\mathbf{\Psi}$, rather than trying to find the distribution of $\mathbf{\Sigma}$.]

Exercise 8.8.26 (Bayesian inference). Exercise 3.7.31 considered Bayesian inference on the normal mean when the covariance matrix is known, and the above Exercise 8.8.25 treated the covariance case with no mean apparent. Here we present a prior to deal with the mean and covariance simultaneously. It is a two-stage prior:

$$\begin{aligned} \boldsymbol{\mu} \mid \mathbf{\Psi} = \psi &\sim N_{p\times q}(\boldsymbol{\mu}_0, \mathbf{K}_0^{-1} \otimes \psi^{-1}), \\ \mathbf{\Psi} &\sim \text{Wishart}_q(\nu_0, \mathbf{\Psi}_0). \end{aligned} \tag{8.87}$$

Here, $\mathbf{K}_0, \boldsymbol{\mu}_0, \mathbf{\Psi}_0$ and ν_0 are known, where $\mathbf{K}_0$ and $\mathbf{\Psi}_0$ are positive definite, and $\nu_0 \geq q$. As above, set $\nu_0 \mathbf{\Psi}_0 = \mathbf{\Sigma}_0^{-1}$. Show that unconditionally, $E[\boldsymbol{\mu}] = \boldsymbol{\mu}_0$ and, using (8.82),

$$Cov[\boldsymbol{\mu}] = \frac{1}{\nu_0 - q - 1}\mathbf{K}_0^{-1} \otimes \mathbf{\Psi}_0^{-1} = \frac{\nu_0}{\nu_0 - q - 1}\mathbf{K}_0^{-1} \otimes \mathbf{\Sigma}_0. \tag{8.88}$$

[Hint: Use the covariance decomposition in (2.74) on $\mathbf{\Psi}$.]

Exercise 8.8.27. This exercise finds the density of the marginal distribution of the $\boldsymbol{\mu}$ in (8.87). (a) Show that the joint density of $\boldsymbol{\mu}$ and $\mathbf{\Psi}$ can be written

$$\begin{aligned} f_{\boldsymbol{\mu},\mathbf{\Psi}}(\mathbf{m}, \psi) = &\frac{1}{(2\pi)^{pq/2} d(\nu_0, q)} |\mathbf{\Psi}_0|^{-\nu_0/2} |\mathbf{K}_0|^{q/2} |\psi|^{(\nu_0+p-q-1)/2} \\ &e^{-\frac{1}{2}\text{trace}(((\mathbf{m}-\boldsymbol{\mu}_0)'\mathbf{K}_0(\mathbf{m}-\boldsymbol{\mu}_0)+\mathbf{\Psi}_0^{-1})\psi)}, \end{aligned} \tag{8.89}$$

for the Wishart constant $d(\nu_0, q)$ given in (8.71). [Hint: Use the pdfs in (8.53) and (8.68).] (b) Argue that the final two terms in (8.89) (the $|\psi|$ term and the exponential term) look like the density of $\mathbf{\Psi}$ if

$$\mathbf{\Psi} \sim \text{Wishart}_q(\nu_0 + p, ((\mathbf{m} - \boldsymbol{\mu}_0)'\mathbf{K}_0(\mathbf{m} - \boldsymbol{\mu}_0) + \mathbf{\Psi}_0^{-1})^{-1}), \tag{8.90}$$

but without the constants, hence integrating over ψ yields the inverse of those constants. Then show that the marginal density of μ is

$$\begin{aligned} f_{\mu}(\mathbf{m}) &= \int f_{\mu,\Psi}(\mathbf{m},\psi)d\psi \\ &= \frac{d(\nu_0+p,q)}{(2\pi)^{pq/2}d(\nu_0,q)} |\Psi_0|^{-\nu_0/2}|\mathbf{K}_0|^{q/2} \, |(\mathbf{m}-\mu_0)'\mathbf{K}_0(\mathbf{m}-\mu_0)+\Psi_0^{-1}|^{-(\nu_0+p)/2} \\ &= \frac{1}{c(\nu_0,p,q)} \frac{|\Psi_0|^{p/2}|\mathbf{K}_0|^{q/2}}{|(\mathbf{m}-\mu_0)'\mathbf{K}_0(\mathbf{m}-\mu_0)\Psi_0+\mathbf{I}_q|^{(\nu_0+p)/2}}, \end{aligned} \tag{8.91}$$

where

$$c(\nu_0,p,q) = (2\pi)^{pq/2}d(\nu_0,q)/d(\nu_0+p,q). \tag{8.92}$$

This density for μ is a type of multivariate t. Hotellings T^2 is another type. (c) Show that if $p = q = 1$, $\mu_0 = 0$, $K_0 = 1/\nu_0$ and $\Psi_0 = 1$, that the pdf (8.91) is that of a Student's t on ν_0 degrees of freedom:

$$f(t\,|\,\nu_0) = \frac{\Gamma((\nu_0+1)/2)}{\sqrt{\nu\pi}\,\Gamma(\nu_0/2)} \frac{1}{(1+t^2/\nu_0)^{(\nu_0+1)/2}}. \tag{8.93}$$

Exercise 8.8.28 (Bayesian inference). Now we add some data to the prior in Exercise 8.8.26. The conditional model for the data is

$$\begin{aligned} \mathbf{Y}\,|\,\mu=\mathbf{m},\Psi=\psi &\sim N_{p\times q}(\mathbf{m},\mathbf{K}^{-1}\otimes\psi^{-1}), \\ \mathbf{W}\,|\,\mu=\mathbf{m},\Psi=\psi &\sim \text{Wishart}_q(\nu,\psi^{-1}), \end{aligned} \tag{8.94}$$

where $\mathbf{Y}$ and $\mathbf{W}$ are independent given μ and Ψ. Note that $\mathbf{W}$'s distribution does not depend on the μ. The conjugate prior is given in (8.87), with the conditions given therebelow. The $\mathbf{K}$ is a fixed positive definite matrix. A curious element is that the prior covariance of the mean and the conditional covariance of $\mathbf{Y}$ have the same ψ, which helps tractability (as in Exercise 6.6.6). (a) Justify the following equations:

$$\begin{aligned} f_{\mathbf{Y},\mathbf{W},\mu,\Psi}(\mathbf{y},\mathbf{w},\mathbf{m},\psi) &= f_{\mathbf{Y}\,|\,\mu,\Psi}(\mathbf{y}\,|\,\mathbf{m},\psi)f_{\mathbf{W}\,|\,\Psi}(\mathbf{w}\,|\,\psi)f_{\mu\,|\,\Psi}(\mathbf{m}\,|\,\psi)f_{\Psi}(\psi) \\ &= f_{\mu\,|\,\mathbf{Y},\Psi}(\mathbf{m}\,|\,\mathbf{y},\psi)f_{\mathbf{Y}\,|\,\Psi}(\mathbf{y}\,|\,\psi)f_{\mathbf{W}\,|\,\Psi}(\mathbf{w}\,|\,\psi)f_{\Psi}(\psi) \end{aligned} \tag{8.95}$$

(b) Show that the conditional distribution of μ given $\mathbf{Y}$ and Ψ is multivariate normal with

$$\begin{aligned} E[\mu\,|\,\mathbf{Y}=\mathbf{y},\Psi=\psi] &= (\mathbf{K}+\mathbf{K}_0)^{-1}(\mathbf{K}\mathbf{y}+\mathbf{K}_0\mu_0), \\ Cov[\mu\,|\,\mathbf{Y}=\mathbf{y},\Psi=\psi] &= (\mathbf{K}+\mathbf{K}_0)^{-1}\otimes\psi^{-1}. \end{aligned} \tag{8.96}$$

[Hint: Follows from Exercise 3.7.31, noting that ψ is fixed (conditioned upon) for this calculation.] (c) Show that

$$\mathbf{Y}\,|\,\Psi=\psi \sim N_{p\times q}(\mu_0,(\mathbf{K}^{-1}+\mathbf{K}_0^{-1})\otimes\psi^{-1}). \tag{8.97}$$

[Hint: See (3.113).] (d) Let $\mathbf{Z} = (\mathbf{K}^{-1} + \mathbf{K}_0^{-1})^{-1/2}(\mathbf{Y} - \boldsymbol{\mu}_0)$, and show that the middle two densities in the last line of (8.95) can be combined into the density of

$$\mathbf{U} = \mathbf{W} + \mathbf{Z}'\mathbf{Z} \mid \boldsymbol{\Psi} = \boldsymbol{\psi} \sim \text{Wishart}_q(\nu + p, \boldsymbol{\psi}^{-1}), \tag{8.98}$$

that is,

$$f_{\mathbf{Y}\mid\boldsymbol{\Psi}}(\mathbf{y} \mid \boldsymbol{\psi}) f_{\mathbf{W}\mid\boldsymbol{\Psi}}(\mathbf{w} \mid \boldsymbol{\psi}) = c^*(\mathbf{u}, \mathbf{w}) f_{\mathbf{U}\mid\boldsymbol{\Psi}}(\mathbf{u} \mid \boldsymbol{\psi}) \tag{8.99}$$

for some constant $c^*(\mathbf{u}, \mathbf{w})$ that does not depend on $\boldsymbol{\psi}$. (e) Now use Exercise 8.8.25 to show that

$$\boldsymbol{\Psi} \mid \mathbf{U} = \mathbf{u} \sim \text{Wishart}_q(\nu + \nu_0 + p, (\mathbf{u} + \boldsymbol{\Psi}_0^{-1})^{-1}). \tag{8.100}$$

(f) Thus the posterior distribution of $\boldsymbol{\mu}$ and $\boldsymbol{\Psi}$ in (8.96) and (8.100) are given in the same two stages as the prior in (8.87). The only differences are in the parameters. The prior parameters are $\boldsymbol{\mu}_0, \mathbf{K}_0, \nu_0$, and $\boldsymbol{\Psi}_0$. What are the corresponding posterior parameters? (g) Using (8.86) to (8.88), show that the posterior means of $\boldsymbol{\mu}$ and $\boldsymbol{\Sigma}$ $(= \boldsymbol{\Psi}^{-1})$ are

$$\begin{aligned} E[\boldsymbol{\mu} \mid \mathbf{Y} = \mathbf{y}, \mathbf{W} = \mathbf{w}] &= (\mathbf{K} + \mathbf{K}_0)^{-1}(\mathbf{K}\mathbf{y} + \mathbf{K}_0\boldsymbol{\mu}_0), \\ E[\boldsymbol{\Sigma} \mid \mathbf{Y} = \mathbf{y}, \mathbf{W} = \mathbf{w}] &= \frac{1}{\nu + \nu_0 + p - q - 1}(\mathbf{u} + \nu_0\boldsymbol{\Sigma}_0), \end{aligned} \tag{8.101}$$

and the posterior covariance of $\boldsymbol{\mu}$ is

$$Cov[\boldsymbol{\mu} \mid \mathbf{Y} = \mathbf{y}, \mathbf{W} = \mathbf{w}] = \frac{1}{\nu + \nu_0 + p - q - 1}(\mathbf{K} + \mathbf{K}_0)^{-1} \otimes (\mathbf{u} + \nu_0\boldsymbol{\Sigma}_0). \tag{8.102}$$

Chapter 9

Likelihood Methods

9.1 Likelihood

For the linear models, we derived estimators of $\boldsymbol{\beta}$ using the least squares principle, and found estimators of $\boldsymbol{\Sigma}_R$ in an obvious manner. Likelihood provides another general approach to deriving estimators, hypothesis tests, and model selection procedures, with a wider scope than least squares. In this chapter we present general likelihood methods, and apply them to the linear models. Chapter 10 considers likelihood analysis of models on covariance matrices.

Throughout this chapter, we assume we have a statistical model consisting of a random object (usually a matrix or a set of matrices) $\mathbf{Y}$ with space $\mathcal{Y}$, and a set of distributions $\{P_{\boldsymbol{\theta}} \mid \boldsymbol{\theta} \in \Theta\}$, where Θ is the parameter space. We assume that these distributions have densities, with $P_{\boldsymbol{\theta}}$ having associated density $f(\mathbf{y} \mid \boldsymbol{\theta})$.

Definition 9.1. *For a statistical model with densities, the* ***likelihood function*** *is defined for each fixed* $\mathbf{y} \in \mathcal{Y}$ *as the function* $L(\cdot\,;\mathbf{y}) : \boldsymbol{\theta} \to [0, \infty)$ *given by*

$$L(\boldsymbol{\theta}\,;\mathbf{y}) = a(\mathbf{y}) f(\mathbf{y} \mid \boldsymbol{\theta}), \tag{9.1}$$

for any positive $a(\mathbf{y})$.

Likelihoods are to be interpreted in only relative fashion, that is, to say the likelihood of a particular $\boldsymbol{\theta}_1$ is $L(\boldsymbol{\theta}_1;\mathbf{y})$ does not mean anything by itself. Rather, meaning is attributed to saying that the relative likelihood of $\boldsymbol{\theta}_1$ to $\boldsymbol{\theta}_2$ (in light of the data $\mathbf{y}$) is $L(\boldsymbol{\theta}_1;\mathbf{y})/L(\boldsymbol{\theta}_2;\mathbf{y})$. Which is why the "$a(\mathbf{y})$" in (9.1) is allowed. There is a great deal of controversy over *what* exactly the relative likelihood means. We do not have to worry about that particularly, since we are just using likelihood as a means to an end. The general idea, though, is that the data supports $\boldsymbol{\theta}$'s with relatively high likelihood.

The next few sections consider maximum likelihood estimation. Subsequent sections look at likelihood ratio tests, and two popular model selection techniques (AIC and BIC).

9.2 Maximum likelihood estimation

Given the data $\mathbf{y}$, it is natural (at least it sounds natural terminologically) to take as estimate of $\boldsymbol{\theta}$ the value that is most likely. Indeed, that is the maximum likelihood estimate.

Definition 9.2. *The **maximum likelihood estimate** (MLE) of the parameter $\boldsymbol{\theta}$ based on the data $\mathbf{y}$ is the unique value, if it exists, $\widehat{\boldsymbol{\theta}}(\mathbf{y}) \in \Theta$ that maximizes the likelihood $L(\boldsymbol{\theta}; \mathbf{y})$.*

It may very well be that the maximizer is not unique, or does not exist at all, in which case there is no MLE for that particular $\mathbf{y}$. The MLE of a function of $\boldsymbol{\theta}$, $g(\boldsymbol{\theta})$, is defined to be the function of the MLE, that is, $\widehat{g(\boldsymbol{\theta})} = g(\widehat{\boldsymbol{\theta}})$. See Exercises 9.6.1 and 9.6.2 for justification.

9.3 The MLE in the both-sides model

9.3.1 Maximizing the likelihood

The basic model for this section is the both-sides model, where we allow setting an arbitrary set of the β_{ij}'s to zero. To specify a submodel of the $p \times l$ matrix $\boldsymbol{\beta}$, we use a pattern $\mathbf{P}$ as in Section 6.5, $\mathbf{P}$ being a $p \times l$ matrix with just 0's and 1's indicating which elements of $\boldsymbol{\beta}$ are allowed to be nonzero. Then given the pattern $\mathbf{P}$, the model is formally defined as

$$\mathbf{Y} \sim N_{n\times q}(\mathbf{x}\boldsymbol{\beta}\mathbf{z}', \mathbf{I}_n \otimes \boldsymbol{\Sigma}_R), \quad (\boldsymbol{\beta}, \boldsymbol{\Sigma}_R) \in \mathcal{B}(\mathbf{P}) \times \mathcal{S}_q, \tag{9.2}$$

where $\mathbf{Y}$ is $n \times q$, $\mathbf{x}$ is $n \times p$, $\mathbf{z}$ is $q \times l$, $\mathcal{B}(\mathbf{P})$ is the set of $\boldsymbol{\beta}$'s with pattern $\mathbf{P}$ as in (6.51), and $\mathcal{S}_q$ is the set of $q \times q$ positive definite symmetric matrices as in (5.51). We also assume that

$$n \geq p + q, \quad \mathbf{x}'\mathbf{x} \text{ is invertible, and } \mathbf{z}'\mathbf{z} \text{ is invertible.} \tag{9.3}$$

It is useful to look at two expressions for the likelihood of $\mathbf{Y}$, corresponding to the equations (8.53) and (8.47). Dropping the constant from the density, we have

$$L(\boldsymbol{\beta}, \boldsymbol{\Sigma}_R \,;\, \mathbf{y}) = \frac{1}{|\boldsymbol{\Sigma}_R|^{n/2}}\, e^{-\frac{1}{2}\,\text{trace}(\boldsymbol{\Sigma}_R^{-1}(\mathbf{y}-\mathbf{x}\boldsymbol{\beta}\mathbf{z}')'(\mathbf{y}-\mathbf{x}\boldsymbol{\beta}\mathbf{z}'))} \tag{9.4}$$

$$= \frac{1}{|\boldsymbol{\Sigma}_R|^{n/2}}\, e^{-\frac{1}{2}(\mathbf{v}-\boldsymbol{\gamma}\mathbf{D}')(\mathbf{I}_n\otimes\boldsymbol{\Sigma}_R^{-1})(\mathbf{v}-\boldsymbol{\gamma}\mathbf{D}')'}, \tag{9.5}$$

where in the latter expression, as in (5.39),

$$\mathbf{v} = \text{row}(\mathbf{y}), \quad \boldsymbol{\gamma} = \text{row}(\boldsymbol{\beta}), \text{ and } \mathbf{D} = \mathbf{x} \otimes \mathbf{z}. \tag{9.6}$$

Now with $\boldsymbol{\beta} \in \mathcal{B}(\mathbf{P})$, certain elements of $\boldsymbol{\gamma}$ are set to zero, indicated by the zeroes in $\text{row}(\mathbf{P})$. Let $\boldsymbol{\gamma}^*$ be the vector $\boldsymbol{\gamma}$, but with the elements assumed zero dropped. Some examples, paralleling the patterns in (6.53) and (6.54):

$$\mathbf{P} = \begin{pmatrix} 1 & 1 & 0 & 0 \\ 1 & 1 & 0 & 0 \end{pmatrix} \Rightarrow \boldsymbol{\gamma}^* = (\beta_{11} \;\; \beta_{12} \;\; \beta_{21} \;\; \beta_{22}),$$

$$\mathbf{P} = \begin{pmatrix} 1 & 1 \\ 0 & 1 \\ 0 & 1 \end{pmatrix} \Rightarrow \boldsymbol{\gamma}^* = (\beta_{11} \;\; \beta_{12} \;\; \beta_{22} \;\; \beta_{32}),$$

$$\text{and } \mathbf{P} = \begin{pmatrix} 1 & 1 \\ 1 & 0 \end{pmatrix} \Rightarrow \boldsymbol{\gamma}^* = (\beta_{11} \;\; \beta_{12} \;\; \beta_{21}). \tag{9.7}$$

Also, let $\mathbf{D}^*$ be the matrix $\mathbf{D}$, but dropping the columns corresponding to zeros in row($\mathbf{P}$), so that

$$\gamma \mathbf{D}' = \gamma^* \mathbf{D}^{*\prime}. \tag{9.8}$$

To start the maximization, suppose first that we know the $\widehat{\mathbf{\Sigma}}_R$, but not the MLE of γ^*. Then maximizing the likelihood (9.4) over $\beta \in \mathcal{B}(\mathbf{P})$ is equivalent to maximizing the likelihood over γ (with specified zeroes) in (9.5), which by (9.6), is equivalent to minimizing

$$(\mathbf{v} - \gamma^* \mathbf{D}^{*\prime})(\mathbf{I}_n \otimes \widehat{\mathbf{\Sigma}}_R^{-1})(\mathbf{v} - \gamma^* \mathbf{D}^{*\prime})' \tag{9.9}$$

over γ^*, where there are no restrictions on the γ^*. This final equation is the objective function in (5.36), which we know is minimized by the weighted least squares estimator (5.35), i.e.,

$$\widehat{\gamma}^* = \mathbf{v}(\mathbf{I}_n \otimes \widehat{\mathbf{\Sigma}}_R^{-1})\mathbf{D}^*(\mathbf{D}^{*\prime}(\mathbf{I}_n \otimes \widehat{\mathbf{\Sigma}}_R^{-1})\mathbf{D}^*)^{-1}. \tag{9.10}$$

Next, turn it around, so that we know $\widehat{\gamma}^*$, but not the MLE of $\mathbf{\Sigma}_R$. We can then reconstruct $\widehat{\beta} \in \mathcal{B}(\mathbf{P})$ from $\widehat{\gamma}^*$, yielding the residual

$$\widehat{\mathbf{R}} = \mathbf{y} - \mathbf{x}\widehat{\beta}\mathbf{z}', \tag{9.11}$$

and can write the likelihood (9.4) as

$$L(\widehat{\beta}, \mathbf{\Sigma}_R\,;\,\mathbf{y}) = \frac{1}{|\mathbf{\Sigma}_R|^{n/2}}\, e^{-\frac{1}{2}\,\text{trace}(\mathbf{\Sigma}_R^{-1}\mathbf{U})}, \text{ where } \mathbf{U} = \widehat{\mathbf{R}}'\widehat{\mathbf{R}}. \tag{9.12}$$

We need to maximize that likelihood over $\mathbf{\Sigma}_R \in \mathcal{S}_q^+$. We appeal to the following lemma, proved in Section 9.3.4.

Lemma 9.1. *Suppose $a > 0$ and $\mathbf{U} \in \mathcal{S}_q^+$. Then*

$$g(\mathbf{\Sigma}) = \frac{1}{|\mathbf{\Sigma}|^{a/2}}\, e^{-\frac{1}{2}\,\text{trace}(\mathbf{\Sigma}^{-1}\mathbf{U})} \tag{9.13}$$

is uniquely maximized over $\mathbf{\Sigma} \in \mathcal{S}_q^+$ by

$$\widehat{\mathbf{\Sigma}} = \frac{1}{a}\,\mathbf{U}, \tag{9.14}$$

and the maximum is

$$g(\widehat{\mathbf{\Sigma}}) = \frac{1}{|\widehat{\mathbf{\Sigma}}|^{a/2}}\, e^{-\frac{aq}{2}}. \tag{9.15}$$

Applying this lemma to (9.12) yields

$$\widehat{\mathbf{\Sigma}}_R = \frac{1}{n}\,\widehat{\mathbf{R}}'\widehat{\mathbf{R}}. \tag{9.16}$$

We have gotten to the circular point that if we know $\widehat{\beta}$, then we can find $\widehat{\mathbf{\Sigma}}_R$, and if we know $\widehat{\mathbf{\Sigma}}_R$, then we can find $\widehat{\beta}$. But in reality we start knowing neither. The solution is to iterate the two steps, that is, start with a guess at $\widehat{\mathbf{\Sigma}}_R$ (such as the identity or the sample covariance matrix), then find the $\widehat{\beta}$, then update the $\widehat{\mathbf{\Sigma}}_R$, then back to the $\widehat{\boldsymbol{\beta}}$, etc., until convergence. In most cases the process does converge, and rather quickly. At the least, at each step, the likelihood does not decrease. This type of algorithm is called **iteratively reweighted least squares**.

Once we have the MLE, we also have from (9.15) that the maximum of the likelihood is

$$L(\widehat{\boldsymbol{\beta}}, \widehat{\boldsymbol{\Sigma}}_R \,;\, \mathbf{y}) = \frac{1}{|\widehat{\boldsymbol{\Sigma}}_R|^{n/2}}\, e^{-\frac{nq}{2}}. \tag{9.17}$$

The distributions of the estimators may not be multivariate normal and Wishart, depending on the pattern $\mathbf{P}$. In the model with no restrictions, we can find exact distributions, as in the next section. In general, we have the asymptotic (as $n \to \infty$, but p, q and l are fixed) results for the MLE, which imply the approximation for the nonzero coefficients,

$$\widehat{\gamma}^* \approx N(\gamma^*, (\mathbf{D}^{*\prime}(\mathbf{I}_n \otimes \widehat{\boldsymbol{\Sigma}}_R^{-1})\mathbf{D}^*)^{-1}), \tag{9.18}$$

should be reasonable for large n. See Lehmann and Casella [2003] for a general review of the asymptotic distribution of MLE's. For small n, the approximate standard errors arising from (9.18) are likely to be distinct underestimates. Section 9.3.3 below has suggestions for improving these estimates, which we incorporate into our R function, bothsidesmodel.mle.

9.3.2 Examples

Continue with the model (6.46) for the mouth size data, in which the growth curves are assumed linear:

$$\mathbf{Y} = \mathbf{x}\boldsymbol{\beta}\mathbf{z}' + \mathbf{R} = \begin{pmatrix} \mathbf{1}_{11} & \mathbf{1}_{11} \\ \mathbf{1}_{16} & \mathbf{0}_{16} \end{pmatrix} \begin{pmatrix} \beta_0 & \beta_1 \\ \delta_0 & \delta_1 \end{pmatrix} \begin{pmatrix} 1 & 1 & 1 & 1 \\ -3 & -1 & 1 & 3 \end{pmatrix} + \mathbf{R}. \tag{9.19}$$

We will use the functions bothsidesmodel and bothsidesmodel.mle to find the least squares and maximum likelihood estimates of $\boldsymbol{\beta}$:

```
x <− cbind(1,mouths[,5])
y <− mouths[,1:4]
z <− cbind(1,c(−3,−1,1,3))
bothsidesmodel(x,y,z) # For LS
bothsidesmodel.mle(x,y,z) # For MLE
```

The next table compares the estimates and their standard errors for the two methods:

	Estimates		Standard errors		
	LS	MLE	LS	MLE	Ratio
$\widehat{\beta}_0$	24.969	24.937	0.486	0.521	0.93
$\widehat{\beta}_1$	0.784	0.827	0.086	0.091	0.95
$\widehat{\delta}_0$	−2.321	−2.272	0.761	0.794	0.96
$\widehat{\delta}_1$	−0.305	−0.350	0.135	0.138	0.98

(9.20)

The two methods give fairly similar results. The last column is the ratio of the standard errors, least squares to maximum likelihood. Least squares is about 2 to 7% better in this example.

Now turn to the leprosy example in Sections 6.4.3 and 6.4.4. We will consider the model with the before effects assumed to be zero, so that the model is

$$\mathbf{Y} = \mathbf{x}\boldsymbol{\beta} + \mathbf{R}$$
$$= \left[\begin{pmatrix} 1 & 1 & 1 \\ 1 & 1 & -1 \\ 1 & -2 & 0 \end{pmatrix} \otimes \mathbf{1}_{10}\right] \begin{pmatrix} \mu_b & \mu_a \\ 0 & \alpha_a \\ 0 & \beta_a \end{pmatrix} + \mathbf{R}. \tag{9.21}$$

This model is multivariate regression, but with restrictions on the $\boldsymbol{\beta}$. The parameters of interest are the after effects α_a and β_a. We use the same functions as above, but utilizing a pattern, the first $\mathbf{P}$ in (6.54):

```
x <− kronecker(cbind(1,c(1,1,−2),c(1,−1,0)),rep(1,10))
y <− leprosy[,1:2]
pattern <− cbind(c(1,0,0),1)
bothsidesmodel(x,y,diag(2),pattern)
bothsidesmodel.mle(x,y,diag(2),pattern)
```

The parameters of interest:

	Estimates		Standard errors		
	LS	MLE	LS	MLE	Ratio
$\widehat{\alpha}_a$	−2.200	−1.130	0.784	0.507	1.544
$\widehat{\beta}_a$	−0.400	−0.054	1.357	0.879	1.544

(9.22)

In this case, the two methods produce somewhat different results, and maximum likelihood yields substantially smaller standard errors. But here, the likelihood method does underestimate the standard errors a bit.

Now compare the estimates here in (9.22) to those in (6.43), which are least squares, but differ on whether we adjusted for covariates. Rearranging (6.43) a bit, we have

	Estimates		Standard errors		
	Original	Covariate-adjusted	Original	Covariate-adjusted	Ratio
$\widehat{\alpha}_a$	−2.200	−1.130	0.784	0.547	1.433
$\widehat{\beta}_a$	−0.400	−0.054	1.357	0.898	1.511

(9.23)

Note that the least squares estimates (and standard errors) in (9.22) are the same as for the model without the covariates in (6.43). That is to be expected, since in both cases, to find the after parameters, we are fitting the regression model with just the $\mathbf{x}$ to the before variable. On the other hand, the maximum likelihood estimates in (9.22) are the same as the covariate-adjusted estimates in (6.43). Exercise 9.6.5 explains why. Note that the standard errors for the covariate-adjusted estimates are larger than those for the MLE. The larger ones are in fact the correct ones, so the likelihood approximation underestimated the two standard errors by about 7% and 2%, respectively. Thus the MLE's are still substantially better than the least squares estimates, just not quite 54% better as suggested by (9.22).

The main difference between least squares and maximum likelihood for these models is that least squares looks at the columns of $\mathbf{Y}$ one-by-one, while maximum likelihood takes into account the correlation between the columns, thus may be able to use information in all the columns to improve the estimation in each column. In particular, maximum likelihood automatically performs the covariate adjustment in the leprosy example. For small samples, it may be that the noise in the cross-column information overwhelms the signal, so that least squares is preferable. But maximum likelihood is known to be asymptotically efficient, so that, for large enough n, it produces the lowest standard errors.

9.3.3 Calculating the estimates

We have seen in (5.46) that if $\mathbf{z}$ is invertible, and there is no restriction on the $\boldsymbol{\beta}$, then the weighted least squares estimator (9.10) of $\boldsymbol{\beta}$ does not depend on the $\widehat{\boldsymbol{\Sigma}}_R$, being the regular least squares estimate. That is, the maximum likelihood and least squares estimators of $\boldsymbol{\beta}$ are the same. Otherwise, the least squares and maximum likelihood estimates will differ.

Consider the both-sides model,

$$\mathbf{Y} = \mathbf{x}\boldsymbol{\beta}\mathbf{z}_a' + \mathbf{R}, \tag{9.24}$$

where $\mathbf{z}_a$ is not square, i.e., it is $q \times l$ with $l < q$, but there are still no restrictions on $\boldsymbol{\beta}$. We can find the MLE of $\boldsymbol{\beta}$ in this model without using iteratively reweighted least squares. In Section 6.2.2, for least squares, we turned this model into a multivariate regression by calculating $\mathbf{Y}_{\mathbf{z}}$ in (6.16). Here we do the same, but first fill out the $\mathbf{z}$ so that it is square. That is, we find an additional $q \times (q - l)$ matrix $\mathbf{z}_b$ so that

$$\mathbf{z} = (\mathbf{z}_a \;\; \mathbf{z}_b) \text{ is invertible, and } \mathbf{z}_a'\mathbf{z}_b = \mathbf{0}. \tag{9.25}$$

Exercise 5.8.29 guarantees that there is such a matrix. Then we can write model (9.24) equivalently as

$$\mathbf{Y} = \mathbf{x}\,(\boldsymbol{\beta} \;\; \mathbf{0}) \begin{pmatrix} \mathbf{z}_a' \\ \mathbf{z}_b' \end{pmatrix} + \mathbf{R} = \mathbf{x}(\boldsymbol{\beta} \;\; \mathbf{0})\mathbf{z}' + \mathbf{R}, \tag{9.26}$$

where the $\mathbf{0}$ matrix is $p \times (q - l)$. Now we move the $\mathbf{z}$ to the other side, so that with $\mathbf{Y}_{\mathbf{z}} = \mathbf{Y}(\mathbf{z}')^{-1}$, the model becomes

$$\mathbf{Y}_{\mathbf{z}} = (\mathbf{Y}_{\mathbf{z}a} \;\; \mathbf{Y}_{\mathbf{z}b}) = \mathbf{x}\,(\boldsymbol{\beta} \;\; \mathbf{0}) + \mathbf{R}_{\mathbf{z}}, \tag{9.27}$$

where

$$\mathbf{R}_{\mathbf{z}} \sim N(\mathbf{0}, \mathbf{I}_n \otimes \boldsymbol{\Sigma}_{\mathbf{z}}). \tag{9.28}$$

Here, $\mathbf{Y}_{\mathbf{z}a}$ is $n \times l$, to conform with $\boldsymbol{\beta}$, and $\mathbf{Y}_{\mathbf{z}b}$ is $n \times (q - l)$.

Compare the model (9.27) to the model with covariates in (6.39). Because $E[\mathbf{Y}_{\mathbf{z}b}] = \mathbf{0}$, we can condition as in (6.40). Partitioning the covariance matrix as

$$\boldsymbol{\Sigma}_{\mathbf{z}} = \begin{pmatrix} \boldsymbol{\Sigma}_{\mathbf{z}aa} & \boldsymbol{\Sigma}_{\mathbf{z}ab} \\ \boldsymbol{\Sigma}_{\mathbf{z}ba} & \boldsymbol{\Sigma}_{\mathbf{z}bb} \end{pmatrix}, \tag{9.29}$$

where $\boldsymbol{\Sigma}_{\mathbf{z}aa}$ is $l \times l$ and $\boldsymbol{\Sigma}_{\mathbf{z}bb}$ is $(q - l) \times (q - l)$, we have that

$$\mathbf{Y}_{\mathbf{z}a} \mid \mathbf{Y}_{\mathbf{z}b} = \mathbf{y}_{\mathbf{z}b} \;\sim\; N(\boldsymbol{\alpha} + \mathbf{y}_{\mathbf{z}b}\boldsymbol{\eta}, \mathbf{I}_n \otimes \boldsymbol{\Sigma}_{\mathbf{z}aa\cdot b}), \tag{9.30}$$

where

$$\begin{aligned} \boldsymbol{\eta} &= (\boldsymbol{\Sigma}_{\mathbf{z}bb})^{-1}\boldsymbol{\Sigma}_{\mathbf{z}ba}, \\ \boldsymbol{\alpha} &= E[\mathbf{Y}_{\mathbf{z}a}] - E[\mathbf{Y}_{\mathbf{z}b}]\boldsymbol{\eta} = \mathbf{x}\boldsymbol{\beta}, \\ \text{and } \boldsymbol{\Sigma}_{\mathbf{z}aa\cdot b} &= \boldsymbol{\Sigma}_{\mathbf{z}aa} - \boldsymbol{\Sigma}_{\mathbf{z}ab}\boldsymbol{\Sigma}_{\mathbf{z}bb}^{-1}\boldsymbol{\Sigma}_{\mathbf{z}ba}. \end{aligned} \tag{9.31}$$

Thus we have the conditional model

$$\mathbf{Y_z}_a = (\mathbf{x}\ \mathbf{y_z}_b)\begin{pmatrix}\boldsymbol{\beta}\\ \boldsymbol{\eta}\end{pmatrix} + \mathbf{R}^*, \tag{9.32}$$

where

$$\mathbf{R}^* \mid \mathbf{Y_z}_b = \mathbf{y_z}_b \sim N(\mathbf{0}, \mathbf{I}_n \otimes \boldsymbol{\Sigma}_{\mathbf{z}aa\cdot b}). \tag{9.33}$$

There is also the marginal model, for $\mathbf{Y_z}_b$, which has mean zero:

$$\mathbf{Y_z}_b \sim N(\mathbf{0}, \mathbf{I}_n \otimes \boldsymbol{\Sigma}_{\mathbf{z}bb}). \tag{9.34}$$

The density of $\mathbf{Y_z}$ can now be factored as

$$f(\mathbf{y_z} \mid \boldsymbol{\beta}, \boldsymbol{\Sigma}_\mathbf{z}) = f(\mathbf{y_z}_a \mid \mathbf{y_z}_b, \boldsymbol{\beta}, \boldsymbol{\eta}, \boldsymbol{\Sigma}_{\mathbf{z}aa\cdot b}) \times f(\mathbf{y_z}_b \mid \boldsymbol{\Sigma}_{\mathbf{z}bb}). \tag{9.35}$$

The set of parameters $(\boldsymbol{\beta}, \boldsymbol{\eta}, \boldsymbol{\Sigma}_{\mathbf{z}aa\cdot b}, \boldsymbol{\Sigma}_{\mathbf{z}bb})$ can be seen to be in one-to-one correspondence with $(\boldsymbol{\beta}, \boldsymbol{\Sigma}_\mathbf{z})$, and has space $\mathbb{R}^{p\times l} \times \mathbb{R}^{(q-l)\times l} \times \mathcal{S}_l^+ \times \mathcal{S}_{q-l}^+$. In particular, the parameters in the conditional density are functionally independent of those in the marginal density, which means that we can find the MLE of $(\boldsymbol{\beta}, \boldsymbol{\eta}, \boldsymbol{\Sigma}_{\mathbf{z}aa\cdot b})$ without having to worry about the $\boldsymbol{\Sigma}_{\mathbf{z}bb}$.

The $\boldsymbol{\beta}$ can be estimated using the model (9.32), which is again a standard multivariate regression, though now with the $\mathbf{y_z}_b$ moved from the left-hand part of the model into the $\mathbf{x}$-part. Thus we can use least squares to calculate the MLE of the coefficients, and their standard errors, then find confidence intervals and t-statistics for the $\widehat{\beta}_{ij}$'s as in Section 6.3. Under the null, the t-statistics are conditionally Student's t, conditioning on $\mathbf{Y_z}_b$, hence unconditionally as well.

To illustrate, consider the both-sides model in (9.19) for the mouth size data. Here, the $\mathbf{z}_a$ in (9.24) is the 2×4 matrix $\mathbf{z}$ in (9.19) used to fit the linear growth curve model. We could use the function fillout in Section A.11 to fill out the matrix as in (9.25), but we already have a candidate for the $\mathbf{z}_b$, i.e., the quadratic and cubic parts of the orthogonal polynomial in (6.29). Thus to find the $\widehat{\boldsymbol{\beta}}$ and its standard errors, we use

```
x <− cbind(1,mouths[,5])
y <− mouths[,1:4]
z <− cbind(1,c(−3,−1,1,3),c(−1,1,1,−1),c(−1,3,−3,1))
yz <− y%*%solve(t(z))
yza <− yz[,1:2]
xyzb <− cbind(x,yz[,3:4])
lm(yza ~xyzb−1) # or bothsidesmodel(xyzb,yza)
```

We obtain the "MLE" results presented in (9.20).

There are submodels of the both-sides models (including multivariate regression) defined by a pattern $\mathbf{P}$ that can be fit analytically, but the function bothsidesmodel.mle uses the iteratively reweighted least squares method in Section 9.3.1 to estimate the $\widehat{\boldsymbol{\beta}}$. We could use the covariance in (9.11) to estimate the standard errors of the estimated coefficients, but we can improve the estimates by first conditioning on any columns $\mathbf{Y}_j$ of $\mathbf{Y}$ for which the corresponding columns $\boldsymbol{\beta}_j$ of $\boldsymbol{\beta}$ are assumed zero. Thus we actually fit a conditional model as in (9.32), where now the $\boldsymbol{\beta}$ may have restrictions, though the $\boldsymbol{\eta}$ does not. Another slight modification we make is to use $n - p^*$ instead of n as the divisor in (9.16), where p^* is the number of rows of $\boldsymbol{\beta}$ with at least one nonzero component.

9.3.4 Proof of the MLE for the Wishart

Because $\mathbf{U}$ is positive definite and symmetric, it has an invertible symmetric square root, $\mathbf{U}^{1/2}$. Let $\boldsymbol{\Psi} = \mathbf{U}^{-1/2}\boldsymbol{\Sigma}\mathbf{U}^{-1/2}$, and from (9.13) write

$$g(\boldsymbol{\Sigma}) = h(\mathbf{U}^{-1/2}\boldsymbol{\Sigma}\mathbf{U}^{-1/2}), \text{ where } h(\boldsymbol{\Psi}) \equiv \frac{1}{|\mathbf{U}|^{a/2}} \frac{1}{|\boldsymbol{\Psi}|^{a/2}} e^{-\frac{1}{2}\text{trace}(\boldsymbol{\Psi}^{-1})} \tag{9.36}$$

is a function of $\boldsymbol{\Psi} \in \mathcal{S}_q^+$. Exercise 9.6.6 shows that (9.36) is maximized by $\widehat{\boldsymbol{\Psi}} = (1/a)\mathbf{I}_q$, hence

$$h(\widehat{\boldsymbol{\Psi}}) = \frac{1}{|\mathbf{U}|^{a/2}|\frac{1}{a}\mathbf{I}_q|^{a/2}} e^{-\frac{1}{2}a\,\text{trace}(\mathbf{I}_q)}, \tag{9.37}$$

from which follows (9.15). Also,

$$\widehat{\boldsymbol{\Psi}} = \mathbf{U}^{-1/2}\widehat{\boldsymbol{\Sigma}}\mathbf{U}^{-1/2} \Rightarrow \widehat{\boldsymbol{\Sigma}} = \mathbf{U}^{1/2}\frac{1}{a}\mathbf{I}_q\mathbf{U}^{1/2} = \frac{1}{a}\mathbf{U}, \tag{9.38}$$

which proves (9.14). □

9.4 Likelihood ratio tests

Again our big model has $\mathbf{Y}$ with space $\mathcal{Y}$ and a set of distributions $\{P_{\boldsymbol{\theta}} \mid \boldsymbol{\theta} \in \Theta\}$ with associated densities $f(\mathbf{y} \mid \boldsymbol{\theta})$. Testing problems we consider are of the form

$$H_0 : \boldsymbol{\theta} \in \Theta_0 \;\; \textit{versus} \;\; H_A : \boldsymbol{\theta} \in \Theta_A, \tag{9.39}$$

where

$$\Theta_0 \subset \Theta_A \subset \Theta. \tag{9.40}$$

Technically, the space in H_A should be $\Theta_A - \Theta_0$, but we take that to be implicit.

The likelihood ratio statistic for problem (9.39) is defined to be

$$LR = \frac{\sup_{\boldsymbol{\theta} \in \Theta_A} L(\boldsymbol{\theta}; \mathbf{y})}{\sup_{\boldsymbol{\theta} \in \Theta_0} L(\boldsymbol{\theta}; \mathbf{y})}, \tag{9.41}$$

where the likelihood L is given in Definition 9.1.

The idea is that the larger LR, the more likely the alternative H_A is relative to the null H_0. For testing, one would either use LR as a basis for calculating a p-value, or find a c_α such that rejecting the null when $LR > c_\alpha$ yields a level of (approximately) α. Either way, the null distribution of LR is needed, at least approximately. The general result we use says that under certain conditions (satisfied by most of what follows), under the null hypothesis,

$$2\log(LR) \longrightarrow^{\mathcal{D}} \chi^2_{df} \text{ where } df = \dim(H_A) - \dim(H_0) \tag{9.42}$$

as $n \to \infty$ (which means there must be some n to go to infinity). The dimension of a hypothesis is the number of free parameters it takes to uniquely describe the associated distributions. This definition is not very explicit, but in most examples the dimension will be "obvious."

9.4.1 The LRT in the both-sides model

In the both-sides model (9.2), we can use the likelihood ratio test to test any two nested submodels that are defined by patterns. That is, suppose we have two patterns, the null $\mathbf{P}_0$ and the alternative $\mathbf{P}_A$, where $\mathbf{P}_0$ is a subpattern of $\mathbf{P}_A$, i.e., $\mathcal{B}(\mathbf{P}_0) \subset \mathcal{B}(\mathbf{P}_A)$. (That is, everywhere $\mathbf{P}_0$ has a one, $\mathbf{P}_A$ also has a one.) Thus we test

$$H_0 : \boldsymbol{\beta} \in \mathcal{B}(\mathbf{P}_0) \ \ versus \ \ H_A : \boldsymbol{\beta} \in \mathcal{B}(\mathbf{P}_A). \tag{9.43}$$

Using (9.17), we can write (9.42) as

$$2\log(LR) = n(\log(|\widehat{\boldsymbol{\Sigma}}_0|) - \log(|\widehat{\boldsymbol{\Sigma}}_A|)) \rightarrow \chi^2_{df}, \ \ df = \sum_{ij}(\mathbf{P}_{Aij} - \mathbf{P}_{0ij}), \tag{9.44}$$

where $\widehat{\boldsymbol{\Sigma}}_0$ and $\widehat{\boldsymbol{\Sigma}}_A$ are the MLE's of $\boldsymbol{\Sigma}_R$ under the null and alternative, respectively. The degrees of freedom in the test statistic (9.42) is the difference in the number of ones in $\mathbf{P}_A$ and $\mathbf{P}_0$.

The chi-squared character of the statistic in (9.44) is asymptotic. In testing block hypotheses as in Section 7.2, we can improve the approximation (or find an exact test) using the techniques in that section. In fact, the likelihood ratio test is a Wilks' Λ test. The only change we make is to use the idea in Section 9.3.3 to obtain a multivariate regression model from the both-sides model.

We start with the both-sides model as in (9.24), and suppose $\mathbf{P}_A$ is a matrix consisting of all ones, so that it provides no restriction, and the zeroes in $\mathbf{P}_0$ form a block, leading to the null exemplified by that in (7.13). By permuting the rows and column appropriately, we can partition the $p \times l$ coefficient matrix so that the model is written

$$\mathbf{Y} = (\mathbf{x}_1 \ \ \mathbf{x}_2) \begin{pmatrix} \boldsymbol{\beta}_{11} & \boldsymbol{\beta}_{21} \\ \boldsymbol{\beta}_{12} & \boldsymbol{\beta}_{22} \end{pmatrix} \mathbf{z}_a' + \mathbf{R}, \tag{9.45}$$

where $\boldsymbol{\beta}_{11}$ is $p^* \times l^*$, $\mathbf{x}_1$ is $n \times p^*$, and $\mathbf{x}_2$ is $n \times (p - p^*)$, and the hypotheses are

$$H_0 : \boldsymbol{\beta}_{11} = 0 \ \ versus \ \ H_A : \boldsymbol{\beta}_{11} \neq 0. \tag{9.46}$$

The other $\boldsymbol{\beta}_{ij}$'s have no restriction. This testing problem is often called the generalized multivariate analysis of variance (GMANOVA) problem.

Repeat the steps in (9.25) through (9.35). Using the partitioning (9.45), the conditional model $\mathbf{Y}_{\mathbf{z}a} | \mathbf{Y}_{\mathbf{z}b} = \mathbf{y}_{\mathbf{z}b}$ in (9.32) is

$$\mathbf{Y}_{\mathbf{z}a} = (\mathbf{x}_1 \ \ \mathbf{x}_2 \ \ \mathbf{y}_{\mathbf{z}b}) \begin{pmatrix} \boldsymbol{\beta}_{11} & \boldsymbol{\beta}_{21} \\ \boldsymbol{\beta}_{12} & \boldsymbol{\beta}_{22} \\ \boldsymbol{\eta}_1 & \boldsymbol{\eta}_2 \end{pmatrix} + \mathbf{R}_{\mathbf{z}a}, \tag{9.47}$$

where $\mathbf{Y}_{\mathbf{z}a}$ is $n \times l$, $\mathbf{Y}_{\mathbf{z}b}$ is $n \times (q - l)$, and $\boldsymbol{\eta}_1$ is $(q - l) \times l^*$. Now we can implement the tests in Section 7.2 based on the model (9.47), where $\boldsymbol{\beta}_{11}$ here plays the role of $\boldsymbol{\beta}^*$ in (7.12), and $(\mathbf{x}_1, \mathbf{x}_2, \mathbf{y}_{\mathbf{z}b})$ plays the role of $\mathbf{x}$. In this case, the $\mathbf{B}$ and $\mathbf{W}$ in (7.20) are again independent, with $\nu = n - p - q + l$,

$$\mathbf{B} \sim \text{Wishart}_{l^*}(p^*, \boldsymbol{\Sigma}_{\mathbf{z}}^*), \ \text{ and } \ \mathbf{W} \sim \text{Wishart}_{l^*}(n - p - q + l, \boldsymbol{\Sigma}_{\mathbf{z}}^*). \tag{9.48}$$

Exercises 9.6.7 through 9.6.9 show that the likelihood ratio (9.41) is Wilks' Λ, i.e.,

$$\Lambda = (LR)^{-2/n} = \frac{|\mathbf{W}|}{|\mathbf{W} + \mathbf{B}|} \sim \text{Wilks}_{l^*}(p^*, n - p - q + l). \tag{9.49}$$

Hence the statistic in (9.44) with the Bartlett correction (7.37) is

$$-\left(n-p-q+l-\frac{l^*-p^*+1}{2}\right)\ \log(\Lambda)\ \approx\ \chi^2_{p^*l^*}. \tag{9.50}$$

9.5 Model selection: AIC and BIC

We often have a number of models we wish to consider, rather than just two as in hypothesis testing. (Note also that hypothesis testing may not be appropriate even when choosing between two models, e.g., when there is no obvious allocation to "null" and "alternative" models.) For example, in the both-sides model, each pattern of zeroes for the $\boldsymbol{\beta}$ defines a different model. Here, we assume there are K models under consideration, labelled $M_1, M_2, \ldots, M_K$. Each model is based on the same data, $\mathbf{Y}$, but has its own density and parameter space:

$$\text{Model } M_k\ \Rightarrow\ \mathbf{Y} \sim f_k(\mathbf{y}\,|\,\boldsymbol{\theta}_k),\ \boldsymbol{\theta}_k \in \Theta_k. \tag{9.51}$$

The densities need not have anything to do with each other, i.e., one could be normal, another uniform, another logistic, etc., although often they will be of the same family. It is possible that the models will overlap, so that several models might be correct at once, e.g., when there are nested models.

Let

$$l_k(\boldsymbol{\theta}_k\,;\,\mathbf{y}) = \log(L_k(\boldsymbol{\theta}_k\,;\,\mathbf{y})) = \log(f_k(\mathbf{y}\,|\,\boldsymbol{\theta}_k)) + C(\mathbf{y}),\ k = 1,\ldots,K, \tag{9.52}$$

be the loglikelihoods for the models. The constant $C(\mathbf{y})$ is arbitrary, being the $\log(a(\mathbf{y}))$ from (9.1). As long as it is the same for each k, it will not affect the outcome of the following procedures. Define the **deviance** of the model M_k at parameter value $\boldsymbol{\theta}_k$ by

$$\text{deviance}(M_k(\boldsymbol{\theta}_k)\,;\,\mathbf{y}) = -2\,l_k(\boldsymbol{\theta}_k\,;\,\mathbf{y}). \tag{9.53}$$

It is a measure of fit of the model to the data; the smaller the deviance, the better the fit. The MLE of $\boldsymbol{\theta}_k$ for model M_k minimizes this deviance, giving us the **observed deviance**,

$$\text{deviance}(M_k(\widehat{\boldsymbol{\theta}}_k)\,;\,\mathbf{y}) = -2\,l_k(\widehat{\boldsymbol{\theta}}_k\,;\,\mathbf{y}) = -2\max_{\boldsymbol{\theta}_k\in\Theta_k} l_k(\boldsymbol{\theta}_k\,;\,\mathbf{y}). \tag{9.54}$$

Note that the likelihood ratio statistic in (9.42) is just the difference in observed deviance of the two hypothesized models:

$$2\log(LR) = \text{deviance}(H_0(\widehat{\boldsymbol{\theta}}_0)\,;\,\mathbf{y}) - \text{deviance}(H_A(\widehat{\boldsymbol{\theta}}_A)\,;\,\mathbf{y}). \tag{9.55}$$

At first blush one might decide the best model is the one with the smallest observed deviance. The problem with that approach is that because the deviances are based on minus the maximum of the likelihoods, the model with the best observed deviance will be the largest model, i.e., one with highest dimension. Instead, we add a penalty depending on the dimension of the parameter space, as for Mallow's C_p in (7.78). The two most popular likelihood-based procedures are the Bayes information criterion (BIC) of Schwarz [1978] and the Akaike information criterion (AIC) of Akaike [1974] (who actually meant for the "A" to stand for "An"):

$$\text{BIC}(M_k\,;\,\mathbf{y}) = \text{deviance}(M_k(\widehat{\boldsymbol{\theta}}_k)\,;\,\mathbf{y}) + \log(n)d_k, \text{ and} \tag{9.56}$$

$$\text{AIC}(M_k\,;\,\mathbf{y}) = \text{deviance}(M_k(\widehat{\boldsymbol{\theta}}_k)\,;\,\mathbf{y}) + 2d_k, \tag{9.57}$$

where

$$d_k = \dim(\Theta_k). \tag{9.58}$$

Whichever criterion is used, it is implemented by finding the value for each model, then choosing the model with the smallest value of the criterion, or looking at the models with the smallest values.

Note that the only difference between AIC and BIC is the factor multiplying the dimension in the penalty component. The BIC penalizes each dimension more heavily than does the AIC, at least if $n > 7$, so tends to choose more parsimonious models. (The corrected version of the AIC in (9.87) for multivariate regression has the penalty depending on (n, p, q).) In more complex situations than we deal with here, the *deviance information criterion* is useful, which uses more general definitions of the deviance. See Spiegelhalter et al. [2002].

The next two sections present some further insight into the two criteria.

9.5.1 BIC: Motivation

The AIC and BIC have somewhat different motivations. The BIC, as hinted at by the "Bayes" in the name, is an attempt to estimate the Bayes posterior probability of the models. More specifically, if the prior probability that model M_k is the true one is π_k, then the BIC-based estimate of the posterior probability is

$$\widehat{P}^{\text{BIC}}[M_k \mid \mathbf{y}] = \frac{e^{-\frac{1}{2}\,\text{BIC}(M_k\,;\,\mathbf{y})}\pi_k}{e^{-\frac{1}{2}\,\text{BIC}(M_1\,;\,\mathbf{y})}\pi_1 + \cdots + e^{-\frac{1}{2}\,\text{BIC}(M_K\,;\,\mathbf{y})}\pi_K}. \tag{9.59}$$

If the prior probabilities are taken to be equal, then because each posterior probability has the same denominator, the model that has the highest posterior probability is indeed the model with the smallest value of BIC. The advantage of the posterior probability form is that it is easy to assess which models are nearly as good as the best, if there are any.

To see where the approximation arises, we first need a prior on the parameter space. In this case, there are several parameter spaces, one for each model under consideration. Thus is it easier to find conditional priors for each $\boldsymbol{\theta}_k$, conditioning on the model:

$$\boldsymbol{\theta}_k \mid M_k \sim \rho_k(\boldsymbol{\theta}_k), \tag{9.60}$$

for some density ρ_k on Θ_k. The marginal probability of each model is the prior probability:

$$\pi_k = P[M_k]. \tag{9.61}$$

The conditional density of $(\mathbf{Y}, \boldsymbol{\theta}_k)$ given M_k is

$$g_k(\mathbf{y}, \boldsymbol{\theta}_k \mid M_k) = f_k(\mathbf{y} \mid \boldsymbol{\theta}_k)\rho_k(\boldsymbol{\theta}_k). \tag{9.62}$$

To find the density of $\mathbf{Y}$ given M_k, we integrate out the $\boldsymbol{\theta}_k$:

$$\begin{aligned} \mathbf{Y} \mid M_k \sim g_k(\mathbf{y} \mid M_k) &= \int_{\Theta_k} g_k(\mathbf{y}, \boldsymbol{\theta}_k \mid M_k) d\boldsymbol{\theta}_k \\ &= \int_{\Theta_k} f_k(\mathbf{y} \mid \boldsymbol{\theta}_k)\rho_k(\boldsymbol{\theta}_k) d\boldsymbol{\theta}_k. \end{aligned} \tag{9.63}$$

With the parameters hidden, it is straightforward to find the posterior probabilities of the models using Bayes theorem, Theorem 2.2:

$$P[M_k \mid \mathbf{y}] = \frac{g_k(\mathbf{y} \mid M_k)\pi_k}{g_1(\mathbf{y} \mid M_1)\pi_1 + \cdots + g_K(\mathbf{y} \mid M_K)\pi_K}. \tag{9.64}$$

Comparing (9.64) to (9.59), we see that the goal is to approximate $g_k(\mathbf{y} \mid M_k)$ by $e^{-\frac{1}{2}BIC(M_k;\mathbf{y})}$. To do this, we use the Laplace approximation, as in Schwarz [1978]. The following requires a number of regularity assumptions, not all of which we will detail. One is that the data $\mathbf{y}$ consists of n iid observations, another that n is large. Many of the standard likelihood-based assumptions needed can be found in Chapter 6 of Lehmann and Casella [2003], or any other good mathematical statistics text. For convenience we drop the "k", and from (9.63) consider

$$\int_\Theta f(\mathbf{y} \mid \boldsymbol{\theta})\rho(\boldsymbol{\theta})d\boldsymbol{\theta} = \int_\Theta e^{l(\boldsymbol{\theta};\mathbf{y})}\rho(\boldsymbol{\theta})d\boldsymbol{\theta}. \tag{9.65}$$

The Laplace approximation expands $l(\boldsymbol{\theta};\mathbf{y})$ around its maximum, the maximum occurring at the maximum likelihood estimator $\widehat{\boldsymbol{\theta}}$. Then, assuming all the derivatives exist,

$$l(\boldsymbol{\theta};\mathbf{y}) \approx l(\widehat{\boldsymbol{\theta}};\mathbf{y}) + (\boldsymbol{\theta}-\widehat{\boldsymbol{\theta}})'\nabla(\widehat{\boldsymbol{\theta}}) + \frac{1}{2}\,(\boldsymbol{\theta}-\widehat{\boldsymbol{\theta}})'\mathbf{H}(\widehat{\boldsymbol{\theta}})(\boldsymbol{\theta}-\widehat{\boldsymbol{\theta}}), \tag{9.66}$$

where $\nabla(\widehat{\boldsymbol{\theta}})$ is the $d \times 1$ ($\boldsymbol{\theta}$ is $d \times 1$) vector with

$$\nabla_i(\widehat{\boldsymbol{\theta}}) = \frac{\partial}{\partial\boldsymbol{\theta}_i}\,l(\boldsymbol{\theta};\mathbf{y})\mid_{\boldsymbol{\theta}=\widehat{\boldsymbol{\theta}}'} \tag{9.67}$$

and $\mathbf{H}$ is the $d \times d$ matrix with

$$\mathbf{H}_{ij} = \frac{\partial^2}{\partial\boldsymbol{\theta}_i\partial\boldsymbol{\theta}_j}\,l(\boldsymbol{\theta};\mathbf{y})\mid_{\boldsymbol{\theta}=\widehat{\boldsymbol{\theta}}}\,. \tag{9.68}$$

Because $\widehat{\boldsymbol{\theta}}$ is the MLE, the derivative of the loglikelihood at the MLE is zero, i.e., $\nabla(\widehat{\boldsymbol{\theta}}) = \mathbf{0}$. Also, let

$$\widehat{\mathbf{F}} = -\frac{1}{n}\,\mathbf{H}(\widehat{\boldsymbol{\theta}}), \tag{9.69}$$

which is called the observed Fisher information contained in one observation. Then (9.65) and (9.66) combine to show that

$$\int_\Theta f(\mathbf{y} \mid \boldsymbol{\theta})\rho(\boldsymbol{\theta})d\boldsymbol{\theta} \approx e^{l(\widehat{\boldsymbol{\theta}};\mathbf{y})}\int_\Theta e^{-\frac{1}{2}(\boldsymbol{\theta}-\widehat{\boldsymbol{\theta}})'n\widehat{\mathbf{F}}(\boldsymbol{\theta}-\widehat{\boldsymbol{\theta}})}\rho(\boldsymbol{\theta})d\boldsymbol{\theta}. \tag{9.70}$$

If n is large, the exponential term in the integrand drops off precipitously when $\boldsymbol{\theta}$ is not close to $\widehat{\boldsymbol{\theta}}$, and assuming that the prior density $\rho(\boldsymbol{\theta})$ is fairly flat for $\boldsymbol{\theta}$ near $\widehat{\boldsymbol{\theta}}$, we have

$$\int_\Theta e^{-\frac{1}{2}(\boldsymbol{\theta}-\widehat{\boldsymbol{\theta}})'n\widehat{\mathbf{F}}(\boldsymbol{\theta}-\widehat{\boldsymbol{\theta}})}\rho(\boldsymbol{\theta})d\boldsymbol{\theta} \approx \int_\Theta e^{-\frac{1}{2}(\boldsymbol{\theta}-\widehat{\boldsymbol{\theta}})'n\widehat{\mathbf{F}}(\boldsymbol{\theta}-\widehat{\boldsymbol{\theta}})}d\boldsymbol{\theta}\rho(\widehat{\boldsymbol{\theta}}). \tag{9.71}$$

The integrand in the last term looks like the density (8.47) as if

$$\boldsymbol{\theta} \sim N_d(\widehat{\boldsymbol{\theta}}, (n\widehat{\mathbf{F}})^{-1}), \tag{9.72}$$

but without the constant. Thus the integral is just the reciprocal of that constant, i.e.,

$$\int_{\Theta} e^{-\frac{1}{2}(\boldsymbol{\theta}-\widehat{\boldsymbol{\theta}})' n \widehat{\mathbf{F}}(\boldsymbol{\theta}-\widehat{\boldsymbol{\theta}})} d\boldsymbol{\theta} = (\sqrt{2\pi})^{d/2} |n\widehat{\mathbf{F}}|^{-1/2} = (\sqrt{2\pi})^{d/2} |\widehat{\mathbf{F}}|^{-1/2} n^{-d/2}. \tag{9.73}$$

Putting (9.70) and (9.73) together gives

$$\begin{aligned}
\log\left(\int_{\Theta} f(\mathbf{y} \mid \boldsymbol{\theta}) \rho(\boldsymbol{\theta}) d\boldsymbol{\theta}\right) &\approx l(\widehat{\boldsymbol{\theta}}; \mathbf{y}) - \frac{d}{2}\log(n) \\
&\quad + \log(\rho(\widehat{\boldsymbol{\theta}})) + \frac{d}{2}\log(2\pi) - \frac{1}{2}\log(|\widehat{\mathbf{F}}|) \\
&\approx l(\widehat{\boldsymbol{\theta}}; \mathbf{y}) - \frac{d}{2}\log(n) \\
&= -\frac{1}{2}\,\mathrm{BIC}(M; \mathbf{y}).
\end{aligned} \tag{9.74}$$

Dropping the last three terms in the first equation is justified by noting that as $n \to \infty$, $l(\widehat{\boldsymbol{\theta}}; \mathbf{y})$ is of order n (in the iid case), $\log(n)d/2$ is clearly of order $\log(n)$, and the other terms are bounded. (This step may be a bit questionable since n has to be extremely large before $\log(n)$ starts to dwarf a constant.)

There are a number of approximations and heuristics in this derivation, and indeed the resulting approximation may not be especially good. See Berger, Ghosh, and Mukhopadhyay [2003], for example. A nice property is that under conditions, if one of the considered models is the correct one, then the BIC chooses the correct model as $n \to \infty$.

9.5.2 AIC: Motivation

The Akaike information criterion can be thought of as a generalization of Mallows' C_p from Section 7.5, based on deviance rather than error sum of squares. To evaluate model M_k as in (9.51), we imagine fitting the model based on the data $\mathbf{Y}$, then testing it out on a new (unobserved) variable, $\mathbf{Y}^{New}$, which has the same distribution as and is independent of $\mathbf{Y}$. The measure of discrepancy between the model and the new variable is the deviance in (9.53), where the parameter is estimated using $\mathbf{Y}$. We then take the expected value, yielding the expected predictive deviance,

$$EPredDev_k = E[\mathrm{deviance}(M_k(\widehat{\boldsymbol{\theta}}_k)\,;\,\mathbf{Y}^{New})]. \tag{9.75}$$

The expected value is over $\widehat{\boldsymbol{\theta}}_k$, which depends on only $\mathbf{Y}$, and $\mathbf{Y}^{New}$.

As for Mallows' C_p, we estimate the expected predictive deviance using the observed deviance, then add a term to ameliorate the bias. Akaike argues that for large n, if M_k is the true model,

$$\Delta = EPredDev_k - E[\mathrm{deviance}(M_k(\widehat{\boldsymbol{\theta}}_k)\,;\,\mathbf{Y})] \approx 2d_k, \tag{9.76}$$

where d_k is the dimension of the model as in (9.58), from which the estimate AIC in (9.57) arises. A good model is then one with a small AIC.

Note also that by adjusting the priors $\pi_k = P[M_k]$ in (9.59), one can work it so that the model with the lowest AIC has the highest posterior probability. See Exercise 9.6.13.

Akaike's original motivation was information-theoretic, based on the **Kullback-Leibler** divergence from density f to density g. This divergence is defined as

$$\text{KL}(f \,||\, g) = -\int g(\mathbf{w}) \log\left(\frac{f(\mathbf{w})}{g(\mathbf{w})}\right) d\mathbf{w}. \tag{9.77}$$

For fixed g, equation (1.59) shows that the Kullback-Leibler divergence is positive unless $g = f$, in which case it is zero. For the Akaike information criterion, g is the true density of $\mathbf{Y}$ and $\mathbf{Y}^{New}$, and for model k, f is the density estimated using the maximum likelihood estimate of the parameter, $f_k(\mathbf{w} \mid \widehat{\boldsymbol{\theta}})$, where $\widehat{\boldsymbol{\theta}}$ is based on $\mathbf{Y}$. Write

$$\begin{aligned}\text{KL}(f_k(\mathbf{w} \mid \widehat{\boldsymbol{\theta}}_k) \,||\, g) &= -\int g(\mathbf{w}) \log(f_k(\mathbf{w} \mid \widehat{\boldsymbol{\theta}}))d\mathbf{w} + \int g(\mathbf{w}) \log(g(\mathbf{w}))d\mathbf{w} \\ &= \frac{1}{2}\, E[\text{deviance}(M_k(\widehat{\boldsymbol{\theta}}_k)\,;\, \mathbf{Y}^{New}) \mid \mathbf{Y} = \mathbf{y}] - \text{Entropy}(g). \end{aligned} \tag{9.78}$$

(The $\mathbf{w}$ is representing the $\mathbf{Y}^{New}$, and the dependence on the observed $\mathbf{y}$ is through only $\widehat{\boldsymbol{\theta}}_k$.) Here the g, the true density of $\mathbf{Y}$, does not depend on the model M_k, hence neither does its entropy, defined in (1.44). Thus $EPredDev_k$ from (9.75) is equivalent to (9.78) upon taking the further expectation over $\mathbf{Y}$.

One slight logical glitch in the development is that while the theoretical criterion (9.75) is defined assuming $\mathbf{Y}$ and $\mathbf{Y}^*$ have the true distribution, the approximation in (9.76) assumes the true distribution is contained in the model M_k. Thus it appears that the approximation is valid for all models under consideration only if the true distribution is contained in all the models. Even so, the AIC is a legitimate method for model selection. See the book Burnham and Anderson [2003] for more information.

Rather than justify the result in full generality, we will show the exact value for Δ for multivariate regression, as Hurvich and Tsai [1989] did in the multiple regression model.

9.5.3 AIC: Multivariate regression

The multivariate regression model (4.9) (with no constraints) is

$$\text{Model } M: \; \mathbf{Y} \sim N_{n\times q}\left(\mathbf{x}\boldsymbol{\beta}, \mathbf{I}_n \otimes \boldsymbol{\Sigma}_R\right), \;\; \boldsymbol{\beta} \in \mathbb{R}^{p\times q}, \tag{9.79}$$

where $\mathbf{x}$ is $n \times p$ and $\boldsymbol{\Sigma}_R$ is $q \times q$. Now from (9.4),

$$l(\boldsymbol{\beta}, \boldsymbol{\Sigma}_R; \mathbf{y}) = -\frac{n}{2}\log(|\boldsymbol{\Sigma}_R|) - \frac{1}{2}\,\text{trace}(\boldsymbol{\Sigma}_R^{-1}(\mathbf{y} - \mathbf{x}\boldsymbol{\beta})'(\mathbf{y} - \mathbf{x}\boldsymbol{\beta})). \tag{9.80}$$

The MLE's are then

$$\widehat{\boldsymbol{\beta}} = (\mathbf{x}'\mathbf{x})^{-1}\mathbf{x}'\mathbf{y} \;\text{ and }\; \widehat{\boldsymbol{\Sigma}}_R = \frac{1}{n}\,\mathbf{y}'\mathbf{Q}_\mathbf{x}\mathbf{y}, \tag{9.81}$$

as in (9.16), since by (5.46) the least squares estimate is the weighted least squares estimate (hence MLE) when $\mathbf{z}$ is square and there are no constraints on $\boldsymbol{\beta}$. Using (9.17) and (9.53), we see that the deviances evaluated at the data $\mathbf{Y}$ and the unobserved $\mathbf{Y}^{New}$ are, respectively,

$$\begin{aligned}\text{deviance}(M(\widehat{\boldsymbol{\beta}}, \widehat{\boldsymbol{\Sigma}}_R)\,;\, \mathbf{Y}) &= n\log(|\widehat{\boldsymbol{\Sigma}}_R|) + nq, \;\text{ and} \\ \text{deviance}(M(\widehat{\boldsymbol{\beta}}, \widehat{\boldsymbol{\Sigma}}_R)\,;\, \mathbf{Y}^{New}) &= n\log(|\widehat{\boldsymbol{\Sigma}}_R|) + \text{trace}(\widehat{\boldsymbol{\Sigma}}_R^{-1}(\mathbf{Y}^{New} - \mathbf{x}\widehat{\boldsymbol{\beta}})'(\mathbf{Y}^{New} - \mathbf{x}\widehat{\boldsymbol{\beta}})).\end{aligned} \tag{9.82}$$

The first terms on the right-hand sides in (9.82) are the same, hence the difference in (9.76) is

$$\Delta = E[\text{trace}(\widehat{\mathbf{\Sigma}}_R^{-1}\mathbf{U}'\mathbf{U})] - nq, \quad \text{where } \mathbf{U} = \mathbf{Y}^{New} - \mathbf{x}\widehat{\boldsymbol{\beta}} = \mathbf{Y}^{New} - \mathbf{P}_\mathbf{x}\mathbf{Y}. \tag{9.83}$$

From Theorem 6.1, we know that $\widehat{\boldsymbol{\beta}}$ and $\widehat{\mathbf{\Sigma}}_R$ are independent, and further both are independent of $\mathbf{Y}^{New}$, hence we have

$$E[\text{trace}(\widehat{\mathbf{\Sigma}}_R^{-1}\mathbf{U}'\mathbf{U})] = \text{trace}(E[\widehat{\mathbf{\Sigma}}_R^{-1}]E[\mathbf{U}'\mathbf{U}]). \tag{9.84}$$

Using calculations as in Section 7.5.2, we find (in Exercise 9.6.11) that

$$\Delta = \frac{n}{n-p-q-1}\, 2\, d_M, \tag{9.85}$$

where d_M is the dimension of model M, summing the pq for the β_{ij}'s and $q(q+1)/2$ for the $\mathbf{\Sigma}_R$:

$$d_M = pq + \frac{q(q+1)}{2}. \tag{9.86}$$

Note that the dimension in (7.76) resolves to d_M when there are no zeroes in the $\boldsymbol{\beta}$.

Then from (9.76), the estimate of $EPredDev$ is

$$\text{AICc}(M\,;\,\mathbf{y}) = \text{deviance}(M(\widehat{\boldsymbol{\beta}}, \widehat{\mathbf{\Sigma}}_R)\,;\,\mathbf{y}) + \frac{n}{n-p-q-1}\, 2\, d_M. \tag{9.87}$$

The lower case "c" stands for "corrected." For large n, $\Delta \approx 2d_M$. In univariate regression $q = 1$, and (9.87) is the value given in Hurvich and Tsai [1989].

The exact value for $\boldsymbol{\Delta}$ in (9.85) for submodels of the model M, where a subset of β_{ij}'s is set to zero, is not in general easy to find. If the submodel is itself a multivariate regression model, which occurs if each row of the $\boldsymbol{\beta}$ is either all zero or all nonzero, then the answer is as in (9.87) and (9.86), but substituting the number of nonzero rows for the p. An *ad hoc* solution for model M_k given by the pattern $\mathbf{P}_k$ is to let p_k be the number of rows of $\mathbf{P}_k$ that have at least one "1," then use

$$\Delta_k = \frac{n}{n-p_k-q-1}\, 2\, d_k, \quad d_k = \sum_{ij} p_{k,ij} + \frac{q(q+1)}{2}. \tag{9.88}$$

9.5.4 Example: Skulls

We will look at the data on Egyptian skulls treated previously in Exercises 4.4.2, 6.6.10, 7.6.15, and 7.6.20. There are thirty observations on each of five time periods, and four measurements made on each observation. The $\mathbf{x}$ we consider is an orthogonal polynomial (of degree 4) matrix in the time periods, which we take to be equally spaced. Thus $(n, p, q) = (150, 5, 4)$. The model is multivariate regression, $\mathbf{Y} = \mathbf{x}\boldsymbol{\beta} + \mathbf{R}$, so that $\mathbf{z} = \mathbf{I}_4$. The estimates of the coefficients, and their t-statistics, for the unrestricted model are below:

	Maximal Breadth		Basibregmatic Height		Basialveolar Length		Nasal Height	
	Est.	t	Est.	t	Est.	t	Est.	t
Intercept	133.97	357.12	132.55	334.98	96.46	240.26	50.93	195.78
Linear	1.27	4.80	−0.69	−2.48	−1.59	−5.59	0.34	1.85
Quadratic	−0.12	−0.55	−0.34	−1.43	−0.02	−0.10	0.03	0.21
Cubic	−0.15	−0.55	−0.25	−0.88	0.34	1.20	−0.26	−1.43
Quartic	0.04	0.41	0.10	0.91	−0.08	−0.74	−0.05	−0.72

(9.89)

Looking at the t-statistics, we see that all the intercepts are highly significant, which is not surprising, and all the quadratic, cubic, and quartic coefficients' are under 2 in absolute value. Two of the linear terms are highly significant, one is somewhat significant, and one is somewhat not so.

Hypothesis testing is one approach to comparing pairs of models. Consider testing the null hypothesis that all measurements have linear fits, versus the general model. In terms of patterns for the $\boldsymbol{\beta}$, we test

$$H_0 : \mathbf{P} = \begin{pmatrix} 1111 \\ 1111 \\ 0000 \\ 0000 \\ 0000 \end{pmatrix} \quad versus \quad H_A : \mathbf{P} = \begin{pmatrix} 1111 \\ 1111 \\ 1111 \\ 1111 \\ 1111 \end{pmatrix}. \tag{9.90}$$

We first set up the data matrices.

```
x <− cbind(1,c(−2,−1,0,1,2),c(2,−1,−2,−1,2),c(−1,2,0,−2,1),c(1,−4,6,−4,1))
# or x <− cbind(1,poly(1:5,4))
x <− kronecker(x,rep(1,30))
y <− skulls[,1:4]
z <− diag(4)
```

To find the likelihood ratio test statistic in (9.55), we need to fit the two models, and find their deviances and dimensions.

```
pattern0 <− rbind(c(1,1,1,1),1,0,0,0)
b0 <− bothsidesmodel.mle(x,y,z,pattern0)
bA <− bothsidesmodel.mle(x,y,z) # If the pattern is all 1's, we can omit it
lrstat <− b0$Dev−bA$Dev
b0$Dim
bA$Dim
```

The statistic is lrstat $= 10.4086$. The dimension of the null model is 18, which includes the 8 nonzero β_{ij}'s, plus the 10 parameters in the $\boldsymbol{\Sigma}_R$. For the alternative model, we have $20 + 10 = 30$. Thus the statistic should be compared to a χ^2_ν for $\nu = 30 - 18 = 12$ (which is just the number of 0's in the null's pattern). Clearly, the result is not significant, hence we fail to reject the null hypothesis.

Furthermore, the fourth measurement, nasal height, may not need a linear term, leading us to consider testing

$$H_0 : \mathbf{P} = \begin{pmatrix} 1111 \\ 1110 \\ 0000 \\ 0000 \\ 0000 \end{pmatrix} \quad versus \quad H_A : \mathbf{P} = \begin{pmatrix} 1111 \\ 1111 \\ 0000 \\ 0000 \\ 0000 \end{pmatrix}. \tag{9.91}$$

We already have the results of what is now the alternative hypothesis. For the null, we find

```
pattern0[2,4]<−0 # New null pattern
b00 <− bothsidesmodel.mle(x,y,z,pattern0)
b00$Dev−b0$Dev
[1] 3.431563
```

It is easy to see that the difference in dimensions is 1, so with a statistic of 3.4316, the χ^2_1 yields a p-value of 0.064, not quite significant. Thus this last linear parameter may or may not be zero.

The model selection approach allows a more comprehensive comparison of all possible models. We will consider the submodels that honor the monotonicity of the polynomials for each measurement variable. That is, each column of pattern $\mathbf{P}$ indicates the degree of the polynomial for the corresponding measurement, where a degree p polynomial has 1's in rows 1 through $p+1$, and zeroes elsewhere. Denote a model by four integers indicating the *degree* $+1$ for each variable, so that, for example, M_{2321} is the model with linear, quadratic, linear, and constant models for the four variables, respectively. The hypotheses in (9.90) are M_{2222} and M_{5555}, and the null in (9.91) is M_{2221}.

For model M_{ijkl}, we first set up the pattern, then fit the model. It helps to write a little function to provide the pattern:

```
skullspattern <− function(i,j,k,l) {
        pattern <− matrix(0,ncol=4,nrow=5)
        if(i>0) pattern[1:i,1] <− 1
        if(j>0) pattern[1:j,2] <− 1
        if(k>0) pattern[1:k,3] <− 1
        if(l>0) pattern[1:l,4] <− 1
        pattern
}
```

We then loop over the function, fitting the model each time and collecting the results:

```
results <− NULL
models <− NULL
for(i in 0:5) {
  for(j in 0:5) {
    for(k in 0:5) {
      for (l in 0:5) {
         bothsidesmodel <− bothsidesmodel.mle(x,y,z,skullspattern(i,j,k,l))
         results <− rbind(results,c(bothsidesmodel$Dev,bothsidesmodel$Dim,
                                   bothsidesmodel$AICc,bothsidesmodel$BIC))
         models <− rbind(models,c(i,j,k,l))
}}}}
```

We also find the BIC-based estimates of the posterior probabilities (actually, percentages) of the models from (9.59):

```
bic <− results[,4]
p <− exp(−(bic−max(bic))/2) # Subtract the max to avoid underflow
p <− 100*p/sum(p)
```

Since each column can have from 0 to 5 ones, there are $6^4 = 1296$ possible models. We could reasonably drop all the models with at least one zero in the first row without changing the results. The next table has the results from fitting all 1296 models, though we just show the ten with the best (lowest) BIC's:

Model	Deviance	Dimension	AICc	BIC	$\widehat{p}^{BIC}$
2221	2334.29	17	2369.96	2419.47	32.77
2222	2330.86	18	2368.62	2421.05	14.88
2122	2336.93	17	2372.60	2422.12	8.75
2321	2331.98	18	2370.01	2422.17	8.52
2121	2342.18	16	2375.74	2422.35	7.80
2322	2328.55	19	2368.69	2423.75	3.87
3221	2333.91	18	2371.94	2424.10	3.24
2231	2334.26	18	2372.29	2424.45	2.72
3222	2330.48	19	2370.62	2425.68	1.47
2223	2330.54	19	2370.68	2425.74	1.43

(9.92)

The factor multiplying the dimension d_k for the AICc as in (9.88) here ranges from 2.07 to 2.14, while for the BIC in (9.56), it is $\log(n) = \log(150) = 5.01$. Thus the BIC will tend to choose sparser models. The top four models for BIC are ranked numbers 3, 1, 6, and 4 for AICc. But BIC's third best, 2122, is AICc's 32nd best, and BIC's fifth best, 2121, is AICc's 133rd best. Note that these two models are quite sparse.

Looking at the top ten models, we see they are within one or two parameters of M_{2222}, which is the model with linear fits for all four variables. That model is the best under AICc, and second best under BIC. The best model under BIC shows linear fits for the first three variables, but just a constant fit for the fourth. The BIC-based estimated probabilities for the models are quite spread out, with the top ten models having a total probability of about 85%. The best model has a probability of 33%, which is high but not nearly overwhelming. All the top five have probabilities over 7%.

We can also estimate the probabilities of other events. For example, the probability that the maximum degree of the four polynomials is quartic sums the probabilities of all models for which the maximum of its indices is 5, yielding 0.73%. Similarly, the probabilities that the maximum degree is cubic, quartic, linear, and constant are estimated to be 5.03%, 30.00%, 64.23%, and 0. Thus we can be approximately 94% sure that linear or quadratics are enough. The chance any of the variables has a zero model (i.e., has a zero mean) is zero.

Looking at the individual variables, we can sum the probabilities of models for which variable i has degree j. We obtain the following, where the entries are the estimated percentages:

	Maximal Breadth	Basibregmatic Height	Basialveolar Length	Nasal Height
Constant	0.05	21.53	0.00	61.70
Linear	90.22	60.51	90.48	34.39
Quadratic	8.89	15.64	7.50	3.23
Cubic	0.78	1.96	1.80	0.61
Quartic	0.07	0.35	0.23	0.08

(9.93)

Thus we can be fairly confident that linear fits are good for the first and third variables, maximal breadth and basialveolar length. Basibregmatic height is also likely to be linear, but the constant and quadratic models have nonnegligible probabilities. Nasal height is more likely to be constant than linear.

9.5.5 Example: Histamine in dogs

Now turn to the histamine in dogs example. The data are in Table 4.2, and a both-sides model is presented in (6.44). Notice that the drug effect and interaction effect are approximately opposite, which arises because the two depleted cells have approximately the same after values. That is, when histamine is depleted, the drug does not seem to matter. Thus we try a different model, where the contrasts are the depletion effect, the drug effect within the intact group, and the drug effect within the depleted group. We also use the before measurements as covariates, and orthogonal polynomials for the three after measurements. Thus $\mathbf{Y}_a = (\mathbf{x}, \mathbf{y}_b)\boldsymbol{\beta}\mathbf{z}' + \mathbf{R}$, where

$$\mathbf{x} = \begin{pmatrix} 1 & -1 & -1 & 0 \\ 1 & 1 & 0 & -1 \\ 1 & -1 & 1 & 0 \\ 1 & 1 & 0 & 1 \end{pmatrix} \otimes \mathbf{1}_4 \text{ and } \mathbf{z} = \begin{pmatrix} 1 & -1 & 1 \\ 1 & 0 & -2 \\ 1 & 1 & 1 \end{pmatrix}. \tag{9.94}$$

Here, $\boldsymbol{\beta}$ is 5×3. We are mainly interested in the three contrasts, which are represented by the middle three rows of $\boldsymbol{\beta}$. The first row contains the constants, and the last row contains the effects of the before measurements on the after measurements' coefficients. The models we will consider are then of the form M_{ijklm}, where i, j, k, l, m are the numbers of ones in the five rows, respectively, indicating the *degree* $+\,1$ of the polynomials. If the depletion effect is in the model, then the constant effect should be, and if one of the drug effects is in the model, then the depletion effect should be as well. Thus we have the constraints

$$0 \le i \le j \le k,\ l \le 3,\ 0 \le m \le 3. \tag{9.95}$$

For example, the model with a quadratic constant, a linear depletion effect, a linear drug effect within the intact group, a constant drug effect within the depleted group, and no before effect, has $(i, j, k, l, m) = (3, 2, 2, 1, 0)$:

$$M_{32210} : \ \mathbf{P} = \begin{pmatrix} 111 \\ 110 \\ 110 \\ 100 \\ 000 \end{pmatrix} \tag{9.96}$$

Fitting model M_{ijklm} proceeds by first assigning the values of i, j, k, l, and m, then using the following:

```
x <- cbind(1,c(-1,1,-1,1),c(-1,0,1,0),c(0,-1,0,1))
x <- kronecker(x,rep(1,4))
xyb <- cbind(x,histamine[,1])
ya <- histamine[,2:4]
z <- cbind(1,c(-1,0,1),c(1,-2,1))
pattern <- matrix(0,nrow=5,ncol=3)
if(i>0) pattern[1,1:i] <- 1
if(j>0) pattern[2,1:j] <- 1
if(k>0) pattern[3,1:k] <- 1
if(l>0) pattern[4,1:l] <- 1
if(m>0) pattern[5,1:m] <- 1
bothsidesmodel.mle(xyb,ya,z,pattern)
```

Here, the AICc yields sparser models, because the factor multiplying the dimension d_k ranges from 3.20 to 4.57 ($n = 16, q = 3$ and $p_k = 2, 3, 4, 5$), while for BIC, the $\log(n) = 2.77$. There are 200 models this time. The top ten according to BIC are

Model	Deviance	Dimension	AICc	BIC	$\widehat{p}^{BIC}$
22200	−146.70	12	−104.03	−113.42	12.24
22202	−151.70	14	−95.70	−112.89	9.36
22201	−148.19	13	−96.19	−112.15	6.48
32200	−147.94	13	−101.72	−111.89	5.70
32202	−152.95	15	−92.95	−111.36	4.36
22100	−141.28	11	−102.17	−110.78	3.26
22203	−152.35	15	−92.35	−110.77	3.24
22210	−146.80	13	−94.80	−110.75	3.22
32201	−149.44	14	−93.44	−110.62	3.02
33200	−149.40	14	−99.62	−110.58	2.96

(9.97)

The best model for both AICc and BIC is M_{22200}, which fits a linear model to the depletion effect, a linear model to the drug effect within the intact group, and zeroes out the drug effect within the depleted group and the before effect. As noted above, once the dogs have their histamine depleted, the drugs have no effect. Note that the top three models are the same, except for the before effects. Since we are not really interested in that effect, it is annoying that so much of the ordering of the models depends on it. To clean up the presentation, we can find the marginals for the effects of interest. That is, we sum the probabilities of all models that have a particular jkl. This leaves us with 30 probabilities, the top twelve of which are in the table below:

Model(jkl)	$\widehat{\mathbf{P}}^{BIC}$	Model(jkl)	$\widehat{\mathbf{P}}^{BIC}$
220	45.58	222	3.08
221	12.30	211	2.35
210	8.76	321	1.98
110	7.36	111	1.98
320	7.35	310	1.43
330	4.25	331	1.14

(9.98)

The top amalgamated model now has the same effects for the contrasts of interest as the previous best model, but with a fairly substantial probability of 46%. We can further summarize the results as in (9.93) to find the estimated percentages of the polynomial degrees for the contrasts:

Model	Depletion effect	Drug effect within intact	Drug effect within depleted
Zero	0.12	0.36	75.04
Constant	9.44	22.98	20.18
Linear	72.78	70.92	4.56
Quadratic	17.66	5.75	0.22

(9.99)

Thus the first two effects have over 99.5% probability of being nonzero, with each over a 70% chance the effect is linear. The third effect has a 75% of being zero.

The estimated nonzero coefficients of interest for the best model in (9.97) are

	Constant		Linear	
	Est.	t	Est.	t
Depletion effect (Depleted−Intact)	−0.259	−2.70	0.092	2.14
Drug effect within intact (Trimethaphan−Morphine)	0.426	3.13	−0.140	−2.32

(9.100)

The average depletion effect is negative, which means the depleted dogs have lower histamine levels than the intact ones, which is reasonable. Also, the positive slope means this difference decreases over time. For the drug effect within the intact group, we see that trimethanphan is better on average, but its advantage decreases somewhat over time.

9.6 Exercises

Exercise 9.6.1. Consider the statistical model with space $\mathcal{Y}$ and densities $f(\mathbf{y} \mid \boldsymbol{\theta})$ for $\boldsymbol{\theta} \in \Theta$. Suppose the function $g : \Theta \to \Omega$ is one-to-one and onto, so that a reparametrization of the model has densities $f^*(\mathbf{y} \mid \boldsymbol{\omega})$ for $\boldsymbol{\omega} \in \Omega$, where $f^*(\mathbf{y} \mid \boldsymbol{\omega}) = f(\mathbf{y} \mid g^{-1}(\boldsymbol{\omega}))$. (a) Show that $\widehat{\boldsymbol{\theta}}$ uniquely maximizes $f(\mathbf{y} \mid \boldsymbol{\theta})$ over $\boldsymbol{\theta}$ if and only if $\widehat{\boldsymbol{\omega}} \equiv g(\widehat{\boldsymbol{\theta}})$ uniquely maximizes $f^*(\mathbf{y} \mid \boldsymbol{\omega})$ over $\boldsymbol{\omega}$. [Hint: Show that $f(\mathbf{y} \mid \widehat{\boldsymbol{\theta}}) > f(\mathbf{y} \mid \boldsymbol{\theta})$ for all $\boldsymbol{\theta}' \neq \widehat{\boldsymbol{\theta}}$ implies $f^*(\mathbf{y} \mid \widehat{\boldsymbol{\omega}}) > f^*(\mathbf{y} \mid \boldsymbol{\omega})$ for all $\boldsymbol{\omega} \neq \widehat{\boldsymbol{\omega}}$, and *vice versa*.] (b) Argue that if $\widehat{\boldsymbol{\theta}}$ is the MLE of $\boldsymbol{\theta}$, then $g(\widehat{\boldsymbol{\theta}})$ is the MLE of $\boldsymbol{\omega}$.

Exercise 9.6.2. Again consider the model with space $\mathcal{Y}$ and densities $f(\mathbf{y} \mid \boldsymbol{\theta})$ for $\boldsymbol{\theta} \in \Theta$, and suppose $g : \Theta \to \Omega$ is just onto. Let g^* be any function of $\boldsymbol{\theta}$ such that the joint function $h(\boldsymbol{\theta}) = (g(\boldsymbol{\theta}), g^*(\boldsymbol{\theta}))$, $h : \Theta \to \Lambda$, is one-to-one and onto, and set the reparametrized density as $f^*(\mathbf{y} \mid \boldsymbol{\lambda}) = f(\mathbf{y} \mid h^{-1}(\boldsymbol{\lambda}))$. Exercise 9.6.1 shows that if $\widehat{\boldsymbol{\theta}}$ uniquely maximizes $f(\mathbf{y} \mid \boldsymbol{\theta})$ over Θ, then $\widehat{\boldsymbol{\lambda}} = h(\widehat{\boldsymbol{\theta}})$ uniquely maximizes $f^*(\mathbf{y} \mid \boldsymbol{\lambda})$ over $\boldsymbol{\Lambda}$. Argue that if $\widehat{\boldsymbol{\theta}}$ is the MLE of $\boldsymbol{\theta}$, that it is legitimate to define $g(\widehat{\boldsymbol{\theta}})$ to be the MLE of $\boldsymbol{\omega} = g(\boldsymbol{\theta})$.

Exercise 9.6.3. Give $(\boldsymbol{\beta}, \boldsymbol{\Sigma}_\mathbf{z})$ as a function of $(\boldsymbol{\beta}, \boldsymbol{\gamma}, \boldsymbol{\Sigma}_{\mathbf{z}aa\cdot b}, \boldsymbol{\Sigma}_{\mathbf{z}bb})$ defined in (9.29) and (9.31), and show that the latter set of parameters has space $\mathbb{R}^{p\times l} \times \mathbb{R}^{(q-l)\times l} \times \mathcal{S}_l \times \mathcal{S}_{q-l}$.

Exercise 9.6.4. Verify (9.36).

Exercise 9.6.5. Consider the covariate example in Section 6.4.4 and (9.21), where

$$(\mathbf{Y}_b\ \ \mathbf{Y}_a) = \left[\begin{pmatrix} 1 & 1 & 1 \\ 1 & 1 & -1 \\ 1 & -2 & 0 \end{pmatrix} \otimes \mathbf{1}_{10} \right] \begin{pmatrix} \mu_b & \mu_a \\ 0 & \alpha_a \\ 0 & \beta_a \end{pmatrix} + \mathbf{R}, \tag{9.101}$$

where $Cov[\mathbf{R}] = \mathbf{I}_{10} \otimes \mathbf{\Sigma}_R$ with

$$\mathbf{\Sigma}_R = \begin{pmatrix} \sigma_{bb} & \sigma_{ba} \\ \sigma_{ab} & \sigma_{aa} \end{pmatrix}. \tag{9.102}$$

Then conditionally,

$$\mathbf{Y}_a \,|\, \mathbf{Y}_b = \mathbf{y}_b \;\sim\; N\left((\mathbf{x}\ \mathbf{y}_b) \begin{pmatrix} \mu^* \\ \alpha_a \\ \beta_a \\ \gamma \end{pmatrix}, \sigma_{aa\cdot b}\, \mathbf{I}_n \right), \tag{9.103}$$

where $\mu^* = \mu_a - \gamma\mu_b$ and $\gamma = \sigma_{ab}/\sigma_{bb}$. (a) What is the marginal distribution of $\mathbf{Y}_b$? Note that its distribution depends on only (μ_b, σ_{bb}). Write the joint density of $(\mathbf{Y}_b, \mathbf{Y}_a)$ as a product of the conditional and marginal densities, as in (9.35). (b) Show that the parameter vectors $(\mu_b, \mu_a, \alpha_a, \beta_a, \mathbf{\Sigma}_R)$ and $(\mu^*, \alpha_a, \beta_a, \gamma, \sigma_{aa\cdot b}, \mu_b, \sigma_{bb})$ are in one-to-one correspondence. (c) Show that the parameters for the conditional distribution of $\mathbf{Y}_a|\mathbf{Y}_b$ are functionally independent of those for the marginal distribution of $\mathbf{Y}_b$. That is, the joint space of $(\mu^*, \alpha_a, \beta_a, \gamma, \sigma_{aa\cdot b})$ and (μ_b, σ_{bb}) is a rectangle formed by their individual spaces. (d) Argue that therefore the MLE of (α_a, β_a) in the model (9.101) is the least squares estimate from the conditional model (9.103). (Thus confirming the correspondence found in (9.22) and (9.23).)

Exercise 9.6.6. Consider maximizing $h(\mathbf{\Psi})$ in (9.36) over $\mathbf{\Psi} \in \mathcal{S}_q$. (a) Let $\mathbf{\Psi} = \mathbf{\Gamma}\mathbf{\Lambda}\mathbf{\Gamma}'$ be the spectral decomposition of $\mathbf{\Psi}$, so that the diagonals of $\mathbf{\Lambda}$ are the eigenvalues $\lambda_1 \geq \lambda_2 \geq \cdots \geq \lambda_q \geq 0$. (Recall Theorem 1.1.) Show that

$$\frac{1}{|\mathbf{\Psi}|^{a/2}}\, e^{-\frac{1}{2}\,\mathrm{trace}(\mathbf{\Psi}^{-1})} = \prod_{i=1}^{q} [\lambda_i^{-a/2}\, e^{-1/(2\lambda_i)}]. \tag{9.104}$$

(b) Find $\widehat{\lambda}_i$, the maximizer of $\lambda_i^{-a/2}\exp(-1/(2\lambda_i))$, for each $i = 1, \ldots, q$. (c) Show that these $\widehat{\lambda}_i$'s satisfy the conditions on the eigenvalues of $\mathbf{\Lambda}$. (d) Argue that then $\widehat{\mathbf{\Psi}} = (1/a)\mathbf{I}_q$ maximizes $h(\mathbf{\Psi})$.

Exercise 9.6.7. This exercise is to show that the Wilks' Λ test for the testing situation in Section 7.2.5 is the likelihood ratio test. We have

$$\mathbf{Y} = \mathbf{x}\boldsymbol{\beta} + \mathbf{R} = (\mathbf{x}_1\ \ \mathbf{x}_2) \begin{pmatrix} \boldsymbol{\beta}_1 \\ \boldsymbol{\beta}_2 \end{pmatrix} + \mathbf{R}, \tag{9.105}$$

where $\mathbf{x}_1$ is $n \times p_1$ and $\mathbf{x}_2$ is $n \times p_2$. The hypotheses are

$$H_0 : \boldsymbol{\beta}_2 = 0 \ \ versus\ \ H_A : \boldsymbol{\beta}_2 \neq 0. \tag{9.106}$$

(a) Show that the MLE of $\mathbf{\Sigma}_R$ under the alternative is $\widehat{\mathbf{\Sigma}}_A = \mathbf{W}/n$, where $\mathbf{W} = \mathbf{Y}'\mathbf{Q}_{(\mathbf{x}_1,\mathbf{x}_2)}\mathbf{Y}$. (b) Show that $\widehat{\mathbf{\Sigma}}_0 = (\mathbf{W}+\mathbf{B})/n$, where $\mathbf{W}+\mathbf{B} = \mathbf{Y}'\mathbf{Q}_{\mathbf{x}_1}\mathbf{Y}$. (c) Find the κ so that likelihood ratio statistic LR in (9.41) equals Λ^κ for $\Lambda = |\mathbf{W}|/|\mathbf{B}+\mathbf{W}|$. This Λ is Wilks' Λ as in (7.35).

Exercise 9.6.8. Now consider the model with partitioning

$$\mathbf{Y} = (\mathbf{Y}_1 \ \ \mathbf{Y}_2) = (\mathbf{x}_1 \ \ \mathbf{x}_2) \begin{pmatrix} \boldsymbol{\beta}_{11} & \boldsymbol{\beta}_{21} \\ \boldsymbol{\beta}_{12} & \boldsymbol{\beta}_{22} \end{pmatrix} + \mathbf{R}, \tag{9.107}$$

where $\mathbf{Y}_1$ is $n \times l^*$, $\mathbf{Y}_2$ is $n \times (q - l^*)$, $\boldsymbol{\beta}_{11}$ is $p_1 \times l^*$, $\mathbf{x}_1$ is $n \times p_1$, and $\mathbf{x}_2$ is $n \times p_2$. We test

$$H_0 : \boldsymbol{\beta}_{11} = 0 \ \ versus \ \ H_A : \boldsymbol{\beta}_{11} \neq 0. \tag{9.108}$$

(a) Find the conditional distribution of $\mathbf{Y}_2 | \mathbf{Y}_1 = \mathbf{y}_1$ and the marginal distribution of $\mathbf{Y}_1$. (b) Argue that that conditional distribution is the same under the null and alternative hypotheses, hence we can find the likelihood ratio test by using just the marginal distribution of $\mathbf{Y}_1$. (c) Use Exercise 9.6.7 to show that the likelihood ratio test is equivalent to Wilks' Λ test. What are the parameters for the Λ in this case?

Exercise 9.6.9. Argue that the likelihood ratio test of (9.46) based on (9.45) is equivalent to Wilks' Λ test, and specify the parameters for the Λ. [Hint: Use Exercise 9.6.8 to show that the test can be based on the conditional distribution in (9.47), then make the appropriate identifications of parameters.]

Exercise 9.6.10. Consider the multivariate regression model (4.9), where $\boldsymbol{\Sigma}_R$ is known. (a) Use (9.80) to show that

$$l(\boldsymbol{\beta}; \mathbf{y}) - l(\widehat{\boldsymbol{\beta}}; \mathbf{y}) = -\frac{1}{2}\,\text{trace}(\boldsymbol{\Sigma}_R^{-1}(\widehat{\boldsymbol{\beta}} - \boldsymbol{\beta})'\mathbf{x}'\mathbf{x}(\widehat{\boldsymbol{\beta}} - \boldsymbol{\beta})). \tag{9.109}$$

(b) Show that in this case, (9.66) is actually an equality, and give $\mathbf{H}$, which is a function of $\boldsymbol{\Sigma}_R$ and $\mathbf{x}'\mathbf{x}$.

Exercise 9.6.11. (a) Show that for the $\widehat{\boldsymbol{\Sigma}}_R$ in (9.81),

$$E[\widehat{\boldsymbol{\Sigma}}_R^{-1}] = \frac{n}{n - p - q - 1}\,\boldsymbol{\Sigma}_R^{-1}. \tag{9.110}$$

(b) Show that

$$\mathbf{U} \sim N(\mathbf{0}, (\mathbf{I}_n + \mathbf{P}_{\mathbf{x}}) \otimes \boldsymbol{\Sigma}_R) \tag{9.111}$$

for $\mathbf{U}$ in (9.83), and thus

$$E[\mathbf{U}'\mathbf{U}] = (n + p)\boldsymbol{\Sigma}_R. \tag{9.112}$$

(c) Show that

$$\text{trace}(E[\widehat{\boldsymbol{\Sigma}}_R^{-1}]E[\mathbf{U}'\mathbf{U}]) = nq\,\frac{n + p}{n - p - q - 1}. \tag{9.113}$$

(d) Finally, show that

$$\text{trace}(E[\widehat{\boldsymbol{\Sigma}}_R^{-1}]E[\mathbf{U}'\mathbf{U}]) - nq = 2\frac{n}{n - p - q - 1}\left(pq + \frac{q(q+1)}{2}\right), \tag{9.114}$$

proving (9.85).

Exercise 9.6.12. Show that in (9.85), $\boldsymbol{\Delta} \to 2d_M$ as $n \to \infty$.

Exercise 9.6.13. Show that in (9.59), if we take the prior probabilities as

$$\pi_k \propto \left(\frac{\sqrt{n}}{e}\right)^{d_k}, \tag{9.115}$$

where d_k is the dimension of Model k, then the model that maximizes the estimated posterior probability is the model with the lowest AIC. Note that except for very small n, this prior places relatively more weight on higher-dimensional models.

Exercise 9.6.14 (Prostaglandin). Continue with the prostaglandin data from Exercise 7.6.16. (a) Find the maximum likelihood estimates and standard errors for the coefficients in the model with just one cycle. Compare these results to the least squares estimates and standard errors. Which method produces the larger standard errors? Approximately how much larger are they? (b) Find the BIC's and the corresponding estimated probabilities for the following models: constant, one cycle, one and two cycle, all cycles. Which has the best BIC? Does it have a relatively high estimated probability?

Exercise 9.6.15 (Caffeine). This question continues with the caffeine data in Exercises 4.4.4 and 6.6.11. Start with the both-sides model $\mathbf{Y} = \mathbf{x}\boldsymbol{\beta}\mathbf{z}' + \mathbf{R}$, where as before the $\mathbf{Y}$ is 2×28, the first column being the scores without caffeine, and the second being the scores with caffeine. The $\mathbf{x}$ is a 28×3 matrix for the three grades, with orthogonal polynomials. The linear vector is $(-\mathbf{1}_9', \mathbf{0}_{10}', \mathbf{1}_9')'$ and the quadratic vector is $(\mathbf{1}_9', -1.81\mathbf{1}_{10}', \mathbf{1}_9')'$. The $\mathbf{z}$ looks at the sum and difference of scores:

$$\mathbf{z} = \begin{pmatrix} 1 & -1 \\ 1 & 1 \end{pmatrix}. \tag{9.116}$$

The goal of this problem is to use BIC to find a good model, choosing among the constant, linear and quadratic models for $\mathbf{x}$, and the "overall mean" and "overall mean + difference models" for the scores. (a) For each of the 6 models, find the deviance, number of free parameters, BIC, and estimated probability. (b) Which model has highest probability? (c) What is the chance that the difference effect is in the model? (d) Find the MLE of $\boldsymbol{\beta}$ for the best model.

Exercise 9.6.16 (Leprosy). Continue with the leprosy example from Exercise 9.6.5. We are interested in finding a good model, but are not interested in the parameters μ_b or σ_{bb}. Thus we can base the analysis solely on the conditional model in (9.103). (a) Find the BIC's and corresponding probability estimates for the $2^3 = 8$ models found by setting subsets of $(\alpha_a, \beta_a, \gamma)$ to 0. Which model has the best BIC? What is its estimated probability? (b) What is the probability (in percent) that the drug vs. placebo effect is in the model? The Drug A vs. Drug D effect? The before effect (i.e.,$\gamma \neq 0$)?

Exercise 9.6.17 (Mouth sizes). Section 7.5.1 presents Mallows' C_p for a number of models fit to the mouth size data. Fit the same models, but find their BIC's. (a) Which model has the best BIC? Which has the second best? What is the total estimated probability of those two models? Are they the same as the best two models chosen by C_p in (7.83)? (b) What is the estimated probability that the two curves are different, i.e., that the δ_j's are not all zero? (c) Find the estimated probabilities for each degree of polynomial (zero, constant, linear, quadratic, cubic) for each effect (boys, girls−boys), analogous to the table in (9.99).

Exercise 9.6.18. (This is a discussion question, in that there is no exact answer. Your reasoning should be sound, though.) Suppose you are comparing a number of models using BIC, and the lowest BIC is b_{min}. How much larger than b_{min} would a BIC have to be for you to consider the corresponding model ignorable? That is, what is δ so that models with BIC $> b_{min} + \delta$ don't seem especially viable. Why?

Exercise 9.6.19. Often, in hypothesis testing, people misinterpret the p-value to be the probability that the null is true, given the data. We can approximately compare the two values using the ideas in this chapter. Consider two models, the null (M_0) and alternative (M_A), where the null is contained in the alternative. Let deviance$_0$ and deviance$_A$ be their deviances, and dim$_0$ and dim$_A$ be their dimensions, respectively. Supposing that the assumptions are reasonable, the p-value for testing the null is p-value $= P[\chi^2_\nu > \delta]$, where $\nu = \text{dim}_A - \text{dim}_0$ and $\delta = \text{deviance}_0 - \text{deviance}_A$. (a) Give the BIC-based estimate of the probability of the null for a given ν, δ and sample size n. (b) For each of various values of n and ν (e.g, $n = 1, 5, 10, 25, 100, 1000$ and $\nu = 1, 5, 10, 25$), find the δ that gives a p-value of 5%, and find the corresponding estimate of the probability of the null. (c) Are the probabilities of the null close to 5%? What do you conclude?

Chapter 10

Models on Covariance Matrices

The models so far have been on the means of the variables. In this chapter, we look at some models for the covariance matrix. We start with testing the equality of two or more covariance matrices, then move on to testing independence and conditional independence of sets of variables. Next is factor analysis, where the relationships among the variables are assumed to be determined by latent (unobserved) variables. Principal component analysis is sometimes thought of as a type of factor analysis, although it is more of a decomposition than actual factor analysis. See Section 13.1.5. We conclude with a particular class of structural models, called *invariant normal models*.

We will base our hypothesis tests on Wishart matrices (one, or several independent ones). In practice, these matrices will often arise from the residuals in linear models, especially the $\mathbf{Y}'\mathbf{Q}_\mathbf{x}\mathbf{Y}$ as in (6.14). If $\mathbf{U} \sim \text{Wishart}_q(\nu, \mathbf{\Sigma})$, where $\mathbf{\Sigma}$ is invertible and $\nu \geq q$, then the likelihood is

$$L(\mathbf{\Sigma}; \mathbf{U}) = |\mathbf{\Sigma}|^{-\nu/2}\, e^{-\frac{1}{2}\,\text{trace}(\mathbf{\Sigma}^{-1}\mathbf{U})}. \tag{10.1}$$

The likelihood follows from the density in (8.70). An alternative derivation is to note that by (8.53), $\mathbf{Z} \sim N(0, \mathbf{I}_\nu \otimes \mathbf{\Sigma})$ has likelihood $L^*(\mathbf{\Sigma}; \mathbf{z}) = L(\mathbf{\Sigma}; \mathbf{z}'\mathbf{z})$. Thus $\mathbf{z}'\mathbf{z}$ is a sufficient statistic, and it can be shown that the likelihood for any $\mathbf{X}$ is the same as the likelihood for its sufficient statistic. Since $\mathbf{Z}'\mathbf{Z} =^{\mathcal{D}} \mathbf{U}$, (10.1) is the likelihood for $\mathbf{U}$.

Recall from (5.51) that $\mathcal{S}_q^+$ denotes the set of $q \times q$ positive definite symmetric matrices. Then Lemma 9.1 shows that the MLE of $\mathbf{\Sigma} \in \mathcal{S}_q^+$ based on (10.1) is

$$\widehat{\mathbf{\Sigma}} = \frac{\mathbf{U}}{\nu}, \tag{10.2}$$

and the maximum likelihood is

$$L(\widehat{\mathbf{\Sigma}}\,;\,\mathbf{U}) = \left|\frac{\mathbf{U}}{\nu}\right|^{-\nu/2} e^{-\frac{1}{2}\nu q}. \tag{10.3}$$

10.1 Testing equality of covariance matrices

We first suppose we have two groups, e.g., boys and girls, and wish to test whether their covariance matrices are equal. Let $\mathbf{U}_1$ and $\mathbf{U}_2$ be independent, with

$$\mathbf{U}_i \sim \text{Wishart}_q(\nu_i, \mathbf{\Sigma}_i), \quad i = 1, 2. \tag{10.4}$$

The hypotheses are then

$$H_0 : \mathbf{\Sigma}_1 = \mathbf{\Sigma}_2 \;\; \textit{versus} \;\; H_A : \mathbf{\Sigma}_1 \neq \mathbf{\Sigma}_2, \tag{10.5}$$

where both $\mathbf{\Sigma}_1$ and $\mathbf{\Sigma}_2$ are in $\mathcal{S}_q^+$. (That is, we are not assuming any particular structure for the covariance matrices.) We need the likelihoods under the two hypotheses. Because the $\mathbf{U}_i$'s are independent,

$$L(\mathbf{\Sigma}_1, \mathbf{\Sigma}_2; \mathbf{U}_1, \mathbf{U}_2) = |\mathbf{\Sigma}_1|^{-\nu_1/2} \, e^{-\frac{1}{2}\text{trace}(\mathbf{\Sigma}_1^{-1}\mathbf{U}_1)} |\mathbf{\Sigma}_2|^{-\nu_2/2} \, e^{-\frac{1}{2}\text{trace}(\mathbf{\Sigma}_2^{-1}\mathbf{U}_2)}, \tag{10.6}$$

which, under the null hypothesis, becomes

$$L(\mathbf{\Sigma}, \mathbf{\Sigma}; \mathbf{U}_1, \mathbf{U}_2) = |\mathbf{\Sigma}|^{-(\nu_1+\nu_2)/2} \, e^{-\frac{1}{2}\text{trace}(\mathbf{\Sigma}^{-1}(\mathbf{U}_1+\mathbf{U}_2))}, \tag{10.7}$$

where $\mathbf{\Sigma}$ is the common value of $\mathbf{\Sigma}_1$ and $\mathbf{\Sigma}_2$. The MLE under the alternative hypothesis is found by maximizing (10.5), which results in two separate maximizations:

$$\text{Under } H_A\text{:} \;\; \widehat{\mathbf{\Sigma}}_{A1} = \frac{\mathbf{U}_1}{\nu_1}, \; \widehat{\mathbf{\Sigma}}_{A2} = \frac{\mathbf{U}_2}{\nu_2}. \tag{10.8}$$

Under the null, there is just one Wishart, $\mathbf{U}_1 + \mathbf{U}_2$, so that

$$\text{Under } H_0\text{:} \;\; \widehat{\mathbf{\Sigma}}_{01} = \widehat{\mathbf{\Sigma}}_{02} = \frac{\mathbf{U}_1 + \mathbf{U}_2}{\nu_1 + \nu_2}. \tag{10.9}$$

Thus

$$\sup_{H_A} L = \left|\frac{\mathbf{U}_1}{\nu_1}\right|^{-\nu_1/2} e^{-\frac{1}{2}\nu_1 q} \left|\frac{\mathbf{U}_2}{\nu_2}\right|^{-\nu_2/2} e^{-\frac{1}{2}\nu_2 q}, \tag{10.10}$$

and

$$\sup_{H_0} L = \left|\frac{\mathbf{U}_1 + \mathbf{U}_2}{\nu_1 + \nu_2}\right|^{-(\nu_1+\nu_2)/2} e^{-\frac{1}{2}(\nu_1+\nu_2)q}. \tag{10.11}$$

Taking the ratio, note that the parts in the e cancel, hence

$$LR = \frac{\sup_{H_A} L}{\sup_{H_0} L} = \frac{|\mathbf{U}_1/\nu_1|^{-\nu_1/2} |\mathbf{U}_2/\nu_2|^{-\nu_2/2}}{|(\mathbf{U}_1 + \mathbf{U}_2)/(\nu_1 + \nu_2)|^{-(\nu_1+\nu_2)/2}}. \tag{10.12}$$

And

$$2\log(LR) = (\nu_1 + \nu_2) \log \left|\frac{\mathbf{U}_1 + \mathbf{U}_2}{\nu_1 + \nu_2}\right| - \nu_1 \log \left|\frac{\mathbf{U}_1}{\nu_1}\right| - \nu_2 \log \left|\frac{\mathbf{U}_2}{\nu_2}\right|. \tag{10.13}$$

Under the null hypothesis, $2\log(LR)$ approaches a χ^2 as in (9.41). To figure out the degrees of freedom, we have to find the number of free parameters under each hypothesis. A $\boldsymbol{\Sigma} \in \mathcal{S}_q^+$, unrestricted, has $q(q+1)/2$ free parameters, because of the symmetry. Under the alternative, there are two such sets of parameters. Thus,

$$\dim(H_0) = q(q+1)/2, \; \dim(H_A) = q(q+1) \\ \Rightarrow \dim(H_A) - \dim(H_0) = q(q+1)/2. \tag{10.14}$$

Thus, under H_0,

$$2\log(LR) \longrightarrow \chi^2_{q(q+1)/2}. \tag{10.15}$$

10.1.1 Example: Grades data

Using the grades data on $n = 107$ students in (4.11), we compare the covariance matrices of the men and women. There are 37 men and 70 women, so that the sample covariance matrices have degrees of freedom $\nu_1 = 37 - 1 = 36$ and $\nu_2 = 70 - 1 = 69$, respectively. Their estimates are:

$$\text{Men: } \frac{1}{\nu_1}\mathbf{U}_1 = \begin{pmatrix} 166.33 & 205.41 & 106.24 & 51.69 & 62.20 \\ 205.41 & 325.43 & 206.71 & 61.65 & 69.35 \\ 106.24 & 206.71 & 816.44 & 41.33 & 41.85 \\ 51.69 & 61.65 & 41.33 & 80.37 & 50.31 \\ 62.20 & 69.35 & 41.85 & 50.31 & 97.08 \end{pmatrix}, \tag{10.16}$$

and

$$\text{Women: } \frac{1}{\nu_2}\mathbf{U}_2 = \begin{pmatrix} 121.76 & 113.31 & 58.33 & 40.79 & 40.91 \\ 113.31 & 212.33 & 124.65 & 52.51 & 50.60 \\ 58.33 & 124.65 & 373.84 & 56.29 & 74.49 \\ 40.79 & 52.51 & 56.29 & 88.47 & 60.93 \\ 40.91 & 50.60 & 74.49 & 60.93 & 112.88 \end{pmatrix}. \tag{10.17}$$

These covariance matrices are clearly not equal, but are the differences significant? The pooled estimate, i.e., the common estimate under H_0, is

$$\frac{1}{\nu_1+\nu_2}(\mathbf{U}_1 + \mathbf{U}_2) = \begin{pmatrix} 137.04 & 144.89 & 74.75 & 44.53 & 48.21 \\ 144.89 & 251.11 & 152.79 & 55.64 & 57.03 \\ 74.75 & 152.79 & 525.59 & 51.16 & 63.30 \\ 44.53 & 55.64 & 51.16 & 85.69 & 57.29 \\ 48.21 & 57.03 & 63.30 & 57.29 & 107.46 \end{pmatrix}. \tag{10.18}$$

Then

$$\begin{aligned} 2\log(LR) &= (\nu_1+\nu_2)\log\left|\frac{\mathbf{U}_1+\mathbf{U}_2}{\nu_1+\nu_2}\right| - \nu_1\log\left|\frac{\mathbf{U}_1}{\nu_1}\right| - \nu_2\log\left|\frac{\mathbf{U}_2}{\nu_2}\right| \\ &= 105\log(2.6090\times10^{10}) - 36\log(2.9819\times10^{10}) - 69\log(1.8149\times10^{10}) \\ &= 20.2331. \end{aligned} \tag{10.19}$$

The degrees of freedom for the χ^2 is $q(q+1)/2 = 5\times6/2 = 15$. The p-value is 0.16, which shows that we have not found a significant difference between the covariance matrices.

10.1.2 Testing the equality of several covariance matrices

It is not hard to extend the test to testing the equality of more than two covariance matrices. That is, we have $\mathbf{U}_1, \ldots, \mathbf{U}_m$, independent, $\mathbf{U}_i \sim \text{Wishart}_l(\nu_i, \mathbf{\Sigma}_i)$, and wish to test

$$H_0 : \mathbf{\Sigma}_1 = \cdots = \mathbf{\Sigma}_m \ \ versus \ \ H_A : \ not. \tag{10.20}$$

Then

$$2\log(LR) = (\nu_1 + \cdots + \nu_m)\log\left|\frac{\mathbf{U}_1 + \cdots + \mathbf{U}_m}{\nu_1 + \cdots + \nu_m}\right| - \nu_1 \log\left|\frac{\mathbf{U}_1}{\nu_1}\right| - \cdots - \nu_m \log\left|\frac{\mathbf{U}_m}{\nu_m}\right|, \tag{10.21}$$

and under the null,

$$2\log(LR) \longrightarrow \chi^2_{df}, \ \ df = (m-1)q(q+1)/2. \tag{10.22}$$

This procedure is **Bartlett's test** for the equality of covariances.

10.2 Testing independence of two blocks of variables

In this section, we assume $\mathbf{U} \sim \text{Wishart}_q(\nu, \mathbf{\Sigma})$, and partition the matrices:

$$\mathbf{U} = \begin{pmatrix} \mathbf{U}_{11} & \mathbf{U}_{12} \\ \mathbf{U}_{21} & \mathbf{U}_{22} \end{pmatrix} \ \text{and} \ \mathbf{\Sigma} = \begin{pmatrix} \mathbf{\Sigma}_{11} & \mathbf{\Sigma}_{12} \\ \mathbf{\Sigma}_{21} & \mathbf{\Sigma}_{22} \end{pmatrix}, \tag{10.23}$$

where

$$\mathbf{U}_{11} \text{ and } \mathbf{\Sigma}_{11} \text{ are } q_1 \times q_1, \ \text{ and } \ \mathbf{U}_{22} \text{ and } \mathbf{\Sigma}_{22} \text{ are } q_2 \times q_2; \ q = q_1 + q_2. \tag{10.24}$$

Presuming the Wishart arises from multivariate normals, we wish to test whether the two blocks of variables are independent, which translates to testing

$$H_0 : \mathbf{\Sigma}_{12} = \mathbf{0} \ \ versus \ \ H_A : \mathbf{\Sigma}_{12} \neq \mathbf{0}. \tag{10.25}$$

Under the alternative, the likelihood is just the one in (10.1), hence

$$\sup_{H_A} L(\mathbf{\Sigma}; \mathbf{U}) = \left|\frac{\mathbf{U}}{\nu}\right|^{-\nu/2} e^{-\frac{1}{2}\nu q}. \tag{10.26}$$

Under the null, because $\mathbf{\Sigma}$ is then block diagonal,

$$|\mathbf{\Sigma}| = |\mathbf{\Sigma}_{11}|\,|\mathbf{\Sigma}_{22}| \ \text{ and } \ \text{trace}(\mathbf{\Sigma}^{-1}\mathbf{U}) = \text{trace}(\mathbf{\Sigma}_{11}^{-1}\mathbf{U}_{11}) + \text{trace}(\mathbf{\Sigma}_{22}^{-1}\mathbf{U}_{22}). \tag{10.27}$$

Thus the likelihood under the null can be written

$$|\mathbf{\Sigma}_{11}|^{-\nu/2}\, e^{-\frac{1}{2}\text{trace}(\mathbf{\Sigma}_{11}^{-1}\mathbf{U}_{11})} \times |\mathbf{\Sigma}_{22}|^{-\nu/2}\, e^{-\frac{1}{2}\text{trace}(\mathbf{\Sigma}_{22}^{-1}\mathbf{U}_{22})}. \tag{10.28}$$

The two factors can be maximized separately, so that

$$\sup_{H_0} L(\mathbf{\Sigma}; \mathbf{U}) = \left|\frac{\mathbf{U}_{11}}{\nu}\right|^{-\nu/2} e^{-\frac{1}{2}\nu q_1} \times \left|\frac{\mathbf{U}_{22}}{\nu}\right|^{-\nu/2} e^{-\frac{1}{2}\nu q_2}. \tag{10.29}$$

Taking the ratio of (10.26) and (10.28), the parts in the exponent of the e again cancel, hence

$$2\log(LR) = \nu\,(\log(|\mathbf{U}_{11}/\nu|) + \log(|\mathbf{U}_{22}/\nu|) - \log(|\mathbf{U}/\nu|)). \tag{10.30}$$

(The ν's in the denominators of the determinants cancel, so they can be erased if desired.)

Section 13.3 considers canonical correlations, which are a way to summarize relationships between two sets of variables.

10.2.1 Example: Grades data

Continuing Example 10.1.1, we start with the pooled covariance matrix $\widehat{\mathbf{\Sigma}} = (\mathbf{U}_1 + \mathbf{U}_2)/(\nu_1 + \nu_2)$, which has $q = 5$ and $\nu = 105$. Here we test whether the first three variables (homework, labs, inclass) are independent of the last two (midterms, final), so that $q_1 = 3$ and $q_2 = 2$. Obviously, they should not be independent, but we will test it formally. Now

$$\begin{aligned} 2\log(LR) &= \nu\,(\log(|\mathbf{U}_{11}/\nu|) + \log(|\mathbf{U}_{22}/\nu|) - \log(|\mathbf{U}/\nu|)) \\ &= 28.2299. \end{aligned} \tag{10.31}$$

Here the degrees of freedom in the χ^2 are $q_1 \times q_2 = 6$, because that is the number of covariances we are setting to 0 in the null. Or you can count

$$\begin{aligned} \dim(H_A) &= q(q+1)/2 = 15, \\ \dim(H_0) &= q_1(q_1+1)/2 + q_2(q_2+1)/2 = 6 + 3 = 9, \end{aligned} \tag{10.32}$$

which has $\dim(H_A) - \dim(H_0) = 6$. In either case, the result is clearly significant (the p-value is less than 0.0001), hence indeed the two sets of scores are not independent.

Testing the independence of several block of variables is almost as easy. Consider the three variables homework, inclass, and midterms, which have covariance

$$\mathbf{\Sigma} = \begin{pmatrix} \sigma_{11} & \sigma_{13} & \sigma_{14} \\ \sigma_{31} & \sigma_{33} & \sigma_{34} \\ \sigma_{41} & \sigma_{43} & \sigma_{44} \end{pmatrix}. \tag{10.33}$$

We wish to test whether the three are mutually independent, so that

$$H_0 : \sigma_{13} = \sigma_{14} = \sigma_{34} = 0 \;\; \textit{versus} \;\; H_A : \textit{not}. \tag{10.34}$$

Under the alternative, the estimate of $\mathbf{\Sigma}$ is just the usual one from (10.18), where we pick out the first, third, and fourth variables. Under the null, we have three independent variables, so $\widehat{\sigma}_{ii} = \mathbf{U}_{ii}/\nu$ is just the appropriate diagonal from $\widehat{\mathbf{\Sigma}}$. Then the test statistic is

$$\begin{aligned} 2\log(LR) &= \nu\,(\log(|\mathbf{U}_{11}/\nu|) + \log(|\mathbf{U}_{33}/\nu|) + \log(|\mathbf{U}_{44}/\nu|) - \log(|\mathbf{U}^*/\nu|)) \\ &= 30.116, \end{aligned} \tag{10.35}$$

where $\mathbf{U}^*$ contains just the variances and covariances of the variables 1, 3, and 4. We do not need the determinant notation for the $\mathbf{U}_{ii}$, but leave it in for cases in which the three blocks of variables are not 1×1. The degrees of freedom for the χ^2 is then 3, because we are setting three free parameters to 0 in the null. Clearly that is a significant result, i.e., these three variables are not mutually independent.

10.2.2 Example: Testing conditional independence

Imagine that we have (at least) three blocks of variables, and wish to see whether the first two are conditionally independent given the third. The process is exactly the same as for testing independence, except that we use the conditional covariance matrix. That is, suppose

$$\mathbf{Y} = (\mathbf{Y}_1, \mathbf{Y}_2, \mathbf{Y}_3) \sim N(\mathbf{x}\boldsymbol{\beta}\mathbf{z}', \mathbf{I}_n \otimes \boldsymbol{\Sigma}_R), \ \ \mathbf{Y}_i \text{ is } n \times q_i, \ \ q = q_1 + q_2 + q_3, \tag{10.36}$$

where

$$\boldsymbol{\Sigma}_R = \begin{pmatrix} \boldsymbol{\Sigma}_{11} & \boldsymbol{\Sigma}_{12} & \boldsymbol{\Sigma}_{13} \\ \boldsymbol{\Sigma}_{21} & \boldsymbol{\Sigma}_{22} & \boldsymbol{\Sigma}_{23} \\ \boldsymbol{\Sigma}_{31} & \boldsymbol{\Sigma}_{32} & \boldsymbol{\Sigma}_{33} \end{pmatrix}, \tag{10.37}$$

so that $\boldsymbol{\Sigma}_{ii}$ is $q_i \times q_i$. The null hypothesis is

$$H_0 : \mathbf{Y}_1 \text{ and } \mathbf{Y}_2 \text{ are conditionally independent given } \mathbf{Y}_3. \tag{10.38}$$

The conditional covariance matrix is

$$\begin{aligned} Cov[(\mathbf{Y}_1, \mathbf{Y}_2) \mid \mathbf{Y}_3 = \mathbf{y}_3] &= \begin{pmatrix} \boldsymbol{\Sigma}_{11} & \boldsymbol{\Sigma}_{12} \\ \boldsymbol{\Sigma}_{21} & \boldsymbol{\Sigma}_{22} \end{pmatrix} - \begin{pmatrix} \boldsymbol{\Sigma}_{13} \\ \boldsymbol{\Sigma}_{23} \end{pmatrix} \boldsymbol{\Sigma}_{33}^{-1} \begin{pmatrix} \boldsymbol{\Sigma}_{31} & \boldsymbol{\Sigma}_{32} \end{pmatrix} \\ &= \begin{pmatrix} \boldsymbol{\Sigma}_{11} - \boldsymbol{\Sigma}_{13}\boldsymbol{\Sigma}_{33}^{-1}\boldsymbol{\Sigma}_{31} & \boldsymbol{\Sigma}_{12} - \boldsymbol{\Sigma}_{13}\boldsymbol{\Sigma}_{33}^{-1}\boldsymbol{\Sigma}_{32} \\ \boldsymbol{\Sigma}_{21} - \boldsymbol{\Sigma}_{23}\boldsymbol{\Sigma}_{33}^{-1}\boldsymbol{\Sigma}_{31} & \boldsymbol{\Sigma}_{22} - \boldsymbol{\Sigma}_{23}\boldsymbol{\Sigma}_{33}^{-1}\boldsymbol{\Sigma}_{32} \end{pmatrix} \\ &\equiv \begin{pmatrix} \boldsymbol{\Sigma}_{11\cdot 3} & \boldsymbol{\Sigma}_{12\cdot 3} \\ \boldsymbol{\Sigma}_{21\cdot 3} & \boldsymbol{\Sigma}_{22\cdot 3} \end{pmatrix}. \end{aligned} \tag{10.39}$$

Then the hypotheses are

$$H_0 : \boldsymbol{\Sigma}_{12\cdot 3} = \mathbf{0} \ \ \textit{versus} \ \ H_A : \boldsymbol{\Sigma}_{12\cdot 3} \neq \mathbf{0}. \tag{10.40}$$

Letting $\mathbf{U}/\nu$ be the usual estimator of $\boldsymbol{\Sigma}_R$, where

$$\mathbf{U} = \mathbf{Y}'\mathbf{Q}_\mathbf{x}\mathbf{Y} \sim \text{Wishart}_{(q_1+q_2+q_3)}(\nu, \boldsymbol{\Sigma}_R), \ \ \nu = n - p, \tag{10.41}$$

we know from Proposition 8.1 that the conditional covariance is also Wishart, but loses q_3 degrees of freedom:

$$\begin{aligned} \mathbf{U}_{(1:2)(1:2)\cdot 3} &\equiv \begin{pmatrix} \mathbf{U}_{11\cdot 3} & \mathbf{U}_{12\cdot 3} \\ \mathbf{U}_{21\cdot 3} & \mathbf{U}_{22\cdot 3} \end{pmatrix} \\ &\sim \text{Wishart}_{(q_1+q_2)}\left(\nu - q_3, \begin{pmatrix} \boldsymbol{\Sigma}_{11\cdot 3} & \boldsymbol{\Sigma}_{12\cdot 3} \\ \boldsymbol{\Sigma}_{21\cdot 3} & \boldsymbol{\Sigma}_{22\cdot 3} \end{pmatrix}\right). \end{aligned} \tag{10.42}$$

(The $\mathbf{U}$ is partitioned analogously to the $\boldsymbol{\Sigma}_R$.) Then testing the hypothesis $\boldsymbol{\Sigma}_{12\cdot 3}$ here is the same as (10.30) but after dotting out 3:

$$\begin{aligned} 2\log(LR) = (\nu - q_3)\ (&\log(|\mathbf{U}_{11\cdot 3}/(\nu - q_3)|) + \log(|\mathbf{U}_{22\cdot 3}/(\nu - q_3)|) \\ &- \log(|\mathbf{U}_{(1:2)(1:2)\cdot 3}/(\nu - q_3)|)), \end{aligned} \tag{10.43}$$

which is asymptotically $\chi^2_{q_1 q_2}$ under the null.

An alternative (but equivalent) method for calculating the conditional covariance is to move the conditioning variables $\mathbf{Y}_3$ to the $\mathbf{x}$ matrix, as we did for covariates. Thus, leaving out the $\mathbf{z}$,

$$(\mathbf{Y}_1\ \mathbf{Y}_2) \mid \mathbf{Y}_3 = \mathbf{y}_3 \sim N(\mathbf{x}^*\boldsymbol{\beta}^*, \mathbf{I}_n \otimes \boldsymbol{\Sigma}^*), \tag{10.44}$$

where

$$\mathbf{x}^* = (\mathbf{x}\ \mathbf{y}_3) \text{ and } \boldsymbol{\Sigma}^* = \begin{pmatrix} \boldsymbol{\Sigma}_{11\cdot 3} & \boldsymbol{\Sigma}_{12\cdot 3} \\ \boldsymbol{\Sigma}_{21\cdot 3} & \boldsymbol{\Sigma}_{22\cdot 3} \end{pmatrix}. \tag{10.45}$$

Then

$$\mathbf{U}_{(1:2)(1:2)\cdot 3} = (\mathbf{Y}_1\ \mathbf{Y}_2)'\mathbf{Q}_{\mathbf{x}^*}(\mathbf{Y}_1\ \mathbf{Y}_2). \tag{10.46}$$

See Exercise 10.5.5.

We note that there appears to be an ambiguity in the denominators of the $\mathbf{U}_i$'s for the $2\log(LR)$. That is, if we base the likelihood on the original $\mathbf{Y}$ of (10.36), then the denominators will be n. If we use the original $\mathbf{U}$ in (10.41), the denominators will be $n - p$. And what we actually used, based on the conditional covariance matrix in (10.42), was $n - p - q_3$. All three possibilities are fine in that the asymptotics as $n \to \infty$ are valid. We chose the one we did because it is the most focussed, i.e., there are no parameters involved (e.g., $\boldsymbol{\beta}$) that are not directly related to the hypotheses.

Testing the conditional independence of three or more blocks of variables, given another block, again uses the dotted-out Wishart matrix. For example, consider Example 10.2.1 with variables homework, inclass, and midterms, but test whether those three are conditionally independent given the "block 4" variables, labs and final. The conditional $\mathbf{U}$ matrix is now denoted $\mathbf{U}_{(1:3)(1:3)\cdot 4}$, and the degrees of freedom are $\nu - q_4 = 105 - 2 = 103$, so that the estimate of the conditional covariance matrix is

$$\begin{pmatrix} \widehat{\sigma}_{11\cdot 4} & \widehat{\sigma}_{12\cdot 4} & \widehat{\sigma}_{13\cdot 4} \\ \widehat{\sigma}_{21\cdot 4} & \widehat{\sigma}_{22\cdot 4} & \widehat{\sigma}_{23\cdot 4} \\ \widehat{\sigma}_{31\cdot 4} & \widehat{\sigma}_{32\cdot 4} & \widehat{\sigma}_{33\cdot 4} \end{pmatrix} = \frac{1}{\nu - q_4}\mathbf{U}_{(1:3)(1:3)\cdot 4}$$
$$= \begin{pmatrix} 51.9536 & -18.3868 & 5.2905 \\ -18.3868 & 432.1977 & 3.8627 \\ 5.2905 & 3.8627 & 53.2762 \end{pmatrix}. \tag{10.47}$$

Then, to test

$$H_0 : \sigma_{12\cdot 4} = \sigma_{13\cdot 4} = \sigma_{23\cdot 4} = 0 \ \ versus \ \ H_A : not, \tag{10.48}$$

we use the statistic analogous to (10.35),

$$\begin{aligned} 2\log(LR) = {} & (\nu - q_4)\,(\log(|\mathbf{U}_{11\cdot 4}/(\nu - q_4)|) + \log(|\mathbf{U}_{22\cdot 4}/(\nu - q_4)|) \\ & + \log(|\mathbf{U}_{33\cdot 4}/(\nu - q_4)|) - \log(|\mathbf{U}_{(1:3)(1:3)\cdot 4}/(\nu - q_4)|)) \\ = {} & 2.76. \end{aligned} \tag{10.49}$$

The degrees of freedom for the χ^2 is again 3, so we accept the null: There does not appear to be a significant relationship among these three variables given the labs and final scores. This implies, among other things, that once we know someone's labs and final scores, knowing the homework or inclass will not help in guessing the midterms score. We could also look at the sample correlations, unconditionally (from (10.18)) and conditionally:

	Unconditional			Conditional on Labs, Final		
	HW	InClass	Midterms	HW	InClass	Midterms
HW	1.00	0.28	0.41	1.00	−0.12	0.10
InClass	0.28	1.00	0.24	−0.12	1.00	0.03
Midterms	0.41	0.24	1.00	0.10	0.03	1.00

(10.50)

Notice that the conditional correlations are much smaller than the unconditional ones, and the conditional correlation between homework and inclass scores is negative, though not significantly so. Thus it appears that the labs and final scores explain the relationships among the other variables.

Remark

Graphical models are important and useful models for studying more complicated independence and conditional independence relations among variables. Some references: Whitaker [1990], Jordan [1999], and Koller and Friedman [2009].

10.3 Factor analysis

The example above suggested that the relationship among three variables could be explained by two other variables. The idea behind factor analysis is that the relationships (correlations, to be precise) of a set of variables can be explained by a number of other variables, called factors. The kicker here is that the factors are not observed. Spearman [1904] introduced the idea based on the notion of a "general intelligence" factor. This section gives the very basics of factor analysis. More details can be found in Lawley and Maxwell [1971], Harman [1976] and Basilevsky [1994], as well as many other books.

The model we consider sets $\mathbf{Y}$ to be the $n \times q$ matrix of observed variables, and $\mathbf{X}$ to be the $n \times p$ matrix of factor variables, which we do not observe. Assume

$$\begin{pmatrix} \mathbf{X} & \mathbf{Y} \end{pmatrix} \sim N(\mathbf{D}\begin{pmatrix} \boldsymbol{\delta} & \boldsymbol{\gamma} \end{pmatrix}, \mathbf{I}_n \otimes \boldsymbol{\Sigma}), \quad \boldsymbol{\Sigma} = \begin{pmatrix} \boldsymbol{\Sigma}_{XX} & \boldsymbol{\Sigma}_{XY} \\ \boldsymbol{\Sigma}_{YX} & \boldsymbol{\Sigma}_{YY} \end{pmatrix}, \tag{10.51}$$

where $\mathbf{D}$ is an $n \times k$ design matrix (e.g., to distinguish men from women), and $\boldsymbol{\delta}$ ($k \times p$) and $\boldsymbol{\gamma}$ ($k \times q$) are the parameters for the means of $\mathbf{X}$ and $\mathbf{Y}$, respectively. Factor analysis is not primarily concerned with the means (that is what the linear models are for), but with the covariances. The key assumption is that the variables in $\mathbf{Y}$ are conditionally independent given $\mathbf{X}$, which means the conditional covariance is diagonal:

$$Cov[\mathbf{Y} \mid \mathbf{X} = \mathbf{x}] = \mathbf{I}_n \otimes \boldsymbol{\Sigma}_{YY\cdot X} = \mathbf{I}_n \otimes \boldsymbol{\Psi} = \mathbf{I}_n \otimes \begin{pmatrix} \psi_{11} & 0 & \cdots & 0 \\ 0 & \psi_{22} & \cdots & 0 \\ \vdots & \vdots & \ddots & \vdots \\ 0 & 0 & \cdots & \psi_{qq} \end{pmatrix}. \tag{10.52}$$

Writing out the conditional covariance matrix, we have

$$\boldsymbol{\Psi} = \boldsymbol{\Sigma}_{YY} - \boldsymbol{\Sigma}_{YX}\boldsymbol{\Sigma}_{XX}^{-1}\boldsymbol{\Sigma}_{XY} \Rightarrow \boldsymbol{\Sigma}_{YY} = \boldsymbol{\Sigma}_{YX}\boldsymbol{\Sigma}_{XX}^{-1}\boldsymbol{\Sigma}_{XY} + \boldsymbol{\Psi}, \tag{10.53}$$

so that marginally,

$$\mathbf{Y} \sim N(\mathbf{D}\boldsymbol{\gamma}, \mathbf{I}_n \otimes (\boldsymbol{\Sigma}_{YX}\boldsymbol{\Sigma}_{XX}^{-1}\boldsymbol{\Sigma}_{XY} + \boldsymbol{\Psi})). \tag{10.54}$$

Because $\mathbf{Y}$ is all we observe, we cannot estimate $\boldsymbol{\Sigma}_{XY}$ or $\boldsymbol{\Sigma}_{XX}$ separately, but only the function $\boldsymbol{\Sigma}_{YX}\boldsymbol{\Sigma}_{XX}^{-1}\boldsymbol{\Sigma}_{XY}$. Note that if we replace $\mathbf{X}$ with $\mathbf{X}^* = \mathbf{X}\mathbf{A}$ for some invertible matrix $\mathbf{A}$,

$$\boldsymbol{\Sigma}_{YX}\boldsymbol{\Sigma}_{XX}^{-1}\boldsymbol{\Sigma}_{XY} = \boldsymbol{\Sigma}_{YX^*}\boldsymbol{\Sigma}_{X^*X^*}^{-1}\boldsymbol{\Sigma}_{X^*Y}, \tag{10.55}$$

so that the distribution of $\mathbf{Y}$ is unchanged. See Exercise 10.5.6. Thus in order to estimate the parameters, we have to make some restrictions. Commonly it is assumed that $\boldsymbol{\Sigma}_{XX} = \mathbf{I}_p$, and the mean is zero:

$$\mathbf{X} \sim N(\mathbf{0}, \mathbf{I}_n \otimes \mathbf{I}_p). \tag{10.56}$$

Then, letting $\boldsymbol{\beta} = \boldsymbol{\Sigma}_{XX}^{-1}\boldsymbol{\Sigma}_{XY} = \boldsymbol{\Sigma}_{XY}$,

$$\mathbf{Y} \sim N(\mathbf{D}\boldsymbol{\gamma}, \mathbf{I}_n \otimes (\boldsymbol{\beta}'\boldsymbol{\beta} + \boldsymbol{\Psi})). \tag{10.57}$$

Or, we can write the model as

$$\mathbf{Y} = \mathbf{D}\boldsymbol{\gamma} + \mathbf{X}\boldsymbol{\beta} + \mathbf{R}, \ \ \mathbf{X} \sim N(\mathbf{0}, \mathbf{I}_n \otimes \mathbf{I}_p), \ \ \mathbf{R} \sim N(\mathbf{0}, \mathbf{I}_n \otimes \boldsymbol{\Psi}), \tag{10.58}$$

where $\mathbf{X}$ and $\mathbf{R}$ are independent. The equation decomposes each variable (column) in $\mathbf{Y}$ into the fixed mean plus the part depending on the factors plus the parts unique to the individual variables. The element β_{ij} is called the **loading** of factor i on the variable j. The variance ψ_j is the **unique variance** of variable j, i.e., the part not explained by the factors. Any measurement error is assumed to be part of the unique variance.

There is the statistical problem of estimating the model, meaning the $\boldsymbol{\beta}'\boldsymbol{\beta}$ and $\boldsymbol{\Psi}$ (and $\boldsymbol{\gamma}$, but we already know about that), and the interpretative problem of finding and defining the resulting factors. We will take these concerns up in the next two subsections.

10.3.1 Estimation

We estimate the $\boldsymbol{\gamma}$ using least squares as usual, i.e.,

$$\widehat{\boldsymbol{\gamma}} = (\mathbf{D}'\mathbf{D})^{-1}\mathbf{D}'\mathbf{Y}. \tag{10.59}$$

Then the residual sum of squares matrix is used to estimate the $\boldsymbol{\beta}$ and $\boldsymbol{\Psi}$:

$$\mathbf{U} = \mathbf{Y}'\mathbf{Q}_\mathbf{D}\mathbf{Y} \sim \text{Wishart}_q(\nu, \boldsymbol{\beta}'\boldsymbol{\beta} + \boldsymbol{\Psi}), \ \ \nu = n - k. \tag{10.60}$$

The parameters are still not estimable, because for any $p \times p$ orthogonal matrix $\boldsymbol{\Gamma}$, $(\boldsymbol{\Gamma}\boldsymbol{\beta})'(\boldsymbol{\Gamma}\boldsymbol{\beta})$ yields the same $\boldsymbol{\beta}'\boldsymbol{\beta}$. We can use the QR decomposition from Theorem 5.4. Our $\boldsymbol{\beta}$ is $p \times q$ with $p < q$. Write $\boldsymbol{\beta} = (\boldsymbol{\beta}_1, \boldsymbol{\beta}_2)$, where $\boldsymbol{\beta}_1$ has the first p columns of $\boldsymbol{\beta}$. We apply the QR decomposition to $\boldsymbol{\beta}_1$, assuming the columns are linearly independent. Then $\boldsymbol{\beta}_1 = \mathbf{Q}\mathbf{T}$, where $\mathbf{Q}$ is orthogonal and $\mathbf{T}$ is upper triangular with positive diagonal elements. Thus we can write $\mathbf{Q}'\boldsymbol{\beta}_1 = \mathbf{T}$, or

$$\mathbf{Q}'\boldsymbol{\beta} = \mathbf{Q}' \begin{pmatrix} \boldsymbol{\beta}_1 & \boldsymbol{\beta}_2 \end{pmatrix} = \begin{pmatrix} \mathbf{T} & \mathbf{T}^* \end{pmatrix} \equiv \boldsymbol{\beta}^*, \tag{10.61}$$

where $\mathbf{T}^*$ is some $p \times (q-p)$ matrix. E.g., with $p = 3, q = 5$,

$$\boldsymbol{\beta}^* = \begin{pmatrix} \beta_{11}^* & \beta_{12}^* & \beta_{13}^* & \beta_{14}^* & \beta_{15}^* \\ 0 & \beta_{22}^* & \beta_{23}^* & \beta_{24}^* & \beta_{25}^* \\ 0 & 0 & \beta_{33}^* & \beta_{34}^* & \beta_{35}^* \end{pmatrix}, \tag{10.62}$$

where the β_{ii}^*'s are positive. If we require that $\boldsymbol{\beta}$ satisfies constraints (10.62), then it is estimable. (Exercise 10.5.7.) Note that there are $p(p-1)/2$ non-free parameters (since $\beta_{ij} = 0$ for $i > j$), which means the number of free parameters in the model is $pq - p(p-1)/2$ for the $\boldsymbol{\beta}$ part, and q for $\boldsymbol{\Psi}$. Thus for the p-factor model M_p, the number of free parameters is

$$d_p \equiv \dim(M_p) = q(p+1) - \frac{p(p-1)}{2}. \tag{10.63}$$

(We are ignoring the parameters in the $\boldsymbol{\gamma}$, because they are the same for all the models we consider.) In order to have a hope of estimating the factors, the dimension of the factor model cannot exceed the dimension of the most general model, $\boldsymbol{\Sigma}_{YY} \in \mathcal{S}_q^+$, which has $q(q+1)/2$ parameters. Thus for identifiability we need

$$\frac{q(q+1)}{2} - d_p = \frac{(q-p)^2 - p - q}{2} \geq 0. \tag{10.64}$$

E.g., if there are $q = 10$ variables, at most $p = 6$ factors can be estimated.

There are many methods for estimating $\boldsymbol{\beta}$ and $\boldsymbol{\Psi}$. As in (10.1), the maximum likelihood estimator maximizes

$$L(\boldsymbol{\beta}, \boldsymbol{\Psi}; \mathbf{U}) = \frac{1}{|\boldsymbol{\beta}'\boldsymbol{\beta} + \boldsymbol{\Psi}|^{\nu/2}} \, e^{-\frac{1}{2}\,\mathrm{trace}((\boldsymbol{\beta}'\boldsymbol{\beta}+\boldsymbol{\Psi})^{-1}\mathbf{U})} \tag{10.65}$$

over $\boldsymbol{\beta}$ satisfying (10.62) and $\boldsymbol{\Psi}$ being diagonal. There is not a closed form solution to the maximization, so it must be done numerically. There may be problems, too, such as having one or more of the ψ_j's being driven to 0. It is not obvious, but if $\widehat{\boldsymbol{\beta}}$ and $\widehat{\boldsymbol{\Psi}}$ are the MLE's, then the maximum of the likelihood is, similar to (10.3),

$$L(\widehat{\boldsymbol{\beta}}, \widehat{\boldsymbol{\Psi}}; \mathbf{U}) = \frac{1}{|\widehat{\boldsymbol{\beta}}'\widehat{\boldsymbol{\beta}} + \widehat{\boldsymbol{\Psi}}|^{\nu/2}} \, e^{-\frac{1}{2}\nu q}. \tag{10.66}$$

See Section 9.4 of Mardia, Kent, and Bibby [1979].

Typically one is interested in finding the simplest model that fits. To test whether the p-factor model fits, we use the hypotheses

$$H_0 : \boldsymbol{\Sigma}_{YY} = \boldsymbol{\beta}'\boldsymbol{\beta} + \boldsymbol{\Psi},\ \boldsymbol{\beta} \text{ is } p \times q \ \ \textit{versus} \ \ H_A : \boldsymbol{\Sigma}_{YY} \in \mathcal{S}_q^+. \tag{10.67}$$

The MLE for H_A is $\widehat{\boldsymbol{\Sigma}}_{YY} = \mathbf{U}/\nu$, so that

$$LR = \left(\frac{|\widehat{\boldsymbol{\beta}}'\widehat{\boldsymbol{\beta}} + \widehat{\boldsymbol{\Psi}}|}{|\mathbf{U}/\nu|} \right)^{\nu/2}. \tag{10.68}$$

Now

$$2\log(LR) = \nu(\log(|\widehat{\boldsymbol{\beta}}'\widehat{\boldsymbol{\beta}} + \widehat{\boldsymbol{\Psi}}|) - \log(|\mathbf{U}/\nu|)), \tag{10.69}$$

which is asymptotically χ^2_{df} with df being the difference in (10.64). Bartlett suggests a slight adjustment to the factor ν, similar to the Box approximation for Wilks' Λ, so that under the null,

$$2\log(LR)^* = (\nu - \frac{2q+5}{6} - \frac{2p}{3})(\log(|\widehat{\boldsymbol{\beta}}'\widehat{\boldsymbol{\beta}} + \widehat{\boldsymbol{\Psi}}|) - \log(|\mathbf{U}/\nu|)) \longrightarrow \chi^2_{df}, \tag{10.70}$$

where from (10.64),

$$df = \frac{(q-p)^2 - p - q}{2}. \tag{10.71}$$

Alternatively, one can use AIC (9.56) or BIC (9.57) to assess M_p for several p. Because νq is the same for all models, we can take

$$\text{deviance}(M_p(\widehat{\boldsymbol{\beta}}, \widehat{\boldsymbol{\Psi}})\ ;\ \mathbf{y}) = \nu \log(|\widehat{\boldsymbol{\beta}}'\widehat{\boldsymbol{\beta}} + \widehat{\boldsymbol{\Psi}}|), \tag{10.72}$$

so that

$$\text{BIC}(M_p) = \nu \log(|\widehat{\boldsymbol{\beta}}'\widehat{\boldsymbol{\beta}} + \widehat{\boldsymbol{\Psi}}|) + \log(\nu)\ (q(p+1) - p(p-1)/2), \tag{10.73}$$

$$\text{AIC}(M_p) = \nu \log(|\widehat{\boldsymbol{\beta}}'\widehat{\boldsymbol{\beta}} + \widehat{\boldsymbol{\Psi}}|) + 2\ (q(p+1) - p(p-1)/2). \tag{10.74}$$

10.3.2 Describing the factors

Once you have decided on the number p of factors in the model and the estimate $\widehat{\boldsymbol{\beta}}$, you have a choice of rotations. That is, since $\boldsymbol{\Gamma}\widehat{\boldsymbol{\beta}}$ for any $p \times p$ orthogonal matrix $\boldsymbol{\Gamma}$ has exactly the same fit, you need to choose the $\boldsymbol{\Gamma}$. There are a number of criteria. The **varimax** criterion tries to pick a rotation so that the loadings ($\widehat{\beta}_{ij}$'s) are either large in magnitude, or close to 0. The hope is that it is then easy to interpret the factors by seeing which variables they load heavily upon. Formally, the varimax rotation is that which maximizes the sum of the variances of the squares of the elements in each row. That is, if $\mathbf{F}$ is the $p \times q$ matrix consisting of the squares of the elements in $\boldsymbol{\Gamma}\widehat{\boldsymbol{\beta}}$, then the varimax rotation is the $\boldsymbol{\Gamma}$ that maximizes trace$(\mathbf{F}\mathbf{H}_q\mathbf{F}')$, $\mathbf{H}_q$ being the centering matrix (1.12). There is nothing preventing you from trying as many $\boldsymbol{\Gamma}$'s as you wish. It is an art to find a rotation and interpretation of the factors.

The matrix $\mathbf{X}$, which has the scores of the factors for the individuals, is unobserved, but can be estimated. The joint distribution is, from (10.51) with the assumptions (10.56),

$$\begin{pmatrix} \mathbf{X} & \mathbf{Y} \end{pmatrix} \sim N(\begin{pmatrix} \mathbf{0} & \mathbf{D}\boldsymbol{\gamma} \end{pmatrix}, \mathbf{I}_n \otimes \boldsymbol{\Sigma}), \quad \boldsymbol{\Sigma} = \begin{pmatrix} \mathbf{I}_p & \boldsymbol{\beta} \\ \boldsymbol{\beta}' & \boldsymbol{\beta}'\boldsymbol{\beta} + \boldsymbol{\Psi} \end{pmatrix}. \tag{10.75}$$

Then given the observed $\mathbf{Y}$:

$$\mathbf{X} \,|\, \mathbf{Y} = \mathbf{y} \sim N(\boldsymbol{\alpha}^* + \mathbf{y}\boldsymbol{\beta}^*, \mathbf{I}_n \otimes \boldsymbol{\Sigma}_{XX\cdot Y}), \tag{10.76}$$

where

$$\boldsymbol{\beta}^* = (\boldsymbol{\beta}'\boldsymbol{\beta} + \boldsymbol{\Psi})^{-1}\boldsymbol{\beta}', \quad \boldsymbol{\alpha}^* = -\mathbf{D}\boldsymbol{\gamma}\boldsymbol{\beta}^*, \tag{10.77}$$

and

$$\boldsymbol{\Sigma}_{XX\cdot Y} = \mathbf{I}_p - \boldsymbol{\beta}(\boldsymbol{\beta}'\boldsymbol{\beta} + \boldsymbol{\Psi})^{-1}\boldsymbol{\beta}'. \tag{10.78}$$

An estimate of $\mathbf{X}$ is the estimate of $E[\mathbf{X} \mid \mathbf{Y} = \mathbf{y}]$:

$$\widehat{\mathbf{X}} = (\mathbf{y} - \mathbf{D}\widehat{\gamma})(\widehat{\beta}'\widehat{\beta} + \widehat{\Psi})^{-1}\widehat{\beta}'. \tag{10.79}$$

10.3.3 Example: Grades data

Continue with the grades data in Section 10.2.2, where the $\mathbf{D}$ in

$$E[\mathbf{Y}] = \mathbf{D}\gamma \tag{10.80}$$

is a 107×2 matrix that distinguishes men from women. The first step is to estimate $\mathbf{\Sigma}_{YY}$:

$$\widehat{\mathbf{\Sigma}}_{YY} = \frac{1}{\nu}\,\mathbf{Y}'\mathbf{Q}_\mathbf{D}\mathbf{Y}, \tag{10.81}$$

where here $\nu = 107 - 2$ (since $\mathbf{D}$ has two columns), which is the pooled covariance matrix in (10.18).

We illustrate with the R program factanal. The input to the program can be a data matrix or a covariance matrix or a correlation matrix. In any case, the program will base its calculations on the correlation matrix. Unless $\mathbf{D}$ is just a column of 1's, you shouldn't give it $\mathbf{Y}$, but $\mathbf{S} = \mathbf{Y}'\mathbf{Q}_\mathbf{D}\mathbf{Y}/\nu$, where $\nu = n - k$ if $\mathbf{D}$ is $n \times k$. You need to also specify how many factors you want, and the number of observations (actually, $\nu + 1$ for us). We'll start with one factor. The sigmahat is the $\mathbf{S}$, and covmat= indicates to R that you are giving it a covariance matrix. (Do the same if you are giving a correlation matrix.) In such cases, the program does not know what n or k is, so you should set the parameter n.obs. It assumes that $\mathbf{D}$ is $\mathbf{1}_n$, i.e., that $k = 1$, so to trick it into using another k, set n.obs to $n - k + 1$, which in our case is 106. Then the one-factor model is fit to the sigmahat in (10.18) using

```
f <- factanal(covmat=sigmahat,factors=1,n.obs=106)
```

The output includes the uniquenesses (diagonals of $\widehat{\Psi}$), f$uniquenesses, and the (transpose of the) loadings matrix, f$loadings. Here,

diagonals of $\widehat{\Psi}$: (10.82)

HW	Labs	InClass	Midterms	Final
0.247	0.215	0.828	0.765	0.786

and

$\widehat{\beta}$: (10.83)

	HW	Labs	InClass	Midterms	Final
Factor1	0.868	0.886	0.415	0.484	0.463

The given loadings and uniquenesses are based on the correlation matrix, so the fitted correlation matrix can be found using

```
corr0 <- f$loadings%*%t(f$loadings) + diag(f$uniquenesses)
```

The result is

One-factor model	HW	Labs	InClass	Midterms	Final
HW	1.00	0.77	0.36	0.42	0.40
Labs	0.77	1.00	0.37	0.43	0.41
InClass	0.36	0.37	1.00	0.20	0.19
Midterms	0.42	0.43	0.20	1.00	*0.22
Final	0.40	0.41	0.19	0.22	1.00

(10.84)

Compare that to the observed correlation matrix, which is in the matrix f$corr:

$$\tag{10.85}$$

Unrestricted model	HW	Labs	InClass	Midterms	Final
HW	1.00	0.78	0.28	0.41	0.40
Labs	0.78	1.00	0.42	0.38	0.35
InClass	0.28	0.42	1.00	0.24	0.27
Midterms	0.41	0.38	0.24	1.00	*0.60
Final	0.40	0.35	0.27	0.60	1.00

The fitted correlations are reasonably close to the observed ones, except for the midterms/final correlation: The actual is 0.60, but the estimate from the one-factor model is only 0.22. It appears that this single factor is more focused on other correlations.

For a formal goodness-of-fit test, we have

$$H_0 : \text{One-factor model} \quad \textit{versus} \quad H_A : \text{Unrestricted.} \tag{10.86}$$

We can use either the correlation or covariance matrices, as long as we are consistent, and since factanal gives the correlation, we might as well use that. The MLE under H_A is then corrA, the correlation matrix obtained from $\mathbf{S}$, and under H_0 is corr0. Then

$$2\log(LR) = \nu(\log(|\widehat{\boldsymbol{\beta}}'\widehat{\boldsymbol{\beta}} + \widehat{\boldsymbol{\Psi}}|) - \log(|\mathbf{S}|)) \tag{10.87}$$

is found in R using

```
105*log(det(corr0)/det(f$corr))
```

yielding the value 37.65. It is probably better to use Bartlett's refinement (10.70),

```
(105 − (2*5+5)/6−2/3)*log(det(corr0)/det(f$corr))
```

which gives 36.51. This value can be found in f$STATISTIC, or by printing out f. The degrees of freedom for the statistic in (10.71) is $((q-p)^2 - p - q)/2 = 5$, since $p = 1$ and $q = 5$. Thus H_0 is rejected: The one-factor model does not fit.

Two factors

With small q, we have to be careful not to ask for too many factors. By (10.64), two is the maximum when $q = 5$. In R, we just need to set factors=2 in the factanal function. The χ^2 for goodness-of-fit is 2.11, on one degree of freedom, hence the two-factor model fits fine. The estimated correlation matrix is now

$$\tag{10.88}$$

Two-factor model	HW	Labs	InClass	Midterms	Final
HW	1.00	0.78	0.35	0.40	0.40
Labs	0.78	1.00	0.42	0.38	0.35
InClass	0.35	0.42	1.00	0.24	0.25
Midterms	0.40	0.38	0.24	1.00	0.60
Final	0.40	0.35	0.25	0.60	1.00

which is quite close to the observed correlation matrix (10.85) above. Only the InClass/HW correlation is a bit off, but not by much.

The uniquenesses and loadings for this model are

diagonals of $\widehat{\Psi}$: (10.89)

HW	Labs	InClass	Midterms	Final
0.36	0.01	0.80	0.48	0.30

and

$\widehat{\beta}$: (10.90)

	HW	Labs	InClass	Midterms	Final
Factor 1	0.742	0.982	0.391	0.268	0.211
Factor 2	0.299	0.173	0.208	0.672	0.807

The routine gives the loadings using the varimax criterion.

Looking at the uniquenesses, we notice that inclass's is quite large, which suggests that it has a factor unique to itself, e.g., being able to get to class. It has fairly low loadings on both factors. We see that the first factor loads highly on homework and labs, especially labs, and the second loads heavily on the exams, midterms and final. (These results are not surprising given the example in Section 10.2.2, where we see homework, inclass, and midterms are conditionally independent given labs and final.) So one could label the factors "diligence" and "test taking ability."

The exact same fit can be achieved by using other rotations $\mathbf{\Gamma}\beta$, for a 2×2 orthogonal matrix $\mathbf{\Gamma}$. Consider the rotation

$$\mathbf{\Gamma} = \frac{1}{\sqrt{2}} \begin{pmatrix} 1 & 1 \\ 1 & -1 \end{pmatrix}. \tag{10.91}$$

Then the loadings become

$\mathbf{\Gamma}\widehat{\beta}$: (10.92)

	HW	Labs	InClass	Midterms	Final
Factor* 1	0.736	0.817	0.424	0.665	0.720
Factor* 2	0.314	0.572	0.129	−0.286	−0.421

Now Factor* 1 could be considered an overall ability factor, and Factor* 2 a contrast of HW+Lab and Midterms+Final.

Any rotation is fine — whichever you can interpret easiest is the one to take.

Using the BIC to select the number of factors

We have the three models: One-factor (M_1), two-factor (M_2), and unrestricted (M_{Big}). The deviances (10.72) are $\nu \log(|\widehat{\mathbf{\Sigma}}|)$, where here we take the correlation form of the $\widehat{\mathbf{\Sigma}}$'s. The relevant quantities are next:

(10.93)

Model	Deviance	d	BIC	BIC	$\widehat{P}^{BIC}$
M_1	−156.994	10	−110.454	16.772	0
M_2	−192.382	14	−127.226	0	0.768
M_{Big}	−194.640	15	−124.831	2.395	0.232

The only difference between the two BIC columns is that the second one has 127.226 added to each element, making it easier to compare them. These results conform to what we had before. The one-factor model is untenable, and the two-factor model is fine, with 77% estimated probability. The full model has a decent probability as well.

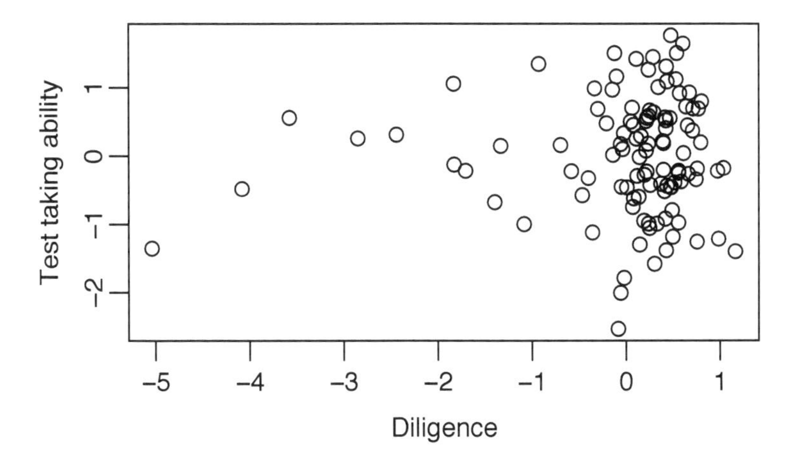

Figure 10.1: Plot of factor scores for the two-factor model.

Estimating the score matrix X

The score matrix is estimated as in (10.79). You have to be careful, though, to use consistently the correlation form or covariance form. That is, if the $(\widehat{\boldsymbol{\beta}}, \widehat{\boldsymbol{\Psi}})$ is estimated from the correlation matrix, then the residuals $\mathbf{y} - \mathbf{D}\widehat{\gamma}$ must be rescaled so that the variances are 1. Or you can let R do it, by submitting the residuals and asking for the "regression" scores:

```
x <- cbind(1,grades[,1])
gammahat <- solve(t(x)%*%x,t(x)%*%grades[,2:6])
resids <- grades[,2:6]-x%*%gammahat
xhat <- factanal(resids,factors=2,scores='regression')$scores
```

The xhat is then 107×2:

$$\widehat{\mathbf{X}} = \begin{pmatrix} -5.038 & -1.352 \\ -4.083 & -0.479 \\ -0.083 & -2.536 \\ \vdots & \vdots \\ 0.472 & 1.765 \end{pmatrix}. \tag{10.94}$$

Now we can use the factor scores in scatter plots. For example, Figure 10.1 contains a scatter plot of the estimated factor scores for the two-factor model. They are by construction uncorrelated, but one can see how diligence has a much longer lower tail (lazy people?).

We also calculated box plots to compare the women's and men's distribution on the factors:

```
par(mfrow=c(1,2))
yl <- range(xhat) # To obtain the same y-scales
```

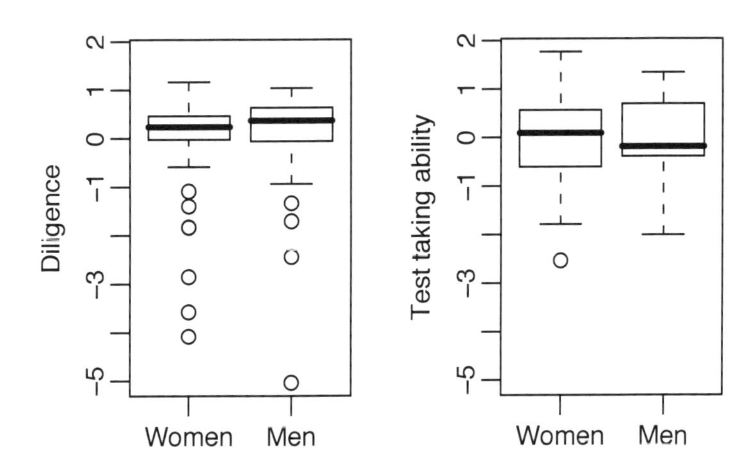

Figure 10.2: Box plots comparing the women and men on their factor scores.

```
w <- (x[,2]==1) # Whether women (T) or not.
boxplot(list(Women=xhat[w,1],Men=xhat[!w,1]),main='Factor 1',ylim=yl)
boxplot(list(Women=xhat[w,2],Men=xhat[!w,2]),main='Factor 2',ylim=yl)
```

See Figure 10.2. There do not appear to be any large overall differences.

10.4 Some symmetry models

Some structural models on covariances matrices, including testing independence, can be defined through group symmetries. The advantage of such models is that the likelihood estimates and tests are very easy to implement. The ones we present are called **invariant normal models** as defined in Andersson [1975]. We will be concerned with these models' restrictions on the covariance matrices. More generally, the models are defined on the means as well. Basically, the models are ones that specify certain linear constraints among the elements of the covariance matrix.

The model starts with

$$\mathbf{Y} \sim N_{n\times q}(\mathbf{0}, \mathbf{I}_n \otimes \mathbf{\Sigma}), \tag{10.95}$$

and a $q \times q$ group $\mathcal{G}$ (see (5.75)), a subgroup of the group of $q \times q$ orthogonal matrices $\mathcal{O}_q$. The model demands that the distribution of $\mathbf{Y}$ be invariant under multiplication on the right by elements of $\mathcal{G}$, that is,

$$\mathbf{Yg} =^{\mathcal{D}} \mathbf{Y} \text{ for all } \mathbf{g} \in \mathcal{G}. \tag{10.96}$$

Now because $Cov[\mathbf{Yg}] = \mathbf{I}_n \otimes \mathbf{g}'\mathbf{\Sigma}\mathbf{g}$, (10.95) and (10.96) imply that

$$\mathbf{\Sigma} = \mathbf{g}'\mathbf{\Sigma}\mathbf{g} \text{ for all } \mathbf{g} \in \mathcal{G}. \tag{10.97}$$

Thus we can define the **mean zero invariant normal model** based on $\mathcal{G}$ to be (10.95) with

$$\boldsymbol{\Sigma} \in \mathcal{S}_q^+(\mathcal{G}) \equiv \{\boldsymbol{\Sigma} \in \mathcal{S}_q^+ \mid \boldsymbol{\Sigma} = \mathbf{g}'\boldsymbol{\Sigma}\mathbf{g} \text{ for all } \mathbf{g} \in \mathcal{G}\}. \tag{10.98}$$

A few examples are in order at this point. Typically, the groups are fairly simple groups.

10.4.1 Some types of symmetry

Independence and block independence

Partition the variables in $\mathbf{Y}$ so that

$$\mathbf{Y} = (\mathbf{Y}_1, \ldots, \mathbf{Y}_K), \text{ where } \mathbf{Y}_k \text{ is } n \times q_k, \tag{10.99}$$

and

$$\boldsymbol{\Sigma} = \begin{pmatrix} \boldsymbol{\Sigma}_{11} & \boldsymbol{\Sigma}_{12} & \cdots & \boldsymbol{\Sigma}_{1K} \\ \boldsymbol{\Sigma}_{21} & \boldsymbol{\Sigma}_{22} & \cdots & \boldsymbol{\Sigma}_{2K} \\ \vdots & \vdots & \ddots & \vdots \\ \boldsymbol{\Sigma}_{K1} & \boldsymbol{\Sigma}_{K2} & \cdots & \boldsymbol{\Sigma}_{KK} \end{pmatrix}, \ \boldsymbol{\Sigma}_{kl} \text{ is } q_k \times q_l. \tag{10.100}$$

Independence of one block of variables from the others entails setting the covariance to zero, that is, $\boldsymbol{\Sigma}_{kl} = \mathbf{0}$ means $\mathbf{Y}_k$ and $\mathbf{Y}_l$ are independent. Invariant normal models can specify a block being independent of all the other blocks. For example, suppose $K = 3$. Then the model that $\mathbf{Y}_1$ is independent of $(\mathbf{Y}_2, \mathbf{Y}_3)$ has

$$\boldsymbol{\Sigma} = \begin{pmatrix} \boldsymbol{\Sigma}_{11} & \mathbf{0} & \mathbf{0} \\ \mathbf{0} & \boldsymbol{\Sigma}_{22} & \boldsymbol{\Sigma}_{23} \\ \mathbf{0} & \boldsymbol{\Sigma}_{32} & \boldsymbol{\Sigma}_{33} \end{pmatrix}. \tag{10.101}$$

The group that gives rise to that model consists of two elements:

$$\mathcal{G} = \left\{ \begin{pmatrix} \mathbf{I}_{q_1} & \mathbf{0} & \mathbf{0} \\ \mathbf{0} & \mathbf{I}_{q_2} & \mathbf{0} \\ \mathbf{0} & \mathbf{0} & \mathbf{I}_{q_3} \end{pmatrix}, \begin{pmatrix} -\mathbf{I}_{q_1} & \mathbf{0} & \mathbf{0} \\ \mathbf{0} & \mathbf{I}_{q_2} & \mathbf{0} \\ \mathbf{0} & \mathbf{0} & \mathbf{I}_{q_3} \end{pmatrix} \right\}. \tag{10.102}$$

(The first element is just $\mathbf{I}_q$, of course.) It is easy to see that $\boldsymbol{\Sigma}$ of (10.101) is invariant under $\mathcal{G}$ of (10.102). Lemma 10.1 below can be used to show any $\boldsymbol{\Sigma}$ in $\mathcal{S}^+(\mathcal{G})$ is of the form (10.101).

If the three blocks $\mathbf{Y}_1, \mathbf{Y}_2$, and $\mathbf{Y}_3$ are mutually independent, then $\boldsymbol{\Sigma}$ is block diagonal,

$$\boldsymbol{\Sigma} = \begin{pmatrix} \boldsymbol{\Sigma}_{11} & \mathbf{0} & \mathbf{0} \\ \mathbf{0} & \boldsymbol{\Sigma}_{22} & \mathbf{0} \\ \mathbf{0} & \mathbf{0} & \boldsymbol{\Sigma}_{33} \end{pmatrix}, \tag{10.103}$$

and the corresponding $\mathcal{G}$ consists of the eight matrices

$$\mathcal{G} = \left\{ \begin{pmatrix} \pm\mathbf{I}_{q_1} & \mathbf{0} & \mathbf{0} \\ \mathbf{0} & \pm\mathbf{I}_{q_2} & \mathbf{0} \\ \mathbf{0} & \mathbf{0} & \pm\mathbf{I}_{q_3} \end{pmatrix} \right\}. \tag{10.104}$$

An extreme case is when all variables are mutually independent, so that $q_k = 1$ for each k (and $K = q$), $\boldsymbol{\Sigma}$ is diagonal, and $\mathcal{G}$ consists of all diagonal matrices with ± 1's down the diagonal.

Intraclass correlation structure

The **intraclass correlation structure** arises when the variables are interchangeable. For example, the variables may be similar measurements (such as blood pressure) made several times, or scores on sections of an exam, where the sections are all measuring the same ability. In such cases, the covariance matrix would have equal variances, and equal covariances:

$$\mathbf{\Sigma} = \sigma^2 \begin{pmatrix} 1 & \rho & \cdots & \rho & \rho \\ \rho & 1 & \cdots & \rho & \rho \\ \vdots & \vdots & \ddots & \vdots & \vdots \\ \rho & \rho & \cdots & \rho & 1 \end{pmatrix}. \tag{10.105}$$

The $\mathcal{G}$ for this model is the group of $q \times q$ permutation matrices $\mathcal{P}_q$. (A permutation matrix has exactly one 1 in each row and one 1 in each column, and zeroes elsewhere. If $\mathbf{g}$ is a permutation matrix and $\mathbf{x}$ is a $q \times 1$ vector, then $\mathbf{gx}$ contains the same elements as $\mathbf{x}$, but in a different order.)

Compound symmetry

Compound symmetry is an extension of intraclass symmetry, where there are groups of variables, and the variables within each group are interchangeable. Such models might arise, e.g., if students are given three interchangeable batteries of math questions, and two interchangeable batteries of verbal questions. The covariance matrix would then have the form

$$\mathbf{\Sigma} = \begin{pmatrix} a & b & b & c & c \\ b & a & b & c & c \\ b & b & a & c & c \\ c & c & c & d & e \\ c & c & c & e & d \end{pmatrix}. \tag{10.106}$$

In general, the group would consist of block diagonal matrices, with permutation matrices as the blocks. That is, with $\mathbf{\Sigma}$ partitioned as in (10.100),

$$\mathcal{G} = \left\{ \begin{pmatrix} \mathbf{G}_1 & \mathbf{0} & \cdots & \mathbf{0} \\ \mathbf{0} & \mathbf{G}_2 & \cdots & \mathbf{0} \\ \vdots & \vdots & \ddots & \vdots \\ \mathbf{0} & \mathbf{0} & & \cdots \mathbf{G}_K \end{pmatrix} \,\middle|\, \mathbf{G}_1 \in \mathcal{P}_{q_1}, \mathbf{G}_2 \in \mathcal{P}_{q_2}, \ldots, \mathbf{G}_K \in \mathcal{P}_{q_K} \right\}. \tag{10.107}$$

IID, or spherical symmetry

Combining independence and intraclass correlation structure yields $\mathbf{\Sigma} = \sigma^2 \mathbf{I}_q$, so that the variables are independent and identically distributed. The group for this model is the set of permutation matrices augmented with $\pm$ signs on the 1's. (Recall Exercise 8.8.6.)

The largest group possible for these models is the group of $q \times q$ orthogonal matrices. When (10.96) holds for all orthogonal $\mathbf{g}$, the distribution of $\mathbf{Y}$ is said to be *spherically symmetric*. It turns out that this choice also yields $\mathbf{\Sigma} = \sigma^2 \mathbf{I}_q$. This result is a reflection of the fact that iid and spherical symmetry are the same for the multivariate normal distribution. If $\mathbf{Y}$ has some other distribution, then the two models are distinct, although they still have the same covariance structure.

10.4.2 Characterizing the structure

It is not always obvious given a structure for the covariance matrix to find the corresponding $\mathcal{G}$, or even to decide whether there is a corresponding $\mathcal{G}$. But given the group, there is a straightforward method for finding the structure. We will consider just finite groups $\mathcal{G}$, but the idea extends to general groups, in which case we would need to introduce uniform (Haar) measure on these groups.

For given finite group $\mathcal{G}$ and general $\mathbf{\Sigma}$, define the $\mathcal{G}$-average of $\mathbf{\Sigma}$ by

$$\overline{\mathbf{\Sigma}} = \frac{\sum_{\mathbf{g}\in\mathcal{G}} \mathbf{g}'\mathbf{\Sigma}\mathbf{g}}{\#\mathcal{G}}. \tag{10.108}$$

It should be clear that if $\mathbf{\Sigma} \in \mathcal{S}_q^+(\mathcal{G})$, then $\overline{\mathbf{\Sigma}} = \mathbf{\Sigma}$. The next lemma shows that all $\mathcal{G}$-averages are in $\mathcal{S}_q^+(\mathcal{G})$.

Lemma 10.1. *For any* $\mathbf{\Sigma} \in \mathcal{S}_q^+$, $\overline{\mathbf{\Sigma}} \in \mathcal{S}_q^+(\mathcal{G})$.

Proof. For any $\mathbf{h} \in \mathcal{G}$,

$$\begin{aligned}\mathbf{h}'\overline{\mathbf{\Sigma}}\mathbf{h} &= \frac{\sum_{\mathbf{g}\in\mathcal{G}} \mathbf{h}'\mathbf{g}'\mathbf{\Sigma}\mathbf{g}\mathbf{h}}{\#\mathcal{G}} \\ &= \frac{\sum_{\mathbf{g}^*\in\mathcal{G}} \mathbf{g}^{*\prime}\mathbf{\Sigma}\mathbf{g}^*}{\#\mathcal{G}} \\ &= \overline{\mathbf{\Sigma}}.\end{aligned} \tag{10.109}$$

The second line follows by setting $\mathbf{g}^* = \mathbf{gh}$, and noting that as $\mathbf{g}$ runs over $\mathcal{G}$, so does $\mathbf{g}^*$. (This is where the requirement that $\mathcal{G}$ is a group is needed.) But (10.109) implies that $\overline{\mathbf{\Sigma}} \in \mathcal{G}$. □

The lemma shows that

$$\mathcal{S}_q^+(\mathcal{G}) = \{\overline{\mathbf{\Sigma}} \mid \mathbf{\Sigma} \in \mathcal{S}_q^+\}, \tag{10.110}$$

so that one can discover the structure of covariance matrices invariant under a particular group by averaging a generic $\mathbf{\Sigma}$. That is how one finds the structures in (10.101), (10.103), (10.106), and (10.107) from their respective groups.

10.4.3 Maximum likelihood estimates

The maximum likelihood estimate of $\mathbf{\Sigma}$ in (10.98) is the $\widehat{\mathbf{\Sigma}} \in \mathcal{S}_q^+(\mathcal{G})$ that maximizes

$$L(\mathbf{\Sigma};\mathbf{y}) = \frac{1}{|\mathbf{\Sigma}|^{n/2}}\, e^{-\frac{1}{2}\,\text{trace}(\mathbf{\Sigma}^{-1}\mathbf{u})}, \mathbf{\Sigma} \in \mathcal{S}^+(\mathcal{G}), \quad \text{where } \mathbf{u} = \mathbf{y}'\mathbf{y}. \tag{10.111}$$

The requirement $\mathbf{\Sigma} \in \mathcal{S}_q^+(\mathcal{G})$ means that

$$\mathbf{\Sigma}^{-1} = (\mathbf{g}'\mathbf{\Sigma}\mathbf{g})^{-1} = \mathbf{g}'\mathbf{\Sigma}^{-1}\mathbf{g} \tag{10.112}$$

for any $\mathbf{g} \in \mathcal{G}$, that is, $\mathbf{\Sigma}^{-1} \in \mathcal{S}_q^+(\mathcal{G})$, hence

$$\begin{aligned}
\text{trace}(\mathbf{\Sigma}^{-1}\mathbf{u}) &= \text{trace}\left(\frac{\sum_{\mathbf{g}\in\mathcal{G}} \mathbf{g}'\mathbf{\Sigma}^{-1}\mathbf{g}}{\#\mathcal{G}}\mathbf{u}\right) \\
&= \text{trace}\left(\frac{\sum_{\mathbf{g}\in\mathcal{G}} \mathbf{g}'\mathbf{\Sigma}^{-1}\mathbf{g}\mathbf{u}}{\#\mathcal{G}}\right) \\
&= \text{trace}\left(\frac{\sum_{\mathbf{g}\in\mathcal{G}} \mathbf{\Sigma}^{-1}\mathbf{g}\mathbf{u}\mathbf{g}'}{\#\mathcal{G}}\right) \\
&= \text{trace}\left(\mathbf{\Sigma}^{-1}\frac{\sum_{\mathbf{g}\in\mathcal{G}} \mathbf{g}\mathbf{u}\mathbf{g}'}{\#\mathcal{G}}\right) \\
&= \text{trace}(\mathbf{\Sigma}^{-1}\overline{\mathbf{u}}).
\end{aligned} \tag{10.113}$$

Thus

$$L(\mathbf{\Sigma}; \mathbf{y}) = \frac{1}{|\mathbf{\Sigma}|^{n/2}} e^{-\frac{1}{2}\text{trace}(\mathbf{\Sigma}^{-1}\overline{\mathbf{u}})}. \tag{10.114}$$

We know from Lemma 9.1 that the maximizer of L in (10.114) over $\mathbf{\Sigma} \in \mathcal{S}_q^+$ is

$$\widehat{\mathbf{\Sigma}} = \frac{\overline{\mathbf{u}}}{n}, \tag{10.115}$$

but since that maximizer is in $\mathcal{S}_q^+(\mathcal{G})$ by Lemma 10.1, and $\mathcal{S}_q^+(\mathcal{G}) \subset \mathcal{S}_q^+$, it must be the maximizer over $\mathcal{S}_q^+(\mathcal{G})$. That is, (10.115) is indeed the maximum likelihood estimate for (10.111).

To illustrate, let $\mathbf{S} = \mathbf{U}/n$. Then if $\mathcal{G}$ is as in (10.104), so that the model is that three sets of variables are independent (10.103), the maximum likelihood estimate is the sample analog

$$\widehat{\mathbf{\Sigma}}(\mathcal{G}) = \begin{pmatrix} \mathbf{S}_{11} & \mathbf{0} & \mathbf{0} \\ \mathbf{0} & \mathbf{S}_{22} & \mathbf{0} \\ \mathbf{0} & \mathbf{0} & \mathbf{S}_{33} \end{pmatrix}. \tag{10.116}$$

In the intraclass correlation model (10.105), the group is the set of $q \times q$ permutation matrices, and the maximum likelihood estimate has the same form,

$$\widehat{\mathbf{\Sigma}}(\mathcal{G}) = \widehat{\sigma}^2 \begin{pmatrix} 1 & \widehat{\rho} & \cdots & \widehat{\rho} \\ \widehat{\rho} & 1 & \cdots & \widehat{\rho} \\ \vdots & \vdots & \ddots & \vdots \\ \widehat{\rho} & \cdots & \widehat{\rho} & 1 \end{pmatrix}, \tag{10.117}$$

where

$$\widehat{\sigma}^2 = \frac{1}{q}\sum_{i=1}^{q} s_{ii}, \text{ and } \widehat{\rho}\widehat{\sigma}^2 = \frac{\sum\sum_{1\le i<j\le q} s_{ij}}{\binom{q}{2}}. \tag{10.118}$$

That is, the common variance is the average of the original variances, and the common covariance is the average of the original covariances.

10.4.4 Hypothesis testing and model selection

The deviance for the model defined by the group $\mathcal{G}$ is, by (10.3),

$$\text{deviance}(M(\mathcal{G})) = n\log(|\widehat{\mathbf{\Sigma}}(\mathcal{G})|), \tag{10.119}$$

where we drop the exponential term since nq is the same for all models. We can then use this deviance in finding AIC's or BIC's for comparing such models, once we figure out the dimensions of the models, which is usually not too hard. E.g., if the model is that $\mathbf{\Sigma}$ is unrestricted, so that $\mathcal{G}_A = \{\mathbf{I}_q\}$, the trivial group, the dimension for H_A is $q(q+1)/2$. The dimension for the independence model in (10.103) and (10.104) sums the dimensions for the diagonal blocks: $q_1(q_1+1)/2 + q_2(q_2+1)/2 + q_3(q_3+1)/2$. The dimension for the intraclass correlation model (10.105) is 2 (for the variance and covariance).

Also, the likelihood ratio statistic for testing two nested invariant normal models is easy to find. These testing problems use two nested groups, $\mathcal{G}_A \subset \mathcal{G}_0$, so that the hypotheses are

$$H_0 : \mathbf{\Sigma} \in \mathcal{S}_q^+(\mathcal{G}_0) \;\; versus \;\; H_A : \mathbf{\Sigma} \in \mathcal{S}_q^+(\mathcal{G}_A). \tag{10.120}$$

Note that the larger $\mathcal{G}$, the smaller $\mathcal{S}_q^+(\mathcal{G})$, since fewer covariance matrices are invariant under a larger group. Then the likelihood ratio test statistic, $2\log(LR)$, is the difference of the deviances, as in (9.55).

The mean is not zero

So far this subsection assumed the mean of $\mathbf{Y}$ is zero. In the more general regression case where $\mathbf{Y} = \mathbf{x}\boldsymbol{\beta} + \mathbf{R}$ and $\mathbf{R} \sim N_{n\times q}(\mathbf{0}, \mathbf{I}_n \otimes \mathbf{\Sigma})$, we assume that $\mathbf{R}$ is invariant, so that (10.96) holds for $\mathbf{R}$ instead of $\mathbf{Y}$. We estimate $\mathbf{\Sigma}$ restricted to $\mathcal{S}_q^+(\mathcal{G})$ by finding $\mathbf{U} = \mathbf{Y}'\mathbf{Q}_\mathbf{x}\mathbf{Y}$, then taking

$$\widehat{\mathbf{\Sigma}} = \frac{\overline{\mathbf{U}}}{n} \;\text{ or }\; \frac{\overline{\mathbf{U}}}{n-p}, \tag{10.121}$$

(where $\mathbf{x}$ is $n \times p$), depending on whether you want the maximum likelihood estimate or an unbiased estimate. In testing, I would suggest taking the unbiased versions, then using

$$\text{deviance}(M(\mathcal{G})) = (n-p)\log(|\widehat{\mathbf{\Sigma}}(\mathcal{G})|). \tag{10.122}$$

10.4.5 Example: Mouth sizes

Continue from Section 7.3.1 with the mouth size data, using the model (7.51). Because the measurements within each subject are of the same mouth, a reasonable question to ask is whether the residuals within each subject are exchangeable, i.e., whether $\mathbf{\Sigma}_R$ has the intraclass correlation structure (10.105). Let $\mathbf{U} = \mathbf{Y}'\mathbf{Q}_\mathbf{x}\mathbf{Y}$ and the unrestricted estimate be $\widehat{\mathbf{\Sigma}}_A = \mathbf{U}/\nu$ for $\nu = n - 2 = 25$. Then $\widehat{\mathbf{\Sigma}}_A$ and the estimate under the intraclass correlation hypothesis $\widehat{\mathbf{\Sigma}}_0$, given in (10.117) and (10.118), are

$$\widehat{\mathbf{\Sigma}}_A = \begin{pmatrix} 5.415 & 2.717 & 3.910 & 2.710 \\ 2.717 & 4.185 & 2.927 & 3.317 \\ 3.910 & 2.927 & 6.456 & 4.131 \\ 2.710 & 3.317 & 4.131 & 4.986 \end{pmatrix} \tag{10.123}$$

and

$$\widehat{\Sigma}_0 = \begin{pmatrix} 5.260 & 3.285 & 3.285 & 3.285 \\ 3.285 & 5.260 & 3.285 & 3.285 \\ 3.285 & 3.285 & 5.260 & 3.285 \\ 3.285 & 3.285 & 3.285 & 5.260 \end{pmatrix}. \tag{10.124}$$

To test the null hypothesis that the intraclass correlation structure holds, versus the general model, we have from (10.119)

$$2\log(LR) = 25\,(\log(|\widehat{\Sigma}_0|) - \log(|\widehat{\Sigma}_A|)) = 9.374. \tag{10.125}$$

The dimension for the general model is $d_A = q(q+1)/2 = 10$, and for the null is just $d_0 = 2$, thus the degrees of freedom for this statistic is $df = d_A - d_0 = 8$. The intraclass correlation structure appears to be plausible.

We can exploit this structure (10.105) on the $\boldsymbol{\Sigma}_R$ to more easily test hypotheses about the $\boldsymbol{\beta}$ in both-sides models like (7.51). First, we transform the matrix $\boldsymbol{\Sigma}_R$ into a diagonal matrix with two distinct variances. Notice that we can write this covariance as

$$\boldsymbol{\Sigma}_R = \sigma^2(1-\rho)\mathbf{I}_q + \sigma^2\rho\mathbf{1}_q\mathbf{1}_q'. \tag{10.126}$$

Let $\boldsymbol{\Gamma}$ be any $q \times q$ orthogonal matrix whose first column is proportional to $\mathbf{1}_q$, i.e., $\mathbf{1}_q/\sqrt{q}$. Then

$$\begin{aligned} \boldsymbol{\Gamma}'\boldsymbol{\Sigma}_R\boldsymbol{\Gamma} &= \sigma^2(1-\rho)\boldsymbol{\Gamma}'\boldsymbol{\Gamma} + \sigma^2\rho\boldsymbol{\Gamma}'\mathbf{1}_q\mathbf{1}_q'\boldsymbol{\Gamma} \\ &= \sigma^2(1-\rho)\mathbf{I}_q + \sigma^2\rho \begin{pmatrix} \sqrt{q} \\ 0 \\ \vdots \\ 0 \end{pmatrix} \begin{pmatrix} \sqrt{q} & 0 & \cdots & 0 \end{pmatrix} \\ &= \sigma^2 \begin{pmatrix} 1+(q-1)\rho & \mathbf{0}' \\ \mathbf{0} & (1-\rho)\mathbf{I}_{q-1} \end{pmatrix} \equiv \boldsymbol{\Lambda}. \end{aligned} \tag{10.127}$$

We used the fact that because all columns of $\boldsymbol{\Gamma}$ except the first are orthogonal to $\mathbf{1}_q$, $\boldsymbol{\Gamma}'\mathbf{1}_q = \sqrt{q}(1,0,\ldots,0)'$. As suggested by the notation, this $\boldsymbol{\Lambda}$ is indeed the eigenvalue matrix for $\boldsymbol{\Sigma}_R$ (though the diagonals are in a nonincreasing order if $\rho < 0$), and $\boldsymbol{\Gamma}$ contains a corresponding set of eigenvectors.

In the model (7.51), the $\mathbf{z}$ is almost an appropriate $\boldsymbol{\Gamma}$:

$$\mathbf{z} = \begin{pmatrix} 1 & -3 & 1 & -3 \\ 1 & -1 & -1 & 1 \\ 1 & 1 & -1 & -1 \\ 1 & 3 & 1 & 3 \end{pmatrix}. \tag{10.128}$$

The columns are orthogonal, and the first is $\mathbf{1}_4$, so we just have to divide each column by its length to obtain orthonormal columns. The squared lengths of the columns are the diagonals of $\mathbf{z}'\mathbf{z}$: $(4, 20, 4, 20)$. Let $\boldsymbol{\Delta}$ be the square root of $\mathbf{z}'\mathbf{z}$,

$$\boldsymbol{\Delta} = \begin{pmatrix} 2 & 0 & 0 & 0 \\ 0 & \sqrt{20} & 0 & 0 \\ 0 & 0 & 2 & 0 \\ 0 & 0 & 0 & \sqrt{20} \end{pmatrix}, \tag{10.129}$$

and set

$$\mathbf{\Gamma} = \mathbf{z}\mathbf{\Delta}^{-1} \text{ and } \boldsymbol{\beta}^* = \boldsymbol{\beta}\mathbf{\Delta}, \tag{10.130}$$

so that the both-sides model can be written

$$\mathbf{Y} = \mathbf{x}\boldsymbol{\beta}\mathbf{\Delta}\mathbf{\Delta}^{-1}\mathbf{z}' + \mathbf{R} = \mathbf{x}\boldsymbol{\beta}^*\mathbf{\Gamma}' + \mathbf{R}. \tag{10.131}$$

Multiplying everything on the right by $\mathbf{\Gamma}$ yields

$$\mathbf{Y}^* \equiv \mathbf{Y}\mathbf{\Gamma} = \mathbf{x}\boldsymbol{\beta}^* + \mathbf{R}^*, \tag{10.132}$$

where

$$\mathbf{R}^* \equiv \mathbf{R}\mathbf{\Gamma} \sim N(\mathbf{0}, \mathbf{I}_n \otimes \mathbf{\Gamma}'\mathbf{\Sigma}_R\mathbf{\Gamma}) = N(\mathbf{0}, \mathbf{I}_n \otimes \mathbf{\Lambda}). \tag{10.133}$$

The estimate of $\boldsymbol{\beta}^*$ is straightforward:

$$\widehat{\boldsymbol{\beta}}^* = (\mathbf{x}'\mathbf{x})^{-1}\mathbf{x}'\mathbf{Y}^* = \begin{pmatrix} 49.938 & 3.508 & 0.406 & -0.252 \\ -4.642 & -1.363 & -0.429 & 0.323 \end{pmatrix}. \tag{10.134}$$

These estimates are the same as those for model (6.29), multiplied by $\mathbf{\Delta}$ as in (10.130). The difference is in their covariance matrix:

$$\widehat{\boldsymbol{\beta}}^* \sim N(\boldsymbol{\beta}^*, \mathbf{C}_\mathbf{x} \otimes \mathbf{\Lambda}). \tag{10.135}$$

To estimate the standard errors of the estimates, we look at the sum of squares and cross products of the estimated residuals,

$$\mathbf{U}^* = \mathbf{Y}^{*\prime}\mathbf{Q}_\mathbf{x}\mathbf{Y}^* \sim \text{Wishart}_q(\nu, \mathbf{\Lambda}), \tag{10.136}$$

where $\nu = \text{trace}(\mathbf{Q}_\mathbf{x}) = n - p = 27 - 2 = 25$. Because the $\mathbf{\Lambda}$ in (10.127) is diagonal, the diagonals of $\mathbf{U}^*$ are independent scaled χ^2_ν's:

$$U^*_{11} \sim \tau_0^2\chi^2_\nu, \; U^*_{jj} \sim \tau_1^2\chi^2_\nu, \; j = 2, \ldots, q = 4. \tag{10.137}$$

Unbiased estimates are

$$\widehat{\tau}_0^2 = \frac{U_{11}}{\nu} \sim \frac{\tau_0^2}{\nu}\chi^2_\nu \text{ and } \widehat{\tau}_1^2 = \frac{U_{22} + \cdots U_{qq}}{(q-1)\nu} \sim \frac{\tau_1^2}{(q-1)\nu}\chi^2_{(q-1)\nu}. \tag{10.138}$$

For our data,

$$\widehat{\tau}_0^2 = \frac{377.915}{25} = 15.117 \text{ and } \widehat{\tau}_1^2 = \frac{59.167 + 26.041 + 62.919}{75} = 1.975. \tag{10.139}$$

The estimated standard errors of the $\widehat{\beta}^*_{ij}$'s from (10.135) are found from

$$\mathbf{C}_\mathbf{x} \otimes \widehat{\mathbf{\Lambda}} = \begin{pmatrix} 0.0625 & -0.0625 \\ -0.0625 & 0.1534 \end{pmatrix} \otimes \begin{pmatrix} 15.117 & 0 & 0 & 0 \\ 0 & 1.975 & 0 & 0 \\ 0 & 0 & 1.975 & 0 \\ 0 & 0 & 0 & 1.975 \end{pmatrix}, \tag{10.140}$$

being the square roots of the diagonals:

	Standard errors			
	Constant	Linear	Quadratic	Cubic
Boys	0.972	0.351	0.351	0.351
Girls−Boys	1.523	0.550	0.550	0.550

(10.141)

The t-statistics divide (10.134) by their standard errors:

t-statistics (10.142)

	Constant	Linear	Quadratic	Cubic
Boys	51.375	9.984	1.156	−0.716
Girls−Boys	−3.048	−2.477	−0.779	0.586

These statistics are not much different from what we found in Section 6.4.1, but the degrees of freedom for all but the first column are now 75, rather than 25. The main impact is in the significance of δ_1, the difference between the girls' and boys' slopes. Previously, the p-value was 0.033 (the $t = -2.26$ on 25 df). Here, the p-value is 0.016, a bit stronger suggestion of a difference.

10.5 Exercises

Exercise 10.5.1. Verify the likelihood ratio statistic (10.21) for testing the equality of several covariance matrices as in (10.20).

Exercise 10.5.2. Verify that $\text{trace}(\mathbf{\Sigma}^{-1}\mathbf{U}) = \text{trace}(\mathbf{\Sigma}_{11}^{-1}\mathbf{U}_{11}) + \text{trace}(\mathbf{\Sigma}_{22}^{-1}\mathbf{U}_{22})$, as in (10.27), for $\mathbf{\Sigma}$ being block-diagonal, i.e., $\mathbf{\Sigma}_{12} = \mathbf{0}$ in (10.23).

Exercise 10.5.3. Show that the value of $2\log(LR)$ of (10.30) does not change if the ν's in the denominators are erased.

Exercise 10.5.4. Suppose $\mathbf{U} \sim \text{Wishart}_q(\nu, \mathbf{\Sigma})$, where $\mathbf{\Sigma}$ is partitioned as

$$\mathbf{\Sigma} = \begin{pmatrix} \mathbf{\Sigma}_{11} & \mathbf{\Sigma}_{12} & \cdots & \mathbf{\Sigma}_{1K} \\ \mathbf{\Sigma}_{21} & \mathbf{\Sigma}_{22} & \cdots & \mathbf{\Sigma}_{2K} \\ \vdots & \vdots & \ddots & \vdots \\ \mathbf{\Sigma}_{K1} & \mathbf{\Sigma}_{K2} & \cdots & \mathbf{\Sigma}_{KK} \end{pmatrix}, \tag{10.143}$$

where $\mathbf{\Sigma}_{ij}$ is $q_i \times q_j$, and the q_i's sum to q. Consider testing the null hypothesis that the blocks are mutually independent, i.e.,

$$H_0 : \mathbf{\Sigma}_{ij} = \mathbf{0} \text{ for } 1 \leq i < j \leq K, \tag{10.144}$$

versus the alternative that $\mathbf{\Sigma}$ is unrestricted. (a) Find the $2\log(LR)$, and the degrees of freedom in the χ^2 approximation. (The answer is analogous to that in (10.35).) (b) Let $\mathbf{U}^* = \mathbf{AUA}$ for some diagonal matrix $\mathbf{A}$ with positive diagonal elements. Replace the $\mathbf{U}$ in $2\log(LR)$ with $\mathbf{U}^*$. Show that the value of the statistic remains the same. (c) Specialize to the case that all $q_i = 1$ for all i, so that we are testing the mutual independence of all the variables. Let $\mathbf{C}$ be the sample correlation matrix. Show that $2\log(LR) = -\nu\log(|\mathbf{C}|)$. [Hint: Find the appropriate $\mathbf{A}$ from part (b).]

Exercise 10.5.5. Show that (10.46) holds. [Hint: Apply Exercise 5.8.25(c), with $\mathbf{D}_1 = \mathbf{x}$ and $\mathbf{D}_2 = \mathbf{y}_3$, to obtain

$$\mathbf{Q}_{\mathbf{x}^*} = \mathbf{Q}_\mathbf{x} - \mathbf{Q}_\mathbf{x}\mathbf{y}_3(\mathbf{y}_3'\mathbf{Q}_\mathbf{x}\mathbf{y}_3)^{-1}\mathbf{y}_3'\mathbf{Q}_\mathbf{x}. \tag{10.145}$$

Then show that the right-hand side of (10.46) is the same as $\mathbf{V}'\mathbf{V}$, where $\mathbf{V}$ is found by regressing $\mathbf{Q}_\mathbf{x}\mathbf{y}_3$ out of $\mathbf{Q}_\mathbf{x}(\mathbf{Y}_1, \mathbf{Y}_2)$. This $\mathbf{V}'\mathbf{V}$ is the left-hand side of (10.46).]

Exercise 10.5.6. Show that (10.55) holds.

Exercise 10.5.7. Suppose $\boldsymbol{\beta}$ and $\boldsymbol{\alpha}$ are both $p \times q$, $p < q$, and let their decompositions from (10.61) and (10.62) be $\boldsymbol{\beta} = \mathbf{Q}\boldsymbol{\beta}^*$ and $\boldsymbol{\alpha} = \boldsymbol{\Gamma}\boldsymbol{\alpha}^*$, where $\mathbf{Q}$ and $\boldsymbol{\Gamma}$ are orthogonal, $\alpha^*_{ij} = \beta^*_{ij} = 0$ for $i > j$, and α^*_{ii}'s and β^*_{ii}'s are positive. (We assume the first p columns of $\boldsymbol{\alpha}$, and of $\boldsymbol{\beta}$, are linearly independent.) Show that $\boldsymbol{\alpha}'\boldsymbol{\alpha} = \boldsymbol{\beta}'\boldsymbol{\beta}$ if and only if $\boldsymbol{\alpha}^* = \boldsymbol{\beta}^*$. [Hint: Use the uniqueness of the QR decomposition, in Theorem 5.4.]

Exercise 10.5.8. Show that the conditional parameters in (10.76) are as in (10.77) and (10.78).

Exercise 10.5.9. Show that if the factor analysis is fit using the correlation matrix, then the correlation between variable j and factor i is estimated to be $\widehat{\beta}_{ij}$, the loading of factor i on variable j.

Exercise 10.5.10. What is the factor analysis model with no factors (i.e., erase the $\boldsymbol{\beta}$ in (10.57))? Choose from the following: The covariance of Y is unrestricted; the mean of Y is 0; the Y variables are mutually independent; the covariance matrix of Y is a constant times the identity matrix.

Exercise 10.5.11. Show directly (without using Lemma 10.1) that if $\boldsymbol{\Sigma} \in \mathcal{S}_q^+(\mathcal{G})$ of (10.98), then $\overline{\boldsymbol{\Sigma}}$ in (10.108) is in $\mathcal{S}_q^+(\mathcal{G})$.

Exercise 10.5.12. Verify the steps in (10.113).

Exercise 10.5.13. Show that if $\boldsymbol{\Sigma}$ has intraclass correlation structure (10.105), then $\boldsymbol{\Sigma} = \sigma^2(1-\rho)\mathbf{I}_q + \sigma^2\rho\mathbf{1}_q\mathbf{1}_q'$ as in (10.126).

Exercise 10.5.14. Multivariate complex normals arise in spectral analysis of multiple time series. A q-dimensional complex normal is $\mathbf{Y}_1 + i\,\mathbf{Y}_2$, where $\mathbf{Y}_1$ and $\mathbf{Y}_2$ are $1 \times q$ real normal vectors with joint covariance of the form

$$\boldsymbol{\Sigma} = Cov\left[\begin{pmatrix} \mathbf{Y}_1 & \mathbf{Y}_2 \end{pmatrix}\right] = \begin{pmatrix} \boldsymbol{\Sigma}_1 & \mathbf{F} \\ -\mathbf{F} & \boldsymbol{\Sigma}_1 \end{pmatrix}, \tag{10.146}$$

i.e., $Cov[\mathbf{Y}_1] = Cov[\mathbf{Y}_2]$. Here, "$i$" is the imaginary $i = \sqrt{-1}$. (a) Show that $\mathbf{F} = Cov[\mathbf{Y}_1, \mathbf{Y}_2]$ is *skew-symmetric*, which means that $\mathbf{F}' = -\mathbf{F}$. (b) What is $\mathbf{F}$ when $q = 1$? (c) Show that the set of $\boldsymbol{\Sigma}$'s as in (10.146) is the set $\mathcal{S}_{2q}^+(\mathcal{G})$ in (10.98) with

$$\mathcal{G} = \left\{\mathbf{I}_{2q}, \begin{pmatrix} \mathbf{0} & -\mathbf{I}_q \\ \mathbf{I}_q & \mathbf{0} \end{pmatrix}, \begin{pmatrix} \mathbf{0} & \mathbf{I}_q \\ -\mathbf{I}_p & \mathbf{0} \end{pmatrix}, -\mathbf{I}_{2q}\right\}. \tag{10.147}$$

(See Andersson and Perlman [1984] for more on this model.)

Exercise 10.5.15 (Mouth sizes). For the boys' and girls' mouth size data in Table 4.1, let $\boldsymbol{\Sigma}_B$ be the covariance matrix for the boys' mouth sizes, and $\boldsymbol{\Sigma}_G$ be the covariance matrix for the girls' mouth sizes. Consider testing

$$H_0 : \boldsymbol{\Sigma}_B = \boldsymbol{\Sigma}_G \text{ versus } H_A : \boldsymbol{\Sigma}_B \neq \boldsymbol{\Sigma}_G. \tag{10.148}$$

(a) What are the degrees of freedom for the boys' and girls' sample covariance matrices? (b) Find $|\widehat{\boldsymbol{\Sigma}}_B|$, $|\widehat{\boldsymbol{\Sigma}}_G|$, and the pooled $|\widehat{\boldsymbol{\Sigma}}|$. (Use the unbiased estimates of the $\boldsymbol{\Sigma}_i$'s.) (c) Find $2\log(LR)$. What are the degrees of freedom for the χ^2? What is the p-value? Do you reject the null hypothesis (if $\alpha = .05$)? (d) Look at trace($\widehat{\boldsymbol{\Sigma}}_B$), trace($\widehat{\boldsymbol{\Sigma}}_G$). Also, look at the correlation matrices for the girls and for the boys. What do you see?

Exercise 10.5.16 (Mouth sizes). Continue with the mouth size data from Exercise 10.5.15. (a) Test whether $\mathbf{\Sigma}_B$ has the intraclass correlation structure (versus the general alternative). What are the degrees of freedom for the χ^2? (b) Test whether $\mathbf{\Sigma}_G$ has the intraclass correlation structure. (c) Now assume that both $\mathbf{\Sigma}_B$ and $\mathbf{\Sigma}_G$ have the intraclass correlation structure. Test whether the covariance matrices are equal. What are the degrees of freedom for this test? What is the p-value? Compare this p-value to that in Exercise 10.5.15, part (c). Why is it so much smaller (if it is)?

Exercise 10.5.17 (Grades). This problem considers the grades data. In what follows, use the pooled covariance matrix in (10.18), which has $\nu = 105$. (a) Test the independence of the first three variables (homework, labs, inclass) from the fourth variable, the midterms score. (So leave out the final exams at this point.) Find $l_1, l_2, |\widehat{\mathbf{\Sigma}}_{11}|, |\widehat{\mathbf{\Sigma}}_{22}|$, and $|\widehat{\mathbf{\Sigma}}|$. Also, find $2\log(LR)$ and the degrees of freedom for the χ^2. Do you accept or reject the null hypothesis? (b) Now test the *conditional* independence of the set (homework, labs, inclass) from the midterms, conditioning on the final exam score. What is the ν for the estimated covariance matrix now? Find the new $l_1, l_2, |\widehat{\mathbf{\Sigma}}_{11}|, |\widehat{\mathbf{\Sigma}}_{22}|$, and $|\widehat{\mathbf{\Sigma}}|$. Also, find $2\log(LR)$ and the degrees of freedom for the χ^2. Do you accept or reject the null hypothesis? (c) Find the correlations between the homework, labs and inclass scores and the midterms scores, as well as the conditional correlations given the final exam. What do you notice?

Exercise 10.5.18 (Grades). The table in (10.93) has the BIC's for the one-factor, two-factor, and unrestricted models for the grades data. Find the deviance, dimension, and BIC for the zero-factor model, M_0. (See Exercise 10.5.10.) Find the estimated probabilities of the four models. Compare the results to those without M_0.

Exercise 10.5.19 (Exams). The exams matrix has data on 191 statistics students, giving their scores (out of 100) on the three midterm exams and the final exam. (a) What is the maximum number of factors that can be estimated? (b) Give the number of parameters in the covariance matrices for the 0, 1, 2, and 3 factor models (even if they are not estimable). (c) Plot the data. There are three obvious outliers. Which observations are they? What makes them outliers? For the remaining exercise, eliminate these outliers, so that there are $n = 188$ observations. (d) Test the null hypothesis that the four exams are mutually independent. What are the adjusted $2\log(LR)^*$ (in (10.70)) and degrees of freedom for the χ^2? What do you conclude? (e) Fit the one-factor model. What are the loadings? How do you interpret them? (f) Look at the residual matrix $\mathbf{C} - \widehat{\beta}'\widehat{\beta}$, where $\mathbf{C}$ is the observed correlation matrix of the original variables. If the model fits exactly, what values would the off-diagonals of the residual matrix be? What is the largest off-diagonal in this observed matrix? Are the diagonals of this matrix the uniquenesses? (g) Does the one-factor model fit?

Exercise 10.5.20 (Exams). Continue with the Exams data from Exercise 10.5.19. Again, do not use the outliers found in part (c). Consider the invariant normal model where the group $\mathcal{G}$ consists of 4×4 matrices of the form

$$\mathbf{G} = \begin{pmatrix} \mathbf{G}^* & \mathbf{0}_3' \\ \mathbf{0}_3 & 1 \end{pmatrix} \tag{10.149}$$

where $\mathbf{G}^*$ is a 3×3 permutation matrix. Thus the model is an example of compound symmetry, from Section 10.4.1. The model assumes the three midterms are interchangeable. (a) Give the form of a covariance matrix $\mathbf{\Sigma}$ that is invariant under that $\mathcal{G}$.

(It should be like the upper-left 4×4 block of the matrix in (10.106).) How many free parameters are there? (b) For the exams data, give the MLE of the covariance matrix under the assumption that it is $\mathcal{G}$-invariant. (c) Test whether this symmetry assumption holds, versus the general model. What are the degrees of freedom? For which elements of $\boldsymbol{\Sigma}$ is the null hypothesis least tenable? (d) Assuming $\boldsymbol{\Sigma}$ is $\mathbf{G}$-invariant, test whether the first three variables are independent of the last. (That is, the null hypothesis is that $\boldsymbol{\Sigma}$ is $\mathcal{G}$-invariant *and* $\sigma_{14} = \sigma_{24} = \sigma_{34} = 0$, while the alternative is that $\boldsymbol{\Sigma}$ is $\mathcal{G}$-invariant, but otherwise unrestricted.) What are the degrees of freedom for this test? What do you conclude?

Exercise 10.5.21 (South Africa heart disease). The data for this question comes from a study of heart disease in adult males in South Africa from Rossouw et al. [1983]. (We return to these data in Section 11.8.) The R data frame is SAheart, found in the ElemStatLearn package [Halvorsen, 2012]. The main variable of interest is "chd," congestive heart disease, where 1 indicates the person has the disease, 0 he does not. Explanatory variables include sbc (measurements on blood pressure), tobacco use, ldl (bad cholesterol), adiposity (fat %), family history of heart disease (absent or present), type A personality, obesity, alcohol usage, and age. Here you are to find common factors among the the explanatory variables excluding age and family history. Take logs of the variables sbc, ldl, and obesity, and cube roots of alcohol and tobacco, so that the data look more normal. Age is used as a covariate. Thus $\mathbf{Y}$ is $n \times 7$, and $\mathbf{D} = (\mathbf{1}_n\ \mathbf{x}_{age})$. Here, $n = 462$. (a) What is there about the tobacco and alcohol variables that is distinctly non-normal? (b) Find the sample correlation matrix of the residuals from the $\mathbf{Y} = \mathbf{D}\boldsymbol{\gamma} + \mathbf{R}$ model. Which pairs of variables have correlations over 0.25, and what are their correlations? How would you group these variables? (c) What is the largest number of factors that can be fit for this $\mathbf{Y}$? (d) Give the BIC-based probabilities of the p-factor models for $p = 0$ to the maximum found in part (c), and for the unrestricted model. Which model has the highest probability? Does this model fit, according to the χ^2 goodness of fit test? (e) For the most probable model from part (d), which variables' loadings are highest (over 0.25) for each factor? (Use the varimax rotation for the loadings.) Give relevant names to the two factors. Compare the factors to what you found in part (b). (f) Keeping the same model, find the estimated factor scores. For each factor, find the two-sample t-statistic for comparing the people with heart disease to those without. (The statistics are not actually distributed as Student's t, but do give some measure of the difference.) (g) Based on the statistics in part (f), do any of the factors seem to be important factors in predicting heart disease in these men? If so, which one(s). If not, what are the factors explaining?

Exercise 10.5.22 (Decathlon). Exercise 1.9.22 created a biplot for the decathlon data. The data consist of the scores (number of points) on each of ten events for the top 24 men in the decathlon at the 2008 Olympics. For convenience, rearrange the variables so that the running events come first, then the jumping, then throwing (ignoring the overall total):

```
y <− decathlon[,c(1,5,10,6,3,9,7,2,4,8)]
```

Fit the 1, 2, and 3 factor models. (The chi-squared approximations for the fit might not be very relevant, because the sample size is too small.) Based on the loadings, can you give an interpretation of the factors? Based on the uniquenesses, which events seem to be least correlated with the others?

Chapter 11

Classification

Multivariate analysis of variance seeks to determine whether there are differences among several groups, and what those differences are. Classification is a related area in which one uses observations whose group memberships are known in order to classify new observations whose group memberships are not known. This goal was the basic idea behind the gathering of the Fisher-Anderson iris data (Section 1.3.1). Based on only the petal and sepal measurements of a new iris, can one effectively classify it into one of the three species setosa, virginica, and versicolor? See Figure 1.4 for an illustration of the challenge.

The task is prediction, as in Section 7.5, except that rather than predicting a continuous variable $\mathbf{Y}$, we predict a categorical variable. We will concentrate mainly on linear methods arising from Fisher's methods, logistic regression, and trees. There is a vast array of additional approaches, including using neural networks, support vector machines, boosting, bagging, and a number of other flashy-sounding techniques.

Related to classification is clustering (Chapter 12), in which one assumes that there are groups in the population, but which groups the observations reside in is unobserved, analogous to the factors in factor analysis. In the machine learning community, classification is **supervised learning**, because we know the groups and have some data on group membership, and clustering is **unsupervised learning**, because group membership itself must be estimated. See the book by Hastie, Tibshirani, and Friedman [2009] for a fine statistical treatment of machine learning.

The basic model is a mixture model, presented in the next section.

11.1 Mixture models

The mixture model we consider assumes there are K groups, numbered from 1 to K, and p predictor variables on which to base the classifications. The data then consist of n observations, each a $1 \times (p+1)$ vector,

$$(\mathbf{X}_i, Y_i), i = 1, \ldots, n, \tag{11.1}$$

where $\mathbf{X}_i$ is the $1 \times p$ vector of predictors for observation i, and Y_i is the group number of observation i, so that $Y_i \in \{1, \ldots, K\}$. Marginally, the proportion of the population in group k is

$$P[Y = k] = \pi_k,\ k = 1, \ldots, K. \tag{11.2}$$

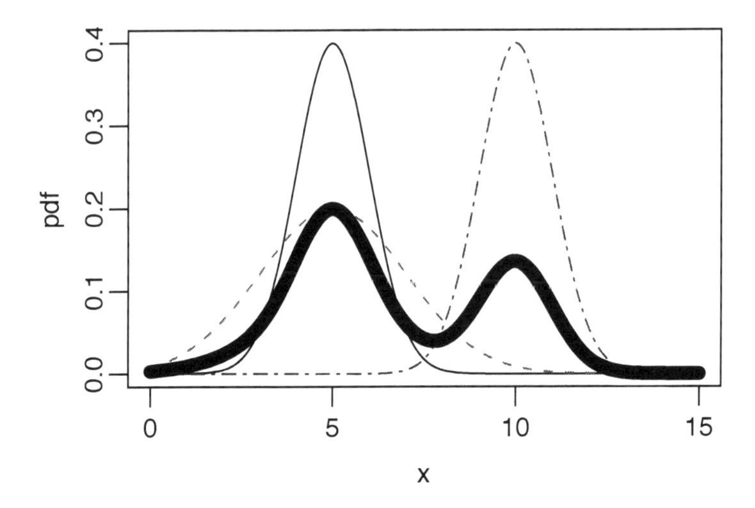

Figure 11.1: Three densities, plus a mixture of the three (the thick line).

Each group then has a conditional distribution P_k:

$$\mathbf{X}_i \mid Y_i = k \sim P_k , \tag{11.3}$$

where the P_k will (almost always) depend on some unknown parameters. Assuming that P_k has density $f_k(\mathbf{x})$, the joint density of $(\mathbf{X}_i, Y_i)$ is a mixed one as in (2.11), with

$$f(\mathbf{x}_i, y_i) = f_{y_i}(\mathbf{x}_i)\pi_{y_i}. \tag{11.4}$$

The marginal pdf of $\mathbf{X}_i$ is found by summing the joint density over the groups:

$$f(\mathbf{x}_i) = \pi_1 f_1(\mathbf{x}_i) + \cdots + \pi_K f_K(\mathbf{x}_i). \tag{11.5}$$

For example, suppose that $K = 3$, $\pi_1 = \pi_2 = \pi_3 = 1/3$, and the three groups are, conditionally,

$$X \mid Y = 1 \sim N(5, 1), \;\; X \mid Y = 2 \sim N(5, 2^2), \;\; X \mid Y = 3 \sim N(10, 1). \tag{11.6}$$

Figure 11.1 exhibits the three pdfs, plus the mixture pdf, which is the thick black line.

The data for classification includes the group index, so that the joint distributions of $(\mathbf{X}_i, Y_i)$ are operative, meaning we can estimate the individual densities. The overall density for the data is then

$$\begin{aligned}
\prod_{i=1}^{n} f(\mathbf{x}_i, y_i) &= \prod_{i=1}^{n} f_{y_i}(\mathbf{x}_i)\pi_{y_i} \\
&= \left[\pi_1^{N_1} \prod_{\{i \mid y_i = 1\}} f_1(\mathbf{x}_i)\right] \cdots \left[\pi_K^{N_K} \prod_{\{i \mid y_i = K\}} f_K(\mathbf{x}_i)\right] \\
&= \left[\prod_{k=1}^{K} \pi_k^{N_k}\right] \times \left[\prod_{k=1}^{K} \prod_{\{i \mid y_i = k\}} f_k(\mathbf{x}_i)\right],
\end{aligned} \tag{11.7}$$

where N_k is the number of observations in group k:

$$N_k = \#\{y_i = k\}. \tag{11.8}$$

The classification task arises when a new observation, $\mathbf{X}^{New}$, arrives without its group identification Y^{New}, so its density is that of the mixture. We have to guess what the group is.

In clustering, the data themselves are without group identification, so we have just the marginal distributions of the $\mathbf{X}_i$. Thus the joint pdf for the data is

$$\Pi_{i=1}^{n} f(\mathbf{x}_i) = \Pi_{i=1}^{n}(\pi_1 f_1(\mathbf{x}_i) + \cdots + \pi_K f_K(\mathbf{x}_i)). \tag{11.9}$$

Thus clustering is similar to classifying new observations, but without having any previous y data to help estimate the π_k's and f_k's. See Section 12.3.

11.2 Classifiers

A **classifier** is a function $\mathcal{C}$ that takes the new observation, and emits a guess at its group:

$$\mathcal{C} : \mathcal{X} \longrightarrow \{1, \ldots, K\}, \tag{11.10}$$

where $\mathcal{X}$ is the space of $\mathbf{X}^{New}$. The classifier may depend on previous data, as well as on the π_k's and f_k's, but not on the Y^{New}. A good classifier is one that is unlikely to make a wrong classification. Thus a reasonable criterion for a classifier is the probability of an error:

$$P[\mathcal{C}(\mathbf{X}^{New}) \neq Y^{New}]. \tag{11.11}$$

We would like to minimize that probability. (This criterion assumes that any type of misclassification is equally bad. If that assumption is untenable, then one can use a weighted probability:

$$\sum_{k=1}^{K}\sum_{l=1}^{K} w_{kl} P[\mathcal{C}(\mathbf{X}^{New}) = k \text{ and } Y^{New} = l], \tag{11.12}$$

where $w_{kk} = 0$.)

Under the (unrealistic) assumption that we know the π_k's and f_k's, the best guess of Y^{New} given $\mathbf{X}^{New}$ is the group that has the highest conditional probability.

Lemma 11.1. *Define the **Bayes classifier** by*

$$\mathcal{C}_B(\mathbf{x}) = k \;\; if \;\; P[Y = k \,|\, \mathbf{X} = \mathbf{x}] > P[Y = l \,|\, \mathbf{X} = \mathbf{x}] \;\; for \;\; l \neq k. \tag{11.13}$$

Then $\mathcal{C}_B$ minimizes (11.11) over classifiers $\mathcal{C}$.

Proof. Let I be the indicator function, so that

$$I[\mathcal{C}(\mathbf{X}^{New}) \neq Y^{New}] = \begin{cases} 1 & if \quad \mathcal{C}(\mathbf{X}^{New}) \neq Y^{New} \\ 0 & if \quad \mathcal{C}(\mathbf{X}^{New}) = Y^{New} \end{cases} \tag{11.14}$$

and

$$P[\mathcal{C}(\mathbf{X}^{New}) \neq Y^{New}] = E[I[\mathcal{C}(\mathbf{X}^{New}) \neq Y^{New}]]. \tag{11.15}$$

As in (2.34), we have that

$$E[I[\mathcal{C}(\mathbf{X}^{New}) \neq Y^{New}]] = E[e_I(\mathbf{X}^{New})], \tag{11.16}$$

where

$$\begin{aligned} e_I(\mathbf{x}^{New}) &= E[I[\mathcal{C}(\mathbf{X}^{New}) \neq Y^{New}] \mid \mathbf{X}^{New} = \mathbf{x}^{New}] \\ &= P[\mathcal{C}(\mathbf{x}^{New}) \neq Y^{New} \mid \mathbf{X}^{New} = \mathbf{x}^{New}] \\ &= 1 - P[\mathcal{C}(\mathbf{x}^{New}) = Y^{New} \mid \mathbf{X}^{New} = \mathbf{x}^{New}]. \end{aligned} \tag{11.17}$$

Thus if we minimize the last expression in (11.17) for each $\mathbf{x}^{New}$, we have minimized the expected value in (11.16). Minimizing (11.17) is the same as maximizing

$$P[\mathcal{C}(\mathbf{x}^{New}) = Y^{New} \mid \mathbf{X}^{New} = \mathbf{x}^{New}], \tag{11.18}$$

but that conditional probability can be written

$$\sum_{l=1}^{K} I[\mathcal{C}(\mathbf{x}^{New}) = l] \; P[Y^{New} = l \mid \mathbf{X}^{New} = \mathbf{x}^{New}]. \tag{11.19}$$

This sum equals $P[Y^{New} = l \mid \mathbf{X}^{New} = \mathbf{x}^{New}]$ for whichever k $\mathcal{C}$ chooses, so to maximize the sum, choose the l with the highest conditional probability, as in (11.13). □

Now the conditional distribution of Y^{New} given $\mathbf{X}^{New}$ is obtained from (11.4) and (11.5) (it is Bayes theorem, Theorem 2.2):

$$P[Y^{New} = k \mid \mathbf{X}^{New} = \mathbf{x}^{New}] = \frac{f_k(\mathbf{x}^{New})\pi_k}{f_1(\mathbf{x}^{New})\pi_1 + \cdots + f_K(\mathbf{x}^{New})\pi_K}. \tag{11.20}$$

Since, given $\mathbf{x}^{New}$, the denominator is the same for each k, we just have to choose the k to maximize the numerator:

$$\mathcal{C}_B(\mathbf{x}) = k \text{ if } f_k(\mathbf{x}^{New})\pi_k > f_l(\mathbf{x}^{New})\pi_l \text{ for } l \neq k. \tag{11.21}$$

We are assuming there is a unique maximum, which typically happens in practice with continuous variables. If there is a tie, any of the top categories will yield the optimum.

Consider the example in (11.6). Because the π_k's are equal, it is sufficient to look at the conditional pdfs. A given x is then classified into the group with highest density, as given in Figure 11.2. Thus the classifications are

$$\mathcal{C}_B(x) = \begin{cases} 1 & \text{if} \quad 3.640 < x < 6.360 \\ 2 & \text{if} \quad x < 3.640 \text{ or } 6.360 < x < 8.067 \text{ or } x > 15.267 \\ 3 & \text{if} \quad 8.067 < x < 15.267. \end{cases} \tag{11.22}$$

In practice, the π_k's and f_k's are not known, but can be estimated from the data. Consider the joint density of the data as in (11.7). The π_k's appear in only the first term. They can be estimated easily (as in a multinomial situation) by

$$\widehat{\pi}_k = \frac{N_k}{n}. \tag{11.23}$$

The parameters for the f_k's can be estimated using the $\mathbf{x}_i$'s that are associated with group k. These estimates are then plugged into the Bayes formula to obtain an approximate Bayes classifier. The next section shows what happens in the multivariate normal case.

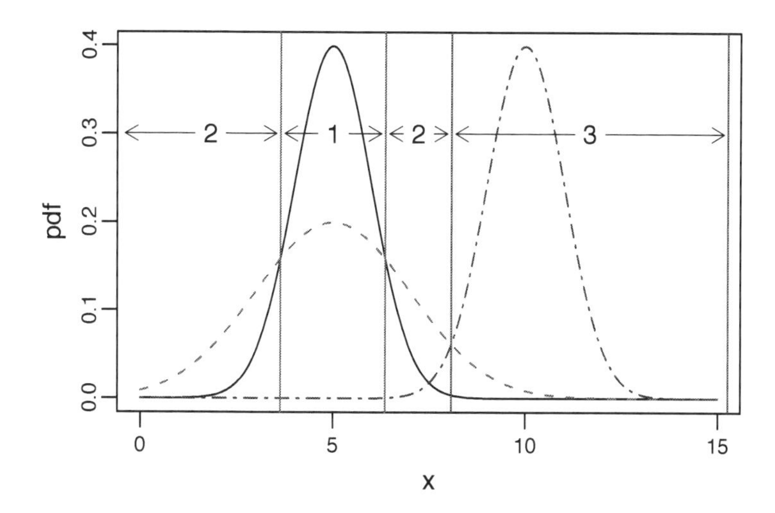

Figure 11.2: Three densities, and the regions in which each is the highest. The densities are 1: N(5,1), solid line; 2: N(5,4), dashed line; 3: N(10,1), dashed/dotted line. Density 2 is also the highest for $x > 15.267$.

11.3 Fisher's linear discrimination

Suppose that the individual f_k's are multivariate normal densities with different means but the same covariance, so that

$$\mathbf{X}_i \mid Y_i = k \sim N_{1\times p}(\boldsymbol{\mu}_k, \boldsymbol{\Sigma}). \tag{11.24}$$

The pdf's (8.47) are then

$$\begin{aligned} f_k(\mathbf{x} \mid \boldsymbol{\mu}_k, \boldsymbol{\Sigma}) &= c\,\frac{1}{|\boldsymbol{\Sigma}|^{1/2}}\, e^{-\frac{1}{2}(\mathbf{x}-\boldsymbol{\mu}_k)\boldsymbol{\Sigma}^{-1}(\mathbf{x}-\boldsymbol{\mu}_k)'} \\ &= c\,\frac{1}{|\boldsymbol{\Sigma}|^{1/2}}\, e^{-\frac{1}{2}\mathbf{x}\boldsymbol{\Sigma}^{-1}\mathbf{x}'} e^{\mathbf{x}\boldsymbol{\Sigma}^{-1}\boldsymbol{\mu}_k' - \frac{1}{2}\boldsymbol{\mu}_k\boldsymbol{\Sigma}^{-1}\boldsymbol{\mu}_k'}. \end{aligned} \tag{11.25}$$

We can ignore the factors that are the same for each group, i.e., that do not depend on k, because we are in quest of the highest pdf$\times\pi_k$. Thus for a given $\mathbf{x}$, we choose the k to maximize

$$\pi_k\, e^{\mathbf{x}\boldsymbol{\Sigma}^{-1}\boldsymbol{\mu}_k' - \frac{1}{2}\boldsymbol{\mu}_k\boldsymbol{\Sigma}^{-1}\boldsymbol{\mu}_k'}, \tag{11.26}$$

or, by taking logs, the k that maximizes

$$d_k^*(\mathbf{x}) \equiv \mathbf{x}\boldsymbol{\Sigma}^{-1}\boldsymbol{\mu}_k' - \frac{1}{2}\boldsymbol{\mu}_k\boldsymbol{\Sigma}^{-1}\boldsymbol{\mu}_k' + \log(\pi_k). \tag{11.27}$$

These d_k^*'s are called the **discriminant functions**. Note that in this case, they are linear in $\mathbf{x}$, hence **linear discriminant functions**. It is often convenient to target one group (say the Kth) as a benchmark, then use the functions

$$d_k(\mathbf{x}) = d_k^*(\mathbf{x}) - d_K^*(\mathbf{x}), \tag{11.28}$$

so that the final function is 0.

We still must estimate the parameters, but that task is straightforward: take the $\widehat{\pi}_k = N_K/n$ as in (11.23), estimate the $\boldsymbol{\mu}_k$'s by the obvious sample means:

$$\widehat{\boldsymbol{\mu}}_k = \frac{1}{N_k} \sum_{\{i \mid y_i = k\}} \mathbf{x}_i, \tag{11.29}$$

and estimate $\boldsymbol{\Sigma}$ by the MLE, i.e., because we are assuming the covariances are equal, the pooled covariance:

$$\widehat{\boldsymbol{\Sigma}} = \frac{1}{n} \sum_{k=1}^{K} \sum_{\{i \mid y_i = k\}} (\mathbf{x}_i - \widehat{\boldsymbol{\mu}}_k)'(\mathbf{x}_i - \widehat{\boldsymbol{\mu}}_k). \tag{11.30}$$

(The numerator equals $\mathbf{X}'(\mathbf{I}_n - \mathbf{P})\mathbf{X}$ for $\mathbf{P}$ being the projection matrix for the design matrix indicating which groups the observations are from. We could divide by $n - K$ to obtain the unbiased estimator, but the classifications would still be essentially the same, exactly so if the π_k's are equal.) Thus the estimated discriminant functions are

$$\widehat{d}_k(\mathbf{x}) = c_k + \mathbf{x}\mathbf{a}_k', \tag{11.31}$$

where

$$\begin{aligned} \mathbf{a}_k &= (\widehat{\boldsymbol{\mu}}_k - \widehat{\boldsymbol{\mu}}_K)\widehat{\boldsymbol{\Sigma}}^{-1} \text{ and} \\ c_k &= -\frac{1}{2}(\widehat{\boldsymbol{\mu}}_k \widehat{\boldsymbol{\Sigma}}^{-1} \widehat{\boldsymbol{\mu}}_k' - \widehat{\boldsymbol{\mu}}_K \widehat{\boldsymbol{\Sigma}}^{-1} \widehat{\boldsymbol{\mu}}_K') + \log(\widehat{\pi}_k / \widehat{\pi}_K). \end{aligned} \tag{11.32}$$

Now we can define the classifier based upon Fisher's linear discrimination function to be

$$\widehat{\mathcal{C}}_{FLD}(\mathbf{x}) = k \;\; if \;\; \widehat{d}_k(\mathbf{x}) > \widehat{d}_l(\mathbf{x}) \;\; for \;\; l \neq k. \tag{11.33}$$

(The hat is there to emphasize the fact that the classifier is estimated from the data.) If $p = 2$, each set $\{\mathbf{x} \mid d_k(\mathbf{x}) = d_l(\mathbf{x})\}$ defines a line in the $\mathbf{x}$-space. These lines divide the space into a number of polygonal regions (some infinite). Each region has the same $\widehat{\mathcal{C}}(\mathbf{x})$. Similarly, for general q, the regions are bounded by hyperplanes. Figure 11.3 illustrates for the iris data when using just the sepal length and width. The solid line is the line for which the discriminant functions for setosas and versicolors are equal. It is basically perfect for these data. The dashed line tries to separate setosas and virginicas. There is one misclassification. The dashed/dotted line tries to separate the versicolors and virginicas. It is not particularly successful. See Section 11.4.1 for a better result using all the variables.

Remark

Fisher's original derivation in Fisher [1936] of the classifier (11.33) did not start with the multivariate normal density. Rather, in the case of two groups, he obtained the $p \times 1$ vector $\mathbf{a}$ that maximized the ratio of the squared difference of means of the variable $\mathbf{X}_i\mathbf{a}'$ for the two groups to the variance:

$$\frac{((\widehat{\boldsymbol{\mu}}_1 - \widehat{\boldsymbol{\mu}}_2)\mathbf{a}')^2}{\mathbf{a}\widehat{\boldsymbol{\Sigma}}\mathbf{a}'}. \tag{11.34}$$

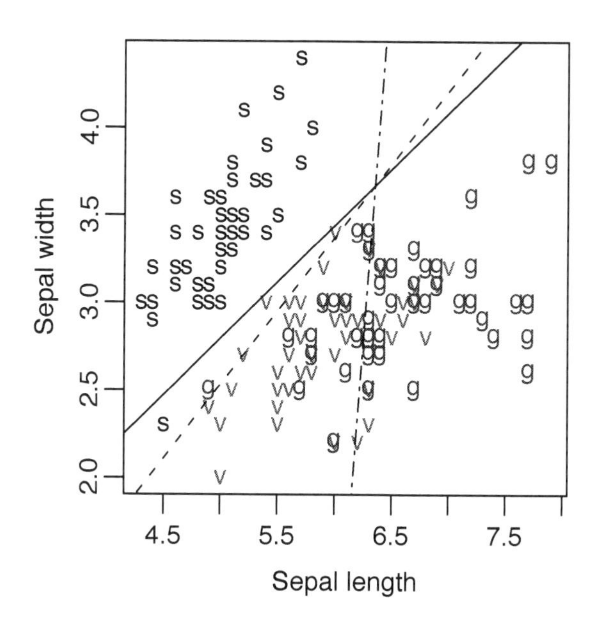

Figure 11.3: Fisher's linear discrimination for the iris data using just sepal length and width. The solid line separates setosa (s) and versicolor (v); the dashed line separates setosa and virginica (g); and the dashed/dotted line separates versicolor and virginica.

The optimal $\mathbf{a}$ is (anything proportional to)

$$\widehat{\mathbf{a}} = (\widehat{\mu}_1 - \widehat{\mu}_2)\widehat{\boldsymbol{\Sigma}}^{-1}, \tag{11.35}$$

which is the $\mathbf{a}_1$ in (11.32). Even though our motivation leading to (11.27) is different than Fisher's, because we end up with his coefficients, we will refer to (11.31) as Fisher's.

11.4 Cross-validation estimate of error

In classification, an error occurs if an observation is misclassified, so one often uses the criterion (11.11) to assess the efficacy of a classifier. Since we are estimating the classifier, we wish to find

$$ClassError = P[\widehat{\mathcal{C}}(\mathbf{X}^{New}) \neq Y^{New}]. \tag{11.36}$$

This probability takes into account the dependence of the estimated classifier on the data, as well as on the randomness of $(\mathbf{X}^{New}, Y^{New})$, assumed to be independent of the data. How does one estimate the error? The obvious approach is to try the classifier on the data, and count the number of misclassifications:

$$ClassError_{Obs} = \frac{1}{n} \sum_{i=1}^{n} I[\widehat{\mathcal{C}}(\mathbf{x}_i) \neq y_i]. \tag{11.37}$$

As in prediction, this error will be an underestimate because we are using the same data to estimate the classifier and test it out. A common approach to a fair estimate is to initially set aside a randomly chosen fraction of the observations (e.g., 10% to 25%) to be **test data**, and use the remaining so-called **training data** to estimate the classifier. Then this estimated classifier is tested on the test data.

Cross-validation is a method that takes the idea one step further, by repeatedly separating the data into test and training data. The "leave-one-out" cross-validation uses single observations as the test data. It starts by setting aside the first observation, $(\mathbf{x}_1, y_1)$, and calculating the classifier using the data $(\mathbf{x}_2, y_2), \ldots, (\mathbf{x}_n, y_n)$. (That is, we find the sample means, covariances, etc., leaving out the first observation.) Call the resulting classifier $\widehat{\mathcal{C}}^{(-1)}$. Then determine whether this classifier classifies the first observation correctly:

$$I[\widehat{\mathcal{C}}^{(-1)}(\mathbf{x}_1) \neq y_1]. \tag{11.38}$$

The $\widehat{\mathcal{C}}^{(-1)}$ and $(\mathbf{X}_1, Y_1)$ are independent, so the function in (11.38) is almost an unbiased estimate of the error in (11.36). The only reason it is not exactly unbiased is that $\widehat{\mathcal{C}}^{(-1)}$ is based on $n-1$ observations, rather than the n for $\widehat{\mathcal{C}}$. This difference should be negligible.

Repeat the process, leaving out each observation in turn, so that $\widehat{\mathcal{C}}^{(-i)}$ is the classifier calculated without observation i. Then the almost unbiased estimate of $ClassError$ in (11.36) is

$$\widehat{ClassError}_{LOOCV} = \frac{1}{n} \sum_{i=1}^{n} I[\widehat{\mathcal{C}}^{(-i)}(\mathbf{x}_i) \neq y_i]. \tag{11.39}$$

If n is large, and calculating the classifier is computationally challenging, then leave-one-out cross-validation can use up too much computer time (especially if one is trying a number of different classifiers). Also, the estimate, though nearly unbiased, might have a high variance. An alternative is to leave out more than one observation each time, e.g., the 10% cross-validation would break the data set into 10 test sets of size $\approx n/10$, and for each set, use the other 90% to estimate the classifier to use on the test set. This approach is much more computationally efficient, and less variable, but does introduce more bias. Kshirsagar [1972] contains a number of other suggestions for estimating the classification error.

11.4.1 Example: Iris data

Turn again to the iris data. Figure 1.3 has the scatter plot matrix. Also see Figures 1.4 and 11.3. In R, the iris data is in the data frame iris. You may have to load the datasets package. The first four columns constitute the $n \times p$ matrix of $\mathbf{x}_i$'s, $n = 150$, $p = 4$. The fifth column has the species, 50 each of setosa, versicolor, and virginica. The basic variables are then

```
x.iris <- as.matrix(iris[,1:4])
y.iris <- rep(1:3,c(50,50,50)) # gets group vector (1,...,1,2,...,2,3,...,3)
```

We will offload many of the calculations to the function lda in Section A.3.1. The following statement calculates the $\mathbf{a}_k$ and c_k in (11.32):

```
ld.iris <- lda(x.iris,y.iris)
```

The $\mathbf{a}_k$ are in the matrix ld.iris$a and the c_k are in the vector ld.iris$c, given below:

k	$\mathbf{a}_k$				c_k
1 (Setosa)	11.325	20.309	−29.793	−39.263	18.428
2 (Versicolor)	3.319	3.456	−7.709	−14.944	32.159
3 (Virginica)	0	0	0	0	0

(11.40)

Note that the final coefficients are zero, because of the way we normalize the functions in (11.28).

To see how well the classifier works on the data, we have to first calculate the $d_k(\mathbf{x}_i)$. The following places these values in an $n \times K$ matrix disc:

```
disc <- x.iris%*%ld.iris$a
disc <- sweep(disc,2,ld.iris$c,'+')
```

The rows corresponding to the first observation from each species are

	k		
i	1	2	3
1	97.703	47.400	0
51	−32.305	9.296	0
101	−120.122	−19.142	0

(11.41)

The classifier (11.33) classifies each observation into the group corresponding to the column with the largest entry. Applied to the observations in (11.41), we have

$$\widehat{C}_{FLD}(\mathbf{x}_1) = 1,\ \widehat{C}_{FLD}(\mathbf{x}_{51}) = 2,\ \widehat{C}_{FLD}(\mathbf{x}_{101}) = 3, \tag{11.42}$$

that is, each of these observations is correctly classified into its group. To find the $\widehat{C}_{FLD}$'s for all the observations, use

```
imax <- function(z) ((1:length(z))[z==max(z)])[1]
yhat <- apply(disc,1,imax)
```

where imax is a little function to give the index of the largest value in a vector. To see how close the predictions are to the observed, use the table command:

```
table(yhat,y.iris)
```

which yields

	y		
$\widehat{y}$	1	2	3
1	50	0	0
2	0	48	1
3	0	2	49

(11.43)

Thus there were 3 observations misclassified — two versicolors were classified as virginica, and one virginica was classified as versicolor. Not too bad. The observed misclassification rate is

$$ClassError_{Obs} = \frac{\#\{\widehat{C}_{FLD}(\mathbf{x}_i) \neq y_i\}}{n} = \frac{3}{150} = 0.02. \tag{11.44}$$

As noted above in Section 11.4, this value is likely to be an optimistic estimate (an underestimate) of $ClassError$ in (11.36), because it uses the same data to find the classifier and to test it out. We will find the leave-one-out cross-validation estimate (11.39) using the code below, where we set varin=1:4 to specify using all four variables.

```
yhat.cv <- NULL
n <- nrow(x.iris)
for(i in 1:n) {
    dcv <- lda(x.iris[-i,varin],y.iris[-i])
    dxi <- x.iris[i,varin]%*%dcv$a+dcv$c
    yhat.cv <- c(yhat.cv,imax(dxi))
}
sum(yhat.cv!=y.iris)/n
```

Here, for each i, we calculate the classifier without observation i, then apply it to that left-out observation i, the predictions placed in the vector yhat.cv. We then count how many observations were misclassified. In this case, $\widehat{ClassError}_{LOOCV} = 0.02$, just the same as the observed classification error. In fact, the same three observations were misclassified.

Subset selection

The above classifications used all four iris variables. We now see if we can obtain equally good or better results using a subset of the variables. We use the same loop as above, setting varin to the vector of indices for the variables to be included. For example, varin = c(1,3) will use just variables 1 and 3, sepal length and petal length.

Below is a table giving the observed error and leave-one-out cross-validation error (in percentage) for 15 models, depending on which variables are included in the classification.

	Classification errors	
Variables	Observed	Cross-validation
1	25.3	25.3
2	44.7	48.0
3	5.3	6.7
4	4.0	4.0
1, 2	20.0	20.7
1, 3	3.3	4.0
1, 4	4.0	4.7
2, 3	4.7	4.7
2, 4	3.3	4.0
3, 4	4.0	4.0
1, 2, 3	3.3	4.0
1, 2, 4	4.0	5.3
1, 3, 4	2.7	2.7
2, 3, 4	2.0	4.0
1, 2, 3, 4	2.0	2.0

(11.45)

Note that the cross-validation error estimates are either the same, or a bit larger, than the observed error rates. The best classifier uses all 4 variables, with an estimated 2% error. Note, though, that variable 4 (petal width) alone has only a 4% error rate. Also, adding variable 1 to variable 4 actual worsens the prediction a little, showing that adding the extra variation is not worth it. Looking at just the observed error, the prediction stays the same.

Figure 11.4 shows the classifications using just petal widths. Because the sample sizes are equal, and the variances are assumed equal, the separating lines between

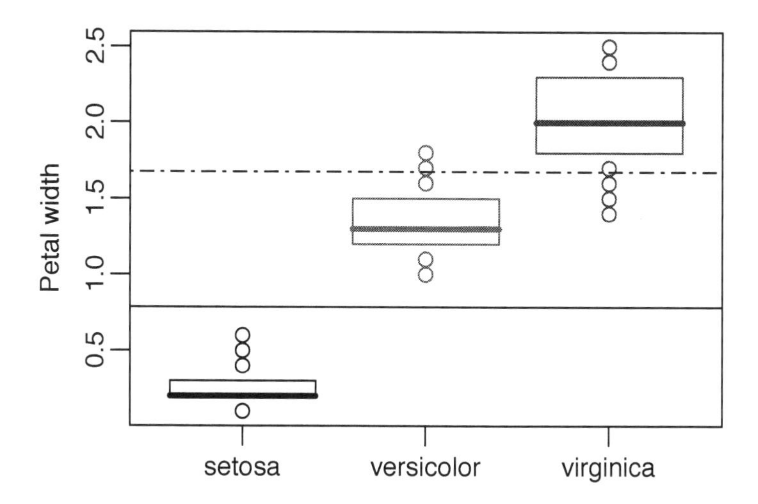

Figure 11.4: Boxplots of the petal widths for the three species of iris. The solid line separates the setosas from the versicolors, and the dashed line separates the versicolors from the virginicas.

two species are just the average of their means. We did not plot the line for setosa versus virginica. There are six misclassifications, two versicolors and four virginicas. (Two of the latter had the same petal width, 1.5.)

11.5 Fisher's quadratic discrimination

When the equality of the covariance matrices is not tenable, we can use a slightly more complicated procedure. Here the conditional probabilities are proportional to

$$\begin{aligned}\pi_k f_k(\mathbf{x} \mid \boldsymbol{\mu}_k, \boldsymbol{\Sigma}_k) &= c\,\pi_k \frac{1}{|\boldsymbol{\Sigma}_k|^{1/2}}\, e^{-\frac{1}{2}(\mathbf{x}-\boldsymbol{\mu}_k)\boldsymbol{\Sigma}_k^{-1}(\mathbf{x}-\boldsymbol{\mu}_k)'} \\ &= c\, e^{-\frac{1}{2}(\mathbf{x}-\boldsymbol{\mu}_k)\boldsymbol{\Sigma}_k^{-1}(\mathbf{x}-\boldsymbol{\mu}_k)' - \frac{1}{2}\log(|\boldsymbol{\Sigma}_k|) + \log(\pi_k)}. \end{aligned} \tag{11.46}$$

Then the discriminant functions can be taken to be the terms in the exponents (times 2, for convenience), or their estimates:

$$\widehat{d}_k^{\,Q}(\mathbf{x}) = -(\mathbf{x} - \widehat{\boldsymbol{\mu}}_k)\widehat{\boldsymbol{\Sigma}}_k^{-1}(\mathbf{x} - \widehat{\boldsymbol{\mu}}_k)' + c_k, \tag{11.47}$$

where

$$c_k = -\log(|\widehat{\boldsymbol{\Sigma}}_k|) + 2\log(N_k/n), \tag{11.48}$$

and $\widehat{\boldsymbol{\Sigma}}_k$ is the sample covariance matrix from the kth group. Now the boundaries between regions are quadratic rather than linear, hence **Fisher's quadratic discrimination function** is defined to be

$$\widehat{\mathcal{C}}_{FQD}(\mathbf{x}) = k \;\; if \;\; \widehat{d}_k^{\,Q}(\mathbf{x}) > \widehat{d}_l^{\,Q}(\mathbf{x}) \;\; for \;\; l \neq k. \tag{11.49}$$

11.5.1 Example: Iris data, continued

We consider the iris data again, but as in Section 11.5 we estimate three separate covariance matrices. Sections A.3.2 and A.3.3 contain the functions qda and predict_qda for calculating the quadratic discriminant functions (11.25) and finding the predictions. Apply these to the iris data as follows:

```
qd.iris <- qda(x.iris,y.iris)
yhat.qd <- NULL
for (i in 1:n) {
        yhat.qd <- c(yhat.qd,imax(predict_qda(qd.iris,x.iris[i,])))
}
table(yhat.qd,y.iris)
```

The resulting table is (11.43), the same as for linear discrimination. The leave-one-out cross-validation estimate of classification error is $4/150 = 0.0267$, which is slightly worse than that for linear discrimination. It does not appear that the extra complication of having three covariance matrices improves the classification rate.

Hypothesis testing, AIC, or BIC can also help decide between the model with equal covariance matrices and the model with three separate covariance matrices. Because we have already calculated the estimates, it is quite easy to proceed. The two models are then

$$\begin{aligned} M_{Same} &\Rightarrow \mathbf{\Sigma}_1 = \mathbf{\Sigma}_2 = \mathbf{\Sigma}_3 \equiv \mathbf{\Sigma}; \\ M_{Diff} &\Rightarrow (\mathbf{\Sigma}_1, \mathbf{\Sigma}_2, \mathbf{\Sigma}_3) \text{ unrestricted.} \end{aligned} \tag{11.50}$$

Both models have the same unrestricted means, and we can consider the π_k's fixed, so we can work with just the sample covariance matrices, as in Section 10.1. Let $\mathbf{U}_1, \mathbf{U}_2$, and $\mathbf{U}_3$ be the sum of squares and cross-products matrices (1.15) for the three species, and $\mathbf{U} = \mathbf{U}_1 + \mathbf{U}_2 + \mathbf{U}_3$ be the pooled version. The degrees of freedom for each species is $\nu_k = 50 - 1 = 49$. Thus from (10.10) and (10.11), we can find the deviances (9.53) to be

$$\begin{aligned} \text{deviance}(M_{Same}) &= (\nu_1 + \nu_2 + \nu_3) \log(|\mathbf{U}/(\nu_1 + \nu_2 + \nu_3)|) \\ &= -1463.905, \\ \text{deviance}(M_{Diff}) &= \nu_1 \log(|\mathbf{U}_1/\nu_1|) + \nu_2 \log(|\mathbf{U}_2/\nu_2|) + \nu_3 \log(|\mathbf{U}_3/\nu_3|) \\ &= -1610.568 \end{aligned} \tag{11.51}$$

Each covariance matrix has $q(q+1)/2 = 10$ parameters, hence

$$d_{Same} = 10 \text{ and } d_{Diff} = 30. \tag{11.52}$$

To test the null hypothesis M_{Same} versus the alternative M_{Diff}, as in (9.55),

$$\begin{aligned} 2\log(LR) &= \text{deviance}(M_{Same}) - \text{deviance}(M_{Diff}) \\ &= -1463.905 + 1610.568 \\ &= 146.663, \end{aligned} \tag{11.53}$$

on $d_{Same} - d_{Diff} = 20$ degrees of freedom. The statistic is highly significant; we reject emphatically the hypothesis that the covariance matrices are the same. The AIC's

(9.56) and BIC's (9.57) are found directly from (11.53):

	AIC	BIC
M_{Same}	−1443.90	−1414.00
M_{Diff}	−1550.57	−1460.86

(11.54)

They, too, favor the separate covariance model. Cross-validation above suggests that the equal-covariance model is slightly better. Thus there seems to be a conflict between AIC/BIC and cross-validation. The conflict can be explained by noting that AIC/BIC are trying to model the distribution of the $(\mathbf{x}_i, y_i)$'s, while cross-validation does not really care about the distribution of the $\mathbf{x}_i$'s, except to the extent it helps in predicting the y_i's.

11.6 Modifications to Fisher's discrimination

The key component in both the quadratic and linear discriminant functions is the quadratic form,

$$q(\mathbf{x}; \boldsymbol{\mu}_k, \boldsymbol{\Sigma}_K) = -\frac{1}{2}\,(\mathbf{x} - \boldsymbol{\mu}_k)\boldsymbol{\Sigma}_k^{-1}(\mathbf{x} - \boldsymbol{\mu}_k)', \tag{11.55}$$

where in the case the $\boldsymbol{\Sigma}_k$'s are equal, the "$\mathbf{x}\boldsymbol{\Sigma}^{-1}\mathbf{x}'$" part is ignored. Without the $-1/2$, (11.55) is a measure of distance (called the **Mahalanobis distance**) between an $\mathbf{x}$ and the mean of the kth group, so that it makes sense to classify an observation into the group to which it is closest (modulo an additive constant). The idea is plausible whether the data are normal or not, and whether the middle component is a general $\boldsymbol{\Sigma}_k$ or not. E.g., when taking the $\boldsymbol{\Sigma}$'s equal, we could take

$$\begin{aligned} \boldsymbol{\Sigma} = \mathbf{I}_p &\Longrightarrow q(\mathbf{x}; \boldsymbol{\mu}_k, \mathbf{I}_p) = -\frac{1}{2}\,\|\mathbf{x} - \boldsymbol{\mu}_k\|^2, \text{ or} \\ \boldsymbol{\Sigma} = \boldsymbol{\Delta}, \text{ diagonal} &\Longrightarrow q(\mathbf{x}; \boldsymbol{\mu}_k, \boldsymbol{\Delta}) = -\frac{1}{2}\,\sum (x_i - \mu_{ki})^2/\delta_{ii}. \end{aligned} \tag{11.56}$$

The first case is regular Euclidean distance. In the second case, one would need to estimate the δ_{ii}'s by the pooled sample variances. These alternatives may be better when there are not many observations per group, and a fairly large number of variables p, so that estimating a full $\boldsymbol{\Sigma}$ introduces enough extra random error into the classification to reduce its effectiveness.

Another modification is to use functions of the individual variables. E.g., in the iris data, one could generate quadratic boundaries by using the variables

$$\textit{Sepal Length}, (\textit{Sepal Length})^2, \textit{Petal Length}, (\textit{Petal Length})^2 \tag{11.57}$$

in the $\mathbf{x}$. The resulting set of variables certainly would not be multivariate normal, but the classification based on them may still be reasonable. See the next section for another method of incorporating such functions.

11.7 Conditioning on X: Logistic regression

Based on the conditional densities of $\mathbf{X}$ given $Y = k$ and priors π_k, Lemma 11.1 shows that the Bayes classifier in (11.13) is optimal. In Section 11.3, we saw that if the conditional distributions of the $\mathbf{X}$ are multivariate normal, with the same covariance matrix for each group, then the classifier devolved to a linear one (11.31) in $\mathbf{x}$. The linearity is not specific to the normal, but is a consequence of the normal being an **exponential family density**, which means the density has the form

$$f(\mathbf{x} \mid \boldsymbol{\theta}) = a(\mathbf{x}) e^{\mathbf{t}(\mathbf{x})\boldsymbol{\theta}' - \psi(\boldsymbol{\theta})} \tag{11.58}$$

for some $1 \times m$ parameter $\boldsymbol{\theta}$, $1 \times m$ function $\mathbf{t}(\mathbf{x})$ (the sufficient statistic), and function $a(\mathbf{x})$, where $\psi(\boldsymbol{\theta})$ is the normalizing constant.

Suppose that the conditional density of $\mathbf{X}$ given $Y = k$ is $f(\mathbf{x} \mid \boldsymbol{\theta}_k)$, that is, each group has the same form of the density, but a different parameter value. Then the analog to equations (11.27) and (11.28) yields discriminant functions like those in (11.31),

$$d_k(\mathbf{x}) = \gamma_k + \mathbf{t}(\mathbf{x})\boldsymbol{\alpha}_k', \tag{11.59}$$

a linear function of $\mathbf{t}(\mathbf{x})$, where $\boldsymbol{\alpha}_k = \boldsymbol{\theta}_k - \boldsymbol{\theta}_K$, and γ_k is a constant depending on the parameters. (Note that $d_K(\mathbf{x}) = 0$.) To implement the classifier, we need to estimate the parameters $\boldsymbol{\theta}_k$ and π_k, usually by finding the maximum likelihood estimates. (Note that Fisher's quadratic discrimination in Section 11.5 also has discriminant functions (11.47) of the form (11.59), where the $\mathbf{t}$ is a function of the $\mathbf{x}$ and its square, $\mathbf{x}\mathbf{x}'$.) In such models the conditional distribution of Y given $\mathbf{X}$ is given by

$$P[Y = k \mid \mathbf{X} = \mathbf{x}] = \frac{e^{d_k(\mathbf{x})}}{e^{d_1(\mathbf{x})} + \cdots + e^{d_{K-1}(\mathbf{x})} + 1} \tag{11.60}$$

for the d_k's in (11.59). This conditional model is called the **logistic regression model**. Then an alternative method for estimating the γ_k's and $\boldsymbol{\alpha}_k$'s is to find the values that maximize the *conditional* likelihood,

$$L((\gamma_1, \boldsymbol{\alpha}_1), \ldots, (\gamma_{K-1}, \boldsymbol{\alpha}_{K-1}); (\mathbf{x}_1, y_1), \ldots, (\mathbf{x}_n, y_n)) = \prod_{i=1}^{n} P[Y = y_i \mid \mathbf{X}_i = \mathbf{x}_i]. \tag{11.61}$$

(We know that $\boldsymbol{\alpha}_K = \mathbf{0}$ and $\gamma_K = 0$.) There is no closed-form solution for solving the likelihood equations, so one must use some kind of numerical procedure like Newton-Raphson. Note that this approach estimates the slopes and intercepts of the discriminant functions directly, rather than (in the normal case) estimating the means and variances, and the π_k's, then finding the slopes and intercepts as functions of those estimates.

Whether using the exponential family model unconditionally or the logistic model conditionally, it is important to realize that both lead to the exact same theoretical classifier. The difference is in the way the slopes and intercepts are estimated in (11.59). One question is then which gives the better estimates. Note that the joint distribution of the $(\mathbf{X}, Y)$ is the product of the conditional of Y given $\mathbf{X}$ in (11.60) and the marginal of $\mathbf{X}$ in (11.5), so that for the entire data set,

$$\begin{aligned}\prod_{i=1}^{n} f(y_i, \mathbf{x}_i \mid \boldsymbol{\theta}_k)\pi_{y_i} = & \left[\prod_{i=1}^{n} P[Y = y_i \mid \mathbf{X}_i = \mathbf{x}_i, \boldsymbol{\theta}_k]\right] \\ & \times \left[\prod_{i=1}^{n} (\pi_1 f(\mathbf{x}_i \mid \boldsymbol{\theta}_1) + \cdots + \pi_K f(\mathbf{x}_i \mid \boldsymbol{\theta}_K))\right].\end{aligned} \tag{11.62}$$

Thus using just the logistic likelihood (11.61), which is the first term on the right-hand side in (11.62), in place of the complete likelihood on the left, leaves out the information about the parameters that is contained in the mixture likelihood (the second term on the right). As we will see in Chapter 12, there is information in the mixture likelihood. One would then expect that the complete likelihood gives better estimates in the sense of asymptotic efficiency of the estimates. It is not clear whether that property always translates to yielding better classification schemes, but maybe.

On the other hand, the conditional logistic model is more general in that it yields valid estimates even when the exponential family assumption does not hold. We can entertain the assumption that the conditional distributions in (11.60) hold for any statistics $\mathbf{t}(\mathbf{x})$ we wish to use, without trying to model the marginal distributions of the $\mathbf{X}$'s at all. This realization opens up a vast array of models to use, that is, we can contemplate any functions $\mathbf{t}$ we wish.

In what follows, we restrict ourselves to having $K = 2$ groups, and renumber the groups $\{0, 1\}$, so that Y is conditionally Bernoulli:

$$Y_i \mid \mathbf{X}_i = \mathbf{x}_i \sim \text{Bernoulli}(\rho(\mathbf{x}_i)), \tag{11.63}$$

where

$$\rho(\mathbf{x}) = P[Y = 1 \mid \mathbf{X} = \mathbf{x}]. \tag{11.64}$$

The modeling assumption from (11.60) can be translated to the **logit** (log odds) of ρ, $\text{logit}(\rho) = \log(\rho/(1-\rho))$. Then

$$\text{logit}(\rho(\mathbf{x})) = \text{logit}(\rho(\mathbf{x} \mid \gamma, \boldsymbol{\alpha})) = \gamma + \mathbf{x}\boldsymbol{\alpha}'. \tag{11.65}$$

(We have dropped the $\mathbf{t}$ from the notation. You can always define $\mathbf{x}$ to be whatever functions of the data you wish.) The form (11.65) exhibits the reason for calling the model "logistic regression." Letting

$$\text{logit}(\boldsymbol{\rho}) = \begin{pmatrix} \text{logit}(\rho(\mathbf{x}_1 \mid \gamma, \boldsymbol{\alpha})) \\ \text{logit}(\rho(\mathbf{x}_2 \mid \gamma, \boldsymbol{\alpha})) \\ \vdots \\ \text{logit}(\rho(\mathbf{x}_n \mid \gamma, \boldsymbol{\alpha})) \end{pmatrix}, \tag{11.66}$$

we can set up the model to look like the regular linear model,

$$\text{logit}(\boldsymbol{\rho}) = \begin{pmatrix} 1 & \mathbf{x}_1 \\ 1 & \mathbf{x}_2 \\ \vdots & \vdots \\ 1 & \mathbf{x}_n \end{pmatrix} \begin{pmatrix} \gamma \\ \boldsymbol{\alpha}' \end{pmatrix} = \mathbf{X}\boldsymbol{\beta}. \tag{11.67}$$

We turn to examples.

11.7.1 Example: Iris data

Consider the iris data, restricting to classifying the virginicas versus versicolors. The next table has estimates of the linear discrimination functions' intercepts and slopes using the multivariate normal with equal covariances, and the logistic regression model:

	Intercept	Sepal Length	Sepal Width	Petal Length	Petal Width
Normal	17.00	3.63	5.69	−7.11	−12.64
Logistic	42.64	2.47	6.68	−9.43	−18.29

(11.68)

The two estimates are similar, the logistic giving more weight to the petal widths, and having a large intercept. It is interesting that the normal-based estimates have an observed error of 3/150, while the logistic has 2/150.

11.7.2 Example: Spam

The Hewlett-Packard spam data was introduced in Exercise 1.9.16. The $n = 4601$ observations are emails to George Forman, at Hewlett-Packard labs. The Y classifies each email as spam ($Y = 1$) or not spam, $Y = 0$. There are $q = 57$ explanatory variables based on the contents of the email. Most of the explanatory variables are frequency variables with many zeroes, hence are not at all normal, so Fisher's discrimination may not be appropriate. One could try to model the variables using Poissons or multinomials. Fortunately, if we use the logistic model, we do not need to model the explanatory variables at all, but only decide on the x_j's to use in modeling the logit in (11.67).

The $\rho(\mathbf{x} \mid \gamma, \alpha)$ is the probability an email with message statistics $\mathbf{x}$ is spam. We start by throwing in all 57 explanatory variables linearly, so that in (11.67), the design matrix contains all the explanatory variables, plus the $\mathbf{1}_n$ vector. This fit produces an observed misclassification error rate of 6.9%.

A number of the coefficients are not significant, hence it makes sense to try subset logistic regression, that is, find a good subset of explanatory variables to use. It is computationally much more time consuming to fit a succession of logistic regression models than regular linear regression models, so that it is often infeasible to do an all-subsets exploration. **Stepwise** procedures can help, though are not guaranteed to find the best model. Start with a given criterion, e.g., AIC, and a given subset of explanatory variables, e.g., the full set or the empty set. At each step, one has an "old" model with some subset of the explanatory variables, and tries every possible model that either adds one variable or removes one variable from that subset. Then the "new" model is the one with the lowest AIC. The next step uses that new model as the old, and adds and removes one variable from that. This process continues until at some step the new model and old model are the same.

The table in (11.69) shows the results when using AIC and BIC. (The R code is below.) The BIC has a stronger penalty, hence ends up with a smaller model, 30 variables (including the $\mathbf{1}_n$) versus 44 for the AIC. For those two "best" models as well as the full model, the table also contains the 46-fold cross-validation estimate of the error, in percent. That is, we randomly cut the data set into 46 blocks of 100 observations, then predict each block of 100 from the remaining 4501. For the latter two models, cross-validation involves redoing the entire stepwise procedure for each reduced data set. A computationally simpler, but maybe not as defensible, approach would be to use cross-validation on the actual models chosen when applying stepwise to the full data set. Here, we found the estimated errors for the best AIC and BIC models were 7.07% and 7.57%, respectively, approximately the same as for the more complicated procedure.

	p	Deviance	AIC	BIC	Obs. error	CV error	CV se
Full	58	1815.8	1931.8	2305.0	6.87	7.35	0.34
Best AIC	44	1824.9	1912.9	2196.0	6.78	7.15	0.35
Best BIC	30	1901.7	1961.7	2154.7	7.28	7.59	0.37

(11.69)

The table shows that all three models have essentially the same cross-validation error, with the best AIC's model being best. The standard errors are the standard deviations of the 46 errors divided by $\sqrt{46}$, so give an idea of how variable the error estimates are. The differences between the three errors are not large relative to these standard errors, so one could arguably take either the best AIC or best BIC model.

The best AIC model has $p = 44$ parameters, one of which is the intercept. The table (11.70) categorizes the 41 frequency variables (word or symbol) in this model, according to the signs of their coefficients. The ones with positive coefficients tend to indicate spam, while the others indicate non-spam. Note that the latter tend to be words particular to someone named "George" who works at a lab at HP, while the spam indicators have words like "credit", "free", "money", and exciting symbols like "!" and "$". Also with positive coefficients are the variables that count the number of capital letters, and the length of the longest run of capitals, in the email.

Positive	Negative
3d our over remove internet order mail addresses free business you credit your font 000 money 650 technology ! $ #	make address will hp hpl george lab data 85 parts pm cs meeting original project re edu table conference ;

(11.70)

Computational details

In R, logistic regression models with two categories can be fit using the generalized linear model function, glm. The spam data is in the data matrix Spam. The indicator variable, Y_i, for spam is called spam. We first must change the data matrix into a data frame for glm: Spamdf <− data.frame(Spam). The full logistic regression model is fit using

```
spamfull <- glm(spam ~.,data=Spamdf,family=binomial)
```

The "spam ~." tells the program that the spam variable is the Y, and the dot means use all the variables except for spam in the $\mathbf{X}$. The "family = binomial" tells the program to fit logistic regression. The summary command, summary(spamfull), will print out all the coefficients, which I will not reproduce here, and some other statistics, including

```
    Null deviance: 6170.2 on 4600 degrees of freedom
Residual deviance: 1815.8 on 4543 degrees of freedom
AIC: 1931.8
```

The "residual deviance" is the regular deviance in (9.53). The full model uses 58 variables, hence

$$\text{AIC} = \text{deviance} + 2p = 1815.8 + 2 \times 58 = 1931.8, \tag{11.71}$$

which checks. The BIC is found by substituting log(4601) for the 2.

We can find the predicted classifications from this fit using the function predict, which returns the estimated linear $\mathbf{X}\widehat{\boldsymbol{\beta}}$ from (11.67) for the fitted model. The $\widehat{Y}_i$'s are then 1 or 0 as the $\rho(\mathbf{x}^{(i)} \mid \widehat{c}, \widehat{\mathbf{a}})$ is greater than or less than $1/2$, i. e., as the $\mathbf{X}\widehat{\boldsymbol{\beta}}$ is greater or less than 0. Thus to find the predictions and overall error rate, do

```
yhat <- ifelse(predict(spamfull)>0,1,0)
sum(yhat!=Spamdf[,'spam'])/4601
```

We find the observed classification error to be 6.87%.

Cross-validation

We will use 46-fold cross-validation to estimate the classification error. We randomly divide the 4601 observations into 46 groups of 100, leaving one observation who doesn't get to play. First, permute the indices from 1 to n:

```
o <- sample(1:4601)
```

Then the first hundred are the indices for the observations in the first leave-out-block, the second hundred in the second leave-out-block, etc. The loop is next, where the err collects the number of classification errors in each block of 100.

```
err <- NULL
for(i in 1:46) {
    oi <- o[(1:100)+(i-1)*100]
    yfiti <- glm(spam ~., family = binomial,data = Spamdf,subset=(1:4601)[-oi])
    dhati <- predict(yfiti,newdata=Spamdf[oi,])
    yhati <- ifelse(dhati>0,1,0)
    err <- c(err,sum(yhati!=Spamdf[oi,'spam']))
}
```

In the loop for cross-validation, the oi is the vector of indices being left out. We then fit the model without those by using the keyword subset=(1:4601)[-oi], which indicates using all indices except those in oi. The dhati is then the vector of discriminant functions evaluated for the left out observations (the newdata). The mean of err is the estimated error, which for us is 7.35%. See the entry in table in (11.69).

Stepwise

The command to use for stepwise regression is step. To have the program search through the entire set of variables, use one of the two statements

```
spamstepa <- step(spamfull,scope=list(upper= ~.,lower = ~1))
spamstepb <- step(spamfull,scope=list(upper= ~.,lower = ~1),k=log(4601))
```

The first statement searches on AIC, the second on BIC. The first argument in the step function is the return value of glm for the full data. The upper and lower inputs refer to the formulas of the largest and smallest models one wishes to entertain. In our case, we wish the smallest model to have just the $\mathbf{1}_n$ vector (indicated by the "~1"), and the largest model to contain all the vectors (indicated by the "~ .").

These routines may take a while, and will spit out a lot of output. The end result is the best model found using the given criterion. (If using the BIC version, while calculating the steps, the program will output the BIC values, though calling them

"AIC." The summary output will give the AIC, calling it "AIC." Thus if you use just the summary output, you must calculate the BIC for yourself.)

To find the cross-validation estimate of classification error, we need to insert the stepwise procedure after fitting the model leaving out the observations, then predict those left out using the result of the stepwise procedure. So for the best BIC model, use the following:

```
errb <- NULL
for(i in 1:46) {
    oi <- o[(1:100)+(i-1)*100]
    yfiti <- glm(spam ~., family = binomial, data = Spamdf,subset=(1:4601)[-oi])
    stepi <- step(yfiti,scope=list(upper= ~.,lower = ~1),k=log(4501))
    dhati <- predict(stepi,newdata=Spamdf[oi,])
    yhati <- ifelse(dhati>0,1,0)
    errb <- c(errb,sum(yhati!=Spamdf[oi,'spam']))
}
```

The estimate for the best AIC model uses the same statements but with k = 2 in the step function. This routine will take a while, because each stepwise procedure is time consuming. Thus one might consider using cross-validation on the model chosen using the BIC (or AIC) criterion for the full data.

The neural networks R package nnet [Venables and Ripley, 2002] can be used to fit logistic regression models for $K > 2$.

11.8 Trees

The presentation here will also use just $K = 2$ groups, labeled 0 and 1, but can be extended to any number of groups. In the logistic regression model (11.60), we modeled $P[Y = 1 \mid \mathbf{X} = \mathbf{x}] \equiv \rho(\mathbf{x})$ using a particular parametric form. In this section we use a simpler, nonparametric form, where $\rho(\mathbf{x})$ is constant over rectangular regions of the $\mathbf{X}$-space.

To illustrate, we will use the South African heart disease data from Rossouw et al. [1983], which was used in Exercise 10.5.21. The Y is congestive heart disease (chd), where 1 indicates the person has the disease, 0 he does not. Explanatory variables include various health measures. Hastie et al. [2009] apply logistic regression to these data. Here we use trees. Figure 11.5 plots the chd variable for the age and adiposity (fat percentage) variables. Consider the vertical line. It splits the data according to whether age is less than 31.5 years. The splitting point 31.5 was chosen so that the proportions of heart disease in each region would be very different. Here, $10/117 =$ 8.85% of the men under age 31.5 had heart disease, while $150/345 = 43.48\%$ of those above 31.5 had the disease.

The next step is to consider just the men over age 31.5, and split them on the adiposity variable. Taking the value 25, we have that $41/106 = 38.68\%$ of the men over age 31.5 but with adiposity under 25 have heart disease; $109/239 = 45.61\%$ of the men over age 31.5 and with adiposity over 25 have the disease. We could further split the younger men on adiposity, or split them on age again. Subsequent steps split the resulting rectangles, each time with either a vertical or horizontal segment.

There are also the other variables we could split on. It becomes easier to represent the splits using a tree diagram, as in Figure 11.6. There we have made several splits, at the *nodes*. Each node needs a variable and a cutoff point, such that people for

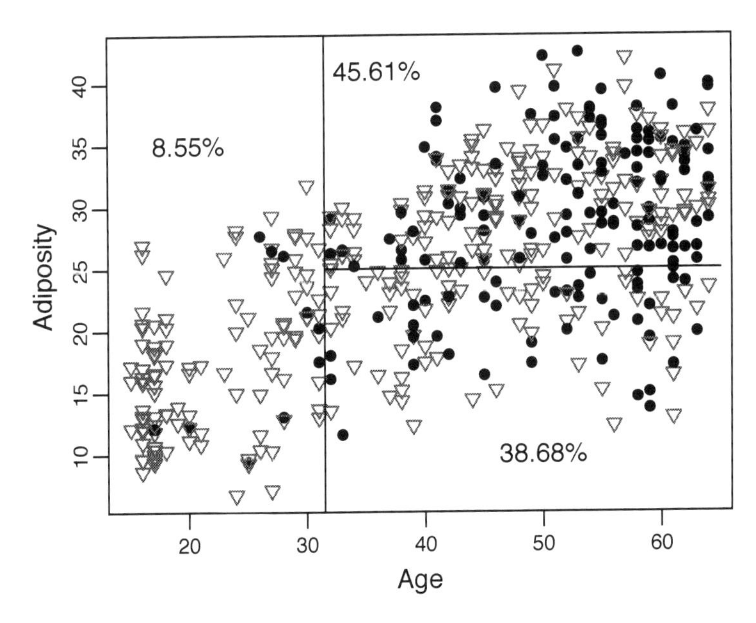

Figure 11.5: Splitting on age and adiposity. The open triangles indicate no heart disease, the solid discs indicate heart disease. The percentages are the percentages of men with heart disease in each region of the plot.

which the variable is less than the cutoff are placed in the left branch, and the others go to the right. The ends of the branches are *terminal nodes* or *leaves*. This plot has 15 leaves. At each leaf, there are a certain number of observations. The plot shows the proportion of 0's (the top number) and 1's (the bottom number) at each leaf.

For classification, we place a 0 or 1 at each leaf, depending on whether the proportion of 1's is less than or greater than 1/2. Figure 11.7 shows the results. Note that for some splits, both leaves have the same classification, because although their proportions of 1's are quite different, they are both on the same side of 1/2. For classification purposes, we can *snip* some of the branches off. Further analysis (Section 11.8.1) leads us to the even simpler tree in Figure 11.8. The tree is very easy to interpret, hence popular among people (e.g., doctors) who need to use them. The tree also makes sense, showing age, type A personality, tobacco use, and family history are important factors in predicting heart disease among these men. The trees also are flexible, incorporating continuous and categorical variables, avoiding having to consider transformations, and automatically incorporating interactions. E.g., the type A variable shows up only for people between the ages of 31.5 and 50.5, and family history and tobacco use show up only for people over 50.5.

Though simple to interpret, it is easy to imagine that finding the "best" tree is a rather daunting prospect, as there is close to an infinite number of possible trees in any large data set (at each stage one can split any variable at any of a number of points), and searching over all the possibilities is a very discrete (versus continuous) process. In the next section, we present a popular, and simple, algorithm to find a

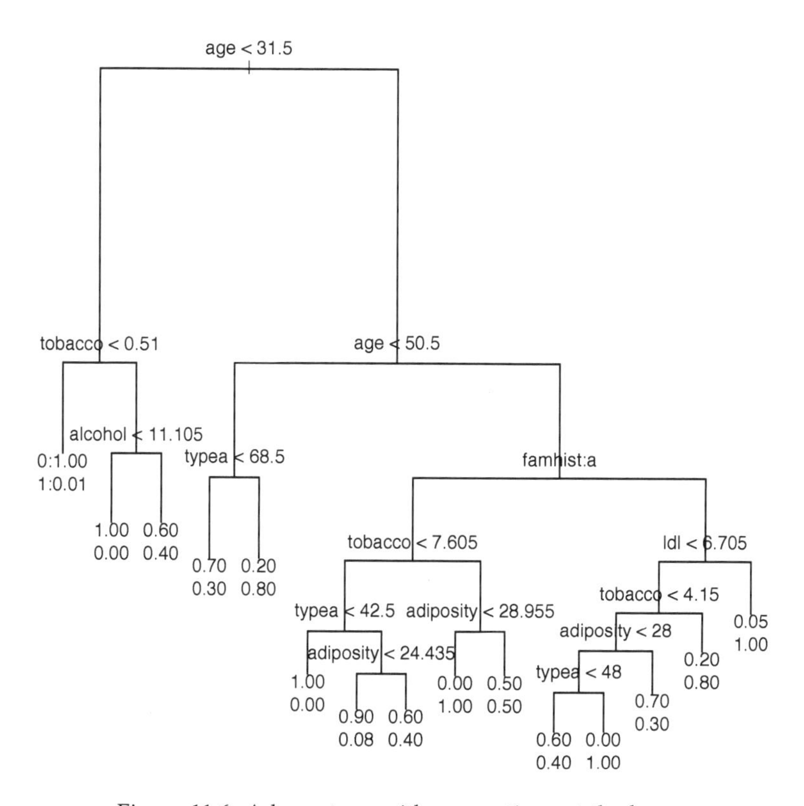

Figure 11.6: A large tree, with proportions at the leaves.

good tree.

11.8.1 CART

Two popular commercial products for fitting trees are Categorization and Regression Trees, CART®, by Breiman et al. [1984], and C5.0, an update of C4.5 by Quinlan [1993]. We will take the CART approach. Both approaches have R versions. It seems that CART would appeal more to statisticians, and C5.0 to data-miners, but I do not think the results of the two methods would differ much.

We first need an objective function to measure the fit of a tree to the data. We will use deviance, although other measures such as the observed misclassification rate are certainly reasonable. For a tree $\mathcal{T}$ with L leaves, each observation is placed in one of the leaves. If observation y_i is placed in leaf l, then that observation's $\rho(\mathbf{x}_i)$ is given by the parameter for leaf l, say p_l. The likelihood for that Bernoulli observation is

$$p_l^{y_i}(1 - p_l)^{1-y_i}. \tag{11.72}$$

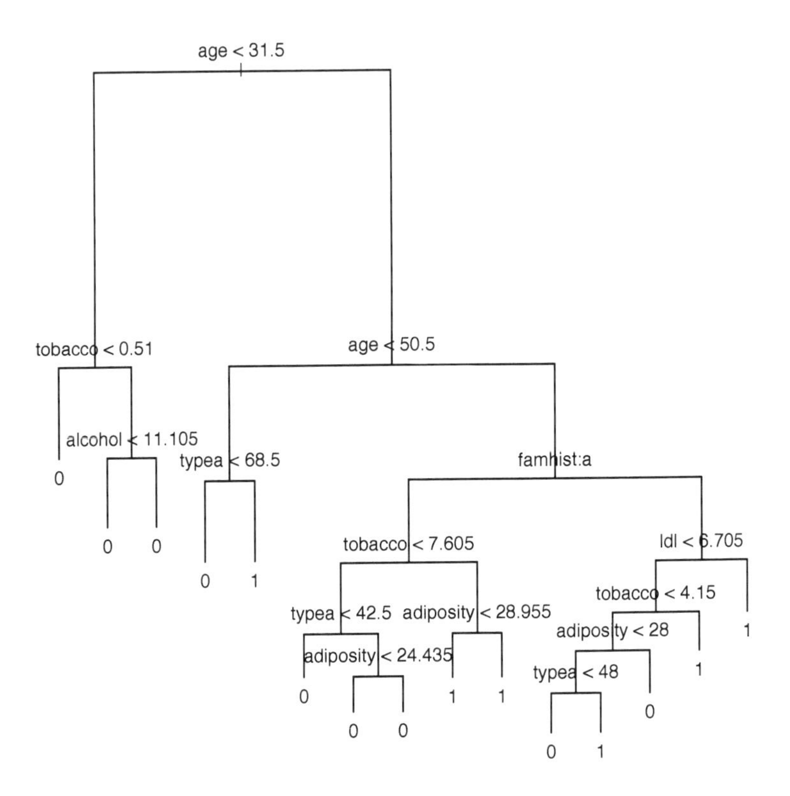

Figure 11.7: A large tree, with classifications at the leaves.

Figure 11.8: A smaller tree, chosen using BIC.

Assuming the observations are independent, at leaf l there is a sample of iid Bernoulli random variables with parameter p_l, hence the overall likelihood of the sample is

$$L(p_1, \ldots, p_L \mid y_1, \ldots, y_n) = \prod_{l=1}^{L} p_l^{w_l} (1 - p_l)^{n_l - w_l}, \tag{11.73}$$

where

$$n_l = \#\{i \text{ at leaf } l\},\ w_l = \#\{y_i = 1 \text{ at leaf } l\}. \tag{11.74}$$

This likelihood is maximized over the p_l's by taking $\widehat{p}_l = w_l / n_l$. Then the deviance (9.53) for this tree is

$$\text{deviance}(\mathcal{T}) = -2 \sum_{l=1}^{L} (w_l \log(\widehat{p}_l) + (n_l - w_l) \log(1 - \widehat{p}_l)). \tag{11.75}$$

The CART method has two main steps: grow the tree, then prune the tree. The tree is grown in a stepwise, greedy fashion, at each stage trying to find the next split that maximally reduces the objective function. We start by finding the single split (variable plus cutoff point) that minimizes the deviance among all such splits. Then the observations at each resulting leaf are optimally split, again finding the variable/cutoff split with the lowest deviance. The process continues until the leaves have just a few observations, e.g., stopping when any split would result in a leaf with fewer than five observations.

To grow the tree for the South African heart disease data in R, we need to install the package called tree [Ripley, 2015]. A good explanation of it can be found in Venables and Ripley [2002]. We use the data frame SAheart in the ElemStatLearn package [Halvorsen, 2012]. The dependent variable is chd. To grow a tree, use

```
basetree <- tree(as.factor(chd)~.,data=SAheart)
```

The as.factor function indicates to the tree function that it should do classification. If the dependent variable is numeric, tree will fit a so-called regression tree, not what we want here. To plot the tree, use one of the two statements

```
plot(basetree);text(basetree,label='yprob',digits=1)
plot(basetree);text(basetree)
```

The first gives the proportions of 0's and 1's at each leaf, and the second gives the classifications of the leaves, yielding the trees in Figures 11.6 and 11.7, respectively.

This basetree is now our base tree, and we consider only subtrees, that is, trees obtainable by snipping branches off this tree. As usual, we would like to balance observed deviance with the number of parameters in the model, in order to avoid overfitting. To whit, we add a penalty to the deviance depending on the number of leaves in the tree. To use AIC or BIC, we need to count the number of parameters for each subtree, conditioning on the structure of the base tree. That is, we assume that the nodes and the variable at each node are given, so that the only free parameters are the cutoff points and the p_l's. The task is one of subset selection, that is, deciding which nodes to snip away. If the subtree has L leaves, then there are $L - 1$ cutoff points (there are $L - 1$ nodes), and L p_l's, yielding $2L - 1$ parameters. Thus the BIC criterion for a subtree $\mathcal{T}$ with L leaves is

$$\text{BIC}(\mathcal{T}) = \text{deviance}(\mathcal{T}) + \log(n)(2L - 1). \tag{11.76}$$

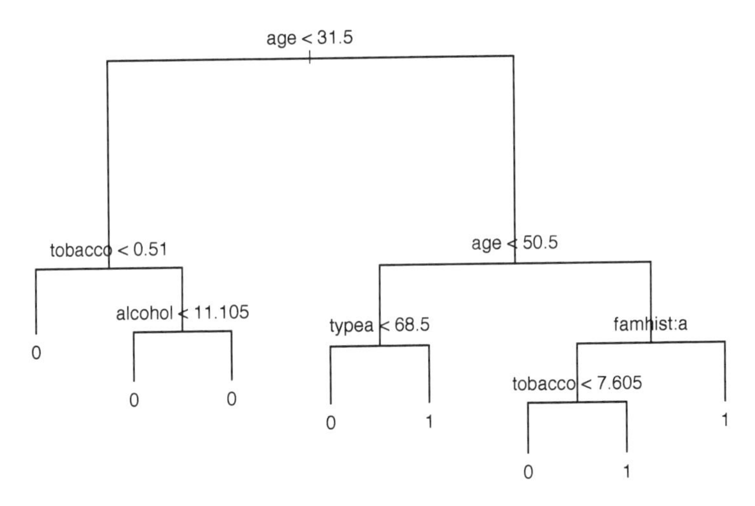

Figure 11.9: The best subtree using the BIC criterion, before snipping redundant leaves.

The prune.tree function can be used to find the subtree with the lowest BIC. It takes the base tree and a value k as inputs, then finds the subtree that minimizes

$$\text{obj}_k(\mathcal{T}) = \text{deviance}(\mathcal{T}) + kL. \tag{11.77}$$

Thus for the best AIC subtree we would take $k = 4$, and for BIC we would take $k = 2\log(n)$:

```
aictree <- prune.tree(basetree,k=4)
bictree <- prune.tree(basetree,k=2*log(462)) # n = 462 here
```

If the k is not specified, then the routine calculates the numbers of leaves and deviances of best subtrees for all values of k. The best AIC subtree is in fact the full base tree, as in Figure 11.7. Figure 11.9 exhibits the best BIC subtree, which has eight leaves. There are also routines in the tree package that use cross-validation to choose a good factor k to use in pruning.

Note that the tree in Figure 11.9 has some redundant splits. Specifically, all leaves to the left of the first split (age < 31.5) lead to classification "0." To snip at that node, we need to determine its index in basetree. One approach is to print out the tree, resulting in the output in Listing 11.1. We see that node #2 is "age < 31.5," which is where we wish to snip, hence we use

```
bictree.2 <- snip.tree(bictree,nodes=2)
```

Plotting the result yields Figure 11.8. It is reasonable to stick with the pre-snipped tree, in case one wished to classify using a cutoff point for $\widehat{p}_l$'s other than 1/2.

There are some drawbacks to this tree-fitting approach. Because of the stepwise nature of the growth, if we start with the wrong variable, it is difficult to recover. That is, even though the best single split may be on age, the best two-variable split may

Listing 11.1: Text representation of the output of tree for the tree in Figure 11.9

```
node), split, n, deviance, yval, (yprob)
      * denotes terminal node

 1) root 462 596.10 0 ( 0.65368 0.34632 )
   2) age < 31.5 117 68.31 0 ( 0.91453 0.08547 )
     4) tobacco < 0.51 81 10.78 0 ( 0.98765 0.01235 ) *
     5) tobacco > 0.51 36 40.49 0 ( 0.75000 0.25000 )
      10) alcohol < 11.105 16 0.00 0 ( 1.00000 0.00000 ) *
      11) alcohol > 11.105 20 27.53 0 ( 0.55000 0.45000 ) *
   3) age > 31.5 345 472.40 0 ( 0.56522 0.43478 )
     6) age < 50.5 173 214.80 0 ( 0.68786 0.31214 )
      12) typea < 68.5 161 188.90 0 ( 0.72671 0.27329 ) *
      13) typea > 68.5 12 10.81 1 ( 0.16667 0.83333 ) *
     7) age > 50.5 172 236.10 1 ( 0.44186 0.55814 )
      14) famhist: Absent 82 110.50 0 ( 0.59756 0.40244 )
        28) tobacco < 7.605 58 68.32 0 ( 0.72414 0.27586 ) *
        29) tobacco > 7.605 24 28.97 1 ( 0.29167 0.70833 ) *
      15) famhist: Present 90 110.00 1 ( 0.30000 0.70000 ) *
```

be on type A and alcohol. There is inherent instability, because having a different variable at a given node can completely change the further branches. Additionally, if there are several splits, the sample sizes for estimating the p_l's at the farther-out leaves can be quite small. Boosting, bagging, and random forests are among the techniques proposed that can help ameliorate some of these problems and lead to better classifications. They are more black-box-like, though, losing some of the simplicity of the simple trees. See Hastie et al. [2009].

Estimating misclassification rate

The observed misclassification rate for any tree is easily found using the summary command. Below we find the 10-fold cross-validation estimates of the classification error. The results are in (11.78). Note that the BIC had the lowest estimate, though by only about 0.01. The base tree was always chosen by AIC. It is interesting that the BIC trees were much smaller, averaging 5 leaves versus 22 for the AIC/base trees.

	Obs. error	CV error	CV se	Average L
Base tree	0.208	0.328	0.057	22
Best AIC	0.208	0.328	0.057	22
Best BIC	0.229	0.317	0.063	5

(11.78)

The following finds the cross-validation estimate for the BIC chosen tree:

```
o <- sample(1:462) # Reorder the indices
err <- NULL # To collect the errors
for(i in 1:10) {
    oi <- o[(1:46)+46*(i-1)] # Left-out indices
    basetreei <- tree(as.factor(chd)~.,data=SAheart,subset=(1:462)[-oi])
    bictreei <- prune.tree(basetreei,k=2*log(416)) # BIC tree w/o left-out data
```

```
    yhati <- predict(bictreei,newdata=SAheart[oi,],type='class')
    err <- c(err,sum(yhati!=SAheart[oi,'chd']))
}
```

For each of the left-out observations, the predict statement with type='class' gives the tree's classification of the left-out observations. The estimate of the error is then mean(err)/46, and the standard error is sd(err)/46.

11.9 Exercises

Exercise 11.9.1. Show that (11.19) follows from (11.18).

Exercise 11.9.2. Compare the statistic in (11.34) and its maximum using the $\widehat{\mathbf{a}}$ in (11.35) to the motivation for Hotelling's T^2 presented in Section 8.4.1.

Exercise 11.9.3. Write the γ_k in (11.59) as a function of the $\boldsymbol{\theta}_i$'s and π_i's.

Exercise 11.9.4 (Spam). Consider the spam data from Section 11.7.2 and Exercise 1.9.16. Here we simplify it a bit, and just look at four of the 0/1 predictors: Whether or not the email contains the words "free" or "remove" or the symbols "!" or "$". The following table summarizes the data, where the first four columns indicate the presence (1) or absence (0) of the word or symbol, and the last two columns give the numbers of corresponding emails that are spam or not spam. E.g., there are 98 emails containing "remove" and "!", but not "free" nor "$", 8 of which are not spam, 90 are spam.

free	remove	!	$	not spam	spam
0	0	0	0	1742	92
0	0	0	1	157	54
0	0	1	0	554	161
0	0	1	1	51	216
0	1	0	0	15	28
0	1	0	1	4	17
0	1	1	0	8	90
0	1	1	1	5	166
1	0	0	0	94	42
1	0	0	1	28	20
1	0	1	0	81	159
1	0	1	1	38	305
1	1	0	0	1	16
1	1	0	1	0	33
1	1	1	0	2	116
1	1	1	1	8	298

(11.79)

Assuming a multinomial distribution for the $2^5 = 32$ possibilities, find the estimated Bayes classifier of email as "spam" or "not spam" based on the other four variables in the table. What is the observed error rate?

Exercise 11.9.5 (Crabs). This problem uses data on 200 crabs, categorized into two species, orange and blue, and two sexes. It is in the MASS R package [Venables and Ripley, 2002]. The data is in the data frame crabs. There are 50 crabs in each species×sex category; the first 50 are blue males, then 50 blue females, then 50 orange males, then 50 orange females. The five measurements are frontal lobe size, rear width, carapace length, carapace width, and body depth, all in millimeters. The goal here is to find linear discrimination procedures for classifying new crabs into species and sex categories. (a) The basic model is that $\mathbf{Y} \sim N(\mathbf{x}\boldsymbol{\beta}, \mathbf{I}_{200} \otimes \boldsymbol{\Sigma})$, where $\mathbf{x}$ is any analysis of variance design matrix ($n \times 4$) that distinguishes the four groups. Find the MLE of $\boldsymbol{\Sigma}$, $\widehat{\boldsymbol{\Sigma}}$. (b) Find the c_k's and $\mathbf{a}_k$'s in Fisher's linear discrimination for classifying all four groups, i.e., classifying on species and sex simultaneously. (Take $\widehat{\pi}_k = 1/4$ for all four groups.) Use the version wherein $d_K = 0$. (c) Using the procedure in part (b) on the observed data, how many crabs had their species misclassified? How many had their sex misclassified? What was the overall observed misclassification rate (for simultaneous classification of color and sex)? (d) Use leave-one-out cross-validation to estimate the overall misclassification rate. What do you get? Is it higher than the observed rate in part (c)?

Exercise 11.9.6 (Crabs). Continue with the crabs data from Exercise 11.9.5, but use classification trees to classify the crabs by just species. (a) Find the base tree using the command

```
crabtree <- tree(sp ~FL+RW+CL+CW+BD,data=crabs)
```

How many leaves does the tree have? Snip off redundant nodes. How many leaves does the snipped tree have? What is its observed misclassification rate? (b) Find the BIC for the subtrees found using prune.tree. Give the number of leaves, deviance, and dimension for the subtree with the best BIC. (c) Consider the subtree with the best BIC. What is its observed misclassification rate? What two variables figure most prominently in the tree? Which variables do not appear? (d) Now find the leave-one-out cross-validation estimate of the misclassification error rate for the best model using BIC. How does this rate compare with the observed rate?

Exercise 11.9.7 (South African heart disease). This question uses the South African heart disease study discussed in Section 11.8. The objective is to use logistic regression to classify people on the presence of heart disease, variable chd. (a) Use the logistic model that includes all the explanatory variables to do the classification. (b) Find the best logistic model using the stepwise function, with BIC as the criterion. Which variables are included in the best model from the stepwise procedure? (c) Use the model with just the variables suggested by the factor analysis of Exercise 10.5.21: tobacco, ldl, adiposity, obesity, and alcohol. (d) Find the BIC, observed error rate, and leave-one-out cross-validation rate for the three models in parts (a), (b) and (c). (e) True or false: (i) The full model has the lowest observed error rate; (ii) The factor-analysis-based model is generally best; (iii) The cross-validation-based error rates are somewhat larger than the corresponding observed error rates; (iv) The model with the best observed error rate has the best cross-validation-based error rate as well; (v) The best model of these three is the one chosen by the stepwise procedure; (vi) Both adiposity and obesity seem to be important factors in classifying heart disease.

Exercise 11.9.8 (Zipcode). The objective here is to classify handwritten numerals $(0, 1, \ldots, 9)$, so that machines can read people's handwritten zipcodes. The data set consists of 16×16 grayscale images, that is, each numeral has been translated to a 16×16 matrix, where the elements of the matrix indicate the darkness (from -1 to 1) of the image at 256 grid points. The data set is from LeCun [1989], and can be found in the R package StatElemLearn [Halvorsen, 2012]. This question will use just the 7's, 8's and 9's, for which there are $n = 1831$ observations. We put the data in three matrices, one for each digit, called train7, train8, and train9. Each row contains first the relevant digit, then the 256 grayscale values, for one image. The task is to use linear discrimination to distinguish between the digits, even though it is clear that the data are not multivariate normal. First, create the three matrices from the large zip.train matrix:

```
train7 <- zip.train[zip.train[,1]==7,-1]
train8 <- zip.train[zip.train[,1]==8,-1]
train9 <- zip.train[zip.train[,1]==9,-1]
```

(a) Using the image, contour, and matrix functions in R, reconstruct the images of some of the 7's, 8's and 9's from their grayscale values. (Or explore the zip2image function in the StatElemLearn package.) (b) Use linear discrimination to classify the observations based on the 256 variables under the three scenarios below. In each case, find both the observed misclassification rate and the estimate using cross-validation. (i) Using $\mathbf{\Sigma} = \mathbf{I}_{256}$. (ii) Assuming $\mathbf{\Sigma}$ is diagonal, using the pooled estimates of the individual variances. (iii) Using the pooled covariance matrix as an estimate of $\mathbf{\Sigma}$. (d) Which method had the best error rate, estimated by cross-validation? (e) Create a data set of digits (7's, 8's, and 9', as well as 5's) to test classifiers as follows:

```
test5 <- zip.test[zip.test[,1]==5,-1]
test7 <- zip.test[zip.test[,1]==7,-1]
test8 <- zip.test[zip.test[,1]==8,-1]
test9 <- zip.test[zip.test[,1]==9,-1]
```

Using the discriminant functions from the original data for the best method from part (b), classify these new observations. What is the error rate for the 7's, 8's, and 9's? How does it compare with the cross-validation estimate? How are the 5's classified?

Exercise 11.9.9 (Spam). Use classification trees to classify the spam data. It is best to start as follows:

```
Spamdf <- data.frame(Spam)
spamtree <- tree(as.factor(spam)~.,data=Spamdf)
```

Turning the matrix into a data frame makes the labeling on the plots simpler. (a) Find the BIC's for the subtrees obtained using prune.tree. How many leaves are in the best model? What is its BIC? What is its observed error rate? (b) Use 46-fold cross-validation (so you leave out 100 observations each time) to estimate the error rate, as in the procedure on page 239. (c) Repeat parts (a) and (b), but using the first ten principal components of the spam explanatory variables as the predictors. (Exercise 1.9.17 calculated the principal components.) Repeat again, but this time using the first ten principal components based on the scaled explanatory variables, scale(Spam[,1:57]). Compare the effectiveness of the three approaches.

Exercise 11.9.10. This question develops a Bayes classifier when there is a mix of normal and binomial explanatory variables. Consider the classification problem based on $(Y, \mathbf{X}, Z)$, where Y is the variable to be classified, with values 0 and 1, and $\mathbf{X}$ and Z are predictors. $\mathbf{X}$ is a 1×2 continuous vector, and Z takes the values 0 and 1. The model for $(Y, \mathbf{X}, Z)$ is given by

$$\mathbf{X} \,|\, Y = y, Z = z \sim N(\boldsymbol{\mu}_{yz}, \mathbf{\Sigma}), \tag{11.80}$$

and

$$P[Y = y \;\&\; Z = z] = p_{yz}, \tag{11.81}$$

so that $p_{00} + p_{01} + p_{10} + p_{11} = 1$. (a) Find an expression for $P[Y = y \,|\, \mathbf{X} = \mathbf{x} \,\&\, Z = z]$. (b) Find the 1×2 vector $\boldsymbol{\alpha}_z$ and the constant γ_z (which depend on z and the parameters) so that

$$P[Y = 1 \,|\, \mathbf{X} = \mathbf{x} \,\&\, Z = z] > P[Y = 0 \,|\, \mathbf{X} = \mathbf{x} \,\&\, Z = z] \;\Leftrightarrow\; \mathbf{x}\boldsymbol{\alpha}_z' + \gamma_z > 0. \tag{11.82}$$

(c) Suppose the data are $(Y_i, \mathbf{X}_i, Z_i)$, $i = 1, \ldots, n$, iid, distributed as above. Find expressions for the MLE's of the parameters (the four $\boldsymbol{\mu}_{yz}$'s, the four p_{yz}'s, and $\boldsymbol{\Sigma}$).

Exercise 11.9.11 (South African heart disease). Apply the classification method in Exercise 11.9.10 to the South African heart disease data, with Y indicating heart disease (chd), $\mathbf{X}$ containing the two variables age and type A, and Z being the family history of heart disease variable (history: 0 = absent, 1 = present). Randomly divide the data into two parts: The training data with $n = 362$, and the test data with $n = 100$. E.g., use

```
random.index <− sample(462,100)
sahd.train <− SAheart[−random.index,]
sahd.test <− SAheart[random.index,]
```

(a) Estimate the $\boldsymbol{\alpha}_z$ and γ_z using the training data. Find the observed misclassification rate on the training data, where you classify an observation as $\widehat{Y}_i = 1$ if $\mathbf{x}_i\widehat{\boldsymbol{\alpha}}_z + \widehat{\gamma}_z > 0$, and $\widehat{Y}_i = 0$ otherwise. What is the misclassification rate for the test data (using the estimates from the training data)? Give the 2×2 table showing true and predicted Y's for the test data. (b) Using the same training data, find the classification tree. You don't have to do any pruning. Just take the full tree from the tree program. Find the misclassification rates for the training data and the test data. Give the table showing true and predicted Y's for the test data. (c) Still using the training data, find the classification using logistic regression, with the $\mathbf{X}$ and Z as the explanatory variables. What are the coefficients for the explanatory variables? Find the misclassification rates for the training data and the test data. (d) What do you conclude?

Chapter 12

Clustering

The classification and prediction we have covered in previous chapters were cases of supervised learning. For example, in classification, we try to find a function that classifies individuals into groups using their $\mathbf{X}$ values, where in the training set we know what the proper groups are because we observe their Y's. In clustering, we again wish to classify observations into groups using their $\mathbf{X}$'s, but do not know the correct groups even in the training set, i.e., we do not observe the Y's, nor often even know how many groups there are. Clustering is a case of unsupervised learning.

There are many clustering algorithms. In this chapter we focus on a few of them. Most are reasonably easy to implement given the number K of clusters. The difficult part is deciding what K should be. Unlike in classification, there is no obvious cross-validation procedure to balance the number of clusters with the tightness of the clusters. Only in the model-based clustering do we have direct AIC or BIC criteria. Otherwise, a number of reasonable but *ad hoc* measures have been proposed. We will look at "silhouettes."

In some situations one is not necessarily assuming that there are underlying clusters, but rather is trying to divide the observations into a certain number of groups for other purposes. For example, a teacher in a class of 40 students might want to break up the class into four sections of about ten each based on general ability (to give more focused instruction to each group). The teacher does not necessarily think there will be wide gaps between the groups, but still wishes to divide for pedagogical purposes. In such cases K is fixed, so the task is a bit simpler.

In general, though, when clustering one is looking for groups that are well separated. There is often an underlying model, just as in Chapter 11 on model-based classification. That is, the data are

$$(\mathbf{X}_1, Y_1), \ldots, (\mathbf{X}_n, Y_n), \text{ iid}, \tag{12.1}$$

where $Y_i \in \{1, \ldots, K\}$,

$$\mathbf{X} \mid Y = k \sim f_k(\mathbf{x}) = f(\mathbf{x} \mid \boldsymbol{\theta}_k) \text{ and } P[Y = k] = \pi_k, \tag{12.2}$$

as in (11.2) and (11.3). If the parameters are known, then the clustering proceeds exactly as for classification, where an observation $\mathbf{x}$ is placed into the group

$$\mathcal{C}(\mathbf{x}) = k \text{ that maximizes } \frac{f_k(\mathbf{x})\pi_k}{f_1(\mathbf{x})\pi_1 + \cdots + f_K(\mathbf{x})\pi_K}. \tag{12.3}$$

See (11.13). The fly in the ointment is that we do not observe the y_i's (neither in the training set nor for the new observations), nor do we necessarily know what K is, let alone the parameter values.

The following sections look at some approaches to clustering. The first, K-means, does not explicitly use a model, but has in the back of its mind f_k's being $N(\boldsymbol{\mu}_k, \sigma^2\mathbf{I}_p)$. Hierarchical clustering avoids the problems of number of clusters by creating a tree containing clusterings of all sizes, from $K = 1$ to n. Finally, the model-based clustering explicitly assumes the f_k's are multivariate normal (or some other given distribution), with various possibilities for the covariance matrices.

12.1 K-means

For a given number K of groups, K-means assumes that each group has a mean vector $\boldsymbol{\mu}_k$. Observation $\mathbf{x}_i$ is assigned to the group with the closest mean. To estimate these means, we minimize the sum of the squared distances from the observations to their group means:

$$\text{obj}(\boldsymbol{\mu}_1, \ldots, \boldsymbol{\mu}_K) = \sum_{i=1}^{n} \min_{\boldsymbol{\mu}_1, \ldots, \boldsymbol{\mu}_K} \|\mathbf{x}_i - \boldsymbol{\mu}_k\|^2. \tag{12.4}$$

An algorithm for finding the clusters starts with a random set of means $\widehat{\boldsymbol{\mu}}_1, \ldots, \widehat{\boldsymbol{\mu}}_K$ (e.g., randomly choose K observations from the data), then iterate the following two steps:

1. Having estimates of the means, assign observations to the group corresponding to the closest mean,

$$C(\mathbf{x}_i) = k \text{ that minimizes } \|\mathbf{x}_i - \widehat{\boldsymbol{\mu}}_k\|^2 \text{ over } k. \tag{12.5}$$

2. Having individuals assigned to groups, find the group means,

$$\widehat{\boldsymbol{\mu}}_k = \frac{1}{\#\{C(\mathbf{x}_i) = k\}} \sum_{i \mid C(\mathbf{x}_i) = k} x_i. \tag{12.6}$$

The algorithm is guaranteed to converge (if ties among the closest means are handled properly), but not necessarily to the global minimum. It is a good idea to try several random starts, then take the one that yields the lowest obj in (12.4). The resulting means and assignments are the K-means and their clustering.

12.1.1 Example: Sports data

Recall the data on people ranking seven sports presented in Section 1.6.2. Using the K-means algorithm for $K = 1, \ldots, 4$, we find the following means (where $K = 1$ gives the overall mean):

$$(12.7)$$

$K=1$	BaseB	FootB	BsktB	Ten	Cyc	Swim	Jog
Group 1	3.79	4.29	3.74	3.86	3.59	3.78	4.95

$K=2$	BaseB	FootB	BsktB	Ten	Cyc	Swim	Jog
Group 1	5.01	5.84	4.35	3.63	2.57	2.47	4.12
Group 2	2.45	2.60	3.06	4.11	4.71	5.21	5.85

$K=3$	BaseB	FootB	BsktB	Ten	Cyc	Swim	Jog
Group 1	2.33	2.53	3.05	4.14	4.76	5.33	5.86
Group 2	4.94	5.97	5.00	3.71	2.90	3.35	2.13
Group 3	5.00	5.51	3.76	3.59	2.46	1.90	5.78

$K=4$	BaseB	FootB	BsktB	Ten	Cyc	Swim	Jog
Group 1	5.10	5.47	3.75	3.60	2.40	1.90	5.78
Group 2	2.30	2.10	2.65	5.17	4.75	5.35	5.67
Group 3	2.40	3.75	3.90	1.85	4.85	5.20	6.05
Group 4	4.97	6.00	5.07	3.80	2.80	3.23	2.13

Look at the $K = 2$ means. Group 1 likes swimming and cycling, while group 2 likes the team sports, baseball, football, and basketball. If we compare these to the $K = 3$ clustering, we see group 1 appears to be about the same as the team sports group from $K = 2$, while groups 2 and 3 both like swimming and cycling. The difference is that group 3 does not like jogging, while group 2 does. For $K = 4$, it looks like the team-sports group has split into one that likes tennis (group 3), and one that doesn't (group 2). At this point it may be more useful to try to decide what number of clusters is "good." (Being able to interpret the clusters is one good characteristic.)

12.1.2 Silhouettes

One measure of clustering efficacy is Rousseeuw's [1987] notion of **silhouettes**. The silhouette of an observation i measures how well it fits in its own cluster versus how well it fits in its next closest cluster. For K-means, we have changed the original definition a little. Let

$$a(i) = \|\mathbf{x}_i - \widehat{\boldsymbol{\mu}}_k\|^2 \text{ and } b(i) = \|\mathbf{x}_i - \widehat{\boldsymbol{\mu}}_l\|^2, \tag{12.8}$$

where observation i is assigned to group k, and group l has the next-closest group mean to $\mathbf{x}_i$. Then its silhouette is

$$\text{silhouette}(i) = \frac{b(i) - a(i)}{\max\{a(i), b(i)\}}. \tag{12.9}$$

By construction, $b(i) \geq a(i)$, hence the denominator is $b(i)$, and the silhouette takes values between 0 and 1. If the observation is equal to its group mean, its silhouette is 1. If it is halfway between the two group means, its silhouette is 0. For other clusterings (K-medoids, as in Section 12.2, for example), the silhouettes can range from -1 to 1, but usually stay above 0, or at least do not go much below.

Figure 12.1 contains the silhouettes for K's from 2 to 5 for the sports data. The observations (along the horizontal axis) are arranged by group and, within group,

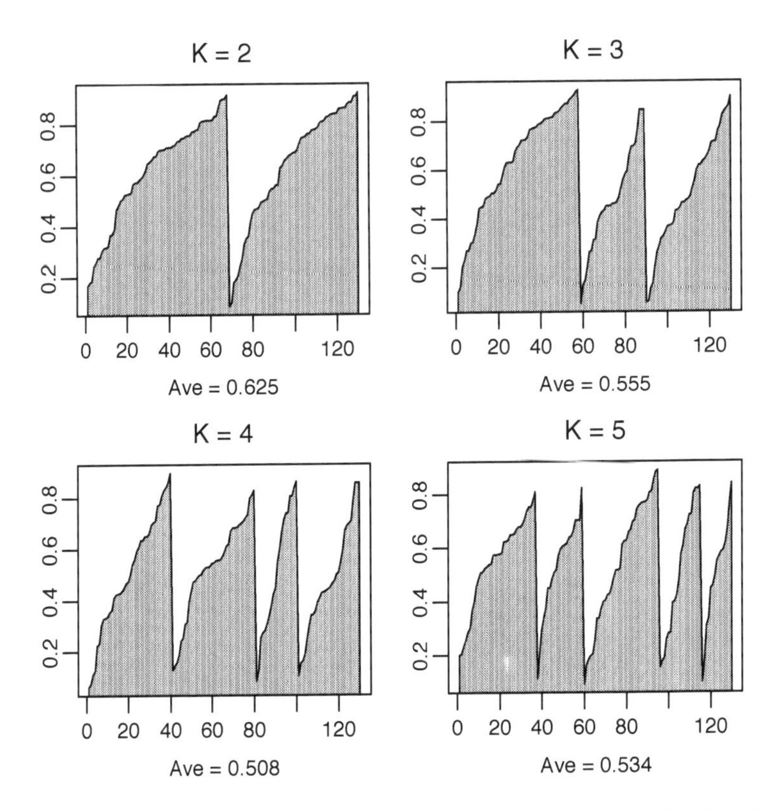

Figure 12.1: The silhouettes for $K = 2, \ldots, 5$ clusters. The horizontal axis indexes the observations. The vertical axis exhibits the values of the silhouettes.

by silhouettes. This arrangement allows one to compare the clusters. In the first plot ($K = 2$ groups), the two clusters have similar silhouettes, and the silhouettes are fairly "full." High silhouettes are good, so that the average silhouette is a measure of goodness for the clustering. In this case, the average is 0.625. For $K = 3$, notice that the first silhouette is still full, while the two smaller clusters are a bit frail. The $K = 4$ and 5 silhouettes are not as full, either, as indicated by their averages.

Figure 12.2 plots the average silhouette versus K. It is clear that $K = 2$ has the highest silhouette, hence we would take $K = 2$ as the best cluster size.

12.1.3 Plotting clusters in one and two dimensions

With two groups, we have two means in $p(= 7)$-dimensional space. To look at the data, we can project the observations to the line that runs through the means. This

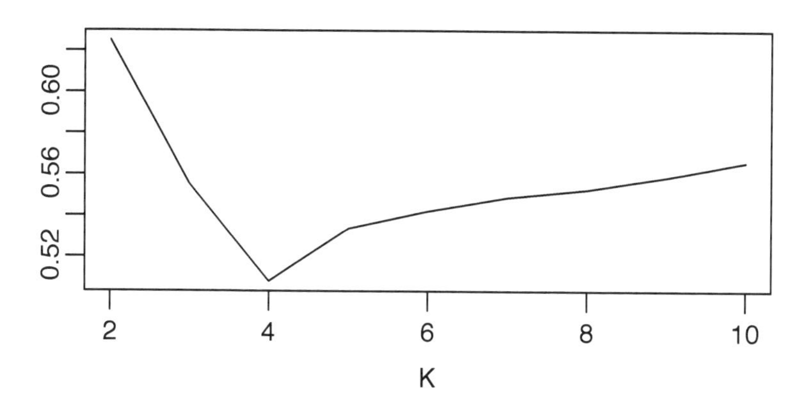

Figure 12.2: The average silhouettes for $K = 2, \ldots, 10$ clusters.

projection is where the clustering is taking place. Let

$$\mathbf{z} = \frac{\widehat{\boldsymbol{\mu}}_1 - \widehat{\boldsymbol{\mu}}_2}{\|\widehat{\boldsymbol{\mu}}_1 - \widehat{\boldsymbol{\mu}}_2\|}, \tag{12.10}$$

the unit vector pointing from $\widehat{\boldsymbol{\mu}}_2$ to $\widehat{\boldsymbol{\mu}}_1$. Then using $\mathbf{z}$ as an axis, the projections of the observations onto $\mathbf{z}$ have coordinates

$$w_i = \mathbf{x}_i \mathbf{z}', \ \ i = 1, \ldots, N. \tag{12.11}$$

Figure 12.3 is the histogram for the w_i's, where group 1 has $w_i > 0$ and group 2 has $w_i < 0$. We can see that the clusters are well-defined in that the bulk of each cluster is far from the center of the other cluster.

We have also plotted the sports, found by creating a "pure" ranking for each sport. Thus the pure ranking for baseball would give baseball the rank of 1, and the other sports the rank of 4.5, so that the sum of the ranks, 28, is the same as for the other rankings. Adding these sports to the plot helps aid in interpreting the groups: team sports on the left, individual sports on the right, with tennis on the individual-sport side, but close to the border.

If $K = 3$, then the three means lie in a plane, hence we would like to project the observations onto that plane. One approach is to use principal components (Section 1.6) on the means. Because there are three, only the first two principal components will have positive variance, so that all the action will be in the first two. Letting

$$\mathbf{Z} = \begin{pmatrix} \widehat{\boldsymbol{\mu}}_1 \\ \widehat{\boldsymbol{\mu}}_2 \\ \widehat{\boldsymbol{\mu}}_3 \end{pmatrix}, \tag{12.12}$$

we apply the spectral decomposition (1.33) in Theorem 1.1 to the sample covariance matrix of $\mathbf{Z}$:

$$\frac{1}{3}\,\mathbf{Z}'\mathbf{H}_3\mathbf{Z} = \mathbf{G}\mathbf{L}\mathbf{G}', \tag{12.13}$$

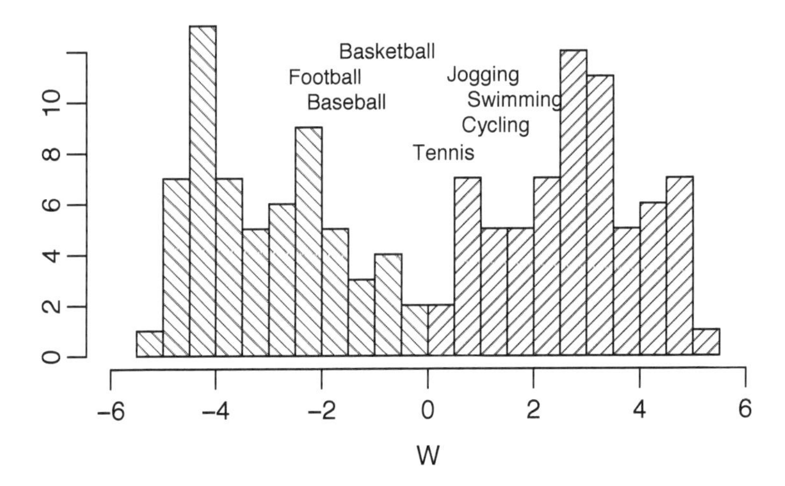

Figure 12.3: The histogram for the observations along the line connecting the two means for $K = 2$ groups.

where **G** is orthogonal and **L** is diagonal. The diagonals of **L** here are 11.77, 4.07, and five zeros. We then rotate the data and the means using **G**,

$$\mathbf{W} = \mathbf{XG} \text{ and } \mathbf{W}^{(means)} = \mathbf{ZG}. \tag{12.14}$$

Figure 12.4 plots the first two variables for **W** and $\mathbf{W}^{(means)}$, along with the seven pure rankings. We see the people who like team sports to the right, and the people who like individual sports to the left, divided into those who can and those who cannot abide jogging. Compare this plot to the biplot that appears in Figure 1.6.

12.1.4 Example: Sports data, using R

The sports data is in the R matrix sportsranks. The K-means clustering uses the function kmeans. This routine uses the method in Hartigan and Wong [1979], which tweaks the basic algorithm in (12.5) and (12.6) by some intelligent reassigning of individual points. We create a list whose K^{th} component contains the results for $K = 2, \ldots, 10$ groups:

```
kms <- vector('list',10)
for(K in 2:10) {
    kms[[K]] <- kmeans(sportsranks,centers=K,nstart=10)
}
```

The centers input specifies the number of groups desired, and nstart=10 means randomly start the algorithm ten times, then use the one with lowest within sum of squares. The output in kms[[K]] for the K-group clustering is a list with centers, the $K \times p$ of estimated cluster means; cluster, an n-vector that assigns each observation to its cluster (i.e., the $\widehat{y}_i$'s); withinss, the K-vector of SS_k's (so that $SS(K)$ is found

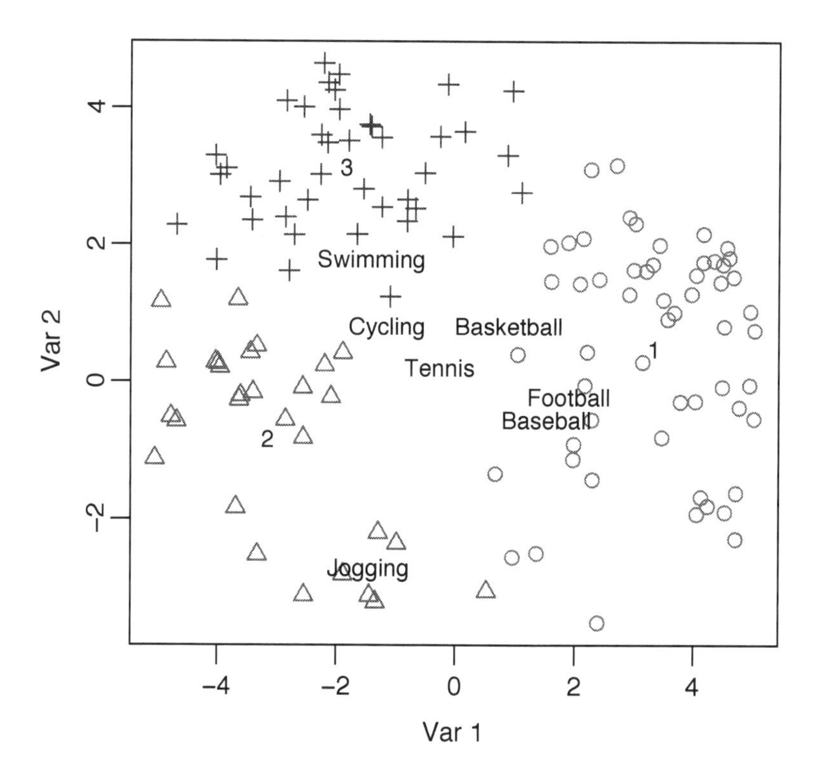

Figure 12.4: The scatter plot for the data projected onto the plane containing the means for $K = 3$.

by sum(kms[[K]]$withinss)); and size, the K-vector giving the numbers of observations assigned to each group.

Silhouettes

Section A.4.1 contains a simple function for calculating the silhouettes in (12.9) for a given K-means clustering. The sort_silhouette function in Section A.4.2 sorts the silhouette values for plotting. The following statements produce Figure 12.1:

```
sil.ave <- NULL # To collect silhouette's means for each K
par(mfrow=c(3,3))
for(K in 2:10) {
    sil <- silhouette.km(sportsranks,kms[[K]]$centers)
    sil.ave <- c(sil.ave,mean(sil))
    ssil <- sort_silhouette(sil,kms[[K]]$cluster)
    plot(ssil,type='h',xlab='Observations',ylab='Silhouettes')
    title(paste('K =',K))
}
```

The sil.ave calculated above can then be used to obtain Figure 12.2:

```
plot(2:10,sil.ave,type='l',xlab='K',ylab='Average silhouette width')
```

Plotting the clusters

Finally, we make plots as in Figures 12.3 and 12.4. For $K = 2$, we have the one $\mathbf{z}$ as in (12.10) and the w_i's as in (12.11):

```
z <- kms[[2]]$centers[1,]-kms[[2]]$centers[2,]
z <- z/sqrt(sum(z^2))
w <- sportsranks%*%z
xl <- c(-6,6); yl <- c(0,13) # Fix the x- and y-ranges
hist(w[kms[[2]]$cluster==1],col=2,xlim=xl,ylim=yl,main='K=2',xlab='W')
par(new=TRUE) # To allow two histograms on the same plot
hist(w[kms[[2]]$cluster==2],col=3,xlim=xl,ylim=yl,main=' ',xlab=' ')
```

To add the sports' names:

```
y <- matrix(4.5,7,7)-3.5*diag(7)
ws <- y%*%z
text(ws,c(10,11,12,8,9,10,11),labels=dimnames(sportsranks)[[2]])
```

The various placement numbers were found by trial and error.

For $K = 3$, or higher, we can use R's eigenvector/eigenvalue function, eigen, to find the $\mathbf{G}$ used in (12.14):

```
z <- kms[[3]]$centers
g <- eigen(var(z))$vectors[,1:2] # Just need the first two columns
w <- sportsranks%*%g # For the observations
ws <- y%*%g # For the sports' names
wm <- z%*%g # For the groups' means
cl <- kms[[3]]$cluster
plot(w,xlab='Var 1',ylab='Var 2',pch=cl)
text(wc,labels=1:3)
text(ws,dimnames(sportsranks)[[2]])
```

12.2 K-medoids

Clustering with medoids [Kaufman and Rousseeuw, 1990] works directly on distances between objects. Suppose we have n objects, $\mathbf{o}_1, \ldots, \mathbf{o}_n$, and a dissimilarity measure $d(\mathbf{o}_i, \mathbf{o}_j)$ between pairs. This d satisfies

$$d(\mathbf{o}_i, \mathbf{o}_j) \geq 0, \;\; d(\mathbf{o}_i, \mathbf{o}_j) = d(\mathbf{o}_j, \mathbf{o}_i), \text{ and } d(\mathbf{o}_i, \mathbf{o}_i) = 0, \tag{12.15}$$

but it may not be an actual metric in that it need not satisfy the triangle inequality. Note that one cannot necessarily impute distances between an object and another vector, e.g., a mean vector. Rather than clustering around means, the clusters are then built around some of the objects. That is, K-medoids finds K of the objects $(\mathbf{c}_1, \ldots, \mathbf{c}_K)$ to act as centers (or medoids), the objective being to find the set that minimizes

$$\text{obj}(\mathbf{c}_1, \ldots, \mathbf{c}_K) = \sum_{i=1}^{N} \min_{\{\mathbf{c}_1, \ldots, \mathbf{c}_K\}} d(\mathbf{o}_i, \mathbf{c}_k). \tag{12.16}$$

Each object is then assigned to the group corresponding to the closest of the $\mathbf{c}_k$'s. Silhouettes are defined as in (12.9), except that here, for each observation i,

$$a(i) = \frac{1}{\#\text{ Group } k} \sum_{j \,\in\, \text{Group } k} d(\mathbf{o}_i, \mathbf{o}_j) \text{ and } b(i) = \frac{1}{\#\text{ Group } l} \sum_{j \,\in\, \text{Group } l} d(\mathbf{o}_i, \mathbf{o}_j), \tag{12.17}$$

where group k is object i's group, and group l is its next closest group.

In R, one can use the package cluster [Maechler et al., 2015], which implements K-medoids clustering in the function pam, which stands for **partitioning around medoids**. Consider the grades data in Section 4.2.1. We will cluster the five variables, homework, labs, inclass, midterms, and final, not the 107 people. A natural measure of similarity between two variables is their correlation. Instead of using the usual Pearson coefficient, we will use Kendall's τ, which is more robust. For $n \times 1$ vectors $\mathbf{x}$ and $\mathbf{y}$, Kendall's τ is

$$T(\mathbf{x}, \mathbf{y}) = \frac{\sum_{1 \le i < j \le n} \text{Sign}(x_i - x_j)\text{Sign}(y_i - y_j)}{\binom{n}{2}}. \tag{12.18}$$

The numerator looks at the line segment connecting each pair of points (x_i, y_i) and (x_j, y_j), counting $+1$ if the slope is positive and -1 if it is negative. The denominator normalizes the statistic so that it is between ± 1. Then $T(\mathbf{x}, \mathbf{y}) = +1$ means that the x_i's and y_i's are exactly monotonically increasingly related, and -1 means they are exactly monotonically decreasingly related, much as the correlation coefficient. The T's measure similarities, so we subtract each T from 1 to obtain the dissimilarity matrix:

	HW	Labs	InClass	Midterms	Final
HW	0.00	0.56	0.86	0.71	0.69
Labs	0.56	0.00	0.80	0.68	0.71
InClass	0.86	0.80	0.00	0.81	0.81
Midterms	0.71	0.68	0.81	0.00	0.53
Final	0.69	0.71	0.81	0.53	0.00

(12.19)

Using R, we find the dissimilarity matrix:

```
x <- grades[,2:6]
dx <- matrix(nrow=5,ncol=5) # To hold the dissimilarities
for(i in 1:5)
    for(j in 1:5)
        dx[i,j] <- 1-cor.test(x[,i],x[,j],method='kendall')$est
```

This matrix is passed to the pam function, along with the desired number of groups K. Thus for $K = 3$, say, use

```
pam3 <- pam(as.dist(dx),k=3)
```

The average silhouette for this clustering is in pam3$silinfo$avg.width. The results for $K = 2, 3$ and 4 are

K	2	3	4
Average silhouette	0.108	0.174	0.088

(12.20)

We see that $K = 3$ has the best average silhouette. The assigned groups for this clustering can be found in pam3$clustering, which is (1,1,2,3,3), meaning the groupings are, reasonably enough,

$$\{\text{HW, Labs}\}\ \{\text{InClass}\}\ \{\text{Midterms, Final}\}. \tag{12.21}$$

The medoids, i.e., the objects chosen as centers, are in this case labs, inclass, and midterms, respectively.

12.3 Model-based clustering

In model-based clustering [Fraley and Raftery, 2002], we assume that the model in (12.2) holds, just as for classification. We then estimate the parameters, which includes the $\boldsymbol{\theta}_k$'s and the π_k's, and assign observations to clusters as in (12.3):

$$\widehat{\mathcal{C}}(\mathbf{x}_i) = k \text{ that maximizes } \frac{f(\mathbf{x}_i \mid \widehat{\boldsymbol{\theta}}_k)\widehat{\pi}_k}{f(\mathbf{x}_i \mid \widehat{\boldsymbol{\theta}}_1)\widehat{\pi}_1 + \cdots + f(\mathbf{x}_i \mid \widehat{\boldsymbol{\theta}}_K)\widehat{\pi}_K}. \tag{12.22}$$

As opposed to classification situations, in clustering we do not observe the y_i's, hence cannot use the joint distribution of $(Y, \mathbf{X})$ to estimate the parameters. Instead, we need to use the marginal of $\mathbf{X}$, which is the denominator in the $\mathcal{C}$:

$$\begin{aligned} f(\mathbf{x}_i) &= f(\mathbf{x}_i \mid \boldsymbol{\theta}_1, \ldots, \boldsymbol{\theta}_K, \pi_1, \ldots, \pi_K) \\ &= f(\mathbf{x}_i \mid \boldsymbol{\theta}_1)\pi_1 + \cdots + f(\mathbf{x}_i \mid \boldsymbol{\theta}_K)\pi_K. \end{aligned} \tag{12.23}$$

The density is a mixture density, as in (11.5).

The likelihood for the data is then

$$L(\boldsymbol{\theta}_1, \ldots, \boldsymbol{\theta}_K, \pi_1, \ldots, \pi_K\,; \mathbf{x}_1, \ldots, \mathbf{x}_n) = \prod_{i=1}^{n} \left(f(\mathbf{x}_i \mid \boldsymbol{\theta}_1)\pi_1 + \cdots + f(\mathbf{x}_i \mid \boldsymbol{\theta}_K)\pi_K \right). \tag{12.24}$$

The likelihood can be maximized for any specific model (specifying the f's as well as K), and models can compared using the BIC (or AIC). The likelihood (12.24) is not always easy to maximize due to its being a product of sums. Often the EM algorithm (see Section 12.4) is helpful.

We will present the multivariate normal case, as we did in (11.24) and (11.25) for classification. The general model assumes for each k that

$$\mathbf{X} \mid Y = k \sim N_{1\times p}(\boldsymbol{\mu}_k, \boldsymbol{\Sigma}_k). \tag{12.25}$$

We will assume the $\boldsymbol{\mu}_k$'s are free to vary, although models in which there are equalities among some of the elements are certainly reasonable. There are also a variety of structural and equality assumptions on the $\boldsymbol{\Sigma}_k$'s used.

12.3.1 Example: Automobile data

The R function we use is in the package mclust, Fraley et al. [2014]. Our data consist of size measurements on 111 automobiles. The variables include length, wheelbase, width, height, front and rear head room, front leg room, rear seating, front and rear shoulder room, and luggage area. The data are in the file cars, from Consumers' Union [1990], and can be found in the S-Plus® [TIBCO Software Inc., 2009] data frame cu.dimensions. The variables in cars have been normalized to have medians of 0 and median absolute deviations (MAD) of 1.4826 (the MAD for a $N(0, 1)$).

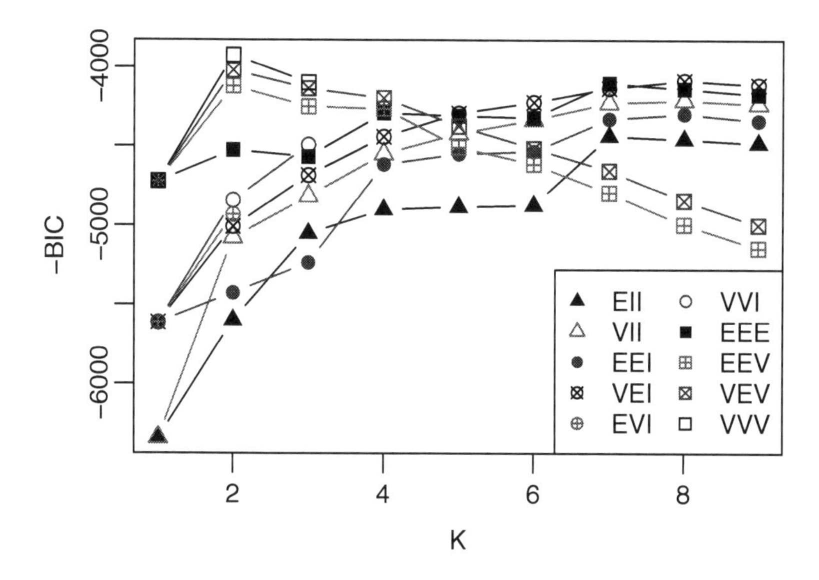

Figure 12.5: −BIC's for fitting the entire data set.

The routine we'll use is Mclust (be sure to capitalize the M). It will try various forms of the covariance matrices and group sizes, and pick the best based on the BIC. To use the default options and have the results placed in mcars, use

```
mcars <− Mclust(cars)
```

There are many options for plotting in the package. To see a plot of the BIC's, use

```
plot(mcars,cars,what='BIC')
```

You have to click on the graphics window, or hit enter, to reveal the plot. The result is in Figure 12.5. The horizontal axis specifies the K, and the vertical axis gives the BIC values, although these are the negatives of our BIC's. The symbols plotted on the graph are codes for various structural hypotheses on the covariances. See (12.31). In this example, the best model is Model "VVV" with $K = 2$, which means the covariance matrices are arbitrary and unequal.

Some pairwise plots (length versus height, width versus front head room, and rear head room versus luggage) are given in Figure 12.6. The plots include ellipses to illustrate the covariance matrices. Indeed we see that the two ellipses in each plot are arbitrary and unequal. To plot variable 1 (length) versus variable 4 (height), use

```
plot(mcars,cars,what='classification',dimens=c(1,4))
```

We also plot the first two principal components (Section 1.6). The matrix of eigenvectors, $\mathbf{G}$ in (1.33), is given by eigen(var(cars))$vectors:

```
carspc <− cars%*%eigen(var(cars))$vectors # Principal components
```

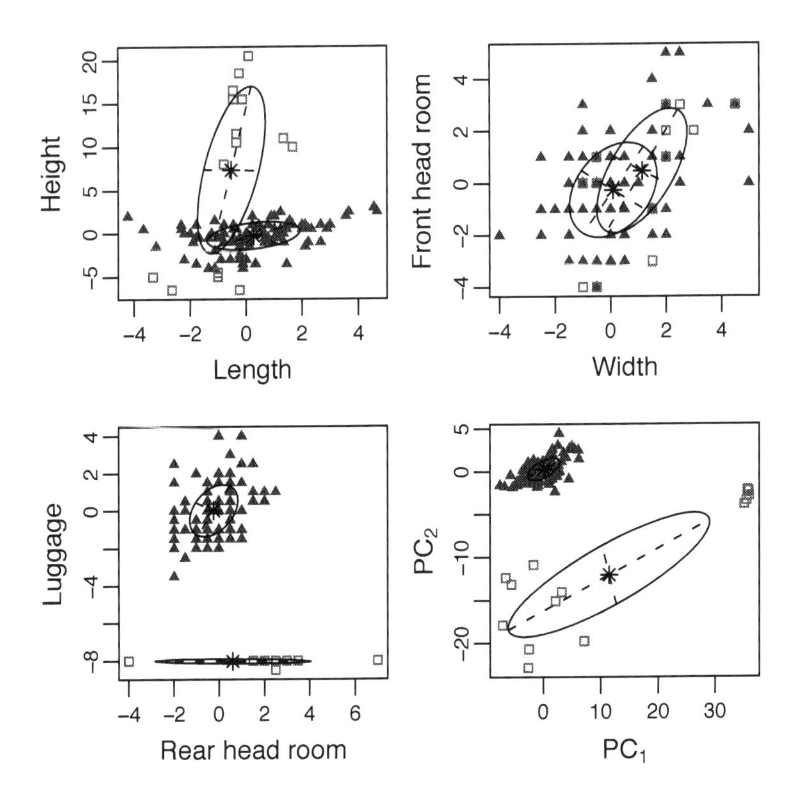

Figure 12.6: Some two-variable plots of the clustering produced by Mclust. The solid triangles indicate group 1, and the open squares indicate group 2. The fourth graph plots the first two principal components of the data.

To obtain the ellipses, we redid the clustering using the principal components as the data, and specifying G=2 groups in Mclust.

Look at the plots. The lower left graph shows that group 2 is almost constant on the luggage variable. In addition, the upper left and lower right graphs indicate that group 2 can be divided into two groups, although the BIC did not pick up the difference. The Table 12.1 exhibits four of the variables for the 15 automobiles in group 2.

We have divided this group as suggested by the principal component plot. Note that the first group of five are all sports cars. They have no back seats or luggage areas, hence the values in the data set for the corresponding variables are coded somehow. The other ten automobiles are minivans. They do not have specific luggage areas, i.e., trunks, either, although in a sense the whole vehicle is a big luggage area. Thus this group really is a union of two smaller groups, both of which are quite a bit different than group 1.

We now redo the analysis on just the group 1 automobiles:

	Rear Head	Rear Seating	Rear Shoulder	Luggage
Chevrolet Corvette	−4.0	−19.67	−28.00	−8.0
Honda Civic CRX	−4.0	−19.67	−28.00	−8.0
Mazda MX5 Miata	−4.0	−19.67	−28.00	−8.0
Mazda RX7	−4.0	−19.67	−28.00	−8.0
Nissan 300ZX	−4.0	−19.67	−28.00	−8.0
Chevrolet Astro	2.5	0.33	−1.75	−8.0
Chevrolet Lumina APV	2.0	3.33	4.00	−8.0
Dodge Caravan	2.5	−0.33	−6.25	−8.0
Dodge Grand Caravan	2.0	2.33	3.25	−8.0
Ford Aerostar	1.5	1.67	4.25	−8.0
Mazda MPV	3.5	0.00	−5.50	−8.0
Mitsubishi Wagon	2.5	−19.00	2.50	−8.0
Nissan Axxess	2.5	0.67	1.25	−8.5
Nissan Van	3.0	−19.00	2.25	−8.0
Volkswagen Vanagon	7.0	6.33	−7.25	−8.0

Table 12.1: The automobiles in group 2 of the clustering of all the data.

```
cars1 <- cars[mcars$classification==1,]
mcars1 <- Mclust(cars1)
```

The model chosen by BIC is "XXX with 1 components" which means the best clustering is one large group, where the $\mathbf{\Sigma}$ is arbitrary. See Figure 12.7 for the BIC plot. The EEE models (equal but arbitrary covariance matrices) appear to be quite good, and similar BIC-wise, for K from 1 to 4. To get the actual BIC values, look at the vector mcars1$BIC[,"EEE"]. The next table has the BIC's and corresponding estimates of the posterior probabilities for the first five model, where we shift the BIC's so that the best is 0:

K	1	2	3	4	5
BIC	0	28.54	9.53	22.09	44.81
$\widehat{p}^{BIC}$	99.15	0	0.84	0	0

(12.26)

Indeed, it looks like one group is best, although three groups may be worth looking at. It turns out the three groups are basically large, middle-sized, and small cars. Not profound, perhaps, but reasonable.

12.3.2 Some of the models in mclust

The mclust package considers several models for the covariance matrices. Suppose that the covariance matrices for the groups are $\mathbf{\Sigma}_1, \ldots, \mathbf{\Sigma}_K$, where each has its spectral decomposition (1.33)

$$\mathbf{\Sigma}_k = \mathbf{\Gamma}_k \mathbf{\Lambda}_k \mathbf{\Gamma}_k', \tag{12.27}$$

and the eigenvalue matrix is decomposed as

$$\mathbf{\Lambda}_k = c_k \mathbf{\Delta}_k \text{ where } |\mathbf{\Delta}_k| = 1 \text{ and } c_k = [\prod_{j=1}^{p} \lambda_j]^{1/p}, \tag{12.28}$$

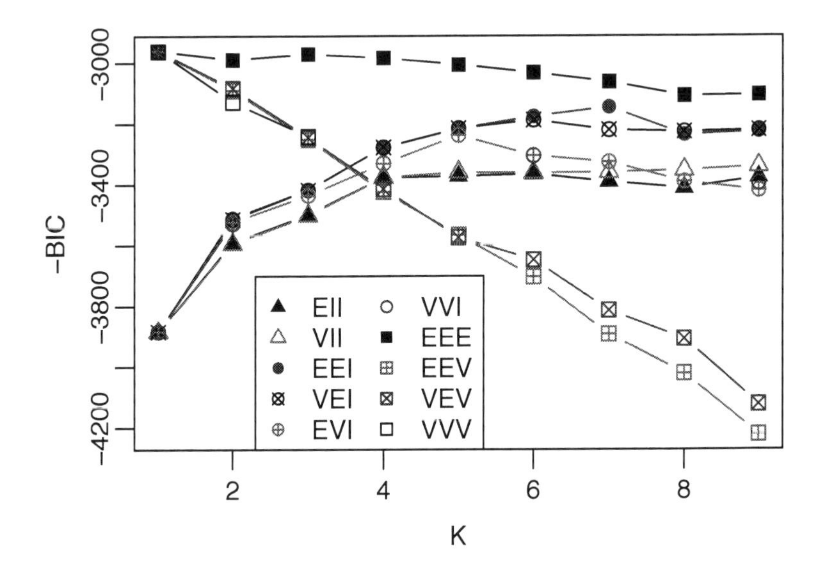

Figure 12.7: −BIC's for the data set without the sports cars or minivans.

the geometric mean of the eigenvalues. A covariance matrix is then described by shape, volume, and orientation:

$$\begin{aligned} \text{Shape}(\mathbf{\Sigma}_k) &= \mathbf{\Delta}_k; \\ \text{Volume}(\mathbf{\Sigma}_k) &= |\mathbf{\Sigma}_k| = c_k^p; \\ \text{Orientation}(\mathbf{\Sigma}_k) &= \mathbf{\Gamma}_k. \end{aligned} \tag{12.29}$$

The covariance matrices are then classified into spherical, diagonal, and ellipsoidal:

$$\begin{aligned} \text{Spherical} &\Rightarrow \mathbf{\Delta}_k = \mathbf{I}_p \ \Rightarrow \ \mathbf{\Sigma}_k = c_k \mathbf{I}_p; \\ \text{Diagonal} &\Rightarrow \mathbf{\Gamma}_k = \mathbf{I}_p \ \Rightarrow \ \mathbf{\Sigma}_k = c_k \mathbf{D}_k; \\ \text{Ellipsoidal} &\Rightarrow \mathbf{\Sigma}_k \text{ is arbitrary.} \end{aligned} \tag{12.30}$$

The various models are defined by the type of covariances, and what equalities there are among them. I haven't been able to crack the code totally, but the descriptions tell the story. When $K \geq 2$ and $p \geq 2$, the following table may help translate the descriptions into restrictions on the covariance matrices through (12.29) and(12.30):

Code	Description	$\boldsymbol{\Sigma}_k$
EII	spherical, equal volume	$\sigma^2 \mathbf{I}_p$
VII	spherical, unequal volume	$\sigma_k^2 \mathbf{I}_p$
EEI	diagonal, equal volume and shape	$\boldsymbol{\Lambda}$
VEI	diagonal, varying volume, equal shape	$c_k \boldsymbol{\Delta}$
EVI	diagonal, equal volume, varying shape	$c \boldsymbol{\Delta}_k$
VVI	diagonal, varying volume and shape	$\boldsymbol{\Lambda}_k$
EEE	ellipsoidal, equal volume, shape, and orientation	$\boldsymbol{\Sigma}$
EEV	ellipsoidal, equal volume and equal shape	$\boldsymbol{\Gamma}_k \boldsymbol{\Lambda} \boldsymbol{\Gamma}_k'$
VEV	ellipsoidal, equal shape	$c_k \boldsymbol{\Gamma}_k \boldsymbol{\Delta} \boldsymbol{\Gamma}_k'$
VVV	ellipsoidal, varying volume, shape, and orientation	arbitrary

(12.31)

Here, $\boldsymbol{\Lambda}$'s are diagonal matrices with positive diagonals, $\boldsymbol{\Delta}$'s are diagonal matrices with positive diagonals whose product is 1 as in (12.28), $\boldsymbol{\Gamma}$'s are orthogonal matrices, $\boldsymbol{\Sigma}$'s are arbitrary nonnegative definite symmetric matrices, and c's are positive scalars. A subscript k on an element means the groups can have different values for that element. No subscript means that element is the same for each group.

If there is only one variable, but $K \geq 2$, then the only two models are "E," meaning the variances of the groups are equal, and "V," meaning the variances can vary. If there is only one group, then the models are as follows:

Code	Description	$\boldsymbol{\Sigma}$
X	one-dimensional	σ^2
XII	spherical	$\sigma^2 \mathbf{I}_p$
XXI	diagonal	$\boldsymbol{\Lambda}$
XXX	ellipsoidal	arbitrary

(12.32)

12.4 An example of the EM algorithm

The aim of this section is to give the flavor of an implementation of the EM algorithm. We assume K groups with the multivariate normal distribution as in (12.25), with different arbitrary $\boldsymbol{\Sigma}_k$'s. The idea is to iterate two steps:

1. Having estimates of the parameters, find estimates of $P[Y = k \mid \mathbf{X} = \mathbf{x}_i]$'s.

2. Having estimates of $P[Y = k \mid \mathbf{X} = \mathbf{x}_i]$'s, find estimates of the parameters.

Suppose we start with initial estimates of the π_k's, $\boldsymbol{\mu}_k$'s, and $\boldsymbol{\Sigma}_k$'s. E.g., one could first perform a K-means procedure, then use the sample means and covariance matrices of the groups to estimate the means and covariances, and estimate the π_k's by the proportions of observations in the groups. Then, as in (12.22), for step 1 we use

$$\widehat{P}[Y = k \mid \mathbf{X} = \mathbf{x}_i] = \frac{f(\mathbf{x}_i \mid \widehat{\boldsymbol{\mu}}_k, \widehat{\boldsymbol{\Sigma}}_k)\widehat{\pi}_k}{f(\mathbf{x}_i \mid \widehat{\boldsymbol{\mu}}_1, \widehat{\boldsymbol{\Sigma}}_1)\widehat{\pi}_1 + \cdots + f(\mathbf{x}_i \mid \widehat{\boldsymbol{\mu}}_K, \widehat{\boldsymbol{\Sigma}}_K)\widehat{\pi}_K}$$
$$\equiv w_k^{(i)}, \qquad (12.33)$$

where $\widehat{\boldsymbol{\theta}}_k = (\widehat{\boldsymbol{\mu}}_k, \widehat{\boldsymbol{\Sigma}}_k)$.

Note that for each i, the $w_k^{(i)}$ can be thought of as weights, because their sum over k is 1. Then in Step 2, we find the weighted means and covariances of the $\mathbf{x}_i$'s:

$$\begin{aligned} \widehat{\boldsymbol{\mu}}_k &= \frac{1}{\widehat{n}_k} \sum_{i=1}^{n} w_k^{(i)} \mathbf{x}_i \\ \text{and } \widehat{\boldsymbol{\Sigma}}_k &= \frac{1}{\widehat{n}_k} \sum_{i=1}^{n} w_k^{(i)} (\mathbf{x}_i - \widehat{\boldsymbol{\mu}}_k)'(\mathbf{x}_i - \widehat{\boldsymbol{\mu}}_k), \\ \text{where } \widehat{n}_k &= \sum_{i=1}^{n} w_k^{(i)}. \\ \text{Also, } \widehat{\pi}_k &= \frac{\widehat{n}_k}{n}. \end{aligned} \tag{12.34}$$

The two steps are iterated until convergence. The convergence may be slow, and it may not approach the global maximum likelihood, but it is guaranteed to increase the likelihood at each step. As in K-means, it is a good idea to try different starting points.

In the end, the observations are clustered using the conditional probabilities, where from (12.22),

$$\widehat{\mathcal{C}}(\mathbf{x}_i) = k \text{ that maximizes } w_k^{(i)}. \tag{12.35}$$

12.5 Soft K-means

We note that the K-means procedure in (12.5) and (12.6) is very similar to the EM procedure in (12.33) and (12.34) if we take a hard form of conditional probability, i.e., take

$$w_k^{(i)} = \begin{cases} 1 & \text{if } \mathbf{x}_i \text{ is assigned to group } k \\ 0 & \text{otherwise.} \end{cases} \tag{12.36}$$

Then the $\widehat{\boldsymbol{\mu}}_k$ in (12.34) becomes the sample mean of the observations assigned to cluster k.

A model for which model-based clustering mimics K-means clustering assumes that in (12.25), the covariance matrices $\boldsymbol{\Sigma}_k = \sigma^2 \mathbf{I}_p$ (model "EII" in (12.31)), so that

$$f_k(\mathbf{x}_i) = c \, \frac{1}{\sigma^p} \, e^{-\frac{1}{2\sigma^2}\|\mathbf{x}_i - \widehat{\boldsymbol{\mu}}_k\|^2}. \tag{12.37}$$

If σ is fixed, then the EM algorithm proceeds as above, except that the covariance calculation in (12.34) is unnecessary. If we let $\sigma \to 0$ in (12.33), fixing the means, we have that

$$\widehat{P}[Y = k \mid \mathbf{X} = \mathbf{x}_i] \longrightarrow w_k^{(i)} \tag{12.38}$$

for the $w_k^{(i)}$ in (12.36), at least if all the $\widehat{\pi}_k$'s are positive. Thus for small fixed σ, K-means and model-based clustering are practically the same.

Allowing σ to be estimated as well leads to **soft K-means**, soft because we use a weighted mean, where the weights depend on the distances from the observations to the group means. See Hastie et al. [2009]. In this case, the EM algorithm is as in

(12.33) and (12.34), but with the estimate of the covariance replaced with the pooled estimate of σ^2,

$$\widehat{\sigma}^2 = \frac{1}{n} \sum_{k=1}^{K} \sum_{i=1}^{n} w_k^{(i)} \, \|\mathbf{x}_i - \widehat{\mu}_k\|^2. \tag{12.39}$$

12.5.1 Example: Sports data

In Section 12.1.1, we used K-means to find clusters in the data on peoples' favorite sports. Here we use soft K-means. There are a couple of problems with using this model (12.37): (1) The data are discrete, not continuous as in the multivariate normal; (2) The dimension is actually 6, not 7, because each observation is a permutation of $1, \ldots, 7$, hence sums to the 28. To fix the latter problem, we multiply the data matrix by any orthogonal matrix whose first column is constant, then throw away the first column of the result (since it is a constant). Orthogonal polynomials are easy in R:

```
h <- poly(1:7,6) # Gives all but the constant term.
x <- sportsranks%*%h
```

The clustering can be implemented in Mclust by specifying the "EII" model in (12.31):

```
skms <- Mclust(x,modelNames='EII')
```

The shifted BIC's are

K	1	2	3	4	5
BIC	95.40	0	21.79	32.28	48.27

(12.40)

Clearly $K = 2$ is best, which is what we found using K-means in Section 12.1.1. It turns out the observations are clustered exactly the same for $K = 2$ whether using K-means or soft K-means. When $K = 3$, the two methods differ on only three observations, but for $K = 4$, 35 are differently clustered.

12.6 Hierarchical clustering

A **hierarchical clustering** gives a sequence of clusterings, each one combining two clusters of the previous stage. We assume n objects and their dissimilarities d as in (12.15). To illustrate, consider the five grades' variables in Section 12.2. A possible hierarchical sequence of clusterings starts with each object in its own group, then combines two of those elements, say midterms and final. The next step could combine two of the other singletons, or place one of them with the midterms/final group. Here we combine homework and labs, then combine all but inclass, then finally have one big group with all the objects:

$$\begin{aligned}
&\{\text{HW}\}\ \{\text{Labs}\}\ \{\text{InClass}\}\ \{\text{Midterms}\}\ \{\text{Final}\}\\
\rightarrow&\{\text{HW}\}\ \{\text{Labs}\}\ \{\text{InClass}\}\ \{\text{Midterms, Final}\}\\
\rightarrow&\{\text{HW, Labs}\}\ \{\text{InClass}\}\ \{\text{Midterms, Final}\}\\
\rightarrow&\{\text{InClass}\}\ \{\text{HW, Labs, Midterms, Final}\}\\
\rightarrow&\{\text{InClass, HW, Labs, Midterms, Final}\}.
\end{aligned} \tag{12.41}$$

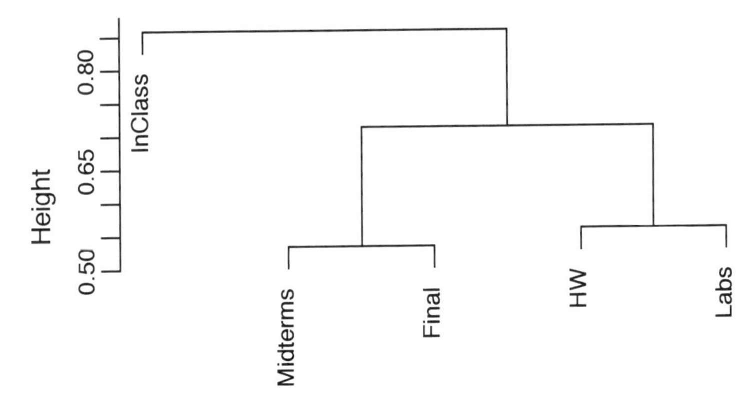

Figure 12.8: Hierarchical clustering of the grades, using complete linkage.

Reversing the steps and connecting, one obtains a tree diagram, or *dendrogram*, as in Figure 12.8.

For a set of objects, the question is which clusters to combine at each stage. At the first stage, we combine the two closest objects, that is, the pair $(\mathbf{o}_i, \mathbf{o}_j)$ with the smallest $d(\mathbf{o}_1, \mathbf{o}_j)$. At any further stage, we may wish to combine two individual objects, or a single object to a group, or two groups. Thus we need to decide how to measure the dissimilarity between any two groups of objects. There are many possibilities. Three popular ones look at the minimum, average, and maximum of the individuals' distances. That is, suppose **A** and **B** are subsets of objects. Then the three distances between the subsets are

$$\begin{aligned} \text{Single linkage: } d(\mathbf{A}, \mathbf{B}) &= \min_{\mathbf{a}\in\mathbf{A}, \mathbf{b}\in\mathbf{B}} d(\mathbf{a}, \mathbf{b}), \\ \text{Average linkage: } d(\mathbf{A}, \mathbf{B}) &= \frac{1}{\#\mathbf{A} \times \#\mathbf{B}} \sum_{\mathbf{a}\in\mathbf{A}} \sum_{\mathbf{b}\in\mathbf{B}} d(\mathbf{a}, \mathbf{b}), \\ \text{Complete linkage: } d(\mathbf{A}, \mathbf{B}) &= \max_{\mathbf{a}\in\mathbf{A}, \mathbf{b}\in\mathbf{B}} d(\mathbf{a}, \mathbf{b}). \end{aligned} \tag{12.42}$$

In all cases, $d(\{\mathbf{a}\}, \{\mathbf{b}\}) = d(\mathbf{a}, \mathbf{b})$. Complete linkage is an example of Hausdorff distance, at least when the d is a distance.

12.6.1 Example: Grades data

Consider the dissimilarities for the five variables of the grades data given in (12.19). The hierarchical clustering using these dissimilarities with complete linkage is given in Figure 12.8. This clustering is not surprising given the results of K-medoids in (12.21). As in (12.41), the hierarchical clustering starts with each object in its own cluster. Next we look for the smallest dissimilarity between two objects, which is the 0.53 between midterms and final. In the dendrogram, we see these two scores being connected at the height of 0.53.

We now have four clusters, with dissimilarity matrix

	HW	Labs	InClass	{Midterms, Final}
HW	0.00	0.56	0.86	0.71
Labs	0.56	0.00	0.80	0.71
InClass	0.86	0.80	0.00	0.81
{Midterms, Final}	0.71	0.71	0.81	0.00

(12.43)

(The dissimilarity between the cluster {midterms, final} and itself is not really zero, but we put zero there for convenience.) Because we are using complete linkage, the dissimilarity between a single object and the cluster with two objects is the maximum of the two individual dissimilarities. For example,

$$\begin{aligned} d(\{\text{HW}\}, \{\text{Midterms}, \text{Final}\}) &= \max\{d(\text{HW}, \text{Midterms}), d(\text{HW}, \text{Final})\} \\ &= \max\{0.71, 0.69\} \\ &= 0.71. \end{aligned} \tag{12.44}$$

The two closest clusters are now the singletons {homework} and {labs}, with a dissimilarity of 0.56. The new dissimilarity matrix is then

	{HW, Labs}	InClass	{Midterms, Final}
{HW, Labs}	0.00	0.86	0.71
InClass	0.86	0.00	0.81
{Midterms, Final}	0.71	0.81	0.00

(12.45)

The next step combines the two two-object clusters, and the final step places inclass with the rest.

To use R, we start with the dissimilarity matrix dx in (12.19). The routine hclust creates the tree, and plot plots it. We need the as.dist there to let the function know we already have the dissimilarities. Then Figure 12.8 is created by the statement

```
plot(hclust(as.dist(dx)))
```

12.6.2 Example: Sports data

Turn to the sports data from Section 12.1.1. Here we cluster the sports, using squared Euclidean distance as the dissimilarity, and compete linkage. To use squared Euclidean distances, use the dist function directly on the data matrix. Figure 12.9 is found using

```
plot(hclust(dist(t(sportsranks))))
```

Compare this plot to the K-means plot in Figure 12.4. We see somewhat similar closenesses among the sports.

Figure 12.10 clusters the individuals using complete linkage and single linkage, created using

```
par(mfrow=c(2,1))
dxs <- dist(sportsranks) # Gets Euclidean distances
lbl <- rep(' ',130) # Prefer no labels for the individuals
plot(hclust(dxs),xlab='Complete linkage',sub=' ',labels=lbl)
plot(hclust(dxs,method='single'),xlab='Single linkage',sub=' ',labels=lbl)
```

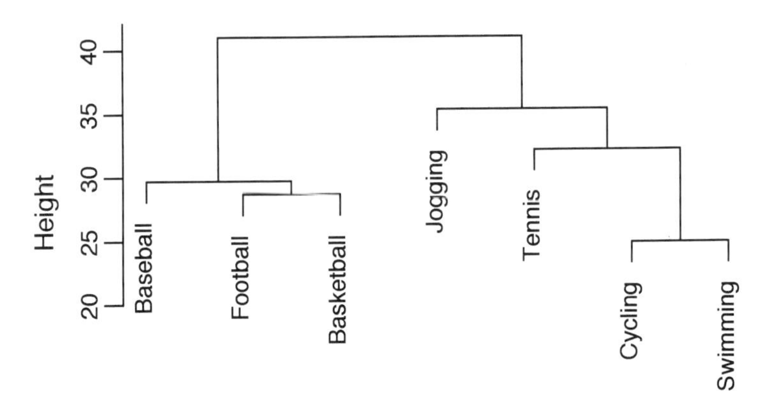

Figure 12.9: Hierarchical clustering of the sports, using complete linkage.

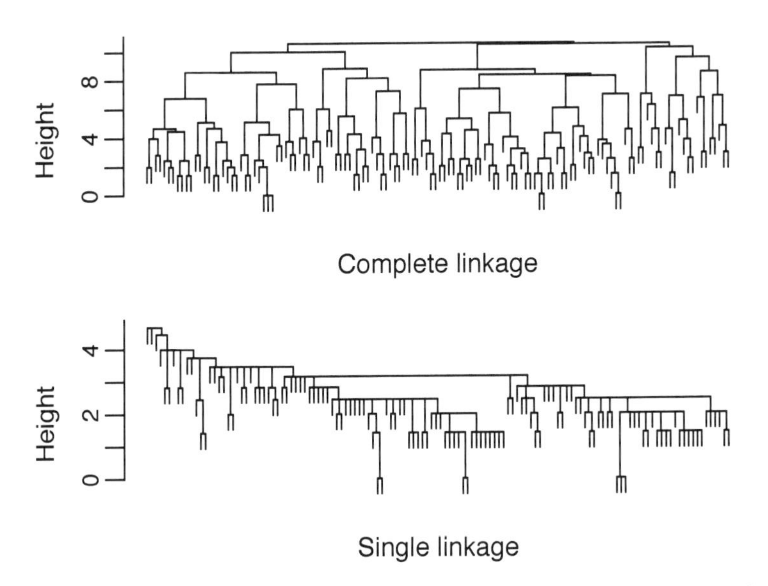

Figure 12.10: Clustering the individuals in the sports data, using complete linkage (top) and single linkage (bottom).

Complete linkage tends to favor similar-sized clusters, because by using the maximum distance, it is easier for two small clusters to get together than anything to attach itself to a large cluster. Single linkage tends to favor a few large clusters, and the rest small, because the larger the cluster, the more likely it will be close to small clusters. These ideas are borne out in the plot, where complete linkage yields a more treey-looking dendrogram.

12.7 Exercises

Exercise 12.7.1. Show that $|\mathbf{\Sigma}_k| = c_k^p$ in (12.29) follows from (12.27) and (12.28).

Exercise 12.7.2. (a) Show that the EM algorithm, where we use the $w_k^{(i)}$'s in (12.36) as the estimate of $\widehat{P}[Y = k \mid \mathbf{X} = \mathbf{x}_i]$, rather than that in (12.33), is the K-means algorithm of Section 12.1. [Note: You have to worry only about the mean in (12.34).] (b) Show that the limit as $\sigma \to 0$ of $\widehat{P}[Y = k \mid \mathbf{X} = \mathbf{x}_i]$ is indeed given in (12.36), if we use the f_k in (12.37) in (12.33).

Exercise 12.7.3 (Grades). This problem is to cluster the students in the grades data based on variables 2–6: homework, labs, inclass, midterms, and final. (a) Use K-means clustering for $K = 2$. (Use nstart=100, which is a little high, but makes sure everyone gets similar answers.) Look at the centers, and briefly characterize the clusters. Compare the men and women (variable 1, 0=male, 1=female) on which clusters they are in. (Be sure to take into account that there are about twice as many women as men.) Any differences? (b) Same question, for $K = 3$. (c) Same question, for $K = 4$. (d) Find the average silhouettes for the $K = 2, 3$ and 4 clusterings from parts (a), (b) and (c). Which K has the highest average silhouette? (e) Use soft K-means to find the $K = 1, 2, 3$ and 4 clusterings. Which K is best according to the BIC's? (Be aware that the BIC's in Mclust are negative what we use.) Is it the same as for the best K-means clustering (based on silhouettes) found in part (d)? (f) For each of $K = 2, 3, 4$, compare the classifications of the data using regular K-means to that of soft K-means. That is, match the clusters produced by both methods for given K, and count how many observations were differently clustered.

Exercise 12.7.4 (Diabetes). The data set diabetes [Reaven and Miller, 1979] is found in the R package mclust. There are $n = 145$ subjects and four variables. The first variable (class) is a categorical variable indicating whether the subject has overt diabetes (symptoms are obvious), chemical diabetes (can be detected only through chemical analysis of the blood), or is normal (no diabetes). The other three variables are blood measurements: glucose, insulin, sspg. (a) First, normalize the three blood measurement variables so that they have means zero and variances 1:

```
data(diabetes)
blood <- scale(diabetes[,2:4])
```

Use K-means to cluster the observations on the three normalized blood measurement variables for $K = 1, 2, \ldots, 9$. (b) Find the average silhouettes for the clusterings found in part (a), except for $K = 1$. Which K would you choose based on this criterion? (c) Use model-based clustering, again with $K = 1, \ldots 9$. Which model and K has the best BIC? (d) For each of the two "best" clusterings in parts (b) and (c), plot each pair of variables, indicating which cluster each point was assigned, as in Figure 12.6. Compare these to the same plots that use the class variable as the indicator. What do you notice? (e) For each of the two best clusterings, find the table comparing the clusters with the class variable. Which clustering was closest to the class variable? Why do you suppose that clustering was closest? (Look at the plots.)

Exercise 12.7.5 (Iris). This question applies model-based clustering to the iris data, pretending we do not know which observations are in which species. (a) Do the model-based clustering without any restrictions (i.e., use the defaults). Which model and number K was best, according to BIC? Compare the clustering for this best model to the actual species. (b) Now look at the BIC's for the model chosen in part (a), but for the various K's from 1 to 9. Calculate the corresponding estimated posterior probabilities. What do you see? (c) Fit the same model, but with $K = 3$. Now compare the clustering to the true species.

Exercise 12.7.6 (Grades). Verify the dissimilarity matrices in (12.43) and (12.45).

Exercise 12.7.7 (Soft drinks). The data set softdrinks has 23 peoples' ranking of 8 soft drinks: Coke, Pepsi, Sprite, 7-up, and their diet equivalents. Do a hierarchical clustering on the drinks, so that the command is

```
hclust(dist(t(softdrinks)))
```

then plot the tree with the appropriate labels. Describe the tree. Does the clustering make sense?

Exercise 12.7.8 (Cereal). Exercise 1.9.21 presented the cereal data (in the R data matrix cereal), finding the biplot. Do hierarchical clustering on the cereals, and on the attributes. Do the clusters make sense? What else would you like to know from these data? Compare the clusterings to the biplot. What is the advantage of the biplot relative to the hierarchical clusterings?

Chapter 13

Principal Components and Related Techniques

Data reduction is a common goal in multivariate analysis — one has too many variables, and wishes to reduce the number of them without losing much information. How to approach the reduction depends of course on the goal of the analysis. For example, in linear models, there are clear dependent variables (in the $\mathbf{Y}$ matrix) that we are trying to explain or predict from the explanatory variables (in the $\mathbf{x}$ matrix, and possibly the $\mathbf{z}$ matrix). Then Mallows' C_p or cross-validation are reasonable approaches. If the correlations between the $\mathbf{Y}$'s are of interest, then factor analysis is appropriate, where the likelihood ratio test is a good measure of how many factors to take. In classification, using cross-validation is a good way to decide on the variables. In model-based clustering, and in fact any situation with a likelihood, one can balance the fit and complexity of the model using something like AIC or BIC.

There are other situations in which the goal is not so clear cut as in those above; one is more interested in exploring the data, using data reduction to get a better handle on the data, in the hope that something interesting will reveal itself. The reduced data may then be used in more formal models, although I recommend first considering targeted reductions as mentioned in the previous paragraph, rather than immediately jumping to principal components.

Below we discuss principal components in more depth, then present multidimensional scaling, and canonical correlations.

13.1 Principal components, redux

Recall way back in Section 1.6 that the objective in principal component analysis was to find linear combinations (with norm 1) with maximum variance. As an exploratory technique, principal components can be very useful, as are other projection pursuit methods. The conceit underlying principal components is that variance is associated with interestingness, which may or may not hold. As long as in an exploratory mood, though, if one finds the top principal components are not particularly interesting or interpretable, then one can go in a different direction.

But be careful not to shift over to the notion that components with low variance can be ignored. It could very well be that they are the most important, e.g., most correlated with a separate variable of interest. Using principal components as the first step in a process, where one takes the first few principal components to use in another procedure, such as clustering or classification, may or may not work out well. In particular, it makes little sense to use principal components to reduce the variables before using them in a linear process such as regression, canonical correlations, or Fisher's linear discrimination. For example, in regression, we are trying to find the linear combination of $\mathbf{x}$'s that best correlates with $\mathbf{y}$. The answer is given by the fit, $\widehat{\mathbf{y}}$. Using principal components first on the $\mathbf{x}$'s will give us a few new variables that are linear combinations of $\mathbf{x}$'s, which we then further take linear combinations of to correlate with the $\mathbf{y}$. What we end up with is a worse correlation than if we just started with the original $\mathbf{x}$'s, since some parts of the $\mathbf{x}$'s are left behind. The same thinking goes when using linear discrimination: We want the linear combination of $\mathbf{x}$'s that best distinguishes the groups, not the best linear combination of a few linear combinations of the $\mathbf{x}$'s. Because factor analysis tries to account for correlations among the variables, if one transforms to principal components, which are uncorrelated, before applying factor analysis, then there will be no common factors. On the other hand, if one is using nonlinear techniques such as classification trees, first reducing by principal components may indeed help.

Even principal components are not unique. E.g., you must choose whether or not to take into account covariates or categorical factors before finding the sample covariance matrix. You also need to decide how to scale the variables, i.e., whether to leave them in their original units, or scale so that all variables have the same sample variance, or scale in some other way. The scaling will affect the principal components, unlike in a factor analysis or linear regression.

13.1.1 Example: Iris data

Recall that the Fisher-Anderson iris data (Section 1.3.1) has $n = 150$ observations and $q = 4$ variables. The measurements of the petals and sepals are in centimeters, so it is reasonable to leave the data unscaled. On the other hand, the variances of the variables do differ, so scaling so that each has unit variance is also reasonable. Furthermore, we could either leave the data unadjusted in the sense of subtracting the overall mean when finding the covariance matrix, or adjust the data for species by subtracting from each observation the mean of its species. Thus we have four reasonable starting points for principal components, based on whether we adjust for species and whether we scale the variables. Figure 13.1 has plots of the first two principal components for each of these possibilities. Note that there is a stark difference between the plots based on adjusted and unadjusted data. The unadjusted plots show a clear separation based on species, while the adjusted plots have the species totally mixed, which would be expected because there are differences in means between the species. Adjusting hides those differences. There are less obvious differences between the scaled and unscaled plots within adjusted/unadjusted pairs. For the adjusted data, the unscaled plot seems to have fairly equal spreads for the three species, while the scaled data has the virginica observations more spread out than the other two species.

The table below shows the sample variances, s^2, and first principal component's loadings (sample eigenvector), PC_1, for each of the four sets of principal components:

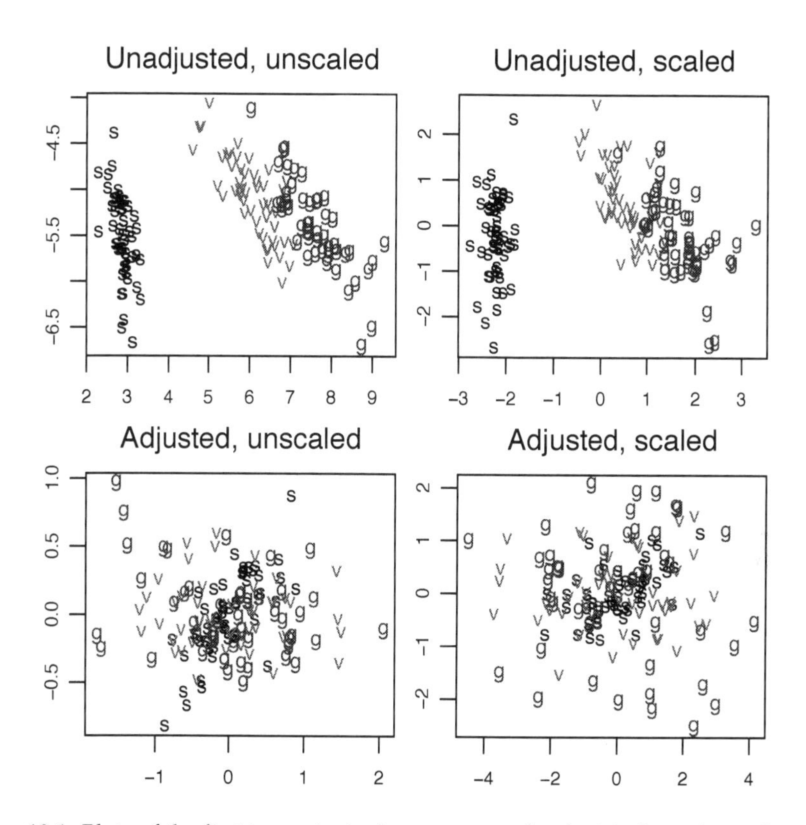

Figure 13.1: Plots of the first two principal components for the iris data, depending on whether adjusting for species and whether scaling the variables to unit variance. For the individual points, "s" indicates setosa, "v" indicates versicolor, and "g" indicates virginica.

	Unadjusted				Adjusted			
	Unscaled		Scaled		Unscaled		Scaled	
	s^2	PC_1	s^2	PC_1	s^2	PC_1	s^2	PC_1
Sepal Length	0.69	0.36	1	0.52	0.26	0.74	1	−0.54
Sepal Width	0.19	−0.08	1	−0.27	0.11	0.32	1	−0.47
Petal Length	3.12	0.86	1	0.58	0.18	0.57	1	−0.53
Petal Width	0.58	0.36	1	0.56	0.04	0.16	1	−0.45

(13.1)

Note that whether adjusted or not, the relative variances of the variables affect the relative weighting they have in the principal component. For example, for the unadjusted data, petal length has the highest variance in the unscaled data, and receives the highest loading in the eigenvector. That is, the first principal component is primarily sepal length. But for the scaled data, all variables are forced to have the same variance, and now the loadings of the variables are much more equal. The

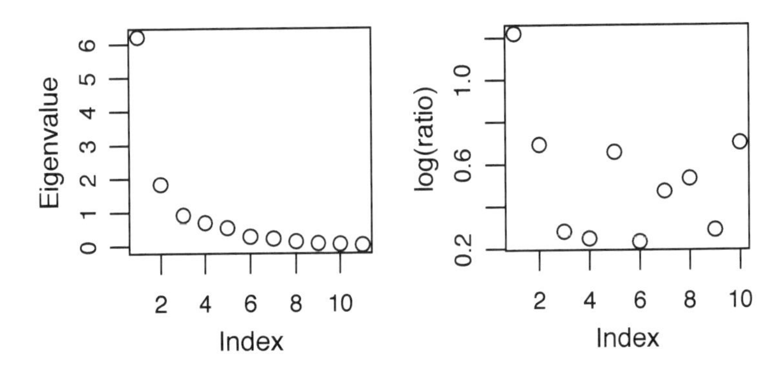

Figure 13.2: The left-hand plot is a scree plot (i versus l_i) of the eigenvalues for the automobile data. The right-hand plot shows i versus $\log(l_i/l_{i+1})$, the successive log-proportional gaps.

opposite holds for sepal width. A similar effect is seen for the adjusted data. The sepal length has the highest unscaled variance and highest loading in PC_1, and petal width the lowest variance and loading. But scaled, the loadings are approximately equal.

Any of the four sets of principal components is reasonable. Which to use depends on what one is interested in, e.g., if wanting to distinguish between species, the unadjusted plots are likely more interesting, while when interested in relations within species, adjusting make sense. In cases where the units are vastly different for the variables, e.g., populations in thousands and areas in square miles of cities, leaving the data unscaled is less defensible.

13.1.2 Choosing the number of principal components

One obvious question is, "How does one choose p?" Unfortunately, there is not any very good answer without some clear objective in mind. There are some reasonable guidelines to use, which can at least suggest which principal components to look at. Of course, nothing prevents you from looking at as many as you have time for.

The most common graphical technique for deciding on p is the **scree plot**, in which the sample eigenvalues are plotted versus their indices. (A scree is a pile of small stones at the bottom of a cliff.) Consider Example 12.3.1 on the automobile data, here using the $n = 96$ autos with trunk space, and all $q = 11$ variables. Scaling the variables so that they are have unit sample variance, we obtain the sample eigenvalues

$$6.210, 1.833, 0.916, 0.691, 0.539, 0.279, 0.221, 0.138, 0.081, 0.061, 0.030. \qquad (13.2)$$

The scree plot is the first one in Figure 13.2. Note that there is a big drop from the first to the second eigenvalue. There is a smaller drop to the third, then the values seem to level off. Other simple plots can highlight the gaps. For example, the second plot in the figure shows the logarithms of the successive proportional drops via

$$\log(ratio_i) \equiv \log\left(\frac{l_i}{l_{i+1}}\right). \tag{13.3}$$

The biggest drops are again from #1 to #2, and #2 to #3, but there are almost as large proportional drops at the fifth and tenth stages.

One may have outside information or requirements that aid in choosing the components. For examples, there may be a reason one wishes a certain number of components (say, three if the next step is a three-dimensional plot), or to have as few components as possible in order to achieve a certain percentage (e.g., 95%) of the total variance. If one has an idea that the measurement error for the observed variables is c, then it makes sense to take just the principal components that have eigenvalue significantly greater than c^2. Or, as in the iris data, all the data is accurate just to one decimal place, so that taking $c = 0.05$ is certainly defensible.

To assess significance, assume that

$$\mathbf{U} \sim \text{Wishart}_q(\nu, \boldsymbol{\Sigma}), \text{ and } \mathbf{S} = \frac{1}{\nu}\mathbf{U}, \tag{13.4}$$

where $\nu > q$ and $\boldsymbol{\Sigma}$ is invertible. Although we do not necessarily expect this distribution to hold in practice, it will help develop guidelines to use. Let the spectral decompositions of $\mathbf{S}$ and $\boldsymbol{\Sigma}$ be

$$\mathbf{S} = \mathbf{GLG}' \text{ and } \boldsymbol{\Sigma} = \boldsymbol{\Gamma\Lambda\Gamma}', \tag{13.5}$$

where $\mathbf{G}$ and $\boldsymbol{\Gamma}$ are each orthogonal matrices, and $\mathbf{L}$ and $\boldsymbol{\Lambda}$ are diagonal with nonincreasing diagonal elements (the eigenvalues), as in Theorem 1.1. The eigenvalues of $\mathbf{S}$ will be distinct with probability 1. If we assume that the eigenvalues of $\boldsymbol{\Sigma}$ are also distinct, then Theorem 13.5.1 in Anderson [2003] shows that for large ν, the sample eigenvalues are approximately independent, and $l_i \approx N(\lambda_i, 2\lambda_i^2/\nu)$. If components with $\lambda_i \leq c^2$ are ignorable, then it is reasonable to ignore the l_i for which

$$\sqrt{\nu}\,\frac{l_i - c^2}{\sqrt{2}\, l_i} < 2, \text{ equivalently, } l_i < \frac{c^2}{1 - 2\sqrt{2}/\sqrt{\nu}}. \tag{13.6}$$

(One may be tempted to take $c = 0$, but if any $\lambda_i = 0$, then the corresponding l_i will be zero as well, so that there is no need for hypothesis testing.) Other test statistics (or really "guidance statistics") can be easily derived, e.g., to see whether the average of the k smallest eigenvalues are less than c^2, or the sum of the first p are greater than some other cutoff.

13.1.3 Estimating the structure of the component spaces

If the eigenvalues of $\boldsymbol{\Sigma}$ are distinct, then the spectral decomposition (13.5) splits the q-dimensional space into q orthogonal one-dimensional spaces. If, say, the first two eigenvalues are equal, then the first two subspaces are merged into one two-dimensional subspace. That is, there is no way to distinguish between the top two dimensions. At the extreme, if all eigenvalues are equal, in which case $\boldsymbol{\Sigma} = \lambda\mathbf{I}_q$, there is no statistically legitimate reason to distinguish any principal components. More generally, suppose there are K distinct values among the λ_i's, say

$$\alpha_1 > \alpha_2 > \cdots > \alpha_K, \tag{13.7}$$

where q_k of the λ_i's are equal to α_k:

$$\begin{aligned}\lambda_1 = \cdots = \lambda_{q_1} &= \alpha_1,\\ \lambda_{q_1+1} = \cdots = \lambda_{q_1+q_2} &= \alpha_2,\\ &\vdots\\ \lambda_{q_1+\cdots+q_{K-1}+1} = \cdots = \lambda_q &= \alpha_K.\end{aligned} \tag{13.8}$$

Then the space is split into K orthogonal subspaces, of dimensions $q_1, \ldots, q_K$, where $q = q_1 + \cdots + q_K$. See Exercise 13.4.3. The vector $(q_1, \ldots, q_K)$ is referred to as the **pattern** of equalities among the eigenvalues. Let $\mathbf{\Gamma}$ be an orthogonal matrix containing eigenvectors as in (13.5), and partition it as

$$\mathbf{\Gamma} = \begin{pmatrix} \mathbf{\Gamma}_1 & \mathbf{\Gamma}_2 & \cdots & \mathbf{\Gamma}_K \end{pmatrix}, \ \mathbf{\Gamma}_k \text{ is } q \times q_k, \tag{13.9}$$

so that $\mathbf{\Gamma}_k$ contains the eigenvectors for the q_k eigenvalues that equal α_k. These are not unique because $\mathbf{\Gamma}_k\mathbf{J}$ for any $q_k \times q_k$ orthogonal matrix $\mathbf{J}$ will also yield a set of eigenvectors for those eigenvalues. The subspaces have corresponding projection matrices $\mathbf{P}_1, \ldots, \mathbf{P}_K$, which *are* unique, and we can write

$$\mathbf{\Sigma} = \sum_{k=1}^{K} \alpha_k \mathbf{P}_k, \ \text{ where } \mathbf{P}_k = \mathbf{\Gamma}_k\mathbf{\Gamma}_k'. \tag{13.10}$$

With this structure, the principal components can be defined only in groups, i.e., the first q_1 of them represent one group, which have higher variance than the next group of q_2 components, etc., down to the final q_K components. There is no distinction within a group, so that one would take either the top q_1 components, or the top $q_1 + q_2$, or the top $q_1 + q_2 + q_3$, etc.

Using the distributional assumption (13.4), we find the Bayes information criterion to choose among the possible patterns (13.8) of equality. The best set can then be used in plots such as in Figure 13.3, where the gaps will be either enhanced (if large) or eliminated (if small). The model (13.8) will be denoted $M_{(q_1,\ldots,q_K)}$. Anderson [1963] (see also Section 13.1.6) shows the following.

Theorem 13.1. *Suppose (13.4) holds, and* $\mathbf{S}$ *and* $\mathbf{\Sigma}$ *have spectral decompositions as in (13.5). Then the MLE of* $\mathbf{\Sigma}$ *under the model* $M_{(q_1,\ldots,q_K)}$ *is given by* $\widehat{\mathbf{\Sigma}} = \mathbf{G}\widehat{\mathbf{\Lambda}}\mathbf{G}'$, *where the* $\widehat{\lambda}_i$*'s are found by averaging the relevant* l_i*'s:*

$$\begin{aligned}\widehat{\lambda}_1 = \ldots = \widehat{\lambda}_{q_1} = \widehat{\alpha}_1 &= \frac{1}{q_1}\,(l_1 + \ldots + l_{q_1}),\\ \widehat{\lambda}_{q_1+1} = \cdots = \widehat{\lambda}_{q_1+q_2} = \widehat{\alpha}_2 &= \frac{1}{q_2}\,(l_{q_1+1} + \cdots + l_{q_1+q_2}),\\ &\vdots\\ \widehat{\lambda}_{q_1+\cdots+q_{K-1}+1} = \cdots = \widehat{\lambda}_q = \widehat{\alpha}_K &= \frac{1}{q_K}\,(l_{q_1+\cdots+q_{K-1}+1} + \cdots + l_q).\end{aligned} \tag{13.11}$$

The number of free parameters is

$$d(q_1, \ldots q_K) = \frac{1}{2}\,(q^2 - \sum_{k=1}^{K} q_k^2) + K. \tag{13.12}$$

The deviance can then be taken to be

$$\text{deviance}(M_{(q_1,\ldots q_K)}(\widehat{\boldsymbol{\Sigma}})\,;\,\mathbf{S}) = \nu \sum_{i=1}^{q} \log(\widehat{\lambda}_i) = \nu \sum_{k=1}^{K} q_k \log(\widehat{\alpha}_k). \tag{13.13}$$

See Exercise 13.4.4. Using (13.12), we have

$$\text{BIC}(M_{(q_1,\ldots q_K)}) = \nu \sum_{k=1}^{K} q_k \log(\widehat{\alpha}_k) + \log(\nu) d(q_1, \ldots q_K). \tag{13.14}$$

13.1.4 Example: Automobile data

Let $\mathbf{S}$ be the scaled covariance matrix for the automobiles with trunks, described in Section 12.3.1. Equation (13.2) and Figure 13.2 exhibit the eigenvalues of $\mathbf{S}$. They are denoted l_j in (13.15). We first illustrate the model (13.8) with pattern $(1,1,3,3,2,1)$. The MLE's of the eigenvalues are then found by averaging the third through fifth, the sixth through eighth, the ninth and tenth, and leaving the others alone, denoted below by the $\widehat{\lambda}_j$'s:

j	1	2	3	4	5
l_j	6.210	1.833	0.916	0.691	0.539
$\widehat{\lambda}_j$	6.210	1.833	0.716	0.716	0.716

j	6	7	8	9	10	11
l_j	0.279	0.221	0.138	0.081	0.061	0.030
$\widehat{\lambda}_j$	0.213	0.213	0.213	0.071	0.071	0.030

(13.15)

With $\nu = n - 1 = 95$,

$$\text{deviance}(M_{(1,1,3,3,2,1)}(\widehat{\boldsymbol{\Sigma}})\,;\,\mathbf{S}) = 95 \sum_{j} \log(\widehat{\lambda}_j) = -1141.398, \tag{13.16}$$

and $d(1,1,3,3,2,1) = 54$, hence

$$\text{BIC}(M_{(1,1,3,3,2,1)}) = -1141.398 + \log(95)\,54 = -895.489. \tag{13.17}$$

Table 13.1.4 contains a number of models, one each for K from 1 to 11. Each pattern after the first was chosen to be the best that is obtained from the previous by summing two consecutive q_k's. The estimated probabilities are among those in the table. Clearly, the preferred is the one with MLE in (13.15). Note that the assumption (13.4) is far from holding here, both because the data are not normal, and because we are using a correlation matrix rather than a covariance matrix. We are hoping, though, that in any case, the BIC is a reasonable balance of the fit of the model on the eigenvalues and the number of parameters.

Figure 13.3 shows the scree plots, using logs, of the sample eigenvalues and the fitted ones from the best model. Note that the latter gives more aid in deciding how many components to choose because the gaps are enhanced or eliminated. That is, taking one or two components is reasonable, but because there is little distinction

Pattern	d	BIC	BIC	$\widehat{P}^{BIC}$
$(1,1,1,1,1,1,1,1,1,1,1)$	66	-861.196	34.292	0.000
$(1,1,1,1,1,2,1,1,1,1)$	64	-869.010	26.479	0.000
$(1,1,1,2,2,1,1,1,1)$	62	-876.650	18.839	0.000
$(1,1,3,2,1,1,1,1)$	59	-885.089	10.400	0.004
$(1,1,3,2,1,2,1)$	57	-892.223	3.266	0.159
$(1,1,3,3,2,1)$	54	-895.489	0.000	0.812
$(1,1,3,3,3)$	51	-888.502	6.987	0.025
$(1,4,3,3)$	47	-870.824	24.665	0.000
$(1,4,6)$	37	-801.385	94.104	0.000
$(5,6)$	32	-657.561	237.927	0.000
11	1	4.554	900.042	0.000

Table 13.1: The BIC's for the sequence of principal component models for the automobile data.

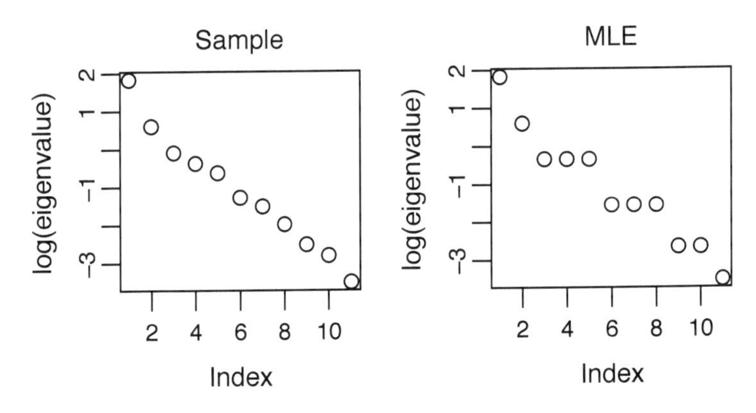

Figure 13.3: Plots of j versus the sample l_j's, and j versus the MLE $\widehat{\lambda}_j$'s, for the chosen model.

among the next three, one may as well take all or none of those three. Similarly with numbers six, seven and eight.

What about interpretations? Below we have the first five principal component loadings, multiplied by 100 then rounded off:

	PC_1	PC_2	PC_3	PC_4	PC_5
Length	36	−23	3	−7	26
Wheelbase	37	−20	−11	6	20
Width	35	−29	19	−11	1
Height	25	41	−41	10	−10
Front Head Room	19	30	68	45	−16
Rear Head Room	25	47	1	28	6
Front Leg Room	10	−49	−30	69	−37
Rear Seating	30	26	−43	2	18
Front Shoulder Room	37	−16	20	−11	10
Rear Shoulder Room	38	−1	7	−13	12
Luggage Area	28	6	−2	−43	−81

(13.18)

The first principal component has fairly equal positive loadings for all variables, indicating an overall measure of bigness. The second component tends to have positive loadings for tallness (height, front head room, rear head room), and negative loadings for the length and width-type variables. This component then measures tallness relative to length and width. The next three may be harder to interpret. Numbers 3 and 4 could be front seat versus back seat measurements, and number 5 is mainly luggage space. But from the analysis in (13.15), we have that from a statistical significance point of view, there is no distinction among the third through fifth components, that is, any rotation of them is equally important. Thus we might try a varimax rotation on the three vectors, to aid in interpretation. (See Section 10.3.2 for a description of varimax.) The R function varimax will do the job. The results are below:

	PC_3^*	PC_4^*	PC_5^*
Length	−4	−20	18
Wheelbase	−11	0	21
Width	13	−17	−6
Height	−32	29	−1
Front Head Room	81	15	7
Rear Head Room	11	18	19
Front Leg Room	4	83	6
Rear Seating	−42	10	18
Front Shoulder Room	12	−22	1
Rear Shoulder Room	0	−19	3
Luggage Area	−4	8	−92

(13.19)

These three components are easy to interpret, weighting heavily on front head room, front leg room, and luggage space, respectively.

Figure 13.4 plots the first two principal components. The horizontal axis represents the size, going from the largest at the left to the smallest at the right. The vertical axis has tall/narrow cars at the top, and short/wide at the bottom. We also performed model-based clustering (Section 12.3) using just these two variables. The best clustering has two groups, whose covariance matrices have the same eigenvalues but different eigenvectors ("EEV" in (12.31)), indicated by the two ellipses, which have the same size and shape, but have different orientations. These clusters are represented in the plot as well. We see the clustering is defined mainly by the tall/wide variable.

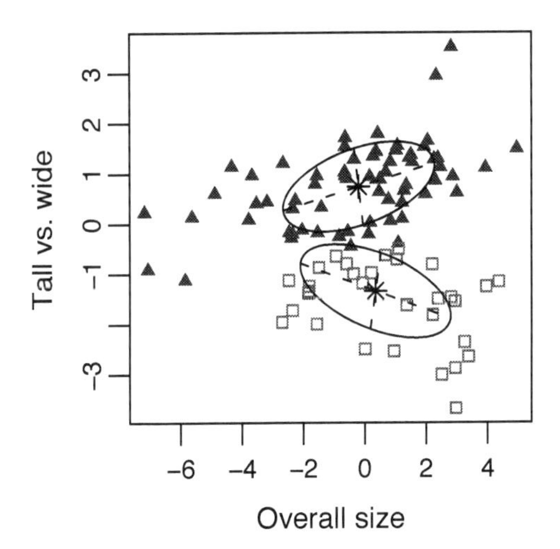

Figure 13.4: The first two principal component variables for the automobile data (excluding sports cars and minivans), clustered into two groups.

Using R

In Section 12.3.1 we created cars1, the reduced data set. To center and scale the data, so that the means are zero and variances are one, use

```
xcars <− scale(cars1)
```

The following obtains eigenvalues and eigenvectors of $\mathbf{S}$:

```
eg <− eigen(var(xcars))
```

The eigenvalues are in eg$values and the matrix of eigenvectors are in eg$vectors. To find the deviance and BIC for the pattern $(1,1,3,3,2,1)$ seen in (13.15) and (13.17), we use the function pcbic (detailed in Section A.5.1):

```
pcbic(eg$values,95,c(1,1,3,3,2,1))
```

In Section A.5.2 we present the function pcbic.stepwise, which uses the stepwise procedure to calculate the elements in Table 13.1.4:

```
pcbic.stepwise(eg$values,95)
```

13.1.5 Principal components and factor analysis

Factor analysis and principal components have some similarities and some differences. Recall the factor analysis model with p factors in (10.58). Taking the mean to be $\mathbf{0}$ for simplicity, we have

$$\mathbf{Y} = \mathbf{X}\beta + \mathbf{R}, \tag{13.20}$$

where $\mathbf{X}$ and $\mathbf{R}$ are independent, with

$$\mathbf{X} \sim N(\mathbf{0}, \mathbf{I}_n \otimes \mathbf{I}_p) \text{ and } \mathbf{R} \sim N(\mathbf{0}, \mathbf{I}_n \otimes \mathbf{\Psi}), \mathbf{\Psi} \text{ diagonal.} \tag{13.21}$$

For principal components, where we take the first p components, partition $\mathbf{\Gamma}$ and $\mathbf{\Lambda}$ in (13.5) as

$$\mathbf{\Gamma} = \begin{pmatrix} \mathbf{\Gamma}_1 & \mathbf{\Gamma}_2 \end{pmatrix} \text{ and } \mathbf{\Lambda} = \begin{pmatrix} \mathbf{\Lambda}_1 & \mathbf{0} \\ \mathbf{0} & \mathbf{\Lambda}_2 \end{pmatrix}. \tag{13.22}$$

Here, $\mathbf{\Gamma}_1$ is $q \times p$, Γ_2 is $q \times (q-p)$, $\mathbf{\Lambda}_1$ is $p \times p$, and $\mathbf{\Lambda}_2$ is $(q-p) \times (q-p)$, the $\mathbf{\Lambda}_k$'s being diagonal. The large eigenvalues are in $\mathbf{\Lambda}_1$, the small ones are in $\mathbf{\Lambda}_2$. Because $\mathbf{I}_q = \mathbf{\Gamma}\mathbf{\Gamma}' = \mathbf{\Gamma}_1\mathbf{\Gamma}_1' + \mathbf{\Gamma}_2\mathbf{\Gamma}_2'$, we can write

$$\mathbf{Y} = \mathbf{Y}\mathbf{\Gamma}_1\mathbf{\Gamma}_1' + \mathbf{Y}\mathbf{\Gamma}_2\mathbf{\Gamma}_2' = \mathbf{X}\boldsymbol{\beta} + \mathbf{R}, \tag{13.23}$$

where

$$\mathbf{X} = \mathbf{Y}\mathbf{\Gamma}_1 \sim N(\mathbf{0}, \mathbf{\Sigma}_X), \ \boldsymbol{\beta} = \mathbf{\Gamma}_1', \ \text{ and } \ \mathbf{R} = \mathbf{Y}\mathbf{\Gamma}_2\mathbf{\Gamma}_2' \sim N(\mathbf{0}, \mathbf{I}_n \otimes \mathbf{\Sigma}_R). \tag{13.24}$$

Because $\mathbf{\Gamma}_1'\mathbf{\Gamma}_2 = \mathbf{0}$, $\mathbf{X}$, and $\mathbf{R}$ are again independent. We also have (Exercise 13.4.5)

$$\mathbf{\Sigma}_X = \mathbf{\Gamma}_1'\mathbf{\Gamma}\mathbf{\Lambda}\mathbf{\Gamma}'\mathbf{\Gamma}_1 = \mathbf{\Lambda}_1, \ \text{ and } \ \mathbf{\Sigma}_R = \mathbf{\Gamma}_2\mathbf{\Gamma}_2'\mathbf{\Gamma}\mathbf{\Lambda}\mathbf{\Gamma}'\mathbf{\Gamma}_2\mathbf{\Gamma}_2' = \mathbf{\Gamma}_2\mathbf{\Lambda}_2\mathbf{\Gamma}_2'. \tag{13.25}$$

Comparing these covariances to the factor analytic ones in (13.20), we see the following:

	$\mathbf{\Sigma}_X$	$\mathbf{\Sigma}_R$
Factor analysis	$\mathbf{I}_p$	$\mathbf{\Psi}$
Principal components	$\mathbf{\Lambda}_1$	$\mathbf{\Gamma}_2\mathbf{\Lambda}_2\mathbf{\Gamma}_2'$

(13.26)

The key difference is in the residuals. Factor analysis chooses the p-dimensional $\mathbf{X}$ so that the residuals are uncorrelated, though not necessarily small. Thus the *correlations* among the $\mathbf{Y}$'s are explained by the factors $\mathbf{X}$. Principal components chooses the p-dimensional $\mathbf{X}$ so that the residuals are small (the variances sum to the sum of the $(q-p)$ smallest eigenvalues), but not necessarily uncorrelated. Much of the *variance* of the $\mathbf{Y}$ is explained by the components $\mathbf{X}$.

A popular model that fits into both frameworks is the factor analytic model (13.20) with the restriction that

$$\mathbf{\Psi} = \sigma^2\mathbf{I}_q, \quad \sigma^2 \text{ "small."} \tag{13.27}$$

The interpretation in principal components is that the $\mathbf{X}$ contains the important information in $\mathbf{Y}$, while the residuals $\mathbf{R}$ contain just random measurement error. For factor analysis, we have that the $\mathbf{X}$ explains the correlations among the $\mathbf{Y}$, and the residuals happen to have the same variances. In this case, we have

$$\mathbf{\Sigma} = \boldsymbol{\beta}\mathbf{\Sigma}_{XX}\boldsymbol{\beta}' + \sigma^2\mathbf{I}_q. \tag{13.28}$$

Because $\boldsymbol{\beta}$ is $q \times p$, there are at most p positive eigenvalues for $\boldsymbol{\beta}\mathbf{\Sigma}_{XX}\boldsymbol{\beta}'$. Call these $\lambda_1^* \geq \lambda_2^* \geq \cdots \geq \lambda_p^*$, and let $\mathbf{\Lambda}_1^*$ be the $p \times p$ diagonal matrix with diagonals λ_i^*. Then the spectral decomposition is

$$\boldsymbol{\beta}\mathbf{\Sigma}_{XX}\boldsymbol{\beta}' = \mathbf{\Gamma}\begin{pmatrix} \mathbf{\Lambda}_1^* & \mathbf{0} \\ \mathbf{0} & \mathbf{0} \end{pmatrix}\mathbf{\Gamma}' \tag{13.29}$$

for some orthogonal $\mathbf{\Gamma}$. But any orthogonal matrix contains eigenvectors for $\mathbf{I}_q = \mathbf{\Gamma}\mathbf{\Gamma}'$, hence $\mathbf{\Gamma}$ is also an eigenvector matrix for $\mathbf{\Sigma}$:

$$\mathbf{\Sigma} = \mathbf{\Gamma}\mathbf{\Lambda}\mathbf{\Gamma}' = \mathbf{\Gamma}\begin{pmatrix} \mathbf{\Lambda}_1^* + \sigma^2\mathbf{I}_p & \mathbf{0} \\ \mathbf{0} & \sigma^2\mathbf{I}_{q-p} \end{pmatrix}\mathbf{\Gamma}'. \tag{13.30}$$

Thus the eigenvalues of $\boldsymbol{\Sigma}$ are

$$\lambda_1^* + \sigma^2 \geq \lambda_2^* + \sigma^2 \geq \cdots \geq \lambda_p^* + \sigma^2 \geq \sigma^2 = \cdots = \sigma^2, \tag{13.31}$$

and the eigenvectors for the first p eigenvalues are the columns of $\boldsymbol{\Gamma}_1$. In this case the factor space and the principal component space are the same. In fact, if the λ_j^* are distinct and positive, the eigenvalues (13.30) satisfy the structural model (13.8) with pattern $(1, 1, \ldots, 1, q-p)$. A common approach to choosing p in principal components is to use hypothesis testing on such models to find the smallest p for which the model fits. See Anderson [1963] or Mardia, Kent, and Bibby [1979]. Of course, AIC or BIC could be used as well.

13.1.6 Justification of the principal component MLE

We first find the MLE of $\boldsymbol{\Sigma}$, and the maximal value of the likelihood, for $\mathbf{U}$ as in (13.4), where the eigenvalues of $\boldsymbol{\Sigma}$ satisfy (13.8). We know from (10.1) that the likelihood for $\mathbf{S}$ is

$$L(\boldsymbol{\Sigma}; \mathbf{S}) = |\boldsymbol{\Sigma}|^{-\nu/2} \, e^{-\frac{\nu}{2} \, \text{trace}(\boldsymbol{\Sigma}^{-1}\mathbf{S})}. \tag{13.32}$$

For the general model, i.e., where there is no restriction among the eigenvalues of $\boldsymbol{\Sigma}$ ($M_{(1,1,\ldots,1)}$), the MLE of $\boldsymbol{\Sigma}$ is $\mathbf{S}$.

Suppose there are nontrivial restrictions (13.8). Write $\boldsymbol{\Sigma}$ and $\mathbf{S}$ in their spectral decomposition forms (13.5) to obtain

$$\begin{aligned} L(\boldsymbol{\Sigma}; \mathbf{S}) &= |\boldsymbol{\Lambda}|^{-\nu/2} \, e^{-\frac{\nu}{2} \, \text{trace}((\boldsymbol{\Gamma}\boldsymbol{\Lambda}\boldsymbol{\Gamma}')^{-1}\mathbf{G}\mathbf{L}\mathbf{G}')} \\ &= (\prod \lambda_i)^{-\nu/2} \, e^{-\frac{1}{2} \, \text{trace}(\boldsymbol{\Lambda}^{-1}\boldsymbol{\Gamma}'\mathbf{G}\mathbf{L}\mathbf{G}'\boldsymbol{\Gamma})}. \end{aligned} \tag{13.33}$$

Because of the multiplicities in the eigenvalues, the $\boldsymbol{\Gamma}$ is not uniquely determined from $\boldsymbol{\Sigma}$, but any orthogonal matrix that maximizes the likelihood is adequate.

We start by fixing $\boldsymbol{\Lambda}$, and maximizing the likelihood over the $\boldsymbol{\Gamma}$. Set

$$\mathbf{F} = \mathbf{G}'\boldsymbol{\Gamma}, \tag{13.34}$$

which is also orthogonal, and note that

$$\begin{aligned} -\frac{1}{2} \text{trace}(\boldsymbol{\Lambda}^{-1}\mathbf{F}'\mathbf{L}\mathbf{F}) &= -\frac{1}{2} \sum_{i=1}^{q} \sum_{j=1}^{q} f_{ij}^2 \, \frac{l_i}{\lambda_j} \\ &= \sum_{i=1}^{q} \sum_{j=1}^{q} f_{ij}^2 \, l_i \beta_j - \frac{1}{2\lambda_q} \sum_{i=1}^{q} l_i, \end{aligned} \tag{13.35}$$

where

$$\beta_j = -\frac{1}{2\lambda_j} + \frac{1}{2\lambda_q}. \tag{13.36}$$

(The $1/(2\lambda_q)$ is added because we need the β_j's to be nonnegative in what follows.) Note that the last term in (13.35) is independent of $\mathbf{F}$. We do summation by parts by letting

$$d_i = l_i - l_{i+1}, \; \delta_i = \beta_i - \beta_{i+1}, \; 1 \leq i \leq q-1, \; d_q = l_q, \; \delta_q = \beta_q, \tag{13.37}$$

so that

$$l_i = \sum_{k=i}^{q} d_k \text{ and } \beta_j = \sum_{m=j}^{q} \delta_m. \tag{13.38}$$

Because the l_i's and λ_i's are nondecreasing in i, the l_i's are positive, and by (13.36) the β_j's are also nonnegative, we have that the δ_i's and d_i's are all nonnegative. Using (13.38) and interchanging the orders of summation, we have

$$\begin{aligned}\sum_{i=1}^{q}\sum_{j=1}^{q} f_{ij}^2\, l_i\beta_j &= \sum_{i=1}^{q}\sum_{j=1}^{q}\sum_{k=i}^{q}\sum_{m=j}^{q} f_{ij}^2\, d_k\delta_m \\ &= \sum_{k=1}^{q}\sum_{m=1}^{q} d_k\delta_m \sum_{i=1}^{k}\sum_{j=1}^{m} f_{ij}^2.\end{aligned} \tag{13.39}$$

Because $\mathbf{F}$ is an orthogonal matrix,

$$\sum_{i=1}^{k}\sum_{j=1}^{m} f_{ij}^2 \le \min\{k, m\}. \tag{13.40}$$

Also, with $\mathbf{F} = \mathbf{I}_q$, the $f_{ii} = 1$, hence expression in (13.40) is an equality. By the nonnegativity of the d_k's and δ_m's, the sum in (13.39), hence in (13.35), is maximized (though not uniquely) by taking $\mathbf{F} = \mathbf{I}_q$. Working back, from (13.34), the (not necessarily unique) maximizer over $\mathbf{\Gamma}$ of (13.35) is

$$\widehat{\mathbf{\Gamma}} = \mathbf{G}. \tag{13.41}$$

Thus the maximum over $\mathbf{\Gamma}$ in the likelihood (13.33) is

$$L(\mathbf{G\Lambda G}'; \mathbf{S}) = (\prod \lambda_i)^{-\nu/2} e^{-\frac{\nu}{2}\sum(l_i/\lambda_i)}. \tag{13.42}$$

Break up the product according to the equalities (13.8):

$$\prod_{k=1}^{K} \alpha_k^{-q_k\nu/2} e^{-\frac{\nu}{2}(t_k/\alpha_k)}, \tag{13.43}$$

where

$$t_1 = \sum_{i=1}^{q_1} l_i \text{ and } t_k = \sum_{i=q_1+\cdots+q_{k-1}+1}^{q_1+\cdots+q_k} l_i \quad \text{for } 2 \le k \le K. \tag{13.44}$$

It is easy to maximize over each α_k in (13.43), which proves that (13.11) is indeed the MLE of the eigenvalues. Thus with (13.41), we have the MLE of $\mathbf{\Sigma}$ as in Theorem 13.1.

We give a heuristic explanation of the dimension (13.12) of the model $M_{(q_1,\ldots q_K)}$ in (13.8). To describe the model, we need the K distinct parameters among the λ_i's, as well as the K orthogonal subspaces that correspond to the distinct values of λ_i. We start by counting the number of free parameters needed to describe an s-dimensional subspace of a t-dimensional space, $s < t$. Any such subspace can be described by a $t \times s$ basis matrix $\mathbf{B}$, that is, the columns of $\mathbf{B}$ comprise a basis for the subspace. (See Section 5.2.) The basis is not unique, in that $\mathbf{BA}$ for any invertible $s \times s$ matrix $\mathbf{A}$ is also a basis matrix, and in fact any basis matrix equals $\mathbf{BA}$ for some such $\mathbf{A}$. Take $\mathbf{A}$

to be the inverse of the top $s \times s$ submatrix of $\mathbf{B}$, so that $\mathbf{BA}$ has $\mathbf{I}_s$ as its top $s \times s$ part. This matrix has $(t-s) \times s$ free parameters, represented in the bottom $(t-s) \times s$ part of it, and is the only basis matrix with $\mathbf{I}_s$ at the top. Thus the dimension is $(t-s) \times s$. (If the top part of $\mathbf{B}$ is not invertible, then we can find some other subset of s rows to use.)

Now for model (13.8), we proceed stepwise. There are $q_1(q_2 + \cdots + q_K)$ parameters needed to specify the first q_1-dimensional subspace. Next, focus on the subspace orthogonal to that first one. It is $(q_2 + \cdots + q_K)$-dimensional, hence to describe the second, q_2-dimensional, subspace within that, we need $q_2 \times (q_3 + \cdots + q_K)$ parameters. Continuing, the total number of parameters is

$$q_1(q_2 + \cdots + q_K) + q_2(q_3 + \cdots + q_K) \; + \cdots + \; q_{K-1}q_k = \frac{1}{2}\,(q^2 - \sum_{k=1}^{K} q_k^2). \tag{13.45}$$

Adding K for the distinct λ_i's, we obtain the dimension in (13.12).

13.2 Multidimensional scaling

Given n objects and defined distances, or dissimilarities, between them, multidimensional scaling tries to mimic the dissimilarities as close as possible by representing the objects in p-dimensional Euclidean space, where p is fairly small. That is, suppose $\mathbf{o}_1, \ldots, \mathbf{o}_n$ are the objects, and $d(\mathbf{o}_i, \mathbf{o}_j)$ is the dissimilarity between the ith and j^{th} objects, as in (12.15). Let $\mathbf{\Delta}$ be the $n \times n$ matrix of $d^2(\mathbf{o}_i, \mathbf{o}_j)$'s. The goal is to find $1 \times p$ vectors $\widehat{\mathbf{x}}_1, \ldots, \widehat{\mathbf{x}}_n$ so that

$$\Delta_{ij} = d^2(\mathbf{o}_i, \mathbf{o}_j) \approx \|\widehat{\mathbf{x}}_i - \widehat{\mathbf{x}}_j\|^2. \tag{13.46}$$

Then the $\widehat{\mathbf{x}}_i$'s are plotted in $\mathbb{R}^p$, giving an approximate visual representation of the original dissimilarities.

There are a number of approaches. Our presentation here follows that of Mardia, Kent, and Bibby [1979], which provides more in-depth coverage. We will start with the case that the original dissimilarities are themselves Euclidean distances, and present the so-called *classical* solution. Next, we exhibit the classical solution when the distances may not be Euclidean. Finally, we briefly mention the *non-metric* approach.

13.2.1 Δ is Euclidean: The classical solution

Here we assume that object $\mathbf{o}_i$ has associated a $1 \times q$ vector $\mathbf{y}_i$, and let $\mathbf{Y}$ be the $n \times q$ matrix with $\mathbf{y}_i$ as the ith row. Of course, now $\mathbf{Y}$ looks like a regular data matrix. It might be that the objects are observations (people), or they are variables, in which case this $\mathbf{Y}$ is really the transpose of the usual data matrix. Whatever the case,

$$d^2(\mathbf{o}_i, \mathbf{o}_j) = \|\mathbf{y}_i - \mathbf{y}_j\|^2. \tag{13.47}$$

For any $n \times p$ matrix $\mathbf{X}$ with rows $\mathbf{x}_i$, define $\mathbf{\Delta}(\mathbf{X})$ to be the $n \times n$ matrix of $\|\mathbf{x}_i - \mathbf{x}_j\|^2$'s, so that (13.47) can be written $\mathbf{\Delta} = \mathbf{\Delta}(\mathbf{Y})$.

The **classical solution** looks for $\widehat{\mathbf{x}}_i$'s in (13.46) that are based on rotations of the $\mathbf{y}_i$'s, much like principal components. That is, suppose $\mathbf{B}$ is a $q \times p$ matrix with orthonormal columns, and set

$$\widehat{\mathbf{x}}_i = \mathbf{y}_i\mathbf{B}, \text{ and } \widehat{\mathbf{X}} = \mathbf{Y}\mathbf{B}. \tag{13.48}$$

The objective is then to choose the $\mathbf{B}$ that minimizes

$$\sum\sum_{1 \le i < j \le n} \left| \|\mathbf{y}_i - \mathbf{y}_j\|^2 - \|\widehat{\mathbf{x}}_i - \widehat{\mathbf{x}}_j\|^2 \right| \tag{13.49}$$

over $\mathbf{B}$. Exercise 13.4.10 shows that $\|\mathbf{y}_i - \mathbf{y}_j\|^2 > \|\widehat{\mathbf{x}}_i - \widehat{\mathbf{x}}_j\|^2$, which means that the absolute values in (13.49) can be removed. Also, note that the sum over the $\|\mathbf{y}_i - \mathbf{y}_j\|^2$ is independent of $\mathbf{B}$, and by symmetry, minimizing (13.49) is equivalent to maximizing

$$\sum_{i=1}^{n}\sum_{j=1}^{n} \|\widehat{\mathbf{x}}_i - \widehat{\mathbf{x}}_j\|^2 \equiv \mathbf{1}_n'\mathbf{\Delta}(\widehat{\mathbf{X}})\mathbf{1}_n. \tag{13.50}$$

The next lemma is useful in relating the $\mathbf{\Delta}(\mathbf{Y})$ to the deviations.

Lemma 13.1. *Suppose* $\mathbf{X}$ *is* $n \times p$. *Then*

$$\mathbf{\Delta}(\mathbf{X}) = \mathbf{a}\mathbf{1}_n' + \mathbf{1}_n\mathbf{a}' - 2\mathbf{H}_n\mathbf{X}\mathbf{X}'\mathbf{H}_n, \tag{13.51}$$

where $\mathbf{a}$ *is the* $n \times 1$ *vector with* $\mathbf{a}_i = \|\mathbf{x}_i - \overline{\mathbf{x}}\|^2$, $\overline{\mathbf{x}}$ *is the mean of the* $\mathbf{x}_i$*'s, and* $\mathbf{H}_n$ *is the centering matrix (1.12).*

Proof. Write

$$\begin{aligned}\|\mathbf{x}_i - \mathbf{x}_j\|^2 &= \|(\mathbf{x}_i - \overline{\mathbf{x}}) - (\mathbf{x}_j - \overline{\mathbf{x}})\|^2 \\ &= \mathbf{a}_i - 2(\mathbf{H}_n\mathbf{X})_i(\mathbf{H}_n\mathbf{X})_j' + \mathbf{a}_j,\end{aligned} \tag{13.52}$$

from which we obtain (13.51). □

By (13.50) and (13.51) we have

$$\begin{aligned}\sum_{i=1}^{n}\sum_{j=1}^{n} \|\widehat{\mathbf{x}}_i - \widehat{\mathbf{x}}_j\|^2 &= \mathbf{1}_n'\mathbf{\Delta}(\widehat{\mathbf{X}})\mathbf{1}_n \\ &= \mathbf{1}_n'(\mathbf{a}\mathbf{1}_n' + \mathbf{1}_n\mathbf{a}' - 2\mathbf{H}_n\mathbf{X}\mathbf{X}'\mathbf{H}_n)\mathbf{1}_n \\ &= 2n\mathbf{1}_n'\mathbf{a} = 2n\sum_{i=1}^{n} \|\widehat{\mathbf{x}}_i - \overline{\widehat{\mathbf{x}}}\|^2,\end{aligned} \tag{13.53}$$

because $\mathbf{H}_n\mathbf{1}_n = \mathbf{0}$. But then

$$\begin{aligned}\sum_{i=1}^{n} \|\widehat{\mathbf{x}}_i - \overline{\widehat{\mathbf{x}}}\|^2 &= \text{trace}(\widehat{\mathbf{X}}'\mathbf{H}_n\widehat{\mathbf{X}}) \\ &= \text{trace}(\mathbf{B}'\mathbf{Y}'\mathbf{H}_n\mathbf{Y}\mathbf{B}).\end{aligned} \tag{13.54}$$

Maximizing (13.54) over $\mathbf{B}$ is a principal components task. That is, as in Proposition 1.1, this trace is maximized by taking $\mathbf{B} = \mathbf{G}_1$, the first p eigenvectors of $\mathbf{Y}'\mathbf{H}_n\mathbf{Y}$. To summarize:

Proposition 13.1. *If $\Delta = \Delta(\mathbf{Y})$, then the classical solution of the multidimensional scaling problem for given p is $\widehat{\mathbf{X}} = \mathbf{Y}\mathbf{G}_1$, where the columns of $\mathbf{G}_1$ consist of the first p eigenvectors of $\mathbf{Y}'\mathbf{H}_n\mathbf{Y}$.*

If one is interested in the distances between *variables*, so that the distances of interest are in $\Delta(\mathbf{Y}')$ (note the transpose), then the classical solution uses the first p eigenvectors of $\mathbf{Y}\mathbf{H}_q\mathbf{Y}'$.

13.2.2 Δ may not be Euclidean: The classical solution

Here, we are given only the $n \times n$ dissimilarity matrix Δ. The dissimilarities may or may not arise from Euclidean distance on vectors $\mathbf{y}_i$, but the solution acts as if they do. That is, we assume there is an $n \times q$ matrix $\mathbf{Y}$ such that $\Delta = \Delta(\mathbf{Y})$, but we do not observe the $\mathbf{Y}$, nor do we know the dimension q. The first step in the process is to derive the $\mathbf{Y}$ from the $\Delta(\mathbf{Y})$, then we apply Proposition 13.1 to find the $\widehat{\mathbf{X}}$.

It turns out that we can assume any value of q as long as it is larger than the values of p we wish to entertain. Thus we are safe taking $q = n$. Also, note that using (13.51), we can see that $\Delta(\mathbf{H}_n\mathbf{X}) = \Delta(\mathbf{X})$, which implies that the sample mean of $\mathbf{Y}$ is indeterminate. Thus we may as well assume the mean is zero, i.e.,

$$\mathbf{H}_n\mathbf{Y} = \mathbf{Y}, \tag{13.55}$$

so that (13.51) yields

$$\Delta = \Delta(\mathbf{Y}) = \mathbf{a}\mathbf{1}_n' + \mathbf{1}_n\mathbf{a}' - 2\mathbf{Y}\mathbf{Y}'. \tag{13.56}$$

To eliminate the $\mathbf{a}$'s, we can pre- and post-multiply by $\mathbf{H}_n$:

$$\mathbf{H}_n\Delta\mathbf{H}_n = \mathbf{H}_n(\mathbf{a}\mathbf{1}_n' + \mathbf{1}_n\mathbf{a}' - 2\mathbf{Y}\mathbf{Y}')\mathbf{H}_n = -2\mathbf{Y}\mathbf{Y}', \tag{13.57}$$

hence

$$\mathbf{Y}\mathbf{Y}' = -\frac{1}{2}\mathbf{H}_n\Delta\mathbf{H}_n. \tag{13.58}$$

Now consider the spectral decomposition (1.33) of $\mathbf{Y}\mathbf{Y}'$,

$$\mathbf{Y}\mathbf{Y}' = \mathbf{J}\mathbf{L}\mathbf{J}', \tag{13.59}$$

where the orthogonal $\mathbf{J} = (\mathbf{j}_1, \ldots, \mathbf{j}_n)$ contains the eigenvectors, and the diagonal $\mathbf{L}$ contains the eigenvalues. Separating the matrices, we can take

$$\mathbf{Y} = \mathbf{J}\mathbf{L}^{1/2} = \begin{pmatrix} \sqrt{l_1}\,\mathbf{j}_1 & \sqrt{l_2}\,\mathbf{j}_2 & \cdots & \sqrt{l_n}\,\mathbf{j}_n \end{pmatrix}. \tag{13.60}$$

Now we are in the setting of Section 13.2.1, hence by Proposition 13.1, we need $\mathbf{G}_1$, the first p eigenvectors of

$$\mathbf{Y}'\mathbf{H}_n\mathbf{Y} = (\mathbf{J}\mathbf{L}^{1/2})'(\mathbf{J}\mathbf{L}^{1/2}) = \mathbf{L}^{1/2}\mathbf{J}'\mathbf{J}\mathbf{L}^{1/2} = \mathbf{L}. \tag{13.61}$$

But the matrix of eigenvectors is just $\mathbf{I}_n$, hence we take the first p columns of $\mathbf{Y}$ in (13.60):

$$\widehat{\mathbf{X}} = \begin{pmatrix} \sqrt{l_1}\,\mathbf{j}_1 & \sqrt{l_2}\,\mathbf{j}_2 & \cdots & \sqrt{l_p}\,\mathbf{j}_p \end{pmatrix}. \tag{13.62}$$

It could be that Δ is not Euclidean, that is, there is no $\mathbf{Y}$ for which $\Delta = \Delta(\mathbf{Y})$. In this case, the classical solution uses the same algorithm as in equations (13.58) to (13.62). A possible glitch is that some of the eigenvalues may be negative, but if p is small, the problem probably won't raise its ugly head.

13.2.3 Non-metric approach

The original approach to multidimensional scaling attempts to find the $\widehat{\mathbf{X}}$ that gives the same ordering of the observed dissimilarities, rather than trying to match the actual values of the dissimilarities. See Kruskal [1964]. That is, take the $t \equiv \binom{n}{2}$ pairwise dissimilarities and order them:

$$d(\mathbf{o}_{i_1}, \mathbf{o}_{j_1}) \leq d(\mathbf{o}_{i_2}, \mathbf{o}_{j_2}) \leq \cdots \leq d(\mathbf{o}_{i_t}, \mathbf{o}_{j_t}). \tag{13.63}$$

The ideal would be to find the $\widehat{\mathbf{x}}_i$'s so that

$$\|\widehat{\mathbf{x}}_{i_1} - \widehat{\mathbf{x}}_{j_1}\|^2 \leq \|\widehat{\mathbf{x}}_{i_2} - \widehat{\mathbf{x}}_{j_2}\|^2 \leq \cdots \leq \|\widehat{\mathbf{x}}_{i_t} - \widehat{\mathbf{x}}_{j_t}\|^2. \tag{13.64}$$

That might not be (actually probably is not) possible for given p, so instead one finds the $\widehat{\mathbf{X}}$ that comes as close as possible, where close is measured by some "stress" function. A popular stress function is given by

$$\text{Stress}^2(\widehat{\mathbf{X}}) = \frac{\sum\sum_{1\leq i<j\leq n}(\|\widehat{\mathbf{x}}_i - \widehat{\mathbf{x}}_j\|^2 - d_{ij}^*)^2}{\sum\sum_{1\leq i<j\leq n}\|\widehat{\mathbf{x}}_i - \widehat{\mathbf{x}}_j\|^4}, \tag{13.65}$$

where the d_{ij}^*'s are constants that have the same ordering as the original dissimilarities $d(\mathbf{o}_i, \mathbf{o}_j)$'s in (13.63), and among such orderings minimize the stress. See Johnson and Wichern [2007] for more details and some examples. The approach is "non-metric" because it does not depend on the actual d's, but just their order.

13.2.4 Examples: Grades and sports

The examples here all start with Euclidean distance matrices, and use the classical solution, so everything is done using principal components.

In Section 12.6.1 we clustered the five variables for the grades data (homework, labs, inclass, midterms, final) for the $n = 107$ students. Here we find the multidimensional scaling plot. The distance between two variables is the sum of squares of the difference in scores the people obtained for them. So we take the transpose of the data. The following code finds the plot.

```
ty <- t(grades[,2:6])
ty <- scale(ty,scale=F)
ev <- eigen(var(ty))$vectors[,1:2]
tyhat <- ty%*%ev
lm <- range(tyhat)*1.1
plot(tyhat,xlim=lm,ylim=lm,xlab='Var 1',ylab='Var 2',type='n')
text(tyhat,labels=dimnames(ty)[[1]])
```

To plot the variables' names rather than points, we first create the plot with no plotting: type='n'. Then text plots the characters in the labels parameters, which we give as the names of the first dimension of ty. The results are in Figure 13.5. Notice how the inclass variable is separated from the others, homework and labs are fairly close, and midterms and final are fairly close together, not surprisingly given the clustering.

Figure 13.5 also has the multidimensional scaling plot of the seven sports from the Louis Roussos sports data in Section 12.1.1, found by substituting sportsranks for grades[,2:6] in the R code. Notice that the first variable orders the sports according

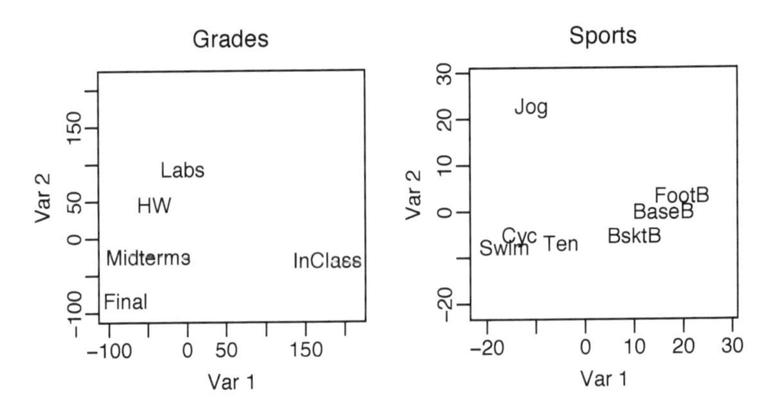

Figure 13.5: Multidimensional scaling plot of the grades' variables (left) and the sports' variables (right).

to how many people typically participate, i.e., jogging, swimming and cycling can be done solo, tennis needs two to four people, basketball has five per team, baseball nine, and football eleven. The second variable serves mainly to separate jogging from the others.

13.3 Canonical correlations

Testing the independence of two sets of variables in the multivariate normal distribution is equivalent to testing their covariance is zero, as in Section 10.2. When there is no independence, one may wish to know where the lack of independence lies. A projection-pursuit approach is to find the linear combination of the first set that is most highly correlated with a linear combination of the second set, hoping to isolate the factors within each group that explain a substantial part of the overall correlations.

The distributional assumption is based on partitioning the $1 \times q$ vector $\mathbf{Y}$ into $\mathbf{Y}_1$ ($1 \times q_1$) and $\mathbf{Y}_2$ ($1 \times q_2$), with

$$\mathbf{Y} = \begin{pmatrix} \mathbf{Y}_1 & \mathbf{Y}_2 \end{pmatrix} \sim N(\boldsymbol{\mu}, \boldsymbol{\Sigma}), \text{ where } \boldsymbol{\Sigma} = \begin{pmatrix} \boldsymbol{\Sigma}_{11} & \boldsymbol{\Sigma}_{12} \\ \boldsymbol{\Sigma}_{21} & \boldsymbol{\Sigma}_{22} \end{pmatrix}, \tag{13.66}$$

$\boldsymbol{\Sigma}_{11}$ is $q_1 \times q_1$, and $\boldsymbol{\Sigma}_{22}$ is $q_2 \times q_2$. If $\boldsymbol{\alpha}$ ($q_1 \times 1$) and $\boldsymbol{\beta}$ ($q_2 \times 1$) are coefficient vectors, then

$$\begin{aligned} Cov[\mathbf{Y}_1\boldsymbol{\alpha}, \mathbf{Y}_2\boldsymbol{\beta}] &= \boldsymbol{\alpha}'\boldsymbol{\Sigma}_{12}\boldsymbol{\beta}, \\ Var[\mathbf{Y}_1\boldsymbol{\alpha}] &= \boldsymbol{\alpha}'\boldsymbol{\Sigma}_{11}\boldsymbol{\alpha}, \text{ and} \\ Var[\mathbf{Y}_2\boldsymbol{\beta}] &= \boldsymbol{\beta}'\boldsymbol{\Sigma}_{22}\boldsymbol{\beta}, \end{aligned} \tag{13.67}$$

hence

$$Corr[\mathbf{Y}_1\boldsymbol{\alpha}, \mathbf{Y}_2\boldsymbol{\beta}] = \frac{\boldsymbol{\alpha}'\boldsymbol{\Sigma}_{12}\boldsymbol{\beta}}{\sqrt{\boldsymbol{\alpha}'\boldsymbol{\Sigma}_{11}\boldsymbol{\alpha}\ \boldsymbol{\beta}'\boldsymbol{\Sigma}_{22}\boldsymbol{\beta}}}. \tag{13.68}$$

Analogous to principal component analysis, the goal in canonical correlation analysis is to maximize the correlation (13.68) over $\boldsymbol{\alpha}$ and $\boldsymbol{\beta}$. Equivalently, we could maximize the covariance $\boldsymbol{\alpha}'\boldsymbol{\Sigma}_{12}\boldsymbol{\beta}$ in (13.68) subject to the two variances equaling one. This pair of linear combination vectors may not explain all the correlations between $\mathbf{Y}_1$ and $\mathbf{Y}_2$, hence we next find the maximal correlation over linear combinations uncorrelated with the first combinations. We continue, with each pair maximizing the correlation subject to being uncorrelated with the previous.

The precise definition is below. Compare it to Definition 1.2 for principal components.

Definition 13.1. *Canonical correlations.* *Assume $(\mathbf{Y}_1, \mathbf{Y}_2)$ are as in (13.66), where $\boldsymbol{\Sigma}_{11}$ and $\boldsymbol{\Sigma}_{22}$ are invertible, and set $m = \min\{q_1, q_2\}$. Let $\boldsymbol{\alpha}_1, \ldots, \boldsymbol{\alpha}_m$ be a set of $q_1 \times 1$ vectors, and $\boldsymbol{\beta}_1, \ldots, \boldsymbol{\beta}_m$ be a set of $q_2 \times 1$ vectors, such that*

$$\begin{aligned}
(\boldsymbol{\alpha}_1, \boldsymbol{\beta}_1) &\text{ is any } (\boldsymbol{\alpha}, \boldsymbol{\beta}) \text{ that maximizes } \boldsymbol{\alpha}'\boldsymbol{\Sigma}_{12}\boldsymbol{\beta} \text{ over } \boldsymbol{\alpha}'\boldsymbol{\Sigma}_{11}\boldsymbol{\alpha} = \boldsymbol{\beta}'\boldsymbol{\Sigma}_{22}\boldsymbol{\beta} = 1;\\
(\boldsymbol{\alpha}_2, \boldsymbol{\beta}_2) &\text{ is any } (\boldsymbol{\alpha}, \boldsymbol{\beta}) \text{ that maximizes } \boldsymbol{\alpha}'\boldsymbol{\Sigma}_{12}\boldsymbol{\beta} \text{ over } \boldsymbol{\alpha}'\boldsymbol{\Sigma}_{11}\boldsymbol{\alpha} = \boldsymbol{\beta}'\boldsymbol{\Sigma}_{22}\boldsymbol{\beta} = 1,\\
&\qquad \boldsymbol{\alpha}'\boldsymbol{\Sigma}_{11}\boldsymbol{\alpha}_1 = \boldsymbol{\beta}'\boldsymbol{\Sigma}_{22}\boldsymbol{\beta}_1 = 0;\\
&\vdots\\
(\boldsymbol{\alpha}_m, \boldsymbol{\beta}_m) &\text{ is any } (\boldsymbol{\alpha}, \boldsymbol{\beta}) \text{ that maximizes } \boldsymbol{\alpha}'\boldsymbol{\Sigma}_{12}\boldsymbol{\beta} \text{ over } \boldsymbol{\alpha}'\boldsymbol{\Sigma}_{11}\boldsymbol{\alpha} = \boldsymbol{\beta}'\boldsymbol{\Sigma}_{22}\boldsymbol{\beta} = 1,\\
&\qquad \boldsymbol{\alpha}'\boldsymbol{\Sigma}_{11}\boldsymbol{\alpha}_i = \boldsymbol{\beta}'\boldsymbol{\Sigma}_{22}\boldsymbol{\beta}_i = 0,\\
&\qquad i = 1, \ldots, m-1.
\end{aligned} \tag{13.69}$$

Then $\delta_i \equiv \boldsymbol{\alpha}_i'\boldsymbol{\Sigma}_{12}\boldsymbol{\beta}_i$ is the ith ***canonical correlation****, and $\boldsymbol{\alpha}_i$ and $\boldsymbol{\beta}_i$ are the associated* ***canonical correlation loading*** *vectors.*

Recall that principal component analysis (Definition 1.2) led naturally to the spectral decomposition theorem (Theorem 1.1). Similarly, canonical correlation analysis will lead to the **singular value decomposition** (Theorem 13.2 below). We begin the canonical correlation analysis with some simplifications. Let

$$\boldsymbol{\psi}_i = \boldsymbol{\Sigma}_{11}^{1/2}\boldsymbol{\alpha}_i \text{ and } \boldsymbol{\gamma}_i = \boldsymbol{\Sigma}_{22}^{1/2}\boldsymbol{\beta}_i \tag{13.70}$$

for each i, so that the $\boldsymbol{\gamma}_i$'s and $\boldsymbol{\psi}_i$'s are sets of orthonormal vectors, and

$$\delta_i = Corr[\mathbf{Y}_1\boldsymbol{\alpha}_i, \mathbf{Y}_2\boldsymbol{\beta}_i] = \boldsymbol{\psi}_i'\boldsymbol{\Xi}\boldsymbol{\gamma}_i, \tag{13.71}$$

where

$$\boldsymbol{\Xi} = \boldsymbol{\Sigma}_{11}^{-1/2}\boldsymbol{\Sigma}_{12}\boldsymbol{\Sigma}_{22}^{-1/2}. \tag{13.72}$$

This matrix $\boldsymbol{\Xi}$ is a multivariate generalization of the correlation coefficient which is useful here, but I don't know exactly how it should be interpreted.

In what follows, we assume that $q_1 \geq q_2 = m$. The $q_1 < q_2$ case can be handled similarly. The matrix $\boldsymbol{\Xi}'\boldsymbol{\Xi}$ is a $q_2 \times q_2$ symmetric matrix, hence by the spectral decomposition in (1.33), there is a $q_2 \times q_2$ orthogonal matrix $\boldsymbol{\Gamma}$ and a $q_2 \times q_2$ diagonal matrix $\boldsymbol{\Lambda}$ with diagonal elements $\lambda_1 \geq \lambda_2 \geq \cdots \geq \lambda_{q_2}$ such that

$$\boldsymbol{\Gamma}'\boldsymbol{\Xi}'\boldsymbol{\Xi}\boldsymbol{\Gamma} = \boldsymbol{\Lambda}. \tag{13.73}$$

Let $\gamma_1, \ldots, \gamma_{q_2}$ denote the columns of $\mathbf{\Gamma}$, so that the ith column of $\Xi\mathbf{\Gamma}$ is $\Xi\gamma_i$. Then (13.73) shows these columns are orthogonal and have squared lengths equal to the λ_i's, i.e.,

$$\|\Xi\gamma_i\|^2 = \lambda_i \text{ and } (\Xi\gamma_i)'(\Xi\gamma_j) = 0 \text{ if } i \neq j. \tag{13.74}$$

Furthermore, because the γ_i's satisfy the equations for the principal components' loading vectors in (1.28) with $\mathbf{S} = \Xi'\Xi$,

$$\|\Xi\gamma_i\| = \sqrt{\lambda_i} \text{ maximizes } \|\Xi\gamma\| \text{ over } \|\gamma\| = 1, \gamma'\gamma_j = 0 \text{ for } j < i. \tag{13.75}$$

Now for the first canonical correlation, we wish to find unit vectors ψ and γ to maximize $\psi'\Xi\gamma$. By Corollary 8.2, for γ fixed, the maximum over ψ is when ψ is proportional to $\Xi\gamma$, hence

$$\psi'\Xi\gamma \leq \|\Xi\gamma\|, \text{ equality achieved with } \psi = \frac{\Xi\gamma}{\|\Xi\gamma\|}. \tag{13.76}$$

But by (13.75), $\|\Xi\gamma\|$ is maximized when $\gamma = \gamma_1$, hence

$$\psi'\Xi\gamma \leq \psi_1'\Xi\gamma_1 = \sqrt{\lambda_1}, \quad \psi_1 = \frac{\Xi\gamma_1}{\|\Xi\gamma_1\|}. \tag{13.77}$$

Thus the first canonical correlation δ_1 is $\sqrt{\lambda_1}$. (The ψ_1 is arbitrary if $\lambda_1 = 0$.)

We proceed for $i = 2, \ldots, k$, where k is the index of the last positive eigenvalue, i.e., $\lambda_k > 0 = \lambda_{k+1} = \cdots = \lambda_{q_2}$. For the ith canonical correlation, we need to find unit vectors ψ orthogonal to $\psi_1, \ldots, \psi_{i-1}$, and γ orthogonal to $\gamma_1, \ldots, \gamma_{i-1}$, that maximize $\psi'\Xi\gamma$. Again by (13.75), the best γ is γ_i, and the best ψ is proportional to $\Xi\gamma_i$, so that

$$\psi'\Xi\gamma \leq \psi_i'\Xi\gamma_i = \sqrt{\lambda_i} \equiv \delta_i, \quad \psi_i = \frac{\Xi\gamma_i}{\|\Xi\gamma_i\|}. \tag{13.78}$$

That this ψ_i is indeed orthogonal to previous ψ_j's follows from the second equation in (13.74).

If $\lambda_i = 0$, then $\psi'\Xi\gamma_i = 0$ for any ψ. Thus the canonical correlations for $i > k$ are $\delta_i = 0$, and the corresponding ψ_i's, $i > k$, can be any set of vectors such that $\psi_1, \ldots, \psi_{q_2}$ are orthonormal.

Finally, to find the $\boldsymbol{\alpha}_i$ and$\boldsymbol{\beta}_i$ in (13.69), we solve the equations in (13.70):

$$\boldsymbol{\alpha}_i = \mathbf{\Sigma}_{11}^{-1/2}\psi_i \text{ and } \boldsymbol{\beta}_i = \mathbf{\Sigma}_{22}^{-1/2}\gamma_i, \quad i = 1, \ldots, m. \tag{13.79}$$

Backing up a bit, we have almost obtained the singular value decomposition of Ξ. Note that by (13.78), $\delta_i = \|\Xi\gamma_i\|$, hence we can write

$$\begin{aligned} \Xi\mathbf{\Gamma} &= \begin{pmatrix} \Xi\gamma_1 & \cdots & \Xi\gamma_{q_2} \end{pmatrix} \\ &= \begin{pmatrix} \psi_1\delta_1 & \cdots & \psi_{q_2}\delta_{q_2} \end{pmatrix} \\ &= \mathbf{\Psi\Delta}, \end{aligned} \tag{13.80}$$

where $\mathbf{\Psi} = (\psi_1, \ldots, \psi_{q_2})$ has orthonormal columns, and $\mathbf{\Delta}$ is diagonal with $\delta_1, \ldots, \delta_{q_2}$ on the diagonal. Shifting the $\mathbf{\Gamma}$ to the other side of the equation, we obtain the following.

Theorem 13.2. ***Singular value decomposition.*** *The $q_1 \times q_2$ matrix Ξ can be written*

$$\Xi = \mathbf{\Psi}\mathbf{\Delta}\mathbf{\Gamma}' \tag{13.81}$$

where $\mathbf{\Psi}$ ($q_1 \times m$) and $\mathbf{\Gamma}$ ($q_2 \times m$) have orthonormal columns, and $\mathbf{\Delta}$ is an $m \times m$ diagonal matrix with diagonals $\delta_1 \geq \delta_2 \geq \cdots \geq \delta_m \geq 0$, where $m = \min\{q_1, q_2\}$.

To summarize:

Corollary 13.1. *Let (13.81) be the singular value decomposition of $\Xi = \mathbf{\Sigma}_{11}^{-1/2}\mathbf{\Sigma}_{12}\mathbf{\Sigma}_{22}^{-1/2}$ for model (13.66). Then for $1 \leq i \leq \min\{q_1, q_2\}$, the ith canonical correlation is δ_i, with loading vectors $\boldsymbol{\alpha}_i$ and $\boldsymbol{\beta}_i$ given in (13.79), where ψ_i (γ_i) is the ith column of $\mathbf{\Psi}$ ($\mathbf{\Gamma}$).*

Next we present an example, where we use the estimate of $\mathbf{\Sigma}$ to estimate the canonical correlations. Theorem 13.3 guarantees that the estimates are the MLE's.

13.3.1 Example: Grades

Return to the grades data. In Section 10.3.3, we looked at factor analysis, finding two main factors: An overall ability factor, and a contrast of homework and labs versus midterms and final. Here we lump in inclass assignments with homework and labs, and find the canonical correlations between the sets (homework, labs, inclass) and (midterms, final), so that $q_1 = 3$ and $q_2 = 2$. The $\mathbf{Y}$ is the matrix of residuals from the model (10.80). In R,

```
x <− cbind(1,grades[,1])
y <− grades[,2:6]−x%∗%solve(t(x)%∗%x,t(x))%∗%grades[,2:6]
s <− t(y)%∗%y/(nrow(y)−2)
corr <− cov2cor(s)
```

The final statement calculates the correlation matrix from the $\mathbf{S}$, yielding

	HW	Labs	InClass	Midterms	Final
HW	1.00	0.78	0.28	0.41	0.40
Labs	0.78	1.00	0.42	0.38	0.35
InClass	0.28	0.42	1.00	0.24	0.27
Midterms	0.41	0.38	0.24	1.00	0.60
Final	0.40	0.35	0.27	0.60	1.00

(13.82)

There are $q_1 \times q_2 = 6$ correlations between variables in the two sets. Canonical correlations aim to summarize the overall correlations by the two $\widehat{\delta}_i$'s. The estimate of the Ξ matrix in (13.69) is given by

$$\begin{aligned}\widehat{\Xi} &= \mathbf{S}_{11}^{-1/2}\mathbf{S}_{12}\mathbf{S}_{22}^{-1/2} \\ &= \begin{pmatrix} 0.236 & 0.254 \\ 0.213 & 0.146 \\ 0.126 & 0.185 \end{pmatrix},\end{aligned} \tag{13.83}$$

found in R using

```
symsqrtinv1 <− symsqrtinv(s[1:3,1:3])
symsqrtinv2 <− symsqrtinv(s[4:5,4:5])
xi <− symsqrtinv1%∗%s[1:3,4:5]%∗%symsqrtinv2
```

where

```
symsqrtinv <- function(x) {
    ee <- eigen(x)
    ee$vectors%*%diag(sqrt(1/ee$values))%*%t(ee$vectors)
}
```

calculates the inverse symmetric square root of an invertible symmetric matrix x. The singular value decomposition function in R is called svd:

```
sv <- svd(xi)
a <- symsqrtinv1%*%sv$u
b <- symsqrtinv2%*%sv$v
```

The component sv$u is the estimate of $\mathbf{\Psi}$ and the component sv$v is the estimate of $\mathbf{\Gamma}$ in (13.81). The matrices of loading vectors are obtained as in (13.79):

$$\mathbf{A} = \mathbf{S}_{11}^{-1/2}\widehat{\mathbf{\Psi}} = \begin{pmatrix} -0.065 & 0.059 \\ -0.007 & -0.088 \\ -0.014 & 0.039 \end{pmatrix},$$

$$\text{and} \quad \mathbf{B} = \mathbf{S}_{22}^{-1/2}\widehat{\mathbf{\Gamma}} = \begin{pmatrix} -0.062 & -0.12 \\ -0.053 & 0.108 \end{pmatrix}. \tag{13.84}$$

The estimated canonical correlations (singular values) are in the vector sv$d, which are

$$d_1 = 0.482 \;\text{ and }\; d_2 = 0.064. \tag{13.85}$$

The d_1 is fairly high, and d_2 is practically negligible. (See the next section.) Thus it is enough to look at the first columns of $\mathbf{A}$ and $\mathbf{B}$. We can change signs, and take the first loadings for the first set of variables to be $(0.065, 0.007, 0.014)$, which is primarily the homework score. For the second set of variables, the loadings are $(0.062, 0.053)$, essentially a straight sum of midterms and final. Thus the correlations among the two sets of variables can be almost totally explained by the correlation between homework and the sum of midterms and final, which correlation is 0.45, versus the optimum of 0.48.

13.3.2 How many canonical correlations are positive?

One might wonder how many of the δ_i's are nonzero. We can use BIC to get an idea. The model is based on the usual

$$\mathbf{S} = \frac{1}{\nu}\,\mathbf{U}, \text{ where } \mathbf{U} \sim \text{Wishart}_q(\nu, \mathbf{\Sigma}), \tag{13.86}$$

$\nu \geq q$, with $\mathbf{\Sigma}$ partitioned as in (13.66), and $\mathbf{S}$ partitioned similarly. Let $\mathbf{\Psi\Delta\Gamma}'$ in (13.81) be the singular value decomposition of Ξ as in (13.72). Then model K $(1 \leq K \leq m \equiv \min\{q_1, q_2\})$ is given by

$$M_K: \; \delta_1 \geq \delta_2 \geq \cdots \geq \delta_K > \delta_{K+1} = \cdots = \delta_m = 0, \tag{13.87}$$

where the δ_i's are the canonical correlations, i.e., diagonals of $\mathbf{\Delta}$. Let

$$\mathbf{S}_{11}^{-1/2}\mathbf{S}_{12}\mathbf{S}_{22}^{-1/2} = \mathbf{PDG}' \tag{13.88}$$

be the sample analog of Ξ (on the left), and its singular value decomposition (on the right).

We first obtain the MLE of $\mathbf{\Sigma}$ under model K. Note that $\mathbf{\Sigma}$ and $(\mathbf{\Sigma}_{11}, \mathbf{\Sigma}_{22}, \Xi)$ are in one-to-one correspondence. Thus it is enough to find the MLE of the latter set of parameters. The next theorem is from Fujikoshi [1974].

Theorem 13.3. *For the above setup, the MLE of $(\mathbf{\Sigma}_{11}, \mathbf{\Sigma}_{22}, \Xi)$ under model M_K in (13.87) is given by*

$$(\mathbf{S}_{11}, \mathbf{S}_{22}, \mathbf{P}\mathbf{D}^{(K)}\mathbf{G}'), \tag{13.89}$$

where $\mathbf{D}^{(K)}$ is the diagonal matrix with diagonals $(d_1, d_2, \ldots, d_K, 0, 0, \ldots, 0)$.

That is, the MLE is obtained by setting to zero the sample canonical correlations that are set to zero in the model. One consequence of the theorem is that the natural sample canonical correlations and accompanying loading vectors are indeed the MLE's. The deviance, for comparing the models M_K, can be expressed as

$$\text{deviance}(M_K) = \nu \sum_{i=1}^{K} \log(1 - d_i^2). \tag{13.90}$$

See Exercise 13.4.15.

For the BIC, we need the dimension of the model. The number of parameters for the $\mathbf{\Sigma}_{ii}$'s we know to be $q_i(q_i + 1)/2$. For Ξ, we look at the singular value decomposition (13.81):

$$\Xi = \mathbf{\Psi}\mathbf{\Delta}^{(K)}\mathbf{\Gamma}' = \sum_{i=1}^{K} \delta_i \psi_i \gamma_i'. \tag{13.91}$$

The dimension for the δ_i's is K. Only the first K of the ψ_i's enter into the equation. Thus the dimension is the same as for principal components with K distinct eigenvalues, and the rest equal at 0, yielding pattern $(1, 1, \ldots, 1, q_1 - K)$, where there are K ones. Similarly, the γ_i's dimension is as for pattern $(1, 1, \ldots, 1, q_2 - K)$. Then by (13.45),

$$\begin{aligned} \dim(\mathbf{\Gamma}) + \dim(\mathbf{\Psi}) + \dim(\mathbf{\Delta}^{(K)}) =& \frac{1}{2}\,(q_1^2 - K - (q_1 - K)^2) \\ &+ \frac{1}{2}\,(q_2^2 - K - (q_2 - K)^2) + K \\ =& K(q - K). \end{aligned} \tag{13.92}$$

Finally, we can take

$$\text{BIC}(M_K) = \nu \sum_{k=1}^{K} \log(1 - d_k^2) + \log(\nu) K(q - K) \tag{13.93}$$

because the $q_i(q_i + 1)/2$ parts are the same for each model.

In the example, we have three models: $K = 0, 1, 2$. $K = 0$ means the two sets of variables are independent, which we already know is not true, and $K = 2$ is the unrestricted model. The calculations, with $\nu = 105$, $d_1 = 0.48226$ and $d_2 = 0.064296$:

K	Deviance	$\dim(\Xi)$	BIC	$\widehat{p}^{BIC}$
0	0	0	0	0.0099
1	−27.7949	4	−9.1791	0.9785
2	−28.2299	6	−0.3061	0.0116

(13.94)

Clearly $K = 1$ is best, which is what we figured above.

13.3.3 Partial least squares

A similar idea is to find the linear combinations of the variables to maximize the covariance, rather than correlation:

$$Cov[\mathbf{Y}_1\mathbf{a}, \mathbf{Y}_2\mathbf{b}] = \mathbf{a}'\mathbf{\Sigma}_{12}\mathbf{b}. \tag{13.95}$$

The process is the same as for canonical correlations, but we use the singular value decomposition of $\mathbf{\Sigma}_{12}$ instead of Ξ. The procedure is called **partial least squares,** but it could have been called **canonical covariances**. It is an attractive alternative to canonical correlations when there are many variables and not many observations, in which cases the estimates of $\mathbf{\Sigma}_{11}$ and $\mathbf{\Sigma}_{22}$ are not invertible.

13.4 Exercises

Exercise 13.4.1. In the model (13.4), find the approximate test for testing the null hypothesis that the average of the last k ($k < q$) eigenvalues is less than the constant c^2.

Exercise 13.4.2. Verify the expression for $\mathbf{\Sigma}$ in (13.10).

Exercise 13.4.3. Consider the $q \times q$ covariance matrix $\mathbf{\Sigma} = \mathbf{\Gamma}\mathbf{\Lambda}\mathbf{\Gamma}'$ as in the discussion from (13.7) to (13.10), where there are q_k eigenvalues equal to α_k. For each k, let $\mathcal{A}_k$ be the set of all eigenvectors corresponding to the eigenvalue α_k, plus the $\mathbf{0}$ vector. That is,

$$\mathcal{A}_k = \{\mathbf{v} \mid \mathbf{\Sigma}\mathbf{v} = \alpha_k\mathbf{v}\}. \tag{13.96}$$

(a) Substituting $\mathbf{u} = \mathbf{\Gamma}'\mathbf{v}$, show that $\mathcal{A}_k = \{\mathbf{\Gamma}\mathbf{u} \mid \mathbf{\Lambda}\mathbf{u} = \alpha_k\mathbf{u}\}$. Argue that $\mathbf{u}$ satisfies the given condition if and only if the only nonzero components of $\mathbf{u}$ are $u_{q_1+\cdots+q_{k-1}+1}$ to $u_{q_1+\cdots+q_k}$, i.e., those with indices i such that $\lambda_i = \alpha_k$. (b) Show that therefore $\mathcal{A}_k = \text{span}\{\text{columns of } \mathbf{\Gamma}_k\}$, and that $\mathbf{P}_k$ in (13.10) is its projection matrix. (c) Suppose $q_k = 1$, i.e., $\lambda_i = \alpha_k$ for a single i. Show that the only eigenvectors corresponding to α_k with norm 1 are $\pm\gamma_i$, the ith column of $\mathbf{\Gamma}$. [This result implies that if all the eigenvalues are distinct, i.e., $q_k = 1$ for each k (and $K = q$), then the matrix $\mathbf{\Gamma}$ is unique up to changing signs of the columns.]

Exercise 13.4.4. Show that the deviance for the model in Theorem 13.1 is given by (13.13). [Hint: Start with the likelihood as in (13.32). Show that

$$\text{trace}(\widehat{\mathbf{\Sigma}}^{-1}\mathbf{S}) = \sum_{i=1}^{q} \frac{l_i}{\widehat{\lambda}_i} = q. \tag{13.97}$$

Argue you can then ignore the part of the deviance that comes from the exponent.]

Exercise 13.4.5. Verify (13.25). [Hint: First, show that $\mathbf{\Gamma}_1'\mathbf{\Gamma} = (\mathbf{I}_p\ \mathbf{0})$ and $\mathbf{\Gamma}_2'\mathbf{\Gamma} = (\mathbf{0}\ \mathbf{I}_{q-p})$.]

Exercise 13.4.6. Show that (13.30) follows from (13.28) and (13.29).

Exercise 13.4.7. Prove (13.40). [Hint: First, explain why $\sum_{i=1}^k f_{ij}^2 \le 1$ and $\sum_{j=1}^m f_{ij}^2 \le 1$.]

Exercise 13.4.8. Verify the equality in (13.42), and show that (13.11) does give the maximizers of (13.43).

Exercise 13.4.9. Verify the equality in (13.45).

Exercise 13.4.10. Show that $\|\mathbf{y}_i - \mathbf{y}_j\|^2 > \|\widehat{\mathbf{x}}_i - \widehat{\mathbf{x}}_j\|^2$ for $\mathbf{y}_i$ and $\widehat{\mathbf{x}}_i$ in (13.48). [Hint: Start by letting $\mathbf{B}_2$ be any $(q-p) \times q$ matrix such that $(\mathbf{B}, \mathbf{B}_2)$ is an orthogonal matrix. Then $\|\mathbf{y}_i - \mathbf{y}_j\|^2 = \|(\mathbf{y}_i - \mathbf{y}_j)(\mathbf{B}, \mathbf{B}_2)\|^2$ (why?), and by expanding equals $\|(\mathbf{y}_i - \mathbf{y}_j)\mathbf{B}\|^2 + \|(\mathbf{y}_i - \mathbf{y}_j)\mathbf{B}_2\|^2$.]

Exercise 13.4.11. Verify (13.52) by expanding the second expression.

Exercise 13.4.12. In (13.80), verify that $\Xi\gamma_i = \psi\delta_i$.

Exercise 13.4.13. For the canonical correlations situation in Corollary 13.1, let $\boldsymbol{\alpha} = (\boldsymbol{\alpha}_1, \ldots, \boldsymbol{\alpha}_m)$ and $\boldsymbol{\beta} = (\boldsymbol{\beta}_1, \ldots, \boldsymbol{\beta}_m)$ be matrices with columns being the loading vectors. Find the covariance matrix of the transformation

$$\begin{pmatrix} \mathbf{Y}_1\boldsymbol{\alpha} & \mathbf{Y}_2\boldsymbol{\beta} \end{pmatrix} = \begin{pmatrix} \mathbf{Y}_1 & \mathbf{Y}_2 \end{pmatrix} \begin{pmatrix} \boldsymbol{\alpha} & \mathbf{0} \\ \mathbf{0} & \boldsymbol{\beta} \end{pmatrix}. \tag{13.98}$$

[It should depend on the parameters only through the δ_i's.]

Exercise 13.4.14. Given the singular value decomposition of Ξ in (13.81), find the spectral decompositions of $\Xi\Xi'$ and of $\Xi'\Xi$. What can you say about the two matrices' eigenvalues? How are these eigenvalues related to the singular values in $\boldsymbol{\Delta}$?

Exercise 13.4.15. This exercise derives the deviance for the canonical correlation model in (13.87). Start with

$$-2\log(L(\widehat{\boldsymbol{\Sigma}}\,;\,\mathbf{S})) = \nu\log(|\widehat{\boldsymbol{\Sigma}}|) + \nu\ \text{trace}(\widehat{\boldsymbol{\Sigma}}^{-1}\mathbf{S}) \tag{13.99}$$

for the likelihood in (13.32), where $\widehat{\boldsymbol{\Sigma}}$ is the estimate given in Theorem 13.3. (a) Show that

$$\boldsymbol{\Sigma} = \begin{pmatrix} \boldsymbol{\Sigma}_{11}^{1/2} & \mathbf{0} \\ \mathbf{0} & \boldsymbol{\Sigma}_{22}^{1/2} \end{pmatrix} \begin{pmatrix} \mathbf{I}_{q_1} & \Xi \\ \Xi' & \mathbf{I}_{q_2} \end{pmatrix} \begin{pmatrix} \boldsymbol{\Sigma}_{11}^{1/2} & \mathbf{0} \\ \mathbf{0} & \boldsymbol{\Sigma}_{22}^{1/2} \end{pmatrix}. \tag{13.100}$$

(b) Suppose $q_1 \geq q_2 = m$. Show that

$$\begin{aligned} \text{trace}(\widehat{\boldsymbol{\Sigma}}^{-1}\mathbf{S}) &= \text{trace}\left(\begin{pmatrix} \mathbf{I}_{q_1} & \mathbf{P}\mathbf{D}^{(K)}\mathbf{G}' \\ \mathbf{G}\mathbf{D}^{(K)}\mathbf{P}' & \mathbf{I}_{q_2} \end{pmatrix}^{-1} \begin{pmatrix} \mathbf{I}_{q_1} & \mathbf{P}\mathbf{D}\mathbf{G}' \\ \mathbf{G}\mathbf{D}\mathbf{P}' & \mathbf{I}_{q_2} \end{pmatrix}\right) \\ &= \text{trace}\left(\begin{pmatrix} \mathbf{I}_{q_1} & \mathbf{P}\mathbf{D}^{(K)} \\ \mathbf{D}^{(K)}\mathbf{P}' & \mathbf{I}_{q_2} \end{pmatrix}^{-1} \begin{pmatrix} \mathbf{I}_{q_1} & \mathbf{P}\mathbf{D} \\ \mathbf{D}\mathbf{P}' & \mathbf{I}_{q_2} \end{pmatrix}\right). \end{aligned} \tag{13.101}$$

[Hint: Note that with $q_1 \geq q_2$, $\mathbf{G}$ is an orthogonal matrix.] (c) Let $\mathbf{D} = (\mathbf{D}_1, \mathbf{D}_2)$, where $\mathbf{D}_1$ contains the first K columns of $\mathbf{D}$, and $\mathbf{D}_2$ the remaining $q_2 - K$ columns. Show that $\mathbf{D}^{(K)} = (\mathbf{D}_1, \mathbf{0})$, and find the $(q_1 + K) \times (q_1 + K)$ matrix $\mathbf{A}$ (a function of $\mathbf{P}\mathbf{D}_1$) such that the trace in (13.101) equals

$$\text{trace}\left(\begin{pmatrix} \mathbf{A} & \mathbf{0} \\ \mathbf{0} & \mathbf{I}_{q_2-K} \end{pmatrix}^{-1} \begin{pmatrix} \mathbf{A} & \mathbf{P}\mathbf{D}_2 \\ \mathbf{D}_2'\mathbf{P}' & \mathbf{I}_{q_2-K} \end{pmatrix}\right). \tag{13.102}$$

From that expression show trace$(\widehat{\mathbf{\Sigma}}^{-1}\mathbf{S}) = q$. (d) Show that $|\widehat{\mathbf{\Sigma}}| = |\mathbf{S}_{11}||\mathbf{S}_{22}||\mathbf{I}_{q_2} - \mathbf{G}(\mathbf{D}^{(K)})^2\mathbf{G}'| = \prod_{i=1}^{K}(1 - d_i^2)$. [Hint: Recall (5.101).] (e) Use parts (c) and (d) to find an expression for (13.99), then argue that for comparing M_K's, we can take the deviance as in (13.90).

Exercise 13.4.16. Verify the calculation in (13.92).

Exercise 13.4.17 (Painters). The biplot for the painters data set (in the MASS package) was analyzed in Exercise 1.9.20. (a) Using the first four variables, without any scaling, find the sample eigenvalues l_i. Which seem to be large, and which small? (b) Find the pattern of the l_i's that has best BIC. What are the MLE's of the λ_i's for the best pattern? Does the result conflict with your answer to part (a)?

Exercise 13.4.18 (Spam). In Exercises 1.9.17 and 11.9.9, we found principal components for the spam data. Here we look for the best pattern of eigenvalues. Note that the data is far from multivariate normal, so the distributional aspects should not be taken too seriously. (a) Using the unscaled spam explanatory variables (1 through 57), find the best pattern of eigenvalues based on the BIC criterion. Plot the sample eigenvalues and their MLE's. Do the same, but for the logs. How many principal components is it reasonable to take? (b) Repeat part (b), but using the scaled data, `scale(Spam[,1:57])`. (c) Which approach yielded the more satisfactory answer? Was the decision to use ten components in Exercise 11.9.9 reasonable, at least for the scaled data?

Exercise 13.4.19 (Iris). This question concerns the relationships between the sepal measurements and petal measurements in the iris data. Let $\mathbf{S}$ be the pooled covariance matrix, so that the denominator is $\nu = 147$. (a) Find the correlation between the sepal length and petal length, and the correlation between the sepal width and petal width. (b) Find the canonical correlation quantities for the two groups of variables {sepal length, sepal width} and {petal length, petal width}. What do the loadings show? Compare the d_i's to the correlations in part (a). (c) Find the BIC's for the three models $K = 0, 1, 2$, where K is the number of nonzero δ_i's. What do you conclude?

Exercise 13.4.20 (Exams). Recall the exams data set (Exercise 10.5.19) has the scores of 191 students on four exams, the three midterms (variables 1, 2, and 3) and the final exam. (a) Find the canonical correlation quantities, with the three midterms in one group, and the final in its own group. Describe the relative weightings (loadings) of the midterms. (b) Apply the regular multiple regression model with the final as the $\mathbf{Y}$ and the three midterms as the $\mathbf{X}$'s. What is the correlation between the $\mathbf{Y}$ and the fit, $\widehat{\mathbf{Y}}$? How does this correlation compare to d_1 in part (a)? What do you get if you square this correlation? (c) Look at the ratios $\widehat{\beta}_i/a_{i1}$ for $i = 1, 2, 3$, where $\widehat{\beta}_i$ is the regression coefficient for midterm i in part (b), and a_{i1} is the first canonical correlation loading. What do you conclude? (d) Run the regression again, with the final still $\mathbf{Y}$, but use just the one explanatory variable $\mathbf{X}\mathbf{a}_1$. Find the correlation of $\mathbf{Y}$ and the $\widehat{\mathbf{Y}}$ for this regression. How does it compare to that in part (b)? (e) Which (if either) yields a linear combination of the midterms that best correlates with the final, canonical correlation analysis or multiple regression. (f) Look at the three midterms' variances. What do you see? Find the regular principal components (without scaling) for the midterms. What are the loadings for the first principal component? Compare them to the canonical correlations' loadings in part (a). (g) Run the regression again, with the final as the $\mathbf{Y}$ again, but with just the first principal component of the midterms as the sole explanatory variable. Find the correlation between $\mathbf{Y}$ and $\widehat{\mathbf{Y}}$ here. Compare to the correlations in parts (b) and (d). What do you conclude?

Exercise 13.4.21 (States)**.** This problems uses the matrix states, which contains several demographic variables on the 50 United States, plus D.C. We are interested in the relationship between crime variables and money variables:

Crime: Violent crimes per 100,000 people
Prisoners: Number of people in prison per 10,000 people.
Poverty: Percentage of people below the poverty line.
Employment: Percentage of people employed
Income: Median household income

Let the first two variables be $\mathbf{Y}_1$, and the other three be $\mathbf{Y}_2$. Scale them to have mean zero and variance one:

```
y1 <- scale(states[,7:8])
y2 <- scale(states[,9:11])
```

Find the canonical correlations between the $\mathbf{Y}_1$ and $\mathbf{Y}_2$. (a) What are the two canonical correlations? How many of these would you keep? (b) Find the BIC's for the $K = 0, 1$ and 2 canonical correlation models. Which is best? (c) Look at the loadings for the first canonical correlation, i.e., $\mathbf{a}_1$ and $\mathbf{b}_1$. How would you interpret these? (d) Plot the first canonical variables: $\mathbf{Y}_1\mathbf{a}_1$ versus $\mathbf{Y}_2\mathbf{b}_1$. Do they look correlated? Which observations, if any, are outliers? (e) Plot the second canonical variables: $\mathbf{Y}_1\mathbf{a}_2$ versus $\mathbf{Y}_2\mathbf{b}_1$. Do they look correlated? (f) Find the correlation matrix of the four canonical variables: $(\mathbf{Y}_1\mathbf{a}_1, \mathbf{Y}_1\mathbf{a}_2, \mathbf{Y}_2\mathbf{b}_1, \mathbf{Y}_2\mathbf{b}_2)$. What does it look like? (Compare it to the result in Exercise 13.4.13.)

Appendix A

Extra R Routines

These functions are very barebones. They do not perform any checks on the inputs, and are not necessarily efficient. You are encouraged to robustify and enhance any of them to your heart's content. They are collected in the R package MSOS [Marden and Balamuta, 2014].

A.1 Entropy

We present a simple method for estimating the best entropy. See Hyvärinen et al. [2001] for a more sophisticated approach, which is implemented in the R package fastICA [Marchini et al., 2013]. First, we need to estimate the negentropy (1.46) for a given univariate sample of n observations. We use the histogram as the density, where we take K bins of equal width d, where K is the smallest integer larger than $\log_2(n)+1$ [Sturges, 1926]. Thus bin i is $(b_{i-1}, b_i]$, $i = 1, \ldots, K$, where $b_i = b_0 + d \times i$, and b_0 and d are chosen so that $(b_0, b_K]$ covers the range of the data. Letting $\widehat{p}_i$ be the proportion of observations in bin i, the histogram estimate of the density g is

$$\widehat{g}(x) = \frac{\widehat{p}_i}{d} \text{ if } b_{i-1} < x \leq b_i. \tag{A.1}$$

From (2.103) in Exercise 2.7.16, we have that the negative entropy (1.46) is

$$\text{Negent}(\widehat{g}) = \frac{1}{2}\left(1 + \log\left(2\pi\left(Var[\mathcal{I}] + 1/12\right)\right)\right) + \sum_{i=1}^{K} \widehat{p}_i \log(\widehat{p}_i), \tag{A.2}$$

where $\mathcal{I}$ is the random variable with $P[\mathcal{I} = i] = \widehat{p}_i$, hence

$$Var[\mathcal{I}] = \sum_{i=1}^{K} i^2 \widehat{p}_i - \left(\sum_{i=1}^{K} i\widehat{p}_i\right)^2. \tag{A.3}$$

See Section A.1.1 for the R function we use to calculate this estimate.

For projection pursuit, we have our $n \times q$ data matrix $\mathbf{Y}$, and wish to find first the $q \times 1$ vector $\mathbf{g}_1$ with norm 1 that maximizes the estimated negentropy of $\mathbf{Y}\mathbf{g}_1$. Next we look for the $\mathbf{g}_2$ with norm 1 orthogonal to $\mathbf{g}_1$ that maximizes the negentropy of $\mathbf{Y}\mathbf{g}_2$, etc. Then our rotation is given by the orthogonal matrix $\mathbf{G} = (\mathbf{g}_1, \mathbf{g}_2, \ldots, \mathbf{g}_q)$.

We need to parametrize the orthogonal matrices somehow. For $q = 2$, we can set

$$\mathbf{G}(\theta) = \mathcal{E}_2(\theta) \equiv \begin{pmatrix} \cos(\theta) & -\sin(\theta) \\ \sin(\theta) & \cos(\theta) \end{pmatrix}. \tag{A.4}$$

Clearly each such matrix is orthogonal. As θ ranges from 0 to 2π, $\mathcal{E}_2(\theta)$ ranges through half of the orthogonal matrices (those with determinant equal to $+1$), the other half obtainable by switching the minus sign from the sine term to one of the cosine terms. For our purposes, we need to take only $0 \leq \theta < \pi$, since the other $\mathbf{G}$'s are obtained from that set by changing the sign on one or both of the columns. Changing signs does not affect the negentropy, nor the graph except for reflection around an axis. To find the best $\mathbf{G}(\theta)$, we perform a simple line search over θ. See Section A.1.2.

For $q = 3$ we use *Euler angles* θ_1, θ_2, and θ_3, so that

$$\begin{aligned} \mathbf{G}(\theta_1, \theta_2, \theta_3) &= \mathcal{E}_3(\theta_1, \theta_2, \theta_3) \\ &\equiv \begin{pmatrix} \mathcal{E}_2(\theta_3) & 0 \\ 0 & 1 \end{pmatrix} \begin{pmatrix} 1 & 0 \\ 0 & \mathcal{E}_2(\theta_2) \end{pmatrix} \begin{pmatrix} \mathcal{E}_2(\theta_1) & 0 \\ 0 & 1 \end{pmatrix}. \end{aligned} \tag{A.5}$$

See Anderson et al. [1987] for similar parametrizations when $q > 3$. The first step is to find the $\mathbf{G} = (\mathbf{g}_1, \mathbf{g}_2, \mathbf{g}_3)$ whose first column, $\mathbf{g}_1$, achieves the maximum negentropy of $\mathbf{Y}\mathbf{g}_1$. Here it is enough to take $\theta_3 = 0$, so that the left-hand matrix is the identity. Because our estimate of negentropy for $\mathbf{Y}\mathbf{g}$ is not continuous in $\mathbf{g}$, we use the *simulated annealing* option in the R function optim to find the optimal $\mathbf{g}_1$. The second step is to find the best further rotation of the remaining variables, $\mathbf{Y}(\mathbf{g}_2, \mathbf{g}_3)$, for which we can use the two-dimensional procedure above. See Section A.1.3.

A.1.1 negent: Estimate negative entropy

Description: Calculates the histogram-based estimate (A.2) of the negentropy (1.46) for a vector of observations. See Listing A.1 for the code.

Usage: negent(x,K=log2(length(x))+1)

Arguments:

x: The n-vector of observations.

K: The number of bins to use in the histogram.

Value: The value of the estimated negentropy.

A.1.2 negent2D: Maximize negentropy for two dimensions

Description: Searches for the rotation that maximizes the estimated negentropy of the first column of the rotated data, for $q = 2$ dimensional data. See Listing A.2 for the code.

Usage: negent2D(y,m=100)

Arguments:

y: The $n \times 2$ data matrix.

m: The number of angles (between 0 and π) over which to search.

Value: A list with the following components:

vectors: The 2×2 orthogonal matrix $\mathbf{G}$ that optimizes the negentropy.

values: Estimated negentropies for the two rotated variables. The largest is first.

A.1.3 negent3D: Maximize negentropy for three dimensions

Description: Searches for the rotation that maximizes the estimated negentropy of the first column of the rotated data, and of the second variable fixing the first, for $q = 3$ dimensional data. The routine uses a random start for the function optim using the simulated annealing option SANN, hence one may wish to increase the number of attempts by setting nstart to an integer larger than 1. See Listing A.3 for the code.

Usage: negent3D(y,nstart=1,m=100,...)

Arguments:

y: The $n \times 3$ data matrix.

nstart: The number of times to randomly start the search routine.

m: The number of angles (between 0 and π) over which to search to find the second variables.

...: Further optional arguments to pass to the optim function to control the simulated annealing algorithm.

Value: A list with the following components:

vectors: The 3×3 orthogonal matrix $\mathbf{G}$ that optimizes the negentropy.

values: Estimated negentropies for the three rotated variables, from largest to smallest.

A.2 Both-sides model

The routines in this section apply to the both-sides model,

$$\mathbf{Y} = \mathbf{x}\boldsymbol{\beta}\mathbf{z}' + \mathbf{R}, \quad \mathbf{R} \sim N(\mathbf{0}, \mathbf{I}_n \otimes \boldsymbol{\Sigma}_R), \tag{A.6}$$

where $\mathbf{Y}$ is $n \times q$, $\mathbf{x}$ is $n \times p$, and $\mathbf{z}$ is $q \times l$. The functions require that $n \geq p$, and $\mathbf{x}'\mathbf{x}$ and $\mathbf{z}'\mathbf{z}$ be invertible.

There are two functions for fitting the model, one for least squares and one for maximum likelihood. They both allow specifying a subset of the parameters in $\boldsymbol{\beta}$ to be zero, using a pattern matrix $\mathbf{P}$. The least squares routine also calculates Mallows' C_p statistic. The MLE routine calculates AICc and BIC.

The reverse.kronecker function is useful when using the linear regression routine in R to find the least squares estimates, as in Section 6.4.2.

There are three functions for testing nested hypotheses. The first two use the least squares estimates. The bothsidesmodel.chisquare finds the approximate χ^2 statistic for testing a arbitrary set of β_{ij}'s equalling zero. The bothsidesmodel.hotelling performs the Hotelling T^2 test and the Wilks' Λ tests, but only for testing blocks of $\boldsymbol{\beta}$ equalling zero, with the alternative being unrestricted. The likelihood ratio test comparing any two nested models uses the routine bothsidesmodel.lrt.

A.2.1 bothsidesmodel: Calculate the least squares estimates

Description: This function fits the model using least squares. It takes an optional pattern matrix **P** as in (6.51) that specifies which β_{ij}'s are zero. See Listing A.4 for the code.

Usage: bothsidesmodel(x,y,z=diag(ncol(y)),pattern=array(1,dim=c(nrow(x),ncol(z))))

Arguments:

x: An $n \times p$ design matrix.

y: The $n \times q$ matrix of observations.

z: A $q \times l$ design matrix.

pattern: An optional $p \times l$ matrix of 0's and 1's.

Value: A list with the following components:

Beta: The least squares estimate of $\boldsymbol{\beta}$.

SE: The $p \times l$ matrix with the ij^{th} element being the standard error of $\widehat{\beta}_{ij}$.

T: The $p \times l$ matrix with the ij^{th} element being the t-statistic based on $\widehat{\beta}_{ij}$.

Covbeta: The estimated covariance matrix of the $\widehat{\beta}_{ij}$'s.

df: A p-dimensional vector of the degrees of freedom for the t-statistics, where the jth component contains the degrees of freedom for the jth column of $\widehat{\boldsymbol{\beta}}$.

Sigmaz: The $q \times q$ matrix $\widehat{\boldsymbol{\Sigma}}_{\mathbf{z}}$.

ResidSS: The $q \times q$ residual sum of squares and cross-products matrix.

Dim: The dimension of the model, counting the nonzero β_{ij}'s and components of $\boldsymbol{\Sigma}_{\mathbf{z}}$.

Cp: Mallow's C_p statistic from Section 7.5.

A.2.2 reverse.kronecker: Reverse the matrices in a Kronecker product

Description: This function takes a matrix that is the Kronecker product $\mathbf{A} \otimes \mathbf{B}$ (Definition 3.5), where $\mathbf{A}$ is $p \times q$ and $\mathbf{B}$ is $n \times m$, and outputs the matrix $\mathbf{B} \otimes \mathbf{A}$. See Listing A.5 for the code.

Usage: reverse.kronecker(ab,p,q)

Arguments:

ab: The $(pn) \times (qm)$ matrix $\mathbf{A} \otimes \mathbf{B}$.

p: The number of rows of $\mathbf{A}$.

q: The number of columns of $\mathbf{A}$.

Value: The $(np) \times (mq)$ matrix $\mathbf{B} \otimes \mathbf{A}$.

A.2.3 bothsidesmodel.mle: Calculate the maximum likelihood estimates

Description: This function fits the model using maximum likelihood. It takes an optional pattern matrix $\mathbf{P}$ as in (6.51) that specifies which β_{ij}'s are zero. See Listing A.6 for the code.

Usage:

```
bothsidesmodel.mle(x,y,z=diag(ncol(y)),pattern=array(1,dim=c(nrow(x),ncol(z))))
```

Arguments:

x: An $n \times p$ design matrix.

y: The $n \times q$ matrix of observations.

z: A $q \times l$ design matrix.

pattern: An optional $p \times l$ matrix of 0's and 1's.

Value: A list with the following components:

Beta: The least squares estimate of $\boldsymbol{\beta}$.

SE: The $p \times l$ matrix with the ij^{th} element being the standard error of $\widehat{\beta}_{ij}$.

T: The $p \times l$ matrix with the ij^{th} element being the t-statistic based on $\widehat{\beta}_{ij}$.

Covbeta: The estimated covariance matrix of the $\widehat{\beta}_{ij}$'s.

df: The degrees of freedom for the t-statistics, $n - p$.

Sigmaz: The $q \times q$ matrix $\widehat{\boldsymbol{\Sigma}}_{\mathbf{z}}$.

Cx: The $q \times q$ matrix $C_{\mathbf{x}} = (\mathbf{x}'\mathbf{x})^{-1}$.

ResidSS: The $q \times q$ residual sum of squares and cross-products matrix.

Deviance: The deviance of the model.

Dim: The dimension of the model, counting the nonzero β_{ij}'s and components of $\mathbf{\Sigma_z}$.

AICc: The corrected AIC criterion from (9.87) and (9.88).

BIC: The BIC criterion from (9.56).

A.2.4 bothsidesmodel.chisquare: Test subsets of β are zero

Description: Tests the null hypothesis that an arbitrary subset of the β_{ij}'s is zero, based on the least squares estimates, using the χ^2 test as in Section 7.1. The null and alternative are specified by pattern matrices $\mathbf{P}_0$ and $\mathbf{P}_A$, respectively. If the $\mathbf{P}_A$ is omitted, then the alternative will be taken to be the unrestricted model. See Listing A.7 for the code.

Usage:
```
bothsidesmodel.chisquare(x,y,z,pattern0,patternA=array(1,dim=c(nrow(x),ncol(z))))
```

Arguments:

x: An $n \times p$ design matrix.

y: The $n \times q$ matrix of observations.

z: A $q \times l$ design matrix.

pattern0: A $p \times l$ matrix of 0's and 1's specifying the null hypothesis.

patternA: An optional $p \times l$ matrix of 0's and 1's specifying the alternative hypothesis.

Value: A list with the following components:

Theta: The vector of estimated parameters of interest.

Covtheta: The estimated covariance matrix of the estimated parameter vector.

df: The degrees of freedom in the test.

chisq: The T^2 statistic in (7.4).

pvalue: The p-value for the test.

A.2.5 bothsidesmodel.hotelling: Test blocks of β are zero

Description: Performs tests of the null hypothesis $H_0 : \boldsymbol{\beta}^* = \mathbf{0}$, where $\boldsymbol{\beta}^*$ is a block submatrix of $\boldsymbol{\beta}$ as in Section 7.2. An example is given in (7.12). The input consists of model matrices, plus vectors giving the rows and columns of $\boldsymbol{\beta}$ to be tested. In the example, we set rows <− c(1,4,5) and cols <− c(1,3). See Listing A.8 for the code.

Usage: bothsidesmodel.test(x,y,z,rows,cols)

Arguments:

x: An $n \times p$ design matrix.

y: The $n \times q$ matrix of observations.

z: A $q \times l$ design matrix.

rows: The vector of rows to be tested.

cols: The vector of columns to be tested.

Value: A list with the following components:

Hotelling: A list with the components of the Lawley-Hotelling T^2 test (7.22):

T2: The T^2 statistic (7.19).
F: The F version (7.22) of the T^2 statistic.
df: The degrees of freedom for the F.
pvalue: The p-value of the F.

Wilks: A list with the components of the Wilks' Λ test (7.37):

lambda: The Λ statistic (7.35).
Chisq: The χ^2 version (7.37) of the Λ statistic, using Bartlett's correction.
df: The degrees of freedom for the χ^2.
pvalue: The p-value of the χ^2.

A.2.6 bothsidesmodel.lrt: Test subsets of β are zero

Description: Tests the null hypothesis that an arbitrary subset of the β_{ij}'s is zero, using the likelihood ratio test as in Section 9.4. The null and alternative are specified by pattern matrices $\mathbf{P}_0$ and $\mathbf{P}_A$, respectively. If the $\mathbf{P}_A$ is omitted, then the alternative will be taken to be the unrestricted model. See Listing A.9 for the code.

Usage: bothsidesmodel.lrt(x,y,z,pattern0,patternA=array(1,dim=c(nrow(x),ncol(z))))

Arguments:

x: An $n \times p$ design matrix.

y: The $n \times q$ matrix of observations.

z: A $q \times l$ design matrix.

pattern0: A $p \times l$ matrix of 0's and 1's specifying the null hypothesis.

patternA: An optional $p \times l$ matrix of 0's and 1's specifying the alternative hypothesis.

Value: A list with the following components:

chisq: The likelihood ratio statistic in (9.44).

df: The degrees of freedom in the test.

pvalue: The p-value for the test.

A.2.7 Helper functions

The functions tr and logdet find the trace and log determinant, respectively, of a square matrix. See Listing A.10. The function fillout takes a $q \times (q - l)$ matrix $\mathbf{z}$ and fills it out as in (9.25) so that it is $q \times q$. See Listings A.11.

The function bothsidesmodel.df is used in bothsidesmodel of Section A.2.1 to find the denominators needed to calculate an unbiased estimator of $\mathbf{\Sigma}_R$, as in Exercise 6.6.8. See Listing A.12. The functions bsm.simple and bsm.fit are used in the function bothsidesmodel.mle of Section A.2.3 to estimate $\boldsymbol{\beta}$. See Listings A.13 and A.14.

A.3 Classification

A.3.1 lda: Linear discrimination

Description: Finds the coefficients $\mathbf{a}_k$ and constants c_k for Fisher's linear discrimination function d_k in (11.31) and (11.32). See Listing A.15 for the code.

Usage: lda(x,y)

Arguments:

x: The $n \times p$ data matrix.

y: The n-vector of group identities, assumed to be given by the numbers $1, \ldots, K$ for K groups.

Value: A list with the following components:

a: A $p \times K$ matrix, where column k contains the coefficients $\mathbf{a}_k$ for (11.31). The final column is all zero.

c: The K-vector of constants c_k for (11.31). The final value is zero.

A.3.2 qda: Quadratic discrimination

Description: The function returns the elements needed to calculate the quadratic discrimination in (11.47). Use the output from this function in predict_qda (Section A.3.2) to find the predicted groups. See Listing A.16 for the code.

Usage: qda(x,y)

Arguments:

x: The $n \times p$ data matrix.

y: The n-vector of group identities, assumed to be given by the numbers $1, \ldots, K$ for K groups.

Value: A list with the following components:

Mean: A $K \times p$ matrix, where row k contains the sample mean vector for group k.

Sigma: A $K \times p \times p$ array, where the Sigma[k,,] contains the sample covariance matrix for group k, $\widehat{\mathbf{\Sigma}}_k$.

c: The K-vector of constants c_k for (11.47).

A.3.3 predict_qda: Quadratic discrimination prediction

Description: The function uses the output from the function qda (Section A.3.2) and a p-vector $\mathbf{x}$, and calculates the predicted group for this $\mathbf{x}$. See Listing A.17 for the code.

Usage: predict_qda(qd,newx)

Arguments:

qd: The output from qda.

newx: A p-vector $\mathbf{x}$ whose components match the variables used in the qda function.

Value: A K-vector of the discriminant values $d_k^Q(\mathbf{x})$ in (11.47) for the given $\mathbf{x}$.

A.4 Silhouettes for K-means clustering

A.4.1 silhouette.km: Calculate the silhouettes

Description: Find the silhouettes (12.9) for K-means clustering from the data and the groups' centers. See Listing A.18 for the code. (This function is a bit different from the silhouette function in the cluster package, Maechler et al. [2015].)

Usage: silhouette.km(x,centers)

Arguments:

x: The $n \times p$ data matrix.

centers: The $K \times p$ matrix of centers (means) for the K clusters, row k being the center for cluster k.

Value: The n-vector of silhouettes, indexed by the observations' indices.

A.4.2 sort_silhouette: Sort the silhouettes by group

Description: Sorts the silhouettes, first by group, then by value, preparatory to plotting. See Listing A.19 for the code.

Usage: sort_silhouette(sil,clusters)

Arguments:

sil: The n-vector of silhouette values.

clusters: The n-vector of cluster indices.

Value: The n-vector of sorted silhouettes.

A.5 Patterns of eigenvalues

We have two main functions, pcbic to find the MLE and BIC for a particular pattern, and pcbic.stepwise, which uses a stepwise search to find a good pattern. The functions pcbic.unite and pcbic.patterns are used by the main functions, and probably not of much interest on their own.

A.5.1 pcbic: BIC for a particular pattern

Description: Find the BIC and MLE from a set of observed eigenvalues for a specific pattern. See Listing A.20 for the code.

Usage: pcbic(eigenvals,n,pattern)

Arguments:

eigenvals: The q-vector of eigenvalues of the covariance matrix, in order from largest to smallest.

n: The degrees of freedom in the covariance matrix.

pattern: The pattern of equalities of the eigenvalues, given by the K-vector $(q_1, \ldots, q_K)$ as in (13.8).

Value: A list with the following components:

lambdaHat: A q-vector containing the MLE's for the eigenvalues.

Deviance: The deviance of the model, as in (13.13).

Dimension: The dimension of the model, as in (13.12).

BIC: The value of the BIC for the model, as in (13.14).

A.5.2 pcbic.stepwise: Choose a good pattern

Description: Uses the stepwise procedure described in Section 13.1.4 to find a pattern for a set of observed eigenvalues with good BIC value. See Listing A.20 for the code.

Usage: pcbic.stepwise(eigenvals,n)

Arguments:

eigenvals: The q-vector of eigenvalues of the covariance matrix, in order from largest to smallest.

n: The degrees of freedom in the covariance matrix.

Value: A list with the following components:

Patterns: A list of patterns, one for each value of length K.

BICs: A vector of the BIC's for the above patterns.

BestBIC: The best (smallest) value among the BIC's in BICs.

BestPattern: The pattern with the best BIC.

lambdaHat: A q-vector containing the MLE's for the eigenvalues for the pattern with the best BIC.

A.5.3 Helper functions

The function pcbic.unite takes as arguments a pattern $(q_1, \ldots, q_K)$, called pattern, and an index i, called index1, where $1 \leq i < K$. It returns the pattern obtained by summing q_i and q_{i+1}. See Listing A.22. The function pcbic.patterns (Listing A.23) takes the arguments eigenvals, n, and pattern0 (as for pcbic in Section A.5.1), and returns the best pattern and its BIC among the patterns obtainable by summing two consecutive terms in pattern0 via pcbic.unite.

A.6 Function listings

Listing A.1: The function negent

```
negent <- function(x,K=ceiling(log2(length(x))+1)) {
        p <- table(cut(x,breaks=K))/length(x)
        sigma2 <- sum((1:K)^2*p)-sum((1:K)*p)^2
        p <- p[(p>0)]
        (1+log(2*pi*(sigma2+1/12)))/2+sum(p*log(p))
}
```

Listing A.2: The function negent2D

```
negent2D <- function(y,m=100) {
        thetas <- (1:m)*pi/m
        ngnt <- NULL
        for(theta in thetas) {
                x <- y%*%c(cos(theta),sin(theta))
                ngnt <- c(ngnt,negent(x))
        }
        i <- imax(ngnt)
        g <- c(cos(thetas[i]),sin(thetas[i]))
        g <- cbind(g,c(-g[2],g[1]))
        list(vectors = g,values = c(ngnt[i],negent(y%*%g[,2])))
}
```

Listing A.3: The function negent3D

```
negent3D <- function(y,nstart=1,m=100,...) {
     f <- function(thetas) {
          cs <- cos(thetas)
          sn <- sin(thetas)
          negent(y%*%c(cs[1],sn[1]*c(cs[2],sn[2])))
     }
     tt <- NULL
     nn <- NULL
     for(i in 1:nstart) {
          thetas <- runif(3)*pi
          o <- optim(thetas,f,method='SANN',control=list(fnscale=-1),...)
          tt <- rbind(tt,o$par)
          nn <- c(nn,o$value)
     }
     i <- imax(nn) # The index of best negentropy
     cs<-cos(tt[i,])
     sn<-sin(tt[i,])
     g.opt <- c(cs[1],sn[1]*cs[2],sn[1]*sn[2])
     g.opt <- cbind(g.opt,c(-sn[1],cs[1]*cs[2],sn[2]*cs[1]))
     g.opt <- cbind(g.opt,c(0,-sn[2],cs[2]))
     x <- y%*%g.opt[,2:3]
     n2 <- negent2D(x,m=m)
     g.opt[,2:3] <- g.opt[,2:3]%*%n2$vectors
     list(vectors=g.opt,values = c(nn[i],n2$values))
}
```

Listing A.4: The function bothsidesmodel

```
bothsidesmodel <- function(x,y,z=diag(qq),pattern=matrix(1,nrow=p,ncol=l)) {
        x <- cbind(x)
        y <- cbind(y)
        n <- nrow(y)
        p <- ncol(x)
        qq <- ncol(y)
        z <- cbind(z)
        l <- ncol(z)

        if((p*nrow(x)==1&&x==0) || (l*nrow(z)==1&&z==0)) {
                sigma <- t(y)%*%y/n
                output <- list(Beta = 0, SE = 0,T = 0, Covbeta = 0,
                                                        df = n, Sigmaz=sigma)
                return(output)
        }

        yz <- t(lm(t(y)~z-1)$coef)
        residb <- lm(yz~x-1)$resid
        sigmab <- t(residb)%*%residb/(n-p)

        if(sum(pattern)==0) {
                residss <- ((n-p-l-1)/(n-p))*tr(solve(sigmab,t(yz)%*%yz))
```

```
            output <- list(Beta = 0, SE = 0,T = 0, Covbeta = 0,
                                        df = n,Sigmaz=sigmab,
            ResidSS=residss,Dim=l*(l+1)/2,Cp=residss+l*(l+1))
            return(output)
      }

      rowse <- rowt <- rowb <- rep(0,p*l)
      rowp <- c(t(pattern))==1
      xx <- t(x)%*%x
      xyz <- t(x)%*%yz
      xyz <- c(t(xyz))[rowp]
      xxzz <- kronecker(xx,diag(l))[rowp,rowp]
      dstarinv <- solve(xxzz)
      g <- xyz%*%dstarinv
      rowb[rowp] <- g
      beta <- matrix(rowb,nrow=p,byrow=T)
      df <- bothsidesmodel.df(xx,n,pattern)
      residz <- yz-x%*%beta
      residsscp <- t(residz)%*%residz
      sigmaz <- residsscp/df
      xxzsz <- kronecker(xx,sigmaz)[rowp,rowp]
      covbeta <- dstarinv%*%xxzsz%*%dstarinv
      covbetab <- matrix(0,p*l,p*l)
      covbetab[rowp,rowp] <- covbeta
      se <- sqrt(diag(covbeta))
      tt <- g/se
      rowse[rowp] <- se
      rowt[rowp] <- tt
      residss <- ((n-p-l-1)/(n-p))*tr(solve(sigmab,residsscp))
      dd <- sum(pattern)+l*(l+1)/2

      list(Beta = beta, SE = matrix(rowse,nrow=p,byrow=T),
            T = matrix(rowt,nrow=p,byrow=T),
            Covbeta = covbetab, df = diag(df),Sigmaz = sigmaz, ResidSS =residss,
            Dim=dd,Cp=residss+2*dd)
}
```

Listing A.5: The function reverse.kronecker

```
reverse.kronecker <- function(ab,p,qq) {
      m <- nrow(ab)/p
      n <- ncol(ab)/qq
      rr <- c(outer((0:(p-1))*m,1:m,"+"))
      cc <- c(outer((0:(qq-1))*n,1:n,"+"))
      ab[rr,cc]
}
```

Listing A.6: The function bothsidesmodel.mle

```
bothsidesmodel.mle <- function(x,y,z=diag(qq),pattern=matrix(1,nrow=p,ncol=l)) {
    x <- cbind(x)
    y <- cbind(y)
    n <- nrow(y)
    p <- ncol(x)
    qq <- ncol(y)
    z <- cbind(z)
    l <- ncol(z)

    if(length(x)==1&&x==0) {
        bsm <- list(Beta=0,SE=0,T=0,Covbeta=0,df=n)
        resid <- y
        psum <- 0
    } else {
        bsm <- list()
        beta <- se <- tt <- matrix(0,nrow=p,ncol=l)
        psum <- sum(pattern)
        if(psum>0) {
            rowin <- apply(pattern,1,max)==1
            colin <- apply(pattern,2,max)==1
            x1 <- x[,rowin,drop=FALSE]
            z1 <- z[,colin,drop=FALSE]
            yz <- cbind(y%*%solve(t(fillout(z1))))
            l1 <- ncol(z1)
            p1 <- ncol(x1)
            yza <- yz[,1:l1,drop=FALSE]
            xyzb <- x[,rowin,drop=FALSE]
            if(l1<qq) xyzb <- cbind(xyzb,yz[,(l1+1):qq])
            pattern1 <- pattern[rowin,colin]
            if(min(pattern1)==1) {
                bsm <- bsm.simple(xyzb,yza,diag(l1))
                bsm$Cx <- bsm$Cx[1:p1,1:p1]
            } else {
                pattern2 <- rbind(pattern1,matrix(1,nrow=qq-l1,ncol=l1))
                bsm <- bsm.fit(xyzb,yza,diag(l1),pattern2)
            }
            beta[rowin,colin] <- bsm$Beta[1:p1,]
            se[rowin,colin] <- bsm$SE[1:p1,]
            tt[rowin,colin] <- bsm$T[1:p1,]
            bsm$Covbeta <- bsm$Covbeta[1:psum,1:psum]
        }
        bsm$Beta <- beta
        bsm$SE <- se
        bsm$T <- tt
        if(psum==0) {
            bsm$Covbeta <- 0
            bsm$df <- n
            p1 <- 0
        }
        resid <- y-x%*%beta%*%t(z)
    }
```

```
        bsm$ResidSS <- t(resid)%*%resid/n
        bsm$Deviance <- n*logdet(bsm$Sigma) + n*qq
        bsm$Dim <- psum+qq*(qq+1)/2
        bsm$AICc <- bsm$Deviance+(n/(n-p1-qq-1))*2*bsm$Dim
        bsm$BIC <- bsm$Deviance + log(n)*bsm$Dim

        bsm
}
```

Listing A.7: The function bothsidesmodel.chisquare

```
bothsidesmodel.chisquare <- function(x,y,z,pattern0,
                        patternA=matrix(1,nrow=ncol(x),ncol=ncol(z))) {
        bsm <- bothsidesmodel(x,y,z,patternA)
        which <- patternA*(1-pattern0)
        which <- c(t(which)) == 1
        theta <- c(t(bsm$Beta))[which]
        covtheta <- bsm$Covbeta[which,which]
        chisq <- theta%*%solve(covtheta,theta)
        df <- sum(which)
        list(Theta=theta,Covtheta = covtheta,df = df,Chisq=chisq,
                                                pvalue=1-pchisq(chisq,df))
}
```

Listing A.8: The function bothsidesmodel.hotelling

```
bothsidesmodel.hotelling <- function(x,y,z,rows,cols) {
        bsm <- bothsidesmodel(x,y,z)
        lstar <- length(cols)
        pstar <- length(rows)
        nu <- bsm$df[1]
        bstar <- bsm$Beta[rows,cols]
        if(lstar==1) bstar <- matrix(bstar,ncol=1)
        if(pstar==1) bstar <- matrix(bstar,nrow=1)
        W.nu <- bsm$Sigmaz[cols,cols]
        cx <- solve(t(x)%*%x)
        B <- t(bstar)%*%solve(cx[rows,rows])%*%bstar
        t2 <- tr(solve(W.nu)%*%B)
        f <- (nu-lstar+1)*t2/(lstar*pstar*nu)
        df <- c(lstar*pstar,nu-lstar+1)
        W <- W.nu*nu
        lambda <- ifelse(lstar==1,W/(W+B),det(W)/det(W+B))
        chis <- -(nu-(lstar-pstar+1)/2)*log(lambda)
        Hotelling <- list(T2 = t2, F = f, df = df,pvalue = 1-pf(f,df[1],df[2]))
        Wilks <- list(Lambda=lambda,Chisq=chis,df=df[1],pvalue=1-pchisq(chis,df[1]))
        list(Hotelling = Hotelling,Wilks = Wilks)
}
```

Listing A.9: The function bothsidesmodel.lrt

```
bothsidesmodel.lrt <- function(x,y,z,pattern0,
                        patternA=matrix(1,nrow=ncol(x),ncol=ncol(z))) {
        bsmA <- bothsidesmodel.mle(x,y,z,patternA)
        bsm0 <- bothsidesmodel.mle(x,y,z,pattern0)
        chisq <- bsm0$Deviance-bsmA$Deviance
        df <- bsmA$Dim - bsm0$Dim
        list(Chisq=chisq,df=df,pvalue=1-pchisq(chisq,df))
}
```

Listing A.10: The functions tr and logdet

```
tr <- function(x) sum(diag(x))

logdet <- function(x) {determinant(x)$modulus[1]}
```

Listing A.11: The function fillout

```
fillout <- function(z) {
        if(is.vector(z)) z <- matrix(z,ncol=1)
        qq <- nrow(z)
        l <- ncol(z)
        if(l<qq) {
                qz <- diag(qq)-z%*%solve(t(z)%*%z,t(z))
                z <- cbind(z,eigen(qz)$vector[,1:(qq-l)])
        }
        if(l>qq) {
                z <- t(fillout(t(z)))
        }
        z
}
```

Listing A.12: The function bothsidesmodel.df

```
bothsidesmodel.df <- function(xx,n,pattern) {
        l <- ncol(pattern)
        tt <- pattern==1
        pj <- apply(pattern,2,sum)
        df <- matrix(0,l,l)
        diag(df) <- n-pj
        a <- vector("list",l)
        a[[1]] <- solve(xx[tt[,1],tt[,1]])
        for(i in 2:l) {
                if(pj[i]>0) a[[i]] <- solve(xx[tt[,i],tt[,i]])
                for(j in 1:(i-1)) {
                        if(pj[i]==0|pj[j]==0) {pij<-0}
                        else {
                                b <- xx[tt[,i],tt[,j]]
                                if(is.vector(b)) b <- matrix(b,nrow=pj[i])
                                pij <- tr(a[[i]]%*%b%*%a[[j]]%*%t(b))
                        }
                        df[j,i] <- df[i,j] <- n-pj[i]-pj[j]+pij
                }
        }
        df
}
```

Listing A.13: The function bsm.simple

```
bsm.simple <- function(x,y,z) {
        yz <- y%*%z%*%solve(t(z)%*%z)
        cx <- solve(t(x)%*%x)
        beta <- cx%*%t(x)%*%yz
        residz <- yz - x%*%beta
        df <- nrow(yz)-ncol(x)
        sigmaz <- t(residz)%*%residz/df
        se <- sqrt(outer(diag(cx),diag(sigmaz),"*"))
        list(Beta = beta, SE = se, T = beta/se, Covbeta = kronecker(cx,sigmaz),df = df,
                Sigmaz = sigmaz, Cx=cx)
}
```

Listing A.14: The function bsm.fit

```
bsm.fit <- function(x,y,z,pattern) {
        n <- nrow(y)
        p <- ncol(x)
        l <- ncol(z)
        xx <- t(x)%*%x
        xy <- t(x)%*%y
        rowp <- c(t(pattern))==1
        lsreg <- lm(y~x-1)
        yqxy <- t(lsreg$resid)%*%lsreg$resid
        beta0 <- lsreg$coef
        sigma <- cov(y)*(n-1)/n
        rowse <- rowt <- rowb <- rep(0,p*l)
        dev0 <- n*logdet(sigma)

        maxiter = 25
        iter = 0;

        repeat {
                sigmainvz <- solve(sigma,z)
                xyz <- c(t(xy%*%sigmainvz))[rowp]
                xxzz <- kronecker(xx,t(z)%*%sigmainvz)[rowp,rowp]
                dstarinv <- solve(xxzz)
                gamma <- xyz%*%dstarinv
                rowb[rowp] <- gamma
                beta <- matrix(rowb,nrow=p,byrow=T)
                betadiff <- beta0-beta%*%t(z)
                sigma <- (yqxy+t(betadiff)%*%xx%*%betadiff)/n
                dev <- n*logdet(sigma)
                iter <- iter+1
                if(abs(dev-dev0) < 0.00001) break;
                if(iter>=maxiter) break;
                dev0 <- dev
        }
        df <- n-p
        covbeta <- solve(xxzz)*(n/df)
        se <- sqrt(diag(covbeta))
        tt <- gamma/se
        rowse[rowp] <- se
        rowt[rowp] <- tt
        se <- matrix(rowse,nrow=p,byrow=T)
        tt <- matrix(rowt,nrow=p,byrow=T)
        list(Beta = beta, SE = se, T = tt, Covbeta = covbeta, df = df)
}
```

Listing A.15: The function lda

```
lda <- function(x,y) {
      if(is.vector(x)) {x <- matrix(x,ncol=1)}
      K <- max(y)
      p <- ncol(x)
      n <- nrow(x)
      m <- NULL
      v <- matrix(0,ncol=p,nrow=p)
      for(k in 1:K) {
            xk <- x[y==k,]
            if(is.vector(xk)) {xk <- matrix(xk,ncol=1)}
            m <- rbind(m,apply(xk,2,mean))
            v <- v + var(xk)*(nrow(xk)-1)
      }
      v <- v/n
      phat <- table(y)/n

      ck <- NULL
      ak <- NULL
      vi <- solve(v)
      for(k in 1:K) {
            c0 <- -(1/2)*(m[k,]%*%vi%*%m[k,]-m[K,]%*%vi%*%m[K,])
                                                   +log(phat[k]/phat[K])
            ck <- c(ck,c0)
            a0 <- vi%*%(m[k,]-m[K,])
            ak <- cbind(ak,a0)
      }
      list(a = ak, c = ck)
}
```

Listing A.16: The function qda

```
qda <- function(x,y) {
      K <- max(y)
      p <- ncol(x)
      n <- nrow(x)
      m <- NULL
      v <- array(0,c(K,p,p))
      ck <- NULL
      phat <- table(y)/n
      for(k in 1:K) {
            xk <- x[y==k,]
            m <- rbind(m,apply(xk,2,mean))
            nk <- nrow(xk)
            v[k,,] <- var(xk)*(nk-1)/nk
            ck <- c(ck,-log(det(v[k,,]))+2*log(phat[k]))
      }

      list(Mean = m,Sigma = v, c = ck)
}
```

Listing A.17: The function predict_qda

```
predict_qda <- function(qd,newx) {
        newx <- c(newx)
    disc <- NULL
    K <- length(qd$c)
    for(k in 1:K) {
        dk <- -t(newx-qd$Mean[k,])%*%
                solve(qd$Sigma[k,,],newx-qd$Mean[k,])+qd$c[k]
        disc <- c(disc,dk)
    }
    disc
}
```

Listing A.18: The function silhouette.km

```
silhouette.km <- function(x,centers) {
    dd <- NULL
    k <- nrow(centers)
    for(i in 1:k) {
        xr <- sweep(x,2,centers[i,],'-')
        dd<-cbind(dd,apply(xr^2,1,sum))
    }
    dd <- apply(dd,1,sort)[1:2,]
    (dd[2,]-dd[1,])/dd[2,]
}
```

Listing A.19: The function sort_silhouette

```
sort_silhouette <- function(sil,cluster) {
    ss <- NULL
    ks <- sort(unique(cluster))
    for(k in ks) {
        ss <- c(ss,sort(sil[cluster==k]))
    }
    ss
}
```

Listing A.20: The function pcbic

```
pcbic <- function(eigenvals,n,pattern) {
      p <- length(eigenvals)
      l <- eigenvals
      k <- length(pattern)
      istart <- 1
      for(i in 1:k) {
            iend <- istart+pattern[i]
            l[istart:(iend-1)] = mean(l[istart:(iend-1)])
            istart <- iend
      }
      dev <- n*sum(log(l))
      dimen <- (p^2-sum(pattern^2))/2 + k
      bic <- dev + log(n)*dimen
      list(lambdaHat = l,Deviance = dev,Dimension = dimen,BIC = bic)
}
```

Listing A.21: The function pcbic.stepwise

```
pcbic.stepwise <- function(eigenvals,n) {
      k <- length(eigenvals)
      p0 <- rep(1,k)
      b <- rep(0,k)
      pb <- vector('list',k)
      pb[[1]] <- p0
      b[1] <- pcbic(eigenvals,n,p0)$BIC
      for(i in 2:k) {
            psb <- pcbic.subpatterns(eigenvals,n,pb[[i-1]])
            b[i] <- min(psb$bic)
            pb[[i]] <- psb$pattern[,psb$bic==b[i]]
      }
      ib <- (1:k)[b==min(b)]
      list(Patterns = pb,BICs = b,
              BestBIC = b[ib],BestPattern = pb[[ib]],
                  LambdaHat = pcbic(eigenvals,n,pb[[ib]])$lambdaHat)
}
```

Listing A.22: The function pcbic.unite

```
pcbic.unite <- function(pattern,index1) {
      k <- length(pattern)
      if(k==1) return(pattern)
      if(k==2) return(sum(pattern))
      if(index1==1) return(c(pattern[1]+pattern[2],pattern[3:k]))
      if(index1==k-1) return(c(pattern[1:(k-2)],pattern[k-1]+pattern[k]))
      c(pattern[1:(index1-1)],pattern[index1]+pattern[index1+1],pattern[(index1+2):k])
}
```

Listing A.23: The function pcbic.subpatterns

```
pcbic.subpatterns <- function(eigenvals,n,pattern0) {
      b <- NULL
      pts <- NULL
      k <- length(pattern0)
      if(k==1) return(F)
      for(i in 1:(k-1)) {
            p1 <- pcbic.unite(pattern0,i)
            b2 <- pcbic(eigenvals,n,p1)
            b <- c(b,b2$BIC)
            pts <- cbind(pts,p1)
      }
      list(bic=b,pattern=pts)
}
```

Bibliography

Hirotugu Akaike. A new look at the statistical model identification. *IEEE Transactions on Automatic Control*, 19:716 – 723, 1974.

E. Anderson. The irises of the Gaspe Peninsula. *Bulletin of the American Iris Society*, 59:2–5, 1935.

E. Anderson. The species problem in iris. *Annals of the Missouri Botanical Garden*, 23: 457–509, 1936.

T. W. Anderson. Asymptotic theory for principal component analysis. *The Annals of Mathematical Statistics*, 34:122–148, 1963.

T. W. Anderson. *An Introduction to Multivariate Statistical Analysis*. Wiley, New York, third edition, 2003.

T. W. Anderson, Ingram Olkin, and L. G. Underhill. Generation of random orthogonal matrices. *SIAM Journal on Scientific and Statistical Computing*, 8:625–629, 1987.

Steen A. Andersson. Invariant normal models. *Annals of Statistics*, 3:132–154, 1975.

Steen A. Andersson and Michael D. Perlman. Two testing problems relating the real and complex multivariate normal distributions. *Journal of Multivariate Analysis*, 15 (1):21 - 51, 1984.

David R. Appleton, Joyce M. French, and Mark P. J. Vanderpump. Ignoring a covariate: An example of Simpson's paradox. *The American Statistician*, 50(4):340–341, 1996.

Robert B. Ash. *Basic Probability Theory*. Wiley, New York, 1970. URL `http://www.math.illinois.edu/~r-ash/BPT.html`.

Daniel Asimov. The grand tour: A tool for viewing multidimensional data. *SIAM Journal on Scientific and Statistical Computing*, 6:128–143, 1985.

M. S. Bartlett. A note on tests of significance in multivariate analysis. *Mathematical Proceedings of the Cambridge Philosophical Society*, 35:180–185, 1939.

M. S. Bartlett. A note on multiplying factors for various χ^2 approximations. *Journal of the Royal Statistical Society B*, 16:296–298, 1954.

Alexander Basilevsky. *Statistical Factor Analysis and Related Methods: Theory and Applications.* Wiley, New York, 1994.

James O. Berger, Jayanta K. Ghosh, and Nitai Mukhopadhyay. Approximations and consistency of Bayes factors as model dimension grows. *Journal of Statistical Planning and Inference*, 112(1-2):241–258, 2003.

Peter J. Bickel and Kjell A. Doksum. *Mathematical Statistics: Basic Ideas and Selected Topics, Volume I.* Pearson Prentice Hall, New Jersey, second edition, 2000.

G. E. P. Box and Mervin E. Muller. A note on the generation of random normal deviates. *Annals of Mathematical Statistics*, 29(2):610–611, 1958.

Leo Breiman, Jerome Friedman, Charles J. Stone, and R. A. Olshen. *Classification and Regression Trees.* Chapman & Hall, New York, 1984.

Andreas Buja and Daniel Asimov. Grand tour methods: An outline. In D. M. Allen, editor, *Computer Science and Statistics: Proceedings of the 17th Symposium on the Interface*, pages 63–67, New York and Amsterdam, 1986. Elsevier/North-Holland.

Kenneth P. Burnham and David R. Anderson. *Model Selection and Multimodel Inference: a Practical Information-theoretic Approach.* Springer-Verlag, New York, second edition, 2003.

T. K. Chakrapani and A. S. C. Ehrenberg. An alternative to factor analysis in marketing research part 2: Between group analysis. *Professional Marketing Research Society Journal*, 1:32–38, 1981.

Herman Chernoff. The use of faces to represent points in k-dimensional space graphically. *Journal of the American Statistical Association*, 68:361–368, 1973.

Vernon M. Chinchilli and Ronald K. Elswick. A mixture of the MANOVA and GMANOVA models. *Communications in Statistics: Theory and Methods*, 14:3075–3089, 1985.

Ronald Christensen. *Plane Answers to Complex Questions: The Theory of Linear Models.* Springer, New York, fourth edition, 2011.

J. W. L. Cole and James E. Grizzle. Applications of multivariate analysis of variance to repeated measurements experiments. *Biometrics*, 22:810–828, 1966.

Consumers' Union. Body dimensions. *Consumer Reports*, 55(5):286–288, 1990.

Dianne Cook and Deborah F. Swayne. *Interactive and Dynamic Graphics for Data Analysis.* Springer, New York, 2007.

DASL Project. *Data and Story Library.* Cornell University, Ithaca, New York, 1996. URL `http://lib.stat.cmu.edu/DASL`.

M. Davenport and G. Studdert-Kennedy. The statistical analysis of aesthetic judgment: An exploration. *Applied Statistics*, 21(3):324–333, 1972.

Bill Davis and J. Jerry Uhl. *Matrices, Geometry & Mathematica.* Math Everywhere, 1999. URL `http://www.matheverywhere.com/`.

Edward J. Dudewicz and Thomas G. Ralley. *The Handbook of Random Number Generation and Testing with TESTRAND Computer Code*. American Sciences Press, Columbus, Ohio, 1981.

R. A. Fisher. The use of multiple measurements in taxonomic problems. *Annals of Eugenics*, 7:179–188, 1936.

Chris Fraley and Adrian E. Raftery. Model-based clustering, discriminant analysis, and density estimation. *Journal of the American Statistical Association*, 97(458):611–631, 2002.

Chris Fraley, Adrian E. Raftery, and Luca Scrucca. mclust: Normal mixture modeling for model-based clustering, classification, and density estimation, 2014. URL `http://cran.r-project.org/web/packages/mclust/`. R package version 4.4.

Yasunori Fujikoshi. The likelihood ratio tests for the dimensionality of regression coefficients. *J. Multivariate Anal.*, 4:327–340, 1974.

K. R. Gabriel. Biplot display of multivariate matrices for inspection of data and diagnosis. In V. Barnett, editor, *Interpreting Multivariate Data*, pages 147–173. Wiley, London, 1981.

Leon J. Gleser and Ingram Olkin. Linear models in multivariate analysis. In *Essays in Probability and Statistics*, pages 276–292. Wiley, New York, 1970.

A. K. Gupta and D. G. Kabe. On Mallows' C_p for the GMANOVA model under double linear restrictions on the regression parameter matrix. *Journal of the Japan Statistical Society*, 30(2):253–257, 2000.

P. R. Halmos. *Measure Theory*. Springer-Verlag, New York, 1950.

Kjetil Halvorsen. ElemStatLearn: Data sets, functions and examples from the book: *The Elements of Statistical Learning, Data Mining, Inference, and Prediction* by Trevor Hastie, Robert Tibshirani and Jerome Friedman., 2012. URL `http://cran.r-project.org/web/packages/ElemStatLearn/`. R package version 2012.04-0.

D. J. Hand and C. C. Taylor. *Multivariate Analysis of Variance and Repeated Measures: A Practical Approach for Behavioural Scientists*. Chapman & Hall, London, 1987.

Harry Horace Harman. *Modern Factor Analysis*. University of Chicago Press, 1976.

J. A. Hartigan and M. A. Wong. Algorithm AS 136: A K-means clustering algorithm. *Journal of the Royal Statistical Society. Series C (Applied Statistics)*, 28(1):pp. 100–108, 1979.

Hikaru Hasegawa. On Mallows' C_p in the multivariate linear regression model under the uniform mixed linear constraints. *Journal of the Japan Statistical Society*, 16:1–6, 1986.

Trevor Hastie, Robert Tibshirani, and Jerome Friedman. *The Elements of Statistical Learning: Data Mining, Inference, and Prediction*. Springer, New York, second edition, 2009. URL `http://www-stat.stanford.edu/~tibs/ElemStatLearn/`.

Claire Henson, Bridget Rogers, and Nadia Reynolds. Always Coca-Cola. Technical report, University Laboratory High School, Urbana, Illinois, 1996.

Arthur E. Hoerl and Robert W. Kennard. Ridge regression: Biased estimation for nonorthogonal problems. *Technometrics*, 12(1):pp. 55–67, 1970.

Robert V. Hogg, Joseph W. McKean, and A. T. Craig. *Introduction to Mathematical Statistics*. Pearson, New Jersey, seventh edition, 2012.

Mark Hopkins, Erik Reeber, George Forman, and Jaap Suermondt. Spam data. Hewlett-Packard Labs, Palo Alto, California, 1999.

Peter J. Huber. Projection pursuit. *The Annals of Statistics*, 13:435–475, 1985. Comments and response, pp. 475–525.

Clifford M. Hurvich and Chih-Ling Tsai. Regression and time series model selection in small samples. *Biometrika*, 76:297–307, 1989.

Aapo Hyvärinen, Juha Karhunen, and Erkki Oja. *Independent Component Analysis*. Wiley, New York, 2001.

IBM. System/360 scientific subroutine package, version III, programmer's manual, program number 360A-CM-03X. In *Manual GH20-0205-4*, White Plains, New York, 1970.

Richard Arnold Johnson and Dean W. Wichern. *Applied Multivariate Statistical Analysis*. Pearson Prentice-Hall, New Jersey, sixth edition, 2007.

Michael I. Jordan, editor. *Learning in Graphical Models*. MIT Press, Cambridge, Massachusetts, 1999.

Takeaki Kariya. *Testing in the Multivariate General Linear Model*. Kinokuniya, Tokyo, 1985.

Leonard Kaufman and Peter J. Rousseeuw. *Finding Groups in Data: An Introduction to Cluster Analysis*. Wiley, New York, 1990.

D. Koller and N. Friedman. *Probabilistic Graphical Models: Principles and Techniques*. MIT Press, Cambridge, Massachusetts, 2009.

J. B. Kruskal. Non-metric multidimensional scaling. *Psychometrika*, 29:1–27, 115–129, 1964.

Anant M. Kshirsagar. Bartlett decomposition and Wishart distribution. *Annals of Mathematical Statistics*, 30(1):239–241, 1959.

Anant M. Kshirsagar. *Multivariate Analysis*. Marcel Dekker, New York, 1972.

S. Kullback and R. A. Leibler. On information and sufficiency. *The Annals of Mathematical Statistics*, 22(1):79–86, 1951.

LaTeX Project Team. LaTeX: A document preparation system, 2015. URL `http://www.latex-project.org/`.

D. N. Lawley and A. E. Maxwell. *Factor Analysis As a Statistical Method*. Butterworths, London, 1971.

Yann LeCun. Generalization and network design strategies. Technical Report CRG-TR-89-4, Department of Computer Science, University of Toronto, 1989.

E. L. Lehmann and George Casella. *Theory of Point Estimation*. Springer, New York, second edition, 2003.

M. Lichman. UCI machine learning repository, 2013. URL `http://archive.ics.uci.edu/ml`.

Lars Madsen and Peter R. Wilson. memoir — typeset fiction, nonfiction and mathematical books, 2015. URL `https://www.ctan.org/tex-archive/macros/latex/contrib/memoir`. Version 3.7c.

Martin Maechler, Peter J. Rousseeuw, Anja Struyf, Mia Hubert, and Kurt Hornik. cluster: Cluster analysis basics and extensions, 2015. URL `http://cran.r-project.org/web/packages/cluster/`. R package version 2.0.1.

C. L. Mallows. Some comments on C_p. *Technometrics*, 15(4):661–675, 1973.

J. L. Marchini, C. Heaton, and B. D. Ripley. fastICA: FastICA algorithms to perform ICA and projection pursuit, 2013. URL `http://cran.r-project.org/web/packages/fastICA/`. R package version 1.2-0.

John Marden and James Balamuta. msos: Datasets and functions from the book: *Multivariate Statistics: Old School* by John I. Marden, 2014. URL `http://cran.r-project.org/web/packages/msos/`. R package version 1.0.1.

K. V. Mardia, J. T. Kent, and J. M. Bibby. *Multivariate Analysis*. Academic Press, London, 1979.

Kenny J. Morris and Robert Zeppa. Histamine-induced hypotension due to morphine and arfonad in the dog. *Journal of Surgical Research*, 3(6):313–317, 1963.

Ingram Olkin and S. N. Roy. On multivariate distribution theory. *Annals of Mathematical Statistics*, 25(2):329–339, 1954.

K. Pearson. On lines and planes of closest fit to systems of points in space. *Philosophical Magazine*, 2(6):559–572, 1901.

Richard F. Potthoff and S. N. Roy. A generalized multivariate analysis of variance model useful especially for growth curve problems. *Biometrika*, 51:313–326, 1964.

J. Ross Quinlan. *C4.5: Programs for Machine Learning*. Morgan Kaufmann, San Francisco, 1993.

R Development Core Team. The Comprehensive R Archive Network, 2015. URL `http://cran.r-project.org/`.

C. R. Rao. *Linear Statistical Inference and Its Applications*. Wiley, New York, 1973.

G. M. Reaven and R. G. Miller. An attempt to define the nature of chemical diabetes using a multidimensional analysis. *Diabetologia*, 16:17–24, 1979.

Brian Ripley. tree: Classification and regression trees, 2015. URL `http://cran.r-project.org/web/packages/tree/`. R package version 1.0-35.

J. E. Rossouw, J. P. du Plessis, A. J. S. Benadé, P. C. J. Jordaan, J. P. Kotzé, P. L. Jooste, and J. J. Ferreira. Coronary risk factor screening in three rural communities: The CORIS baseline study. *South African Medical Journal*, 64(12):430–436, 1983.

Peter J. Rousseeuw. Silhouettes: A graphical aid to the interpretation and validation of cluster analysis. *J. Comput. Appl. Math.*, 20(1):53–65, 1987.

Henry Scheffé. *The Analysis of Variance*. Wiley, New York, 1999.

Gideon Schwarz. Estimating the dimension of a model. *The Annals of Statistics*, 6: 461–464, 1978.

George W. Snedecor and William G. Cochran. *Statistical Methods*. Iowa State University Press, Ames, Iowa, eighth edition, 1989.

C. Spearman. "General intelligence," objectively determined and measured. *American Journal of Psychology*, 15:201–293, 1904.

David J. Spiegelhalter, Nicola G. Best, Bradley P. Carlin, and Angelika van der Linde. Bayesian measures of model complexity and fit. *Journal of the Royal Statistical Society, Series B: Statistical Methodology*, 64(4):583–616, 2002.

H. A. Sturges. The choice of a class interval. *Journal of the American Statistical Association*, 21(153):65–66, 1926.

A. Thomson and R. Randall-MacIver. *Ancient Races of the Thebaid*. Oxford University Press, 1905.

TIBCO Software Inc. *S-Plus*. Palo Alto, California, 2009.

W. N. Venables and B. D. Ripley. *Modern Applied Statistics with S*. Springer, New York, fourth edition, 2002.

J. H. Ware and R. E. Bowden. Circadian rhythm analysis when output is collected at intervals. *Biometrics*, 33(3):566–571, 1977.

Sanford Weisberg. *Applied Linear Regression*. Wiley, New York, fourth edition, 2013.

Joe Whitaker. *Graphical Models in Applied Multivariate Statistics*. Wiley, New York, 1990.

Hadley Wickham, Dianne Cook, Heike Hofmann, and Andreas Buja. tourr: An R package for exploring multivariate data with projections. *Journal of Statistical Software*, 40(2):1–18, 2011.

Wikipedia. Athletics at the 2008 summer olympics: Men's decathlon: Wikipedia, The Free Encyclopedia, 2014a. URL `http://en.wikipedia.org/w/index.php?title=Athletics_at_the_2008_Summer_Olympics_%E2%80%93_Men%27s_decathlon&oldid=620744999`.

Wikipedia. Pterygomaxillary fissure: Wikipedia, The Free Encyclopedia, 2014b. URL `http://en.wikipedia.org/w/index.php?title=Pterygomaxillary_fissure&oldid=635659877`.

Wikipedia. List of breakfast cereals: Wikipedia, The Free Encyclopedia, 2015. URL `http://en.wikipedia.org/w/index.php?title=List_of_breakfast_cereals&oldid=647792889`.

John Wishart. The generalised product moment distribution in samples from a normal multivariate population. *Biometrika*, 20A:32–52, 1928.

Hans Peter Wolf. aplpack: Another Plot PACKage, 2014. URL `http://cran.r-project.org/web/packages/aplpack/`. R package version 1.3.0. [The R package page lists "Uni Bielefeld" as a coauthor. Uni is not Prof. Wolf's colleague, but his university, so I left it off the author list.].

John W. Wright, editor. *The Universal Almanac*. Andrews McMeel Publishing, Kansas City, Missouri, 1997.

Gary O. Zerbe and Richard H. Jones. On application of growth curve techniques to time series data. *Journal of the American Statistical Association*, 75:507–509, 1980.

Author Index

Akaike, Hirotugu 170
Allen, D. M. 9
Anderson, David R. 174
Anderson, E. 5
Anderson, T. W. iii, 130, 133, 271, 272, 278, 296
Andersson, Steen A. 202, 211
Appleton, David R. 46
Ash, Robert B. 35
Asimov, Daniel 9

Balamuta, James iv, 295
Barnett, V. 13
Bartlett, M. S. 130, 151
Basilevsky, Alexander 194
Benadé, A. J. S. 213, 233
Berger, James O. 173
Best, Nicola G. 171
Bibby, J. M. iii, 196, 278, 280
Bickel, Peter J. 27
Bowden, R. E. 82
Box, G. E. P. 67
Breiman, Leo 235
Buja, Andreas 9, 24
Burnham, Kenneth P. 174

Carlin, Bradley P. 171
Casella, George 27, 152, 164, 172
Chakrapani, T. K. 24
Chernoff, Herman 2
Chinchilli, Vernon M. 82
Christensen, Ronald 69
Cochran, William G. 84, 100
Cole, J. W. L. 74
Consumers' Union 24, 254
Cook, Dianne 24
Craig, A. T. iii, 27

DASL Project 122
Davenport, M. 23
Davis, Bill 18
Doksum, Kjell A. 27
du Plessis, J. P. 213, 233
Dudewicz, Edward J. 25

Ehrenberg, A. S. C. 24
Elswick, Ronald K. 82

Ferreira, J. J. 213, 233
Fisher, R. A. 5, 220
Forman, George 22
Fraley, Chris 254
French, Joyce M. 46
Friedman, Jerome 215, 233, 235, 239, 260
Friedman, N. 194
Fujikoshi, Yasunori 289

Gabriel, K. R. 13
Ghosh, Jayanta K. 173
Gleser, Leon J. 82
Grizzle, James E. 74
Gupta, A. K. 136

Halmos, P. R. 153
Halvorsen, Kjetil 22, 213, 237, 241
Hand, D. J. 132
Harman, Harry Horace 194

Hartigan, J. A. 250
Hasegawa, Hikaru 136
Hastie, Trevor 215, 233, 239, 260
Heaton, C. 295
Henson, Claire 83
Hoerl, Arthur E. 121
Hofmann, Heike 24
Hogg, Robert V. iii, 27
Hopkins, Mark 22
Hornik, Kurt 253, 303
Huber, Peter J. 10
Hubert, Mia 253, 303
Hurvich, Clifford M. 174, 175
Hyvärinen, Aapo 295

IBM 25

Johnson, Richard Arnold iii, 84, 283
Jones, Richard H. 82
Jooste, P. L. 213, 233
Jordaan, P. C. J. 213, 233
Jordan, Michael I. 194

Kabe, D. G. 136
Karhunen, Juha 295
Kariya, Takeaki 82
Kaufman, Leonard 252
Kennard, Robert W. 121
Kent, J. T. iii, 196, 278, 280
Koller, D. 194
Kotzé, J. P. 213, 233
Kruskal, J. B. 283
Kshirsagar, Anant M. 151, 222
Kullback, S. 16

LaTeX Project Team ii
Lawley, D. N. 194
LeCun, Yann 241
Lehmann, E. L. 27, 152, 164, 172
Leibler, R. A. 16
Lichman, M. 22

Madsen, Lars ii
Maechler, Martin 253, 303
Mallows, C. L. 135
Marchini, J. L. 295
Marden, John iv, 295
Mardia, K. V. iii, 196, 278, 280
Maxwell, A. E. 194
McKean, Joseph W. iii, 27
Miller, R. G. 265
Morris, Kenny J. 74
Mukhopadhyay, Nitai 173
Muller, Mervin E. 67

Oja, Erkki 295
Olkin, Ingram 82, 151, 296
Olshen, R. A. 235

Pearson, K. 10
Perlman, Michael D. 211
Potthoff, Richard F. 73, 74, 81

Quinlan, J. Ross 235

R Development Core Team iv
Raftery, Adrian E. 254
Ralley, Thomas G. 25
Randall-MacIver, R. 82
Rao, C. R. 87
Reaven, G. M. 265
Reeber, Erik 22
Reynolds, Nadia 83
Ripley, B. D. 23, 233, 237, 240, 295
Rogers, Bridget 83
Rossouw, J. E. 213, 233
Rousseeuw, Peter J. 247, 252, 253, 303
Roy, S. N. 73, 74, 81, 151

Scheffé, Henry 69
Schwarz, Gideon 170, 172
Scrucca, Luca 254
Snedecor, George W. 84, 100
Spearman, C. 194
Spiegelhalter, David J. 171
Stone, Charles J. 235
Struyf, Anja 253, 303
Studdert-Kennedy, G. 23
Sturges, H. A. 295
Suermondt, Jaap 22
Swayne, Deborah F. 24

Taylor, C. C. 132
Thomson, A. 82
TIBCO Software Inc. 24, 254
Tibshirani, Robert 215, 233, 239, 260
Tsai, Chih-Ling 174, 175

Uhl, J. Jerry 18
Underhill, L. G. 296

van der Linde, Angelika 171
Vanderpump, Mark P. J. 46

Venables, W. N. 23, 233, 237, 240

Ware, J. H. 82
Weisberg, Sanford iii, 69
Whitaker, Joe 194
Wichern, Dean W. iii, 84, 283
Wickham, Hadley 24
Wikipedia 24, 73
Wilson, Peter R. ii
Wishart, John 58
Wolf, Hans Peter ii, 2, 24
Wong, M. A. 250
Wright, John W. 2

Zeppa, Robert 74
Zerbe, Gary O. 82

Subject Index

The italic page numbers are references to Exercises.

affine transformation, 40–41
  covariance matrix, 41
  definition, 40
  mean, 41
  multivariate normal distribution, 49–52
    conditional mean, 56
    data matrix, 53–54
Akaike information criterion (AIC), 135, 170–171
  equality of covariance matrices
    iris data, 226–227
  factor analysis, 197
  motivation, 173–174
  multivariate regression, 174–175

Bartlett's decomposition, 152
Bartlett's test, 190
Bayes information criterion (BIC), 135, 170–173
  as posterior probability, 171
  both-sides model
    caffeine data, *184*
    prostaglandin data, *184*
  clustering, 254
  equality of covariance matrices
    iris data, 226–227
  factor analysis, 197
  principal components
    automobile data, 273–276
Bayes theorem, 39–40, 172
  multivariate normal distribution, *65*
Bayesian inference, 39
  binomial, *43–44*
  conjugate prior, *44*

multivariate normal distribution
mean, *65–66*
mean and covariance, *157–159*
regression parameter, *121*
Wishart distribution, *156–157*
biplot, 13–14
cereal data, *24*
decathlon data, *24*
painters data, *23*
sports data, 13–14
both-sides model, 77–82
a.k.a. generalized multivariate analysis of variance (GMANOVA) model, 82
caffeine data, *83*
covariates
histamine data, *123*
estimation of coefficients
bothsidesmodel (R routine), 298
bothsidesmodel.mle (R routine), 299–300
caffeine data, *122*
covariance matrix, 109–111
grades data, *123*
histamine data, 117–118
mouth size data, 112–114
prostaglandin data, *122*
reverse.kronecker (R routine), 299
Student's t, 111–112
estimation of covariance matrix, 110
fits and residuals, 109–111
covariance matrix, 110
expected value, 110
growth curve model
births data, 78–80
mouth size data, 112–114
histamine data, *83–84*
t test, 118
hypothesis testing
approximate χ^2 tests, 125–126
bothsidesmodel.chisquare (R routine), 300
bothsidesmodel.hotelling (R routine), 300–301
bothsidesmodel.lrt (R routine), 301
histamine data, 133–134
Hotelling's T^2 and projection pursuit, 149–150
Hotelling's T^2 test, 128–129, 148–149
Lawley-Hotelling trace test, 126–129
linear restrictions, 134–135
mouth size data, 126, 132–133
Pillai trace test, 130
prostaglandin data, *142*
Roy's maximum root test, 130
t test, 112

Wilks' Λ, 129
iid model, 78–80
intraclass correlation structure
mouth size data, 207–210
leprosy data
t test, 115, 117
Mallows' C_p
caffeine data, *143–144*
mouth size data, 137–139
mouth size data, 81–82, 167
submodel, 119–120
t test, 114, 119
parity data, *84–85*
prostaglandin data, *82*
pseudo-covariates
histamine data, *143*
submodels, 118–120
patterns, 120
Box-Muller transformation, 67

canonical correlations, 191, 284–289
Bayes information criterion, 288–289
grades data, 289
compared to partial least squares, 290
exams data, *292–293*
grades data, 287–288
iris data, *292*
states data, *293*
Cauchy-Schwarz inequality, 145–146, 154
Chernoff's faces, 2–3
iris data, 5
planets data, 2–3
spam data, *22*
sports data, *23*
Choleksy decomposition, 98–99
classification, 215–240, 267
Bayes classifier, 217
spam data, *240*
classifier, 217
estimation, 218
cross-validation, 221–222
leave out more than one, 222
leave-one-out, 222–224
discriminant function, 219
linear, 219
error, 217, 221
cross-validation estimate, 222
observed, 221, 223, 224
Fisher's linear discrimination, 219–225
crabs data, *240–241*

definition, 220
heart disease data, *243*
iris data, 222–225
lda (R routine), 302
modification, 227
zipcode data, *241–242*
Fisher's quadratic discrimination, 225–227
definition, 225
iris data, 226–227
predict_qda (R routine), 303
qda (R routine), 302
illustration, 218
logistic regression, 227–233
heart disease data, *241*
iris data, 229–230
spam data, 230–233
new observation, 217
subset selection
iris data, 224–225
trees, 233–240
Akaike information criterion, 238
Bayes information criterion, 238
C5.0, 235
Categorization and Regression Trees (CART), 235
crabs data, *241*
cross-validation, 239–240
deviance, 237
heart disease data, 233–240
pruning, 237–238
snipping, 234
spam data, *242*
tree (R routine), 237
clustering, 215, 245–265
classifier, 245
density of data, 217
hierarchical, 246, 261–265
average linkage, 262
cereal data, *266*
complete linkage, 262
dissimilarities, 253
grades data, 261–263, *266*
hclust (R routine), 263
single linkage, 262
soft drink data, *266*
sports data, 263–265
K-means, 246–252
algorithm, 246
diabetes data, *265–266*
grades data, *265*
kmeans (R routine), 250–251

objective function, 246
relationship to EM algorithm, 260
silhouette.km (R routine), 303
silhouettes, 247–252
sort_silhouette (R routine), 303
sports data, 246–248, 250–252
K-medoids, 252–254
dissimilarities, 252
grades data, 253–254
objective function, 252
pam (R routine), 253
silhouettes, 247, 253
model-based, 245, 246, 254–260, 267
automobile data, 254–257
Bayes information criterion, 254–255, 257
classifier, 254
EM algorithm, 259–260
iris data, *266*
likelihood, 254
Mclust (R routine), 254
mixture density, 254
multivariate normal model, 254
plotting
sports data, 248–250
soft *K*-means, 260–261
sports data, 261
conditional distributions, *see* probability distributions: conditional distributions
conditional probability, 30
smoking data, *46*
convexity, 20
correlation coefficient, 33
sample, 7
correlation inequality, 146
covariance, 33
sample, 7
covariance (variance-covariance) matrix
collection of random variables, 34
generalized variance, *156*
multivariate normal distribution, 49–51
data matrix, 53
of two vectors, 34–35
principal components, 11
sample, 8
sum of squares and cross-products matrix, 11
covariance models, 187–207
factor analysis, *see separate entry*
invariant normal models, *see* symmetry models
symmetry models, *see separate entry*
testing conditional independence
grades data, *212*

testing equality, 188–190
mouth size data, *211*
testing independence, 190–191
grades data, *212*
testing independence of blocks of variables, 284
covariates, 115–117
histamine data, *123*
cross-validation, 221–222
leave out more than one, 222
leave-one-out, 222, 224
cumulant generating function, 45

data examples
automobiles
entropy and negentropy, *24–25*
model-based clustering, 254–257
principal components, 270–271, 273–276
varimax, 275
births, 78
growth curve models, 78–80
caffeine, *83*
Bayes information criterion, *184*
both-sides model, *83*, *122*, *143–144*
orthogonalization, 106
cereal, *24*
biplot, *24*
hierarchical clustering, *266*
crabs, *240*
classification, *240–241*
decathlon, 24
biplot, *24*
factor analysis, *213*
diabetes, *265*
K-means clustering, *265–266*
elections, *23*
principal components, *23*
exams, *212*
canonical correlations, *292–293*
covariance models, *212–213*
factor analysis, *212*
Fisher-Anderson iris data, 5
Akaike information criterion, 226–227
Bayes information criterion, 226–227
canonical correlations, *292*
Chernoff's faces, 5
classification, 215
Fisher's linear discrimination, 222–225
Fisher's quadratic discrimination, 226–227
logistic regression, 229–230
model-based clustering, *266*

principal components, 12, 268–270
projection pursuit, 17–18
rotations, 10, *24*
scatter plot matrix, 5
subset selection in classification, 224–225
grades, 72
Bayes information criterion, 200
both-sides model, *123*
canonical correlations, 287–289
covariance models, *212*
factor analysis, 198–202, *212*
hierarchical clustering, 261–263, *266*
K-means clustering, *265*
K-medoids clustering, 253–254
multidimensional scaling, 283
multivariate regression, 72–73
testing conditional independence, 193–194
testing equality of covariance matrices, 189
testing independence, 191
Hewlett-Packard spam data, 22
Chernoff's faces, 22
classification, *240*, *242*
logistic regression, 230–233
principal components, *22–23*, *292*
histamine, 74
both-sides model, *83–84*, 117–118, 133–134, *143*
covariates, *123*
model selection, 179–181
multivariate analysis of variance, 74–75
multivariate regression, *142–143*
t test, 118
leprosy, 84
covariates, 115–117
multivariate analysis of variance, *84*
multivariate regression, 115–117, *182*, *184*
orthogonalization, 106–107
t test, 115, 117
Louis Roussos sports data, 13
biplot, 13–14
Chernoff's faces, *23*
hierarchical clustering, 263–265
K-means clustering, 246–248, 250–252
multidimensional scaling, 283–284
plotting clusters, 248–250
principal components, 13–14
soft K-means clustering, 261
mouth sizes, 73
both-sides model, 81–82, 112–114, 126, 132–133, 167, 207–210
covariance models, *211–212*
intraclass correlations, 207–210

Mallows' C_p, 137–139
model selection, 137–139
multivariate regression, 73, *141–142*, *184*
submodel, 119–120
t test, 114, 119
painters, *23*
biplot, *23*
principal components, *292*
parity, *84*
both-sides model, *84–85*
planets, 2
Chernoff's faces, 2–3
star plot, 2–3
prostaglandin, 82
Bayes information criterion, *184*
both-sides model, *82*, *122*, *142*
RANDU
rotations, *25*
skulls, 82
Mallows' C_p, *144*
model selection, 175–179
multivariate regression, *82*, *122*, *142*
orthogonalization, 106
smoking, 46
conditional probability, *46*
soft drinks, *266*
hierarchical clustering, *266*
South African heart disease, 213
classification, *241*, *243*
factor analysis, *213*
trees, 233–240
states, 293
canonical correlations, *293*
zipcode, *241*
classification, *241–242*
data matrix, 2
planets data, 2
data reduction, 267
density, *see* probability distributions: density
deviance, *see* likelihood: deviance
dissimilarities, 252, 253
distributions, *see* probability distributions

eigenvalues and eigenvectors, 12, *see also* principal components
positive and nonnegative definite matrices, *62*
uniqueness, *21*, *290*
EM algorithm, 254, 259–260
entropy and negentropy, 16–18, *44*, 174
automobile data, *24–25*
estimation, 295–296

maximized by normal, 16, 19–20
negent (R routine), 296
negent2D (R routine), 296–297
negent3D (R routine), 297
Euler angles, 296
expected value, 32–33
correlation coefficient, 33
covariance, 33
covariance (variance-covariance) matrix
collection of random variables, 34
of matrix, 34
of two vectors, 34
mean, 33
of collection, 33
of matrix, 34
of vector, 34
moment generating function, 35
variance, 33
with conditioning, 32–33

factor analysis, 187, 194–202, 267
Akaike information criterion, 197
Bartlett's refinement, 199
Bayes information criterion, 197
grades data, 200, *212*
compared to principal components, 276–278
decathlon data, *213*
estimation, 195–196
exams data, *212*
factanal (R routine), 198
goodness-of-fit, 199
grades data, 198–202
heart disease data, *213*
likelihood, 196
model, 194
model selection, 196–197
rotation of factors, 197, 200
scores, 197–198
estimation, 200
structure of covariance matrix, 195
uniquenesses, 198
varimax, 197
Fisher's linear discrimination, *see under* classification
Fisher's quadratic discrimination, *see under* classification
fit, 89

gamma function, 43
$\mathcal{G}$-average of a covariance matrix, 205
generalized multivariate analysis of variance (GMANOVA), 82, 169, *see also* both-sides model

glyphs, 2–3
  Chernoff's faces, 2–3
  star plot, 2–3
grand tour, 9

Hausdorff distance, 262
hypothesis testing, 168
  approximate χ^2 tests, 125–126
  conditional independence, 192–194
    grades data, 193–194
    likelihood ratio statistic, 192, 193
  equality of covariance matrices, 188–190
    Bartlett's test, 190
    grades data, 189
    likelihood ratio statistic, 188, 190
  F test, 128
  Hotelling's T^2 test, 128–129
    null distribution, 128
    proof of distribution, 148–149
  independence of blocks of variables, 190–191
    as symmetry model, 203
    grades data, 191
    likelihood ratio statistic, 191
  Lawley-Hotelling trace test, 126–129
  likelihood ratio test, 168–170
    asymptotic null distribution, 168
  linear restrictions, 134–135
  Pillai trace test, 130
  Roy's maximum root test, 130
  symmetry models, 207
    likelihood ratio statistic, 207
  t test, 112
  Wilks' Λ, 129
    Bartlett's approximation, 130
    beta, *141*

intraclass correlation structure, 78, 204, *see also under symmetry models*
  spectral decomposition, *21*, 208

Jensen's inequality, 20, 22

K-means clustering, *see under* clustering
K-medoids clustering, *see under* clustering
kurtosis, 16, 45

least squares, 55, 63, 90–94
  best linear unbiased estimators (BLUE), 91–94
  both-sides model
    distribution of estimator, 109–120
  definition, 90

equation, 90
Gauss-Markov theorem, 92
normal equations, 90
projection, 91
regression, *44–45*
weighted, 92–93
likelihood, 161–181
definition, 161
deviance
Akaike and Bayes information criteria, 170
canonical correlations, 289
definition, 170
equality of covariance matrices, 226–227
factor analysis, 197
likelihood ratio statistic, 170
multivariate regression, 174
observed, 170
principal components, 273, 276
symmetry models, 207
trees, 237
likelihood ratio test, 168–170
asymptotic null distribution, 168
covariance models, 187
equality of covariance matrices, 188–190
factor analysis, 196–197
independence, 190–191
multivariate regression, 169–170
statistic, 168, 170
symmetry models, 207
maximum likelihood estimation, *see separate entry*
principal components, 278–280
Wishart, 187
linear discrimination, Fisher's, *see* classification: Fisher's linear discrimination
linear models, 69–85, 267
both-sides model, *see separate entry*
covariates
adjusted estimates, 117
leprosy data, 115–117
definition, 94–95
estimation, 109–123
Gram-Schmidt orthogonalization, 95–97
growth curve model, 78
hypothesis testing, 125–144, *see also* both-sides model: hypothesis testing
least squares, *see separate entry*
linear regression, 69–71
analysis of covariance, 71
analysis of variance, 70
cyclic model, 71
model, 69
multiple, 70

polynomial model, 71
simple, 70
model selection, 135–140
multivariate analysis of variance, *see separate entry*
multivariate regression, *see separate entry*
orthogonal polynomials, 99–101
prediction, 135–140
Mallows' C_p, 135–140
repeated measures, 77
ridge regression estimator, *121*
linear subspace
basis
definition, 89
existence, *106*
definition, 87
dimension, 89
least squares, *see separate entry*
linear independence, 88
orthogonal, 89
orthogonal complement, 91
projection, 87–101
definition, 89
properties, 89–90
projection matrix, 90–91, 272
idempotent, 91
properties, 91
span
definition, 87
matrix representation, 88
transformation, 88
linear transformations
sample, 8–9
logistic regression, 227–233
Akaike and Bayes information criteria, 232
iris data, 229–230
likelihood, 228
spam data, 230–233
stepwise, 232
using R, 231
logit, 229

Mallows' C_p
both-sides model, 135–140
caffeine data, *143–144*
prostaglandin data, *142*
marginals, 29
sample, 8
matrix
block diagonal, 53
centering, 7, *20*

decompositions
Bartlett's, 152
Cholesky, 98–99
QR, 97–98, 151–153
singular value, 286–287
spectral, 12, *see also* spectral decomposition
determinant
Cholesky decomposition, *105*
decomposing covariance, *104–105*
group, 202
complex normal, *211*
compound symmetry, 204
definition, 98
orthogonal, 152, 153, 202
permutation, 204
upper triangular, 98
upper unitriangular, 97
idempotent, *20, 21*
definition, 8
in Wishart distribution, 58–59, *67*, 110
Kronecker product, *62*
projection, 91, *102*
spectral decomposition, 58
identity, 7
inverse
block, *104*
Kronecker product
definition, 53
properties, 54, *62–63*
Moore-Penrose inverse, *64*
orthogonal, 202
orthogonal polynomials, 99–101
positive and nonnegative definite, 51
and eigenvalues, *62*
projection, 90–91, 272
idempotent, 91, *102*
properties, 91
square root, 50
Cholesky decomposition, 99
symmetric, 51
maximum likelihood estimation, 161–168
both-sides model, 162–167
covariance matrix, 163
definition, 162
multivariate normal covariance matrix, 163
multivariate regression, 166–167
symmetry models, 205–206
mean, 33
of collection, 33
of matrix, 34

of vector, 34
sample, 6
vector
sample, 7
mixture models, 215–218
conditional probability of predictor, 216
data, 215
density of data, 216
illustration, 216
marginal distribution of predictor, 216
marginal probability of group, 215
model selection, 226–227
Akaike information criterion, *see separate entry*
Bayes information criterion, *see separate entry*
classification, 226–227
factor analysis, 196–197
histamine data
Akaike and Bayes information criteria, 179–181
linear models, 135–140
Mallows' C_p, *see separate entry*
mouth size data
Mallows' C_p, 137–139
penalty, 135, 170
skulls data
Akaike and Bayes information criteria, 175–179
model-based clustering, *see under* clustering
moment generating function, 35
chi-square, *61*
double exponential, *61*
gamma, *61*
multivariate normal, 50
standard normal, 49
standard normal collection, 49
sum of random variables, *45*
uniqueness, 35
multidimensional scaling, 280–284
classical solution, 280–282
Euclidean distances, 280–282
non-Euclidean distances, 282
grades data, 283
non-metric, 283
stress function, 283
principal components, 281
sports data, 283–284
multivariate analysis of variance
between and within matrices, 130–131
histamine data, 74–75
leprosy data, *84*
test statistics, 129–131
multivariate normal distribution, 49–67

affine transformations, 51
Bayesian inference
mean, *65–66*
mean and covariance, *157–159*
regression parameter, *121*
collection, 49
conditional distributions, 55–57
noninvertible covariance, *63–64*
covariance matrix, 49
data matrix, 52–54
affine transformation, 54
conditional distributions, 57
definition, 49
density, 150–151
Kronecker product covariance, 151
independence, 52
marginals, 52
matrix, 51
mean, 49
moment generating function, 50
QR decomposition, 151–153
sample covariance matrix, distribution, 59
sample mean and covariance, independence of, 58
sample mean, distribution, 54
standard normal
collection, 49
univariate, 49
multivariate regression, 72–75
Akaike information criterion, 174–175
Bayes information criterion
leprosy data, *184*
mouth size data, *184*
covariates
histamine data, *142–143*
grades data, 72–73
hypothesis testing
mouth size data, *141–142*
skulls data, *142*
least squares estimation
leprosy data, 115–117
skulls data, *122*
likelihood ratio test, 169–170
Mallows' C_p
skulls data, *144*
maximum likelihood estimation, 166–167
leprosy data, *182*
mouth size data, 73
skulls data, *82*

norm, of a vector, 9

orthogonal, *see also* linear models: Gram-Schmidt orthogonalization
matrix
definition, 10
orthonormal set of vectors, 9
to subspace, 89
vectors, 9
orthogonalization
caffeine data, 106
leprosy data, 106–107
skulls data, 106

partial least squares, 290
Pillai trace test, 130
mouth size data, 133
prediction, 215
linear models, 135–140
Mallows' C_p, 135–140
principal components, 10–15, 187, 267–281
asymptotic distribution of eigenvalues, 271
automobile data
varimax, 275
Bayes information criterion
pcbic (R routine), 304
pcbic.stepwise (R routine), 304–305
best K, 13, 18–19
biplot, 13–14
choosing the number of, 270–276
compared to factor analysis, 276–278
definition, 11
eigenvalues and eigenvectors, 12, 22
election data, *23*
iris data, 12, 268–270
painters data, *292*
pattern of eigenvalues, 271–276
automobile data, 273–276
Bayes information criterion, 273
dimension, 279–280
likelihood, 278–280
maximum likelihood estimate, 272
scree plot, 270
automobile data, 270–271
spam data, *22–23*, *292*
spectral decomposition theorem, 11
sports data, 13–14
structure of component spaces, 271–276
testing size of eigenvalues, 271
uncorrelated property, 11, 18
probability distributions, 27–37
beta
and F, *141*

defined via gammas, *141*
density, *43*
Wilks' Λ, *141*
beta-binomial, *43*
chi-square
and Wisharts, 59
definition, 59
density, *61*
moment generating function, *61*
conditional distributions, 30–31, 37–40
conditional independence, 38
conditional space, 30
continuous, density, 31
covariates, 116
dependence through a function, 38
discrete, 30
independence, 36
iterated expectation, 32
multivariate normal data matrix, 57
multivariate normal distribution, 55–57
notation, 30
plug-in property, 37–38, *42*, 56, 147, 148
variance decomposition, 39
Wishart distribution, 146–147
conditional space, 29
continuous, 28
density, 28–29
mixed-type, 29
probability density function (pdf), 28
probability mass function (pmf), 28
discrete, 28
double exponential, *61–62*
expected values, 32–33
exponential family, 228
F, 128, *155*
gamma, *61*
Haar on orthogonal matrices, 153
Half-Wishart, 152, *156*
Hotelling's T^2, 148–149
projection pursuit, 149–150
independence, 35–37
definition, 36
marginal space, 29
marginals, 29–30
moment generating function, *see separate entry*
multinomial, 43
multivariate normal, *see* multivariate normal distribution
multivariate t, *158*
mutual independence
definition, 37

representations, 27
Student's t, 111, *155*
uniform, 29
Wilks' Λ, *see separate entry*
Wishart, *see* Wishart distribution
projection, *see* linear subspace: projection
projection pursuit, 10, 15–18
entropy and negentropy, 16–18
Hotelling's T^2, 149–150
iris data, 17–18
kurtosis, 16
skewness, 15

QR decomposition, 97–98
multivariate normal, 151–153
quadratic discrimination, Fisher's, *see* classification: Fisher's quadratic discrimination

R routines
both-sides model
bothsidesmodel, 298
bothsidesmodel.chisquare, 300
bothsidesmodel.hotelling, 300–301
bothsidesmodel.lrt, 301
bothsidesmodel.mle, 299–300
helper functions, 302
reverse.kronecker, 299
classification
lda, 302
predict_qda, 303
qda, 302
clustering
silhouette.km, 303
sort_silhouette, 303
entropy and negentropy
negent, 296
negent2D, 296–297
negent3D, 297
principal components
helper functions, 305
pcbic, 304
pcbic.stepwise, 304–305
random variable, 27
collection of, 27
rectangle, 35
representations of random variables, 29–30
residual, 56, 89
ridge regression, 121
rotations, 10
example data sets, *25*
iris data, *24*

RANDU, *25*
Roy's maximum root test, 130
mouth size data, 133

scatter plot
matrix, 3–5
iris data, 5
stars, 3
singular value decomposition, 286–287
skewness, 15, 45
spectral decomposition, 11–12
eigenvalues and eigenvectors, 12, *21*
intraclass correlation structure, *21*
theorem, 12
spherical symmetry, 204
star plot, 2
planets data, 2–3
sum of squares and cross-products matrix, 8
supervised learning, 215, 245
symmetry models, 202–207
a.k.a. invariant normal models, 202
complex normal structure, *211*
compound symmetry, 204
grades data, *212–213*
definition, 202
hypothesis testing, 207
iid, 204–205
independence of blocks of variables, 203
intraclass correlation structure, 204, *211*
and independence, 204
dimension, 207
maximum likelihood estimation, 206
mouth size data, 207–210, *212*
likelihood ratio statistic, 207
maximum likelihood estimation, 205–206
independence, 206
intraclass correlation, 206
spherical symmetry, 204–205
structure from group, 205

total probability formula, 33
trace of a matrix, 18

unsupervised learning, 215, 245

variance, 33, *see also* covariance (variance-covariance) matrix
sample, 6
varimax, 197, 275
R function, 275
vector

norm of, 9
one, 7

Wilks' Λ, 129
beta, *141*
mouth size data, 133
Wishart distribution, 57–60, 187
and chi-squares, 59
Bartlett's decomposition, 152
Bayesian inference, *156–157*
conditional property, 146–147
definition, 58
density, 153–154
expected value of inverse, 129, 147–148
for sample covariance matrix, 59
Half-Wishart, 152, *156*
likelihood, 187
linear transformations, 60
marginals, 60
mean, 59
sum of independent, 59

Made in the USA
Lexington, KY
23 August 2016